Communications Systems Management

HANDBOOK

6TH EDITION

OTHER AUERBACH PUBLICATIONS

Communications Systems Management

HANDBOOK

6TH EDITION

Anura Gurugé *and*
Lisa M. Lindgren

EDITORS

AUERBACH

Boca Raton London New York Washington, D.C.

Library of Congress Cataloging-in-Publication Data

Communications systems management handbook / Anura Gurugé and Lisa Lindgren, editors.
p. cm.
Includes index.
ISBN 0-8493-9826-6 (alk. paper)
1. Telecommunication systems—Management Handbooks, manuals, etc.
2. Computer networks Handbooks, manuals, etc. 3. Business enterprises—Computer networks Handbooks, manuals, etc. I. Gurugé, Anura. II. Lindgren, Lisa.
TK5102.5.C6323 1999
658.4′038—dc21 99-36004
CIP

This book contains information obtained from authentic and highly regarded sources. Reprinted material is quoted with permission, and sources are indicated. A wide variety of references are listed. Reasonable efforts have been made to publish reliable data and information, but the author and the publisher cannot assume responsibility for the validity of all materials or for the consequences of their use.

Direct all inquiries to CRC Press LLC, 2000 N.W. Corporate Blvd., Boca Raton, Florida 33431.

Trademark Notice: Product or corporate names may be trademarks or registered trademarks, and are used only for identification and explanation, without intent to infringe.

Auerbach is an imprint of CRC Press LLC

No claim to original U.S. Government works
International Standard Book Number 0-8493-9826-6
Library of Congress Card Number 99-36004
Printed in the United States of America 1 2 3 4 5 6 7 8 9 0
Printed on acid-free paper

Contributors

RON ANGERZZAME, *Director, Applications Brand Marketing, Attachmate, Bellevue, WA*

ED BAILEY, *Senior Engineer, IBM, Research Triangle Park, NC*

AL BERG, *Director, Strategic Technologies, NETLAN, Inc., New York, NY*

LISA BOBBITT, *Technical Marketing Engineer, Performance and Design Verification Center, Cisco Systems, Inc., Research Triangle Park, NC*

RANDALL CAMPBELL, *Product Line Manager, Token Ring Switching, Cisco Systems, Inc., Research Triangle Park, NC*

JOHN P. CERAOLO, *Information Systems Manager, Coral Springs, FL*

TIM CLARK, *Vice President, Sales, Tavve Software Company, Durham, NC*

JIM FOXWORTHY, *Director, Product Marketing, VM Software Division, Sterling Software, Reston, VA*

ED FRYMOYER, *emf Associates, Mountain View, CA*

DANIEL GONNEAU, *Founder, Zephyr NetWorks, New York, NY*

STEVE GREER, *Product Marketing Manager, Telecommunications Techniques Corp., Germantown, MD*

ANURA GURUGÉ, *Independent Technical Consultant, Meredith, NH*

GILBERT HELD, *Director, 4-Degree Consulting, Macon, GA*

FRANK HORWITZ, *Founder and Chief Executive Officer, SecureNet Technologies Inc., Lynnwood, WA*

DAVID JOHNSON, *Director, Product Marketing, OpenConnect Systems Incorporated, Dallas, TX*

GARY C. KESSLER, *Senior Member, Technical Staff and Chief Information Officer, Hill Associates, Colchester, VT*

DONNA KIDDER, *Product Line Manager, InterWorks Business Unit, Cisco Systems, Inc., Research Triangle Park, NC*

WILLIAM R. KOSS, *Director of Sales, Sonoma Systems, Marlborough, MA*

LISA M. LINDGREN, *Independent Consultant, Meredith, NH*

ANDRES LLANA, JR., *Consultant, Vermont Studies Group, King of Prussia, PA*

PETER LUFF, *Marketing Manager, Telecommunications Techniques Corp., Germantown, MD*

PHILLIP Q. MAIER, *Program Manager, Secure Network Initiative, Lockheed Martin*

COLIN MICK, *Consultant, Palo Alto, CA*

C. KENNETH MILLER, *Founder and Chairman/Chief Technology Officer, StarBurst Communications, Concord, MA*

JIM MORIN, *Director, Corporate Planning and Market Development, CNT, Minneapolis, MN*

NATHAN J. MULLER, *Consultant, The Oxford Group, Huntsville, AL*

GALE PERSIL, *Product Manager, Network Management, Sterling Software, Reston, VA*

RAJ RAJAGOPAL, *Senior Computer Scientist, Illinois Institute of Technology Research Institute, Potomac, MD*

ROBERT ROSENBAUM, *Strategic and Tactical Marketing Consultant, Acton, MA*

BEN ROTHKE, *Consultant, Information Security Services Group, Ernst & Young, LLP, Iselin, NJ*

LARRY SCHESSEL, *Cisco Systems, Inc., Research Triangle Park, NC*

JOHN P. SLONE, *Senior Systems Architect, Lockheed Martin Corp., Orlando, FL*

BILL STACKPOLE, *Network Manager, Olympic Resource Management, Poulsbo, WA*

RALPH R. STAHL, JR., *Chief Global Information Security Officer, Architecture Design and Engineering, Worldwide Information Systems, AT&T, Dayton, OH*

MARTIN TAYLOR, *Vice President, Network Architecture, Madge Networks, San Jose, CA*

JOHN R. VACCA, *Information Technology Consultant, Pomeroy, OH*

DAVID WADSWORTH, *Java Evangelist, Sun Microsystems of Canada, Markham, Ontario, Canada*

LEO A. WROBEL, *President and Chief Executive Officer, Premiere Network Services Inc., Dallas, TX*

SEAN YARBOROUGH, *Product Marketing Engineer, Telecommunications Techniques Corp., Germantown, MD*

Contents

INTRODUCTION xiii

SECTION I PLANNING FOR COMMUNICATIONS SYSTEMS 1

1 The New Enterprise Network 3
Lisa M. Lindgren

SECTION II ENTERPRISE NETWORK ARCHITECTURE 13

2 Gracefully Transitioning from SNA to IP: Why, How, and When? 15
Anura Gurugé

3 One Corporate Network—Dream or Nightmare? 27
Donna Kidder

4 IPv6: The Next-Generation Internet Protocol 37
Gary C. Kessler

5 Moving toward One Network: Practitioner's Steps to Consolidation of Data and Voice 51
Lisa Bobbitt

6 Integrating Data Centers with Intranets 65
Anura Gurugé

7 A Guide to Internet Telephony 77
Nathan J. Muller

8 Implementing and Supporting Extranets 93
Phillip Q. Maier

9 Internet Site Development in a Bandwidth-Hungry World: An Enterprise Dilemma 105
Daniel Gonneau

SECTION III LAN AND CAMPUS NETWORKING 117

10 High-Speed Token Ring 119
Martin Taylor

11 Should You Migrate Your Token Ring Network to Ethernet? 129
Randall Campbell

12 Introduction to 100Base-T: (Fast and Faster) Ethernet 141
Colin Mick

13 Understanding Gigabit Ethernet Performance 155
Gilbert Held

14 Asynchronous Transfer Mode (ATM) Implementation 163
John R. Vacca

15 Virtual LANs 185
Gilbert Held

16 Multicast Networking 193
C. Kenneth Miller

17 Voice and Video on the LAN 205
Martin Taylor

18 Voice and Data Network Integration 215
Larry Schessel

19 Securing Your Router 225
Gilbert Held

SECTION IV REMOTE ACCESS 233

20 Modem Technology Update 235
Gilbert Held

21 Choosing Asymmetrical Digital Subscriber Lines 241
Gilbert Held

22 Preparing for Cable Modems 249
Gilbert Held

23 Virtual Private Networks: Secure Remote Access over the Internet 257
John R. Vacca

24 ATM Access: The Genesis of a New Network 271
William R. Koss

25 Portable Network Access 283
Gilbert Held

26 Cellular Digital Packet Data: An Emerging Mobile Network Service 297
Nathan J. Muller

SECTION V WIDE AREA NETWORK 313

27 An Overview of the Telecommunications Policy Process 315
Leo A. Wrobel

28 The Future of Wireless Communications 325
Robert Rosenbaum

29 Frame Relay Testing and Training 339
Steve Greer, Peter Luff, and Sean Yarborough

30 Transporting Legacy SNA Traffic over Frame Relay 351
Andres Llana, Jr.

31 Maximizing Mainframe-Based WAN Resources 361
Gale Persil

32 SONET for the Future: Satisfying Today's Users with Tomorrow's Technology. 373
Leo A. Wrobel

SECTION VI DATA CENTER 387

33 Data Center Security: Useful Intranet Security Methods and Tools 389
John R. Vacca

34 Why Parallel Sysplex is a Must for Today's IBM Data Centers 407
Anura Gurugé

35 Web-to-Host with VM/ESA: A Viable Platform for the Internet. 419
Jim Foxworthy

36 Java on the Mainframe 429
Anura Gurugé

37 LAN-to-Mainframe Connectivity Choices 437
Lisa M. Lindgren

38 Storage Area Networks 447
Jim Morin

39 Fibre Channel Architecture, Layers, and Services 461
Ed Frymoyer

40 Mainframe Shops Falling Behind on E-Commerce Due to Y2K 469
Anura Gurugé

SECTION VII CLIENTS AND SERVERS 479

41 TN3270 and TN5250 Internet Standards 481
Ed Bailey

42 Y2K on the Desktop: Where the Enterprise Meets Its Waterloo 493
Ron Angerame

43 Network Computers: Promises and Reality 501
Nathan J. Muller

44 Handheld Network Devices to Access Intranets and Enterprise Networks 513
John R. Vacca

45 Networking Features of the Java Programming Language 527
John P. Slone

46 JavaBeans and Java Enterprise Server Platform 539
David Wadsworth

47 A Quick Overview of Linux 549
Raj Rajagopal

48 Evaluating the Performance of NT-Based Systems 559
Gilbert Held

SECTION VIII SECURITY 571

49 Secure Access to Mission-Critical Applications 573
David Johnson

50 Applets and Network Security: A Management Overview 587
Al Berg

51 Intranet Security 595
John R. Vacca

52 A Look at Java Security 615
Ben Rothke

53 Creating Firewall Policy Following a Functional Model 623
John P. Ceraolo

54 Assessing and Eliminating Virus Threats in Distributed Networks 635
Frank Horwitz

55 Mobile User Security 647
Ralph R. Stahl, Jr.

56 An Introduction to IPSec 663
Bill Stackpole

SECTION IX NETWORK MANAGEMENT 677

57 Network Baselining as a Planning Tool 679
Gilbert Held

58 Working with TCP/IP Utilities 687
Gilbert Held

59 Using RMON to Manage Remote Networks 697
Gilbert Held

60 Proactive Performance Management 707
Tim Clark

61 Using Intelligent Agents to Manage Multivendor Environments 717
Nathan J. Muller

INDEX 729

Introduction

AS WE STAND POISED ON THE THRESHOLD OF A NEW MILLENIUM, it is an appropriate time to reflect on the vast changes that worldwide enterprise networks are undergoing. Just a decade ago many industry luminaries were predicting the collapse of the centralized data center and IT structure. In its place would be a more decentralized client/server model built upon the Open Systems Interconnect (OSI) networking architecture. However, client/server never fully realized all of its promises and OSI floundered.

Now, instead of client/server and OSI, we have the Web-based model and TCP/IP. Together, web-oriented technologies (i.e., browsers, web servers, HTML, Java) and TCP/IP are in the process of completely changing the way in which the enterprise views its network. Instead of serving as primarily an internal utility, the enterprise network is becoming a vital means of delivering products and services and of tying an enterprise more closely with its partners and suppliers. The impact on the very structure of the enterprise network could not be more profound. This new edition of the *Handbook of Communications Systems Management* contains approximately 60 articles that will provide you with detailed information on the different facets of change in the enterprise network.

The *Handbook of Communications Systems Management* has been dramatically restructured, to provide a more logical grouping of articles into discrete sections that bring focus to a particular enterprise networking topic. In addition to the new structure, the content of this edition has been substantially updated. Almost three-quarters of the articles are new. The common theme throughout is the change that the enterprise network is undergoing.

The handbook is divided into nine sections. The first section sets the stage for the remainder by documenting the major trends and business drivers facing enterprise network planners and implementers. The remaining eight sections each focus on a different area or element of the enterprise network, providing in-depth information on the changes and technologies.

In section two, Enterprise Network Architecture, the network-wide architectural changes are detailed. Various aspects of implementing TCP/IP on a large scale in enterprise networks are covered, from migration from SNA to TCP/IP to implementing extranets and intranets.

The next four sections examine different areas of the enterprise network. Section three looks at the technologies that enterprises are implementing in the LAN and campus environment. It covers high-speed and switched LAN technologies, virtual LANs, multicast networking, and the integration of voice and video on the LAN. Section four details the vast new variety of new technologies for gaining remote access to enterprise resources. In section five, issues that enterprise managers are facing in analyzing and evaluating Wide Area Networking options are discussed. Section six provides an overview of many of the complex and fundamental networking changes that are facing the data center manager, paying particular attention to the role of the mainframe and the technologies that are keeping it a vital piece of the enterprise infrastructure.

The network today does not stop at the routers and switches that build its backbone, but extends to encompass the client end systems and servers. Section seven discusses some of the new advances in client and server technology, including the introduction of new devices, new server technologies, and new ways of accessing legacy systems and hosts.

Of critical importance to the enterprise network manager considering implementing new web-based electronic commerce initiatives is the issue of security. Section eight discusses some of the critical security considerations and technologies of today. And, as always, network management is another critical consideration, especially when implementing widespread change in the network. Section nine highlights some important utilities and management tools for use in the enterprise network.

The creation of this handbook is an ongoing process since it is published annually. We always appreciate your comments and thoughts on areas important to enterprise network managers that we should explore more deeply. If you have comments or requests for future issues of the handbook, please do not hesitate to send us your feedback either via electronic mail (lindgren@cyberportal.net) or via mail to the publisher:

Auerbach
Handbook of Communications Systems Management
535 Fifth Avenue, Suite 806
New York, NY 10017

Best regards and cheers,
Anura Gurugé
Lisa M. Lindgren
Meredith, NH

Section I

Planning for Communications Systems

Chapter 1
The New Enterprise Network

Lisa M. Lindgren

THE ENTIRE CONCEPT OF WHAT AN ENTERPRISE NETWORK REALLY IS HAS BEEN DRAMATICALLY CHANGING OVER THE PAST FEW YEARS. The sudden, dramatic, and very public success of new firms like eBay and Amazon is causing established Fortune 500 firms to sit up and take notice of the vast potential of the Internet, corporate intranets, and extranets. The impact on the very structure of the enterprise network will be profound.

Just one decade ago, the enterprise network was an internal utility that primarily served as a means of allowing different parts of an enterprise to communicate and share data efficiently. The backbone network of most large enterprises was built upon IBM's Systems Network Architecture (SNA). Departmental computing, however, was on the rise, creating what was referred to in the press and literature of the time as "islands of computing." Industry visionaries were heralding the end of the mainframe era and the beginning of a decentralized, more proletarian era built upon client/server solutions and Open Systems Interconnect (OSI).

Voice and data networks one decade ago were completely separate entities run by different organizations within the IT structure. Most discussion at the time about integrating voice and data assumed that the data traffic, which was the smaller amount, would ride the voice infrastructure. Time division multiplexers were used in some instances to provide low-level integration of data and voice. In the United States, the majority of enterprises maintained private networks based on leased lines.

A decade ago there was a lot of interest in the interconnection of networks for Electronic Data Interchange (EDI). These connections were cumbersome, however, and required a lot of coordination between the two organizations, thus only the closest partners or suppliers were connected. Value-added networks (VANs) like IBM's Advantis, which provided a way for enterprises to link to a network of networks, provided access to a variety of other partners. Early experiments in providing consumers access to enterprise applications (e.g., the Federal Express package tracking application) entailed shipping a special-purpose terminal or a diskette with PC software that the customer would need to install in order to access the application. This method was both expensive and prone to "pilot-error", thus increasing the help desk workload.

The data networks of a decade ago were pretty small by today's standards in terms of total bandwidth capacity. Local area networks were pervasive but were based on 4, 10, or 16 megabits per second shared media for the most part. The 100 Mbps FDDI LANs were expensive but becoming ordinary in data centers and campus environments with high concentrations of servers and LAN traffic. Communication lines linking locations were often limited to 9600 bps, although 56 Kbps lines were becoming more common.

Discussions of security a decade ago centered around how to keep users from one area of the enterprise from accessing private departmental files once the "islands of computing" were interconnected. A common example was the personnel records maintained on the Personnel Department's LAN server. On the enterprise network (i.e., the SNA network), security was handled on the mainframe with tools like RACF and Top Secret. The physical security of the network was "built-in"; in SNA, a specific device, and therefore, specific end user was predefined in the host configuration file and could not be "spoofed" by another device/user.

The new enterprise network is different in almost every way from the enterprise network of a decade ago. Just five years ago enterprise networks were slowly evolving to support both emerging and open technologies. Then the Web appeared and the world changed almost overnight. New firms like eBay and Amazon emerged and built their businesses purely on a web-based electronic commerce model. The five-year IT plans that laid the groundwork for a step-wise evolution of the enterprise network were scrapped as CIOs and CEOs began to demand a response to the Internet.

BUSINESS DRIVERS

The goal of the client/server revolution was to bring more data and better, graphically-oriented applications to the knowledge workers so that information systems users would become more empowered, make better decisions, and create a more effective enterprise as a result. Client/server technology per se was limited in its ability to create this utopian result. Early implementers of client/server found that the costs of client/server systems were much higher than anticipated and the benefits were much smaller. In large part, the barrier to client/server pervasiveness was the cost and complexity of distributing and maintaining application-specific client software on an enterprise's thousands or tens of thousands of desktop systems.

Despite this impediment, the goal of IT organizations to bring data and new GUI-based applications directly to the end user persisted. It wasn't until the World Wide Web, however, that a cost-effective and easy-to-implement model and set of technologies were widely available. The success of the World Wide Web has spurred the wholesale adoption of corporate intranets.

The same web-oriented technologies are creating unprecedented opportunities for enterprises to fulfill the original promises of EDI by tying an enterprise more tightly with its partners, suppliers, distributors, and customers. This creates what Cisco Systems has referred to as the "Networked Enterprise". The common term for the external intranets that extend from one organization to its partners and its distribution chain is "extranet". Consumer-to-business communication and transactions typically take place over the public Internet. Together, extranets and the Internet form the basis for web-based electronic commerce.

Corporate Intranets

The growth of corporate intranets is quite amazing, especially when you consider that many of the key technologies upon which intranets are built are only about five years old. Recently, International Data Corporation (IDC) performed a survey of large enterprises in the United States and Europe and found that, overall, almost 60 percent of organizations had already implemented an intranet. The organizations in the U.S. were ahead of their European counterparts (85 percent versus 43 percent), but the overall message is clear. Most organizations have already implemented an intranet or are in the process of doing so. Exhibit 1.1 depicts the results of the IDC study.

Intranets are pervasive because internal information is easily published and secured in an internal network. New applications are built relatively easily using web servers and the new visual development tools. And there is no barrier to the end users using this information and the new applications since the web-style interface is second nature to many. In fact, the web-style interface is dramatically impacting the bottom line of existing cost centers.

Consider the case of "1-800" customer call centers. Before the emergence of corporate intranets, most of the applications customer care representatives accessed were stand-alone applications, each with its own unique user interface. In some cases these user interfaces were the hostile, "green-on-black", character-oriented interfaces typical of legacy applications. One large telecommunications firm estimated that they spent two weeks training each new customer care representative, and the turnover in those positions averaged one year. This firm was spending almost 5 percent of an average employee's total tenure on the job in training! By creating a common web-

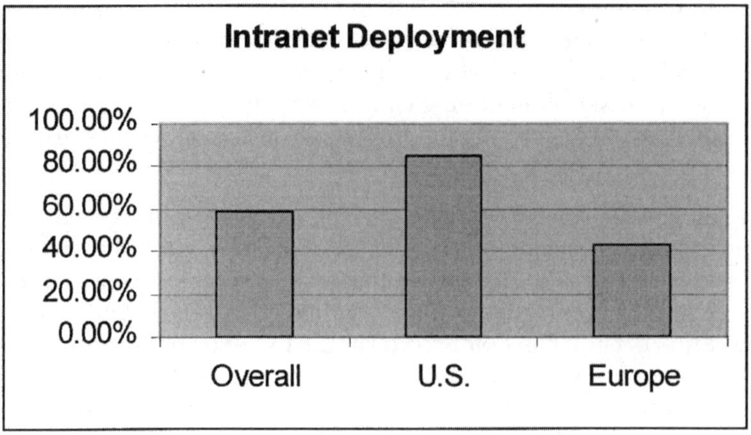

Exhibit 1.1 Intranet Deployment. Source: IDC 500 Report Series: Management Strategies of Worldwide Users, IDC, Sept. 1998

style front end to existing applications, organizations can slash their internal training budgets and dramatically increase productivity.

Today, intranets are being used in many innovative ways, including:

- Publishing internal company information like holiday schedules, policies, job postings, etc.
- Providing a means for employees to electronically submit time sheets, vacation requests, and expense reports
- Posting company-confidential information such as development plans, design documents, business case information, product launch schedules, and meeting agendas and minutes for easy dissemination among appropriate groups of employees
- Creating on-line employee directories and organizational charts
- Providing brand-new applications that draw data from legacy applications and multiple sources

Web-Based Electronic Commerce

One cannot pick up a business publication or listen to a stock market show today without being inundated with the phenomenal growth of the "Internet stocks." Most of the companies included in the list of hot Internet companies (eBay, Amazon, Yahoo) are well-known in part because they are used by consumers to do electronic commerce over the Web. However, business-to-business electronic commerce over the Web is expected to eclipse the size of the consumer-to-business market. As evidence of this, Cisco Systems is currently reporting that 73 percent of its revenue is booked through its web site. A recent study by IDC predicted that the total revenue of web-

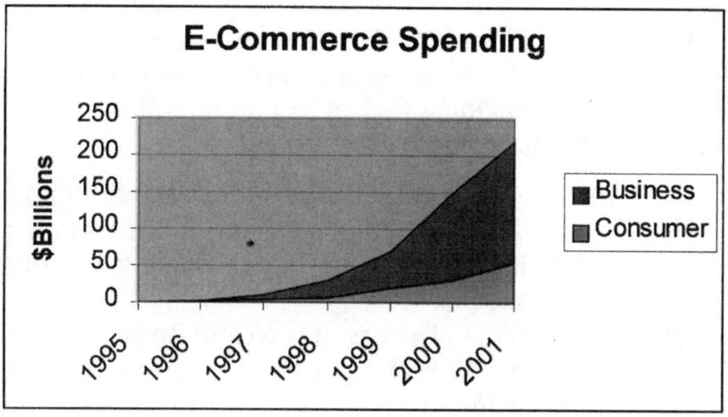

Exhibit 1.2 Web-Based Electronic Commerce Forecast. Source: Lead Electronic Commerce Users: Getting the Enterprise Network Ready (IDC, Sept. 1998)

based electronic commerce would reach approximately $225 billion in 2001. The business-to-business portion of that, however, would be more than 75 percent of that figure. Exhibit 1.2 illustrates IDC's forecast.

Two years ago, delivering products and services over the Web provided the early implementers of web-based electronic commerce with a competitive advantage. Consider FedEx. It was one of the first major commercial enterprises to put a strategic customer service application on the Web with its web-based package tracking. As soon as it did, it experienced rapid growth in its business. While it grew its revenues, it was actually able to lower its costs. The IT department at FedEx estimated that it actually saved between $2 and $5 every time a customer tracked a package via the Web rather than calling its 1-800-GO-FEDEX customer service operation.

Organizations in some industries today are finding that offering products and services over the Web is not a luxury; it is quickly becoming a requirement for competing effectively. One perfect example is the stock brokerage market. Charles Schwab has catapulted itself ahead of some of the most established names in the brokerage industry by extensively investing in its web site and offering web-based trading. The full-service firms are losing business and customers to Schwab, and some are scrambling to offer web-based trading at a competitive rate.

ARCHITECTURAL CHANGES

The enterprise network must change in some pretty dramatic ways to accommodate the new demands of the corporate intranet, extranets, and links to the Internet. The network of a decade ago is not nearly as adaptive, flexible, secure, or scalable as it needs to be in today's web-oriented world. Conceptually, the enterprise network is growing beyond the bounds of the enterprise to encompass partners, suppliers, distributors, and consumers. Enterprise IT organizations must look at key architectural elements of the network with an eye to today's demands:

- *"Re-centralization"*—The hallmark of client/server technology was the downsizing and decentralization of server platforms. Many IT organizations are finding that they need to "recentralize" platforms in order to plan for the scalability required.
- *SNA Migration to TCP/IP*—The urgency to initiate or continue a migration to TCP/IP is heightened with corporate intranet and web-based electronic commerce initiatives.
- *Multiservice Networks*—Enterprises in the forefront are putting in multiservice backbone infrastructures that will incorporate voice, data, and video.
- *Service Providers*—There is an increasing reliance upon service providers, particularly as web-based electronic commerce initiatives take hold.
- *Bandwidth Growth*—The need for faster and faster pipes is growing unabated due to the success of corporate intranets and extranets.
- *Security*—With e-commerce initiatives, one of the primary concerns of unlocking mission-critical systems is with security. IT organizations need to find a way to open the systems while keeping them secure from hackers.

Many IT organizations have already built corporate intranets and/or e-commerce sites. Some of these organizations are finding that the underlying network cannot scale to meet the growth demanded of the new systems. Other organizations are in the planning stages of implementing intranet and e-commerce initiatives and are looking at their networks to determine if they can accommodate the new initiatives. Network planning groups and IT management should review their current network with these architectural elements in mind.

"Recentralization"

Client/server systems were often built upon the model in which a relatively small number of clients would access a specific distributed server for a particular application. A huge number of UNIX servers and NT servers with Microsoft's BackOffice suite were sold to support this model. It scales reasonably well for internal applications, and server clustering technologies have evolved to offer fault tolerance. However, once an enterprise embraces e-commerce, the decentralized model falls apart. An organization that is going to look to extranets and the Internet to deliver its products and services will require the ability to scale to support thousands or tens of thousands of users in a given day. The profile of these sessions or transactions is quite a bit different than for internal applications, but the tolerance for uneven response time is even lower. The total system must be able to scale to support the maximum number of users while providing load balancing and fault tolerance across multiple servers. The technologies now exist, with things like dynamic name servers, to build centralized web server clusters with load balancing and fault tolerance characteristics. These centralized clusters may even be geographically distributed to reach end users spread across the globe, where the closest cluster can be determined based on the end user's location. But these systems are inherently different than simply distributing stand-alone servers as the content and access must be coordinated and administered for high-end scalability and 24x7x52 availability.

SNA Migration to TCP/IP

Some large organizations began to integrate TCP/IP with their mission-critical SNA networks years ago. The early implementers did so for technical reasons (i.e., to achieve redundancy in the network and route around failed components). However, many large enterprises limited this integration to providing TCP/IP in the backbone for redundancy or TCP/IP on the mainframe for file transfers to UNIX systems. They often didn't include desktop conversion to TCP/IP or a wholesale replacement of the network itself.

With the vast success of the web model, IT organizations must now look at the possibility of providing a tighter integration between SNA and TCP/IP resources, or a quickened migration to a complete TCP/IP infrastructure. In order to fully capitalize on the potential of corporate intranets and web-based e-commerce, they need to look at tapping the data and applications that reside on mainframes and other legacy systems. They cannot afford the time or expense of retooling the data and applications to the web model, so they need a way to integrate the legacy data and mission-critical applications with the new model. A variety of products now exist to provide access to legacy data and applications from web browsers.

Multiservice Networks

A sought-after goal for many years has been the consolidation of voice and data into a single network infrastructure. There have been some solutions to this integration in the past, primarily at the lower levels. The difficulty was in the dissimilar nature of the two traffic types. ATM, with its fixed-cell length, promised to integrate voice, data, and even video with quality of service attributes.

ATM is still a viable technology for providing a unified, multiservice network, but some organizations have resisted its widespread implementation for various reasons. With the dominance of TCP/IP, there are new technologies available for providing the multiservice network on top of a TCP/IP infrastructure. Although still early in the life cycle, technologies like Internet telephony and voice-over-IP are being investigated so that organizations can leverage their intranet infrastructure and ties to the Internet. The bottom line is that technologies do exist and are emerging to allow enterprises to build a unified multiservice network to meet their voice, data, and video requirements.

Service Providers

The networks of the past were, at least in the United States, primarily built with leased lines, owned and managed by the enterprise. For some time there has been rising interest in outsourcing the WAN backbone, which has led to the dramatic rise in the popularity of Frame Relay services and of a full suite of managed network services from a variety of different types of service providers. This growth in managed network services is growing unabated and is now reaching down from the very largest corporate customers to the middle tier of companies. But a new model is appearing as well, one that is brought on by the very success of the Internet. Companies of all sizes are looking to manage the costs of supporting a growing number of remote users such as traveling workers and telecommuters. Rather than owning and maintaining their own points of presence with huge modem pools for remote access, they are looking to Internet Service Providers to provide the infrastructure for nationwide or even worldwide remote access at a relatively low cost. This access can be provided securely by implementing Virtual Private Network (VPN) technology, isolating the corporate traffic from the rest of the ISP's traffic. Enterprises should carefully evaluate their existing infrastructure and compute the total cost of ownership, then look for areas that they can outsource to service providers in order to lower costs.

Bandwidth Growth

The shared-media Ethernet LANs of a decade ago were long a bottleneck due to increased traffic within an enterprise. Ethernet switching has been around for a number of years, bringing 10 Mbps dedicated LAN speeds to each individual desktop. Token Ring LANs, with their higher speeds and more deterministic operation, were able to hold out much longer, but even Token Ring users are now demanding switched, dedicated LAN segments.

What almost any network manager who has implemented corporate intranets and/or web-based e-commerce will tell you is that the new networks are sucking up bandwidth almost faster than they can implement solutions. Switched Ethernet has given way to switched Fast Ethernet, and in some cases Gigabit Ethernet. Token Ring manufacturers are providing switched solutions running at 100 Mbps. Backbones that once ran at T1 speeds are being upgraded to support multiple T1 links or OC-3 speeds. The top three sources of bottlenecks today are the LAN backbone, the WAN backbone, and server connections. Enterprises indicate that the top reason for this increased traffic is that the applications are more network-intensive. A vital part of any plan for corporate intranets and web-based e-commerce should be an extensive analysis of current and future bandwidth requirements and an analysis of potential bottlenecks.

Security

In the past, access to mission-critical systems was controlled and limited to known end users. User authentication was usually based on simple user ID and password schemes. Authorization to system resources was predefined (and therefore limited) for each user. Encryption was available but not widely used.

One thing that makes an ordinary data center manager shudder and blanch is the thought of giving access to mission-critical systems to anyone with a web browser. The few but widely publicized "attacks" on these systems merely serve to enhance the fear that was already there. In some cases, the fear of the risks may in fact curtail plans to embrace web-based e-commerce, at least for awhile. However, the security mechanisms available for web-based e-commerce are quite good, and can be integrated with the existing security on legacy systems. IT organizations should evaluate the current set of security solutions available, including: firewalls, encryption, and authentication/authorization systems. In the future, directory services will also play a role in enhancing an organization's ability to provide access to resources easily but with security.

CONCLUSION

The Web is a lot more than a convenient way to chat with others who share our interests, to look up cool information, or to fulfill our voyeuristic fantasies. It has fundamentally changed the relationships between different organizations and between a company and its customers. The technologies upon which the Web is built—the browser, the web server, HTML, HTTP, Java—have finally fulfilled the promises of client/server by bringing information to the end user in a form that is easily understood, easily used, and readily accessible at a very low cost.

The success of the web-based model will have a profound impact on the structure of the enterprise network. The new enterprise network must provide a new degree of scalability with the fault tolerance and availability characteristic of the SNA networks of old. It must seamlessly tie together the new systems and applications while leveraging the vast investment in existing legacy systems and applications at the same time it protects those systems and applications from outside attack. And, if that is not enough of a challenge, the IT organization must implement these changes without disruption to the current, ongoing operation of the network—kind of like changing the tires on a bus as it speeds down the freeway at 60 miles per hour. Luckily, the tools and technologies exist today to make the new enterprise network a reality.

Section II

Enterprise Network Architecture

Chapter 2
Gracefully Transitioning from SNA to IP: Why, How, and When?

Anura Gurugé

WHEN EVALUATING THE MERITS AND IMPLICATIONS OF TRANSITIONING TO A TCP/IP-CENTRIC IT INFRASTRUCTURE, it helps to reflect on the tale of the Great King Canute of England and Denmark (c. 1016) who tried to demonstrate to his adoring subjects that there were powers that even he could not control by showing them that he was powerless to stop the tide from coming ashore. Just as was the case with PCs and LANs, TCP/IP is now an unstemmable technological tide; possibly even a tidal wave. Whether one likes it or not, relishes it or fears it, TCP/IP is here to stay — and will dominate worldwide computing for at least the next two decades, thanks to the endorsement and kudos it gets on a daily basis as the sustaining force behind the Internet miracle.

Mainframe shops today cannot claim unfamiliarity with TCP/IP. Without exception, corporations that use mainframes for their MIS now have a TCP/IP-based intranet in addition to their traditional SNA/APPN or multiprotocol-oriented enterprise network. Most, furthermore, already have a presence on the Internet in the form of a home page, and many are actively exploring the possibilities of using the Internet for electronic commerce, customer support, public relations, product promotions, and global remote access. Not missing out on the tantalizing potential of E-commerce over the Internet, next to that of Y2K concerns, is indubitably the most pressing MIS issue that is being discussed at the highest levels of corporations, starting at the board room. In parallel, intranet-to-intranet communications via extranets are being viewed as the most effective means of streamlining and expediting enterprise-to-enterprise transactions. All of this intranet and Internet (i.e., i·net) activity means that TCP/IP is already being widely used alongside mainframe-based computing systems.

Installing TCP/IP on a mainframe these days is not a difficult, nerve-racking, or laborious undertaking. Extensively proven, extremely efficient, highly scalable, and extremely reliable TCP/IP stacks for mainframes are readily available. IBM claims that more than half of the mainframes running MVS or OS/390 already have TCP/IP installed. Installing TCP/IP on a mainframe facilitates its integration with intranets or the Internet; permits fast, high-speed bulk data transfers with TCP/IP clients or other systems; and, moreover, positions it as a data server for Web-based applications. Once TCP/IP is installed, one could, if required, even have the mainframe acting as a high-capacity Web server. There are companies, such as $9.5B Lafayette Life Insurance (Lafayette, IN), that already have Web servers running on their mainframes — which in the case of Lafayette happens to be an IBM 9672-R24, 3rd generation CMOS-based S/390 Parallel Enterprise Server.

There are significant strategic and tactical advantages to going ahead and installing TCP/IP on a mainframe and moving toward a TCP/IP-centric computing environment. For a start, it provides a solid basis for any and all E-commerce-related initiatives. It can also reduce, sometimes quite significantly, overall capital and operational costs. For example, the browser-based access to SNA solutions that are now readily available from over 40-odd credible vendors for providing unrestricted SNA terminal access across i·nets, totally eliminate the considerable cost associated with installing, managing, and periodically upgrading SNA/3270 emulation software on each and every PC/workstation that needs access to SNA applications.

Using TCP/IP all the way into the mainframe, and then performing SNA conversion at the mainframe per the tn3270(E) standard, also ensures that one no longer needs highly expensive, SNA-oriented communications controllers like the 3745 or the 3746-950. Instead, one can profitably utilize high-performance, low-cost, channel-attached routers such as the IBM 2216-400, Cisco 7500/CIP, or Cisco 7200/CPA as the means of interconnecting the mainframe to the network. Then there are networking-related cost savings. With a TCP/IP-centric infrastructure, one can, albeit with the appropriate security measures (e.g., firewalls), gainfully use the Internet as a way to realize extremely cost-effective remote access for far-flung remote offices, agents, telecommuters, and overseas distributors. Intranets, given that they are based on widely available commodity technology, are also invariably less costly to implement than comparable SNA/APPN or multiprotocol networks. Exhibit 2.1 illustrates a TCP/IP-centric environment.

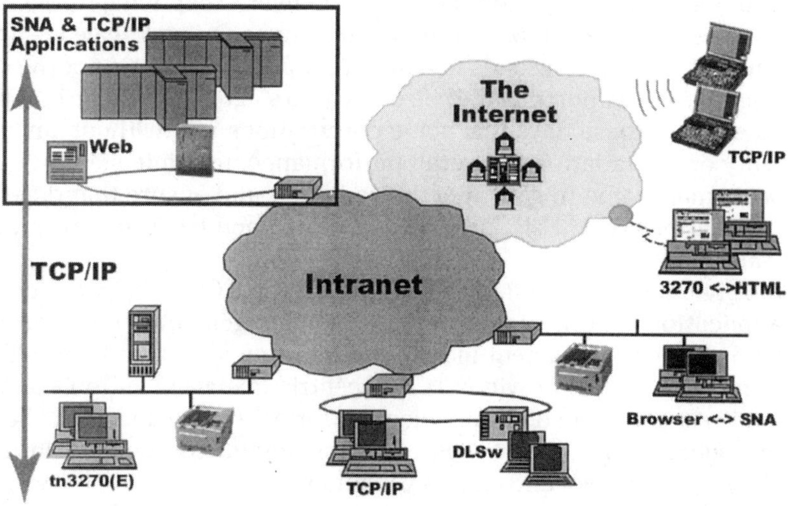

Exhibit 2.1 TCP/IP-Centric Environment. Clients have access to both TCP/IP and SNA resources across an intranet as well as the Internet.

DISPELLING YOUR CONCERNS

Before the advantages of moving to a TCP/IP-centric infrastructure with TCP/IP on the mainframe are articulated any further, it is best to allay any concerns one may have about moving away from SNA.

- Security is no longer the roadblock it used to be with highly proven, bulletproof TCP/IP-specific security solutions for mission-critical commercial computing systems.
- Total unencumbered access to mission-critical SNA/APPN applications running on mainframes or AS/400s is in no way compromised, jeopardized, or even inconvenienced by the installation of TCP/IP on a mainframe and the standardization on a TCP/IP-centric infrastructure. There are a plethora of well-established, standards-based SNA-to-TCP/IP integration technologies, such as tn3270(E) and Data Link Switching (DLSw), that ensure unrestricted SNA access and end-to-end SNA transport across TCP/IP networks.
- Installing TCP/IP on a mainframe and using a TCP/IP-centric i·net for all mainframe access does not prevent one from having ACF/VTAM on that same machine as well. Therefore, one can continue to have the same level of APPN/HPR support that is on one's mainframe today to guarantee that mission-critical SNA/APPN applications will continue to work without any problems.

- Today's TCP/IP stacks for mainframes deliver exceptional throughput and are highly optimized to maximize efficiency and scale easily to support tens of thousands of concurrent users. TCP/IP is not the CPU hog that it was portrayed to be a few years ago. Mainframe TCP/IP is so efficient these days that some corporations run, without any difficulty or degradation in overall performance, multiple stacks on the same mainframe to gain added throughput and ensure that different applications (e.g., FTP and tn3270(E)) can each have their own dedicated stack.
- Incisive, sophisticated, and comprehensive TCP/IP-based network, application, TCP/IP-stack, and system management is now possible with mainframe-resident management systems such as Interlink's e-Control. e-Control provides TCP/IP-centric management tools and facilities for problem determination, performance management, change management (i.e., the configuration and administration of mainframe TCP/IP resources), and capacity planning.
- With today's mature router technology, it is now possible to realize TCP/IP-based networks that are sufficiently resilient and robust to provide high-availability networking with uptimes in excess of 98(+) percent. Today's TCP/IP-centric networks are significantly more reliable and stable than the bridge/router-based multiprotocol networks currently used for transporting SNA/APPN traffic.
- Traffic prioritization between different classes of applications vis-à-vis the TCP/IP network is no longer an issue with today's router software offering functions such as Quality of Service (QoS), Bandwidth Reservation Protocol (RSVP), and highly customizable queuing schemes (e.g., Cisco's Custom Queuing). For those few situations where there is a need to support SNA LU 6.2 Class-of-Service (COS) prioritization on an end-to-end basis, IBM offers a scheme known as Enterprise Extender that permits APPN/HPR routing across IP.
- The continued presence of ACF/VTAM on the mainframe alongside TCP/IP ensures total, uncompromised support for parallel sysplex operation — including multi-node persistent sessions (MNPS), workload balancing, and generic resources.
- High-performance, highly efficient, full-duplex TCP/IP transfers across ESCON channels is not a problem with the TCP/IP-specific CLAW protocol that permits two subchannels to be grouped together for high-throughput and simultaneous bidirectional communications. If anything, TCP/IP channel transfers are significantly faster than SNA/APPN transfers with both IBM or Cisco channel-attached solutions such as the IBM 2216-400 and the Cisco 7500/CIP.
- Mainframe-based TCP/IP printing is not an impediment with tn3270(E) now supporting host print, and with products such as Interlink's very comprehensive Enterprise Print Services (EPS).

CASE AGAINST A LAST-MINUTE SNA REVIVAL

Despite the daily mounting evidence to the contrary, there are still some who believe that IBM will not allow SNA/APPN to succumb to IP, and that there will be a concerted attempt to reestablish SNA/APPN-based networking. IBM recognizes that the role SNA/APPN plays in the future will be restricted to the mainframe in the context of mission-critical applications, and that TCP/IP, unassailably, will be the networking fabric of the future. The following four examples alone should convince the reader that IBM is not just reconciled to, but is in reality one of the greatest advocates of, TCP/IP-centric networking.

- In June 1998, IBM announced an ESCON channel-attachment capability for its flagship 12.8Gbps throughput 8265-17S Nways ATM Switch, which can support 622Mbps ATM uplinks. The only protocol supported across this channel attachment is IP.
- In March 1998, IBM discontinued the 2217 Nways Multiprotocol Concentrator, which was an APPN/HPR-based router that permitted TCP/IP, IPX/SPX, and NetBIOS to be routed end-to-end across an SNA network. The 2217 was the antithesis of a conventional TCP/IP-based router. By discontinuing the 2217, IBM tacitly admitted that there was no call or future for IP-over-SNA routing.
- IBM is avidly promoting the notion of APPN/HPR-over-IP routing with its Enterprise Extender technology, which is now available on the IBM 2216, 2212, and 2210. Cisco, Bay/NT, and others are expected to also support this capability. By promoting the notion of routing APPN/HPR-over-IP, which is the exact opposite of the routing scheme employed by the 2217 that IBM discontinued, IBM is making it very clear that the only WAN networking role it sees for APPN/HPR in the future is within the context of it being used on top of IP.
- The IBM 2216-400 can be attached to an IBM 3746 via an expansion chassis known as the Multi-Access Enclosure (MAE). If one only wants to transfer IP traffic into the mainframe, MAE offers a native, high-speed coupling facility between the 2216-400 and the 3746. If one insists on wanting support for SNA, the best that IBM can offer is a dual Token Ring connection between the 2216-400 and the 3746. When factoring this in with the 8265 IP-only channel-attachment scheme discussed above, it becomes clear that IBM is already positioning itself for an era when most of the mainframe channel traffic is IP based.

THE PRIMARY ADVANTAGES OF MOVING TO MAINFRAME IP

- Enables seamless integration of one's fast-growing intranet with one's mainframe, given that at least 70 percent of the corporate data that one's in-house intranet users require is still on a mainframe rather than on a Web server, NT server, or UNIX system.

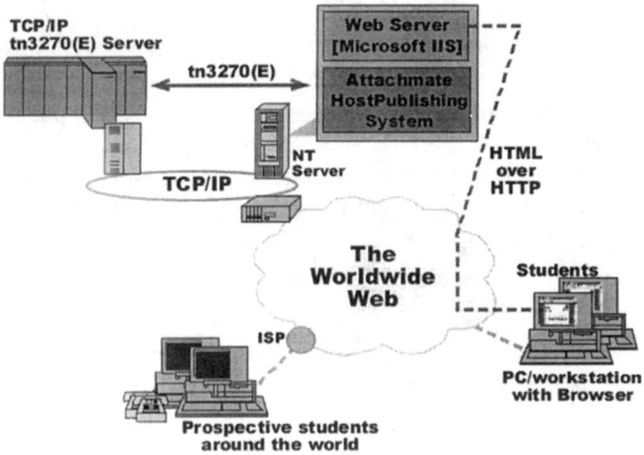

Exhibit 2.2 Ohio State University Diagram. Existing and prospective students are provided with access to mainframe-resident SNA applications over the Web using a totally TCP/IP-centric infrastructure, including TCP/IP and tn3270(E) server on the mainframe.

- Decisively positions one to exploit all the rich potential of E-commerce over the Internet by ensuring that all of the applications and data one may require to enable such commerce is now TCP/IP-ready and can be easily integrated with the necessary Web technology. Business-to-business E-commerce over the Internet is expected to be in excess of $30B by the year 2002.
- Permits one to exploit the Internet as an extremely low-cost means of realizing global remote access to mainframe applications including all mission-critical SNA applications, as shown in Exhibits 2.2 and 2.3. In addition to browser-based access, extremely secure Virtual Private Networking (VPN) solutions — such as those provided by Interlink's NetLOCK V.2.0 — can be used to realize enterprise-specific remote access over the Internet.
- Facilitates and expedites the File Transfer Protocol (FTP)-based file downloads and uploads that one is likely now doing on a daily basis with all of one's distributed servers.
- Allows one to quickly open up mainframe applications for new, Internet-based services such as home banking, online investment, personal travel reservation, and Web-based status checking (e.g., querying the status of an expedited mail item or a cargo shipment), as demonstrated in Exhibit 2.4.
- Greatly minimizes the cost of SNA access by being able to use tn3270(E) or browser-based access to SNA solution. The browser-

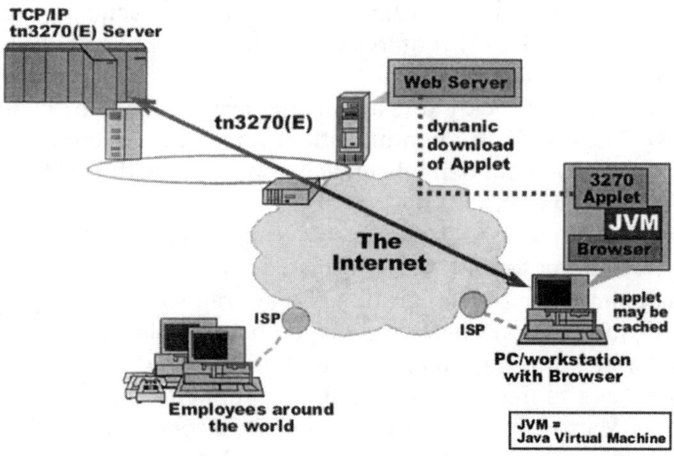

Exhibit 2.3 A Totally TCP/IP-Based System. This system is currently being tried out by a $18B U.S. conglomerate to ensure that telecommuters and mobil users around the world have access to mainframe-resident SNA applications across the Internet.

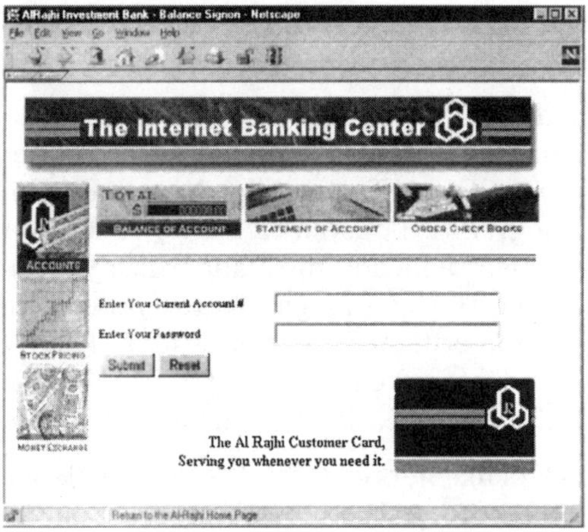

Exhibit 2.4 Actual Screen Shot of a Rejuvenated 3270 User Interface. This was used by a mainframe-centric, Internet-based home banking system realized using browser-based access in the form of 3270-to-HTML conversion.

based access solutions will eliminate the considerable costs associated with installing, managing, and regularly updating SNA/3270 emulation software on individual PCs/workstations by using either an applet-based scheme, where the applet is dynamically downloaded from a Web server, or a 3270-to-HTML conversion scheme, as shown in Exhibit 2.4, which only requires a browser to be present within the client PC/workstation.

- Enables one to quickly phase out the very expensive, SNA-oriented IBM 3745 or IBM 3746 communications controllers in favor of high-performance, low-cost channel gateways such as the IBM 2216-400, Cisco 7500/CIP, or Cisco 7200/CPA.
- Permits one to use the mainframe as a high-capacity, very low-cost-per-user Web server for intranet, extranet, or even Internet applications.
- Greatly simplifies the integration of mainframe data with the new Web applications that are being developed using tools such as NetDynamics 4.0, Bluestone Sapphire/Web, and ColdFusion.
- Eliminates the need for external, low-capacity tn3270(E) gateways such as Microsoft's SNA server by using integrated, highly scalable, mainframe-resident tn3270(E) servers such as the one included within Interlink's e-Access TCPaccess TCP/IP software.
- Gain better channel throughput by using TCP/IP across the channel to a mainframe-resident tn3270(E) server. Phase out the cost and complexity of doing business-to-business transactions using SNA Network Interconnection (SNI) by moving toward a secure, low-cost extranet scheme.

PROVEN TECHNOLOGY TO FACILITATE THE TRANSITION FROM SNA TO IP

The good news is that highly proven and stable technology, from over 40 credible vendors including Interlink, Cisco, OpenConnect Systems, IBM, Attachmate, Wall Data, Eicon Technology, Novell, Farabi Technology, Client/Server Technology, Blue Lobster, etc., is now readily available to facilitate TCP/IP on the mainframe and the standardization on a TCP/IP-centric networking infrastructure — although one still relies on quite a few mainframe-resident, mission-critical SNA applications. The technologies available will enable one to integrate the current SNA/APPN-based environment with the new TCP/IP-centric world in a seamless and synergistic manner.

Athough it may feel like it, one will not be a lone pioneer beating across hitherto uncharted territory. The move from SNA to IP is happening with accelerating pace around the world. In reality, this transition has been happening for the last few years. Enterprises around the world — such as GM, FedEx, Sabre/American Airlines, The Library of Congress, Ohio State University, Royal Jordanian Airlines, Nestles, The Chickering Group, National Van Lines, the state of Idaho, Lafayette Life, Lincoln National Reinsurance, Swiss Air, Al Rajhi Banking & Investment Corp. (Saudi Arabia's largest bank), and Gazprom (a $30B natural gas company in Russia), to name but just a few — have already started to integrate their data center resources with TCP/IP-centric i·nets. Exhibit 2.5 illustrates the solution deployed by Lincoln National Reinsurance.

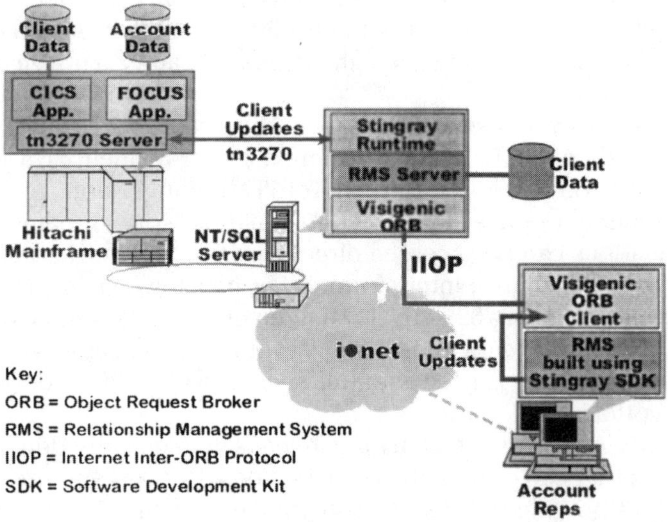

Exhibit 2.5 Mainframe TCP/IP-Based System. This system is being used by Lincoln National Reinsurance Companies, one of the largest reinsurers in the world, to provide its account reps with up-to-date client information across the Web.

To be of use, the technology that enables the transition from SNA to IP needs to be able to accommodate an extremely broad and disparate population of client equipment and functionality. Just some of the entities that need to be dealt with vis-à-vis this transition include PCs; UNIX workstations; coax-attached 3270/5250 terminals; printers; minicomputers; SNA applications that communicate program-to-program using LU 6.2 or LU-LU Session Type 0-based protocols; SNA-only devices (e.g., IBM 4700 Financial Systems); and legacy control units. The PCs, workstations, and printers at remote sites may work in either SNA or TCP/IP mode. Consequently, one will need SNA access technologies to deal with TCP/IP clients, in particular PCs and workstations, and SNA transport technologies to deal with SNA-only clients. This is not a problem. Today, there is a wealth of solid, well-established, field-tested technologies to realize both SNA access and SNA transport in the context of mainframe TCP/IP — and a totally TCP/IP-based network.

Some of the key technologies that will permit an easy transition from SNA to IP include:

- *tn3270(E):* widely used, nearly ten-year-old IETF standard-based access scheme that enables low-cost TCP/IP clients to access SNA applications via a mainframe-resident tn3270(E) server. Today, tn3270(E) is being used by over ten million SNA users. tn3270(E) clients are ubiquitously available from all of the traditional SNA/3270 emulation vendors. All examples shown in Exhibits 2.2, 2.3, and 2.5 utilize tn3270(E) in some form.

- *Browser-based access with 3270-to-HTML conversion:* a thin-client solution (as shown in Exhibit 2.2) where a server-resident SNA-Web gateway performs 3270 datastream-to-HTML conversion, replete with some amount of user interface rejuvenation, so that mainframe SNA applications can be accessed directly from a browser across an i-net. The rejuvenated user interface for home banking (shown in Exhbit 2.4) was realized using 3270-to-HTML conversion. Secure Sockets Layer (SSL)-based authentication and encryption, as available with contemporary browsers, is used with this scheme to provide end-to-end data encryption.
- *Browser-invoked Java or ActiveX applets:* dynamically downloadable applets that can optionally be cached on a PC/workstation hard disk, and that provide tn3270(E) client emulation. This was the technique used in the system shown in Exhibit 2.3. User interface rejuvenation, as well as end-to-end data encryption, is also possible with this technique.
- *Application-specific Web solutions:* such as Interlink ActiveCICX, IBM CICS Web Interface, Interlink ActiveIMX, and Interlink OPEN-IMS, that expeditiously integrate mainframe-resident applications with the Web.
- *Programmatic (or middleware) solutions:* such as IBM MQSeries, Blue Stone Sapphire/Web, or Blue Lobster Stingray SDK, etc., that permit mainframe applications to be interfaced with TCP/IP or Web applications.
- *Data link switching:* like tn3270(E), is a ubiquitous, IETF standards-based encapsulation scheme performed by bridge/routers that permits any kind of SNA/APPN traffic, independent of session type, to be transported end-to-end across a TCP/IP WAN. DLSw ensures that any kind of legacy SNA device or application can be nondisruptively and gracefully accommodated within a TCP/IP-based infrastructure
- *High-performance routing-over-IP:* an alternative to DLSw championed by IBM whereby APPN/HPR-oriented routing is performed across IP. This scheme has the advantage over DLSw in that it can permit APPN-based routing between multiple data centers, and is capable of supporting LU 6.2 COS prioritization on an end-to-end basis over an IP network.

By using one or more of the above technologies, one can gracefully transition from SNA to IP without losing the services of any current mission-critical SNA/APPN applications, sacrificing any functionality, or compromising security or reliability.

THE BOTTOM LINE

With the rapid growth of intranets and the daily increasing significance of the Internet as the next frontier for commerce, the hold that TCP/IP has on commercial sector networking continues to solidify. Even IBM has acknowledged that the role of SNA, APPN, and HPR is going to be relegated to the mainframe as the basis for mission-critical applications. Many of the concerns that MIS professionals had in the past about TCP/IP — such as its security, reliability, and efficiency — are no longer germane. Solid and highly proven TCP/IP solutions are now available from multiple vendors for all aspects of mainframe-oriented computing — whether it be TCP/IP stacks, tn3270(E) servers, security packages, management platforms, applications, channel gateways, or network infrastructures. There really are no impediments to transitioning from SNA to IP. Thousands of companies around the world have already started to standardize on an end-to-end, mainframe-to-PC, TCP/IP fabric. Increasing numbers have already started to use the Internet for remote access and information dissemination. The technology required to successfully and gracefully transition from SNA to IP — such as tn3270(E), browser-based access, and DLSw — is here, is widely available, is cost effective, and is remarkably solid. E-commerce beckons. What are you waiting for?

Chapter 3
One Corporate Network — Dream or Nightmare?

Donna Kidder

FOR THE LAST DECADE, MANY ENTERPRISES HAVE HAD SEPARATE TCP/IP AND SNA NETWORKS. The SNA network handled most mission-critical traffic, while the TCP/IP network carried much of the internal communication — e-mail, file transfers, design specifications, HR data, etc. While maintaining two separate networks is expensive, the benefit of consolidation is always weighed against the cost of degrading the performance of the SNA network. Fear of network degradation and loss of application availability kept many enterprises from consolidating their networks — no one wanted to risk their business on experimental technologies.

Well, it is finally safe to go in the water. The technologies for network consolidation have matured and have been proven in thousands of production networks. These enterprises are enjoying the cost benefits of network consolidation while actually improving SNA network response time and availability.

There are several technologies that allow network consolidation, but the most commonly implemented technology is Data Link Switching (DLSw). However, DLSw lacks some key functionality that is only present in Advanced Peer to Peer Networking (APPN). This article describes both of these technologies, their strengths and weaknesses, and offers recommendations on when and where to use each of them.

DLSw

DLSw is a technology created by the APPN Implementors Workshop (AIW) in 1995 and documented in IETF RFC 1795 and RFC 2166. DLSw provides a standard means for routers to transport SNA data across an IP backbone. The DLSw standard uses TCP as a reliable transport for SNA traffic and calls for local termination of the link layer protocols at either end of the

TCP pipe, as shown in Exhibit 3.1. The standard also defines how DLSw routers communicate with each other to locate resources, establish end-to-end circuits, and address congestion. It defines the frame formats, state machines, and includes media considerations (SDLC and LLC2). The focus of the standard is to ensure interoperability, so it includes a capabilities exchange. Using this capabilities exchange, different vendors' implementations can support additional features without impacting their ability to interoperate with other standard implementations.

What are the benefits of DLSw? First of all, DLSw enhances availability for SNA and NetBIOS traffic. DLSw was the first standard technology to enable nondisruptive rerouting around link failures for SNA traffic. Because the data is transferred in TCP/IP, if a link fails, IP automatically reroutes around the failure. Today's IP routing protocols can reroute within 2 to 15 seconds, so the failure is typically not noticed by an SNA end user. If a frame is dropped in transit, TCP automatically retransmits it, assuring no data is lost. DLSw also eliminates link-layer timeouts. Failure of a DLSw endpoint is disruptive (just as failure of an FEP is disruptive), but recovery is immediate and dynamic.

DLSw simplifies network design by separating the media from the protocol. With DLSw in place, SNA traffic can flow over any combination of campus or carrier technologies, and SNA end systems on disparate media can communicate. DLSw also enables enterprises to leverage state-of-the-art TCP/IP infrastructures and technologies. Once SNA is encapsulated in TCP/IP, intermediate networking gear treats it like TCP/IP and it benefits from all the technology available today. Intermediate Layer 3 switches can process it. It can be compressed, encrypted, accounted for, and even transported over the Internet.[1]

The name "data link switching" is significant. DLSw defines a means to switch SNA traffic between data links. The data link protocols themselves are terminated at each end. This provides two key benefits:

- It eliminates the issue of link-level timeouts and nonproductive polling traffic on wide area links.
- Disparate media can be connected, simplifying network design. With DLSw, one can now enjoy the benefits of attaching via Token Ring at the data center while deploying lower cost Ethernet or maintaining the install base of SDLC controllers at remote sites. This is because DLSw uses a standard media-independent frame format to transfer data between data link switches. The link-layer headings are not added until the data is ready to be dropped on a link. Hence, one can connect remote Ethernet- or SDLC- attached devices to a front-end processor (FEP) over Token Ring. Several enterprises use Token Ring as a means to attach to FEPs because it simplifies FEP configuration

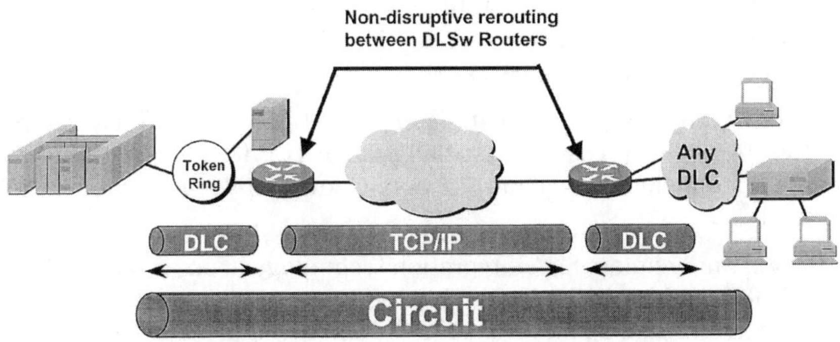

Exhibit 3.1 Conceptual View of DLSw.

and allows for load balancing and automatic backup[2] — either of an FEP or a Token Ring interface coupler (TIC).

DLSw is inherently scalable in a hierarchical SNA network.[3] It relies on the scalability of TCP/IP, which is the same technology used in the largest network in the world — the Internet. And scaling TCP/IP in a hierarchical setting is vastly easier than scaling the Internet. Many networks with thousands of remote sites simply use static routes with defined backup paths. In hierarchical SNA networks, broadcast replication is not an issue. Even if it were, DLSw uses a minimum of broadcasts on the wide area by caching information as it is learned. The key processing load for a DLSw router is LLC2 processing and traffic forwarding. LLC2 timers can be tuned to minimize their impact on central site routers. If the number of SNA physical units (PUs) in the network and the traffic load at the data center are known, then one can determine fairly safely how many central site routers are required to handle the network. As the network grows, one can simply add additional central site routers and balance traffic across them. In general, the number of PUs connected to a single router is more likely based on how many eggs one wants in one basket and not on router capacity.

Finally, DLSw is a relatively simple technology. To implement DLSw, it is key to understand LLC2 and SDLC, but it does not require an extensive knowledge of SNA.

What are the implications on the corporate intranet? When DLSw is implemented, a single TCP/IP infrastructure can support both SNA and IP. All SNA traffic is carried inside of TCP/IP frames; so once the traffic is encapsulated, any intermediate IP router or switch can forward the packet, providing network design flexibility. Routers and switches can distinguish DLSw traffic from other TCP/IP traffic by its unique port number. This is important if SNA traffic needs to be prioritized enroute. Only one routing protocol is required to reroute around failed links, minimizing duplicate and unnecessary routing updates.

Besides technology, one must also look at DLSw from a business perspective. DLSw leverages the vast skill base of TCP/IP — key with the cost of skilled people increasing and the pool of knowledgeable people in SNA decreasing.

DLSw is the premiere SNA integration technology in use today, with well over 450,000 routers deploying it. This fact is key when comparing technologies based on stability, maturity, and the simplicity of integrating two disparate networks after an acquisition. Finally, for most enterprises, the corporate direction for the network is TCP/IP. DLSw allows the network migration to proceed while continuing to provide a transport vehicle for SNA traffic — which is sure to exist for many years to come.

So what are the downsides?

- In parts of the world where bandwidth is at a premium, the overhead introduced by DLSw must be considered. It adds a 40-byte TCP/IP header to every packet, and a 16-byte DLSw header.
- In addition, while DLSw eliminates disruptions caused by link failures, it may add a point of failure in the network (unless the DLSw router is placed in front of an FEP instead of being used as an alternative to an FEP).
- Finally, the DLSw standard does not support SNA class of service (COS).[4] SNA class of service allows prioritization of SNA traffic. For example, interactive SNA traffic can be prioritized ahead of batch SNA traffic. How important is COS? It depends. In a traditional (subarea) SNA network, COS is only supported between SNA nodes — that is, FEPs. Remote SNA controllers and gateways (defined as PU Type 2s) do not understand COS. FEPs can prioritize traffic on their outbound queues based on transmission priority, but remote SNA controllers and gateways cannot prioritize inbound traffic. The bigger issue in most networks today is not prioritizing within SNA traffic, but prioritizing between SNA and TCP/IP.

APPN

APPN is an SNA networking protocol developed by IBM in 1985 (originally for mid-range processors). More recently, APPN has been implemented in VTAM and in all other key communication equipment offered by IBM, including FEPs, 3174s, AS/400s, and the OS/2 Communications Manager. In addition, APPN is implemented in multi-protocol routers from several vendors. APPN is more open than previous SNA protocols. As IBM enhances APPN, the enhancements are offered to the participants in the AIW. Other

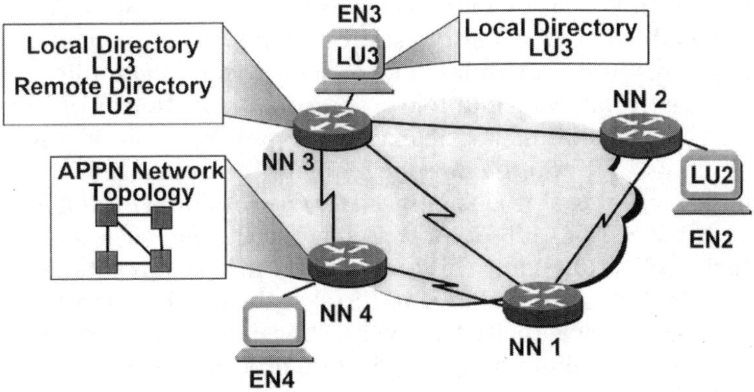

Exhibit 3.2 Conceptual View of APPN.

vendors participating in the AIW can also offer enhancements to APPN. This open architecture approach has enhanced the acceptance of APPN, especially in the data center where it offers strategic advantages over its predecessor, subarea SNA.

APPN natively routes SNA using a link state protocol similar to OSPF to set up routing tables. Routing information is only exchanged among network nodes (NNs), which provide routing function in a similar fashion as IP routers. APPN also has a unique directory function that allows for a central directory in VTAM and uses locators (which are flooded throughout the network) to find resources that are not stored in a local directory or the central directory. Directory functions reside in end nodes (ENs) and NNs, but ENs only maintain a local directory. ENs are similar to IP hosts. Like DLSw, APPN can connect devices on any supported media. Unlike DLSw, APPN supports SNA class of service. Exhibit 3.2 provides a conceptual view of APPN.

The initial implementation of APPN used a packet forwarding protocol called "intermediate session routing," or ISR. ISR performs hop-by-hop error correction and flow control, creating unnecessary overhead and slowing performance. In addition, if any link in the path fails, ISR terminates the SNA sessions using that link, even if an alternate path is available.

For these reasons, high performance routing (HPR) was introduced in the mid-1990s. HPR does error correction at the HPR end points (similar to TCP) and provides stateless, fast packet switching at intermediate HPR nodes (similar to IP). This improves SNA performance in networks with three or more HPR nodes in tandem. More important, however, it enables SNA to natively reroute around link failures without disruption.

APPN, and HPR in particular, have several key advantages. APPN is required to enable use of generic resources in a parallel sysplex complex. The generic resource capability allows one to scale the application processing power by allowing multiple processors to function as one from the perspective of the end user. The end user types in a generic application name (such as CICS), and a coupling facility in conjunction with VTAM finds the processor that is least utilized for the session. The actual session will be with CICS01 or CICS02, for example, but the end user is unaware of the actual application name. HPR is required to take advantage of multinode persistent sessions (MNPS). MNPS takes the parallel sysplex complex to the next level of scalability and availability; it enables nondisruptive rerouting around the failure of the application processor itself.

APPN also gives enterprises a choice of how to connect their mainframes to their networks. Once the mainframe access is via APPN, an enterprise is no longer restricted to using FEPs for mainframe access. It can instead use more versatile, faster, and lower cost channel-attached routers.

With HPR there are no single points of failures other than the HPR end points themselves. If HPR is running in VTAM and in the remote SNA device, then the only way to lose the SNA session is for one of the end points to go away (assuming there is a path between the nodes). This is definitely the key advantage to this technology.

However, APPN and HPR have downsides also. First of all, if a network has a large number of HPR end points, nondisruptive rerouting may be a moot point. A central site error may cause a significant delay in recovering the network and the end users will more than likely hit CTRL-ALT-DEL, terminating their sessions anyway. Most network administrators would opt for short outages rather than long, nondisruptive recovery cycles during which the network is not available.

Most APPN implementations have scalability limitations resulting from directory searches and topology updates. Three-hundred APPN network nodes is about the largest single APPN network one can build. The only way to scale the network beyond 300 is to use a gateway technology known as "border node," which is currently only available from IBM. Border node — once a function of VTAM only — has been introduced to the AIW, so one can expect to see it on other platforms in the near future, but the technology is in its infancy in the marketplace.

APPN networks require careful network design. One needs to worry about the number of network nodes, the number of adjacencies, and the underlying media (not all APPN implementations support all Layer 2 media). APPN takes vastly more SNA knowledge to implement than DLSw.

Finally, in terms of performance, APPN prioritizes SNA traffic by COS, but does not prioritize TCP that is intermingled with SNA. When mixing the two on the same link, APPN performance will tend to degrade. The key reason is that APPN is a well-behaved protocol; in periods of congestion, APPN slows down. When APPN slows down, TCP sees more bandwidth and hence will speed up, resulting in a downward spiral of SNA performance. Even in a pure HPR network, the HPR congestion control mechanism still has kinks that need to be worked out — similar to the TCP slow start problem that was fixed by the IETF a few years ago.

What about the backbone infrastructure? Natively running APPN across your backbone enables physical consolidation as long as all the routers in your backbone support APPN. APPN and TCP/IP can coexist as ships in the night. Each runs its own routing update protocol, causing multiple updates for the same physical failure. Switches in the network have to bridge APPN.

Looking at APPN from a business perspective and putting technical issues aside, APPN has a relatively small installed base; it is less proven in the marketplace and less likely to have all the kinks worked out. The installed base of HPR is even smaller. SNA skills are becoming a hot commodity due to the reducing size of the skilled resource pool. APPN skills are even harder to find. And if the corporate direction is TCP/IP, HPR is a detour.

CONCLUSION

Both APPN and DLSw enable network consolidation onto a router backbone, but each technology addresses certain requirements that the other does not, as illustrated in Exhibit 3.3. Some enterprises will implement one technology or the other, but for many networks the ideal design includes both. APPN is important for many environments — basically, any environment with multiple VTAMs using FEPs for cross domain session routing today. If it is the last SNA enhancement one makes, do it. One can take advantage of the latest parallel sysplex complex capabilities, while at the same time allow for improvement in SNA application availability.

However, it is not necessary to put APPN everywhere to gain the benefits of APPN. The key advantages to APPN require that it be run in the data center. One should also run it in distribution routers — either at the main site or at remote sites where there previously had been FEPs. This minimizes the number of HPR nodes in the network , thereby minimizing HPR recovery time and VTAM processing. It also minimizes the impact of any single failure. In addition, this design provides the highest availability where one needs it, allowing the network to recover quickly and non-disruptively from failures in the data center itself.

Use DLSw to transport SNA traffic from remote sites to distribution routers. DLSw enables a single integrated TCP/IP backbone that carries both IP

Exhibit 3.3 Comparison of APPN and DLSw.

Feature	DLSw Standard	APPN
Nondisruptive recovery from link failures	Yes	Yes, with HPR
Intermediate points of failure	Yes, DLSw routers are intermediate points of failure	With HPR, it is possible to have no intermediate points of failure, although recovery times may be long
Recovery time from any single failure	Implementation dependent, but generally under a minute	With HPR in VTAM, depends on the number of RTP connections; can be in excess of 10 minutes
Consolidated layer 2 backbone	Yes	Yes
Consolidated layer 3 backbone	Yes	No
Routing protocols in consolidated network	Single routing protocol (IP)	Multiple routing protocols (IP and APPN)
SNA routing	No	Yes
Quality of service	Implementation dependent, typically distinguishes DLSw from other traffic by TCP port number	Supports SNA COS to distinguish within SNA, but does not prioritize SNA with other protocols; implementation-dependent extensions are required
Scalability	Up to thousands of DLSw peers in a hierarchical network	Maximum of around 200 NNs in a single network
Installed base	Installed in thousands of networks	Installed in hundreds of networks and/or data centers
Knowledge required to implement	Link layer protocols for SNA, DLSw protocol	Extensive knowledge of both subarea and APPN for most environments

and SNA traffic. Once this infrastructure is in place, costs are minimized (both carrier and management), and migration toward future Web-enabled mainframe applications is simplified. While this design does not eliminate single points of failures, it may in fact speed recovery from failures in large networks. Exhibit 3.4 illustrates a network that leverages the best of both technologies.

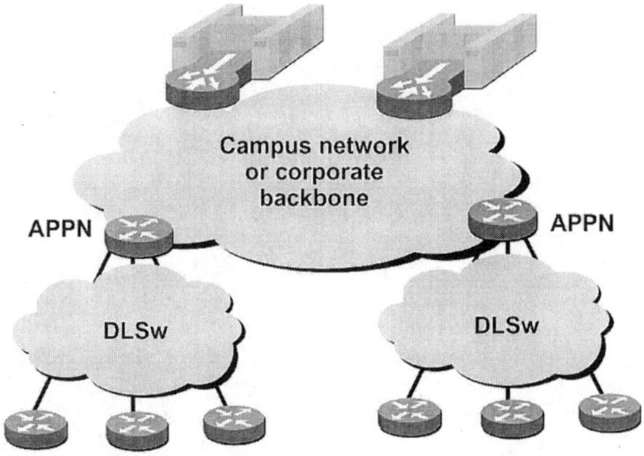

Exhibit 3.4 Using APPN and DLSw to Leverage the Strengths of Each.

Notes

1. One enterprise actually connects a manufacturing site in Australia to a European HQ over the Internet simply to avoid the cost of carrier facilities. In this case, connectivity is the key requirement and the ability to control response time is not cost-justified.

2. Token Ring, in conjunction with source route bridging, allows concurrently active duplicate addresses in the network.

3. Most SNA networks are hierarchical. Traffic goes from remote sites to centrally located data centers.

4. Vendor-specific implementations are available that understand SNA COS and map it to TCP/IP type of service.

Chapter 4
IPv6: The Next-Generation Internet Protocol

Gary C. Kessler

THE ONLY WAY TO COPE WITH THE CHANGES ON THE INTERNET — the number of hosts, types of applications, and growing security concerns — is to implement a new version of the Internet protocol to succeed IPv4. The IETF formed the IPng Working Group to define this transitional protocol, and the result was IPv6 — designed as an evolution from IPv4 rather than as a radical change.

INTRODUCTION

The Internet is historically linked to the ARPANET, the pioneering packet-switched network built for the U.S. Department of Defense in 1969. Starting with four nodes that year, the ARPANET slowly grew to encompass many systems across the US, and connected to hosts in Europe and Asia by the end of the 1970s. By the early 1980s, many regional and national networks across the globe started to become interconnected, and their common communications protocols were based on the TCP/IP suite. By the late 1980s, the number of host systems on these primarily academic and research networks could be counted in the hundreds or thousands. In addition, most of the traffic was supporting simple text-based applications such as E-mail, file transfers, and login.

By the 1990s, however, users discovered the Internet, and commercial use, previously prohibited or constrained on the Internet, was actively encouraged. Since the beginning of this decade, new host systems have been added to the Internet at rates of up to 10 percent per month, and the Internet has been doubling in size every 10-12 months for several years. By January 1997 there were more than 16 million hosts on the Internet, ranging from PC-class systems to supercomputers, on more than 100,000 networks worldwide.

The number of connected hosts is only one measure of the Internet's growth. Another way to quantify the change, however, is in the changing applications. On today's Internet it is common to see hypermedia, audio, video, animation, and other types of traffic that were once thought to be anathema to a packet-switching environment. As the Internet provides better service support, new applications will spark even more growth and changing demographics. In addition, nomadic access has become a major issue with the increased use of laptop computers, and security concerns have grown as a result of the increased amount of sensitive information accessible via the Internet.

IPV6 BACKGROUND AND FEATURES

The IP was introduced in the ARPANET in the mid-1970s. The version of IP commonly used today is IPv4, described in RFC 791.

Although several protocol suites (including OSI) were proposed over the years to replace IPv4, none succeeded because of IPv4's large and continually growing installed base. Nevertheless, IPv4 was never intended for today's Internet in terms of the number of hosts, types of applications, or security concerns.

In the early 1990s, the IETF recognized that the only way to cope with these changes was to design a new version of IP to become the successor to IPv4. The IETF formed the IPng Working Group to define this transitional protocol, ensuring long-term compatibility between the current and new IP versions, and support for current and emerging IP-based applications.

Work started on IPng in 1991, and several IPng proposals were subsequently drafted. The result of this effort was IPv6, described in RFCs 1883 to 1886; these four RFCs were officially entered into the Internet Standards Track in December 1995.

Differences between IPv4 and IPv6

IPv6 is designed as an evolution from IPv4 rather than as a radical change. Useful features of IPv4 were carried over in IPv6, and less useful features were dropped. According to the IPv6 specification, the changes from IPv4 to IPv6 fall primarily into the following categories:

- **Expanded addressing capabilities** The IP address size is increased from 32 bits to 128 bits in IPv6, supporting a much greater number of addressable nodes, more levels of addressing hierarchy, and simpler autoconfiguration of addresses for remote users. The scalability of multicast routing is improved by adding a scope field to multicast addresses. A new type of address, called anycast, is also defined.

- **Header format simplification** Some IPv4 header fields have been dropped or made optional to reduce the necessary amount of packet processing and to limit the bandwidth cost of the IPv6 header.
- **Improved support for extensions and options** IPv6 header options are encoded to allow for more efficient forwarding, less stringent limits on the length of options, and greater flexibility for introducing new options in the future. Some fields of an IPv4 header are optional in IPv6.
- **Flow labeling capability** A new QOS capability has been added to enable the labeling of packets belonging to particular traffic "flows" for which the sender requests special handling, such as real-time service.
- **Authentication and privacy capabilities** Extensions to support security options, such as authentication, data integrity, and data confidentiality, are built into IPv6.

Improved Terminology of IPv6

IPv6 also introduces and formalizes terminology that, in the IPv4 environment, are loosely defined, ill-defined, or undefined. The new and improved terminology includes:

- **Packet** This is an IPv6 PDU, comprising a header and the associated payload. In IPv4, this would have been termed "packet" or "datagram."
- **Node** This is a device that implements IPv6.
- **Router** This is an IPv6 node that forwards packets based on the IP address, not explicitly addressed to itself. In former TCP/IP terminology this device was often referred to as a gateway.
- **Host** This represents any node that is not a router. Hosts are typically end-user systems.
- **Link** This is a medium over which nodes communicate with each other at the data link layer (e.g., an automated teller machine, a frame relay, an SMDS, a WAN, or an Ethernet or Token Ring LAN).
- **Neighbors** These are nodes attached to the same link.

IPV6 HEADER FORMAT

The format of an IPv6 header is shown in Exhibit 4.1. Although IPv6 addresses are four times the size of IPv4 addresses, the basic IPv6 header is only twice the size of an IPv4 header, thus decreasing the impact of the larger address fields. The fields of the IPv6 header are:

- **Version** This represents the IP version number (4 bits). This field's value is 6 for IPv6 and 4 for IPv4. This field is in the same location as the version field in the IPv4 header, making it simple for an IP node to quickly distinguish an IPv4 packet from an IPv6 packet.

Exhibit 4.1 IPv6 Header Format.

Version	Priority	Flow Label
PayloadLength	Next Header	Hop Limit
Source Address		
Destination Address		

Exhibit 4.2 Sample Values of Next Header Field.

Value	Contents of the next header
1	Internet Control Message Protocol (ICMP)
6	Transmission Control Protocol (TCP)
17	User Datagram Protocol (UDP)
43	Routing header
44	Fragment header
58	Internet Control Message Protocol version 6 (ICMPv6)
59	Nothing; this is the final header
60	Destination Options header
89	Open Shortest Path First (OSPF)

- **Priority** This enables a source to identify the desired delivery priority of the packet (4 bits).
- **Flow label** This is used by a source to identify associated packets needing the same type of special handling, such as a real-time service between a pair of hosts (24 bits).
- **Payload length** This is the length of the portion of the packet following the header in octets (16 bits). The maximum value in this field is 65,535; if this field contains zero, it means that the packet contains a payload larger than 64K bytes and the actual payload length value is carried in a jumbo payload hop-by-hop option.
- **Next header** This identifies the type of header immediately following the IPv6 header and uses the same values as the IPv4 protocol field where applicable (8 bits). The next header field can indicate an options header, higher layer protocol, or no protocol above IP. Sample values are listed in Exhibit 4.2.
- **Hop limit** This specifies the maximum number of hops that a packet may take before it is discarded (8 bits). This value is set by the source and decremented by one by each node that forwards the packet; the

packet is discarded if the hop limit reaches zero. The comparable field in IPv4 is the TTL field; it was renamed for IPv6 because the value limits the number of hops, not the amount of time that a packet can stay in the network.

- **Source address** This is the IPv6 address of the originator of the packet (128 bits).
- **Destination address** This is the IPv6 address of the intended recipients of the packet (128 bits).

IPV6 ADDRESSES

To accommodate almost unlimited growth and a variety of addressing formats, IPv6 addresses are 128 bits in length. This address space is probably sufficient to uniquely address every molecule in the solar system.

IPv6 defines three types of addresses:

1. A unicast address specifies a single host.
2. An anycast address specifies a set of hosts, such as a set of FTP servers for a given organization. A packet sent to an anycast address is delivered to one of the hosts identified by that address, usually the "closest" one as defined by the routing protocol.
3. A multicast address also identifies a set of hosts; a packet sent to a multicast address is delivered to all the hosts in the group.

There is no broadcast address in IPv6 as in IPv4, because that function is provided by multicast addresses.

IPv4 addresses are written in dotted decimal notation, where the decimal value of each of the four address bytes is separated by dots. The preferred form of an IPv6 address is to write the hexadecimal value of the eight 16-bit blocks of the address separated by colons (:), such as FF04:19:5:ABD4:187:2C:754:2B1. The leading zeros do not have to be written, and each field must have some value.

IPv6 addresses often contain long strings of zeros because of the way in which addresses are allocated. A compressed address form uses a double colon (::) to indicate multiple 16-bit blocks of zeros; for example, the address FF01:0:0:0:0:0:0:5A could be written as FF01::5A. To avoid ambiguity, the "::" can only appear once in an address.

An alternative hybrid address format has been defined to make it more convenient to represent an IPv4 address in an IPv6 environment. In this scheme, the first 96 address bits (six groups of 16) are represented in the regular IPv6 format, and the remaining 32 address bits are represented in common IPv4 dotted decimal, for example, 0:0:0:0:0:0:199.182.20.17 (or::199.182.20.17).

Exhibit 4.3 Address Prefix Allocation (From RFC 1884).

Allocation	Prefix (Binary)	Fraction of Address Space
Reserved	0000 0000	1/256
Unassigned	0000 0001	1/256
Reserved for NSAP Allocation	0000 001	1/128
Reserved for IPX Allocation	0000 010	1/128
Unassigned	0000 011	1/128
Unassigned	0000 1	1/32
Unassigned	0001	1/16
Unassigned	001	1/8
Provider-Based Unicast Address	010	1/8
Unassigned	011	1/8
Reserved for Geographic-Based Unicast Addresses	100	1/8
Unassigned	101	1/8
Unassigned	110	1/8
Unassigned	1110	1/16
Unassigned	1111 0	1/32
Unassigned	1111 10	1/64
Unassigned	1111 110	1/128
Unassigned	1111 1110 0	1/512
Link Local Use Addresses	1111 1110 10	1/1024
Site Local Use Addresses	1111 1110 11	1/1024
Multicast Addresses	1111 1111	1/256

Address Prefix Allocation

One of the goals of the IPv6 address format is to accommodate many different types of addresses. The beginning of an address contains a 3- to 10-bit format prefix defining the general address type; the remaining bits contain the actual host address in a format specific to the indicated address type. Exhibit 4.3 represents an address prefix allocation (from RFC 1884).

The Provider-Based Unicast Address

The provider-based unicast address is an IPv6 address that might be assigned by an ISP to a customer. Exhibit 4.4 shows a provider-based unicast address format. This type of address contains a number of subfields, including the following:

- **Format prefix** This indicates the type of address as provider-based unicast. It is always 3 bits coded "010."
- **Registry identifier** This identifies the Internet address registry from which the ISP obtains addresses.
- **Provider identifier** This identifies the ISP. This field contains the address block assigned to the ISP by the address registry authority.

Exhibit 4.4 Provider-Based Unicast Address Format.

3 bits	5 bits	n bits	56-n bits	64 bits

- **Subscriber identifier** This identifies the ISP's subscriber. This field contains the address assigned to this subscriber by the ISP. The providerID and subscriberID fields together are 56 bits in length.
- **Intrasubscriber** This contains the portion of the address assigned and managed by the subscriber.

IPv4-Compatible Addresses

Another particularly important address type is the one that indicates an IPv4 address. With more than 16 million hosts using 32-bit addresses, the public Internet must continue to accommodate IPv4 addresses, even as it slowly migrates to IPv6 addressing,

IPv4 addresses are carried in a 128-bit IPv6 address that begins with 80 zeros (0:0:0:0:0). The next 16-bit block contains the compatibility bits, which indicate the way in which the host/router handles IPv4 and IPv6 addresses. If the device can handle either IPv4 or IPv6 addresses, the compatibility bits are all set to zero (0), and this is termed an "IPv4-compatible IPv6 address"; if the address represents an IPv4-only node, the compatibility bits are all set to one (0xFFFF) and the address is termed an "IPv4-mapped IPv6 address." The final 32 bits contain a 32-bit IPv4 address in dotted decimal form.

Multicast Addresses

IPv6 multicast addresses provide an identifier for a group of nodes. A node may belong to any number of multicast groups. Multicast addresses may not be used as a source address in IPv6 packets or appear in any routing.

All multicast addresses, as shown in Exhibit 4.5, begin with eight ones (0xFF). The next four bits are a set of flag bits (flgs); the three high-order bits are set to zero; and the fourth bit (T-bit) indicates a permanently assigned ("well-known") multicast address (T = 0) or a nonpermanently assigned ("transient") multicast address (T = 1). The next four bits indicate the scope of the address, or the part of the network for which this multicast address is relevant; options include node-local (0x1), link-local (0x2), site-local (0x5), organization-local (0x8), or global (0xE).

The remaining 112 bits are the group identifier which identifies the multicast group, either permanent or transient, within the given scope. The interpretation of a permanently assigned multicast address is independent of the scope value. For example, if the World Wide Web (WWW) server group is assigned a permanent multicast address with a group identifier of 0x77, then:

Exhibit 4.5 Multicast Address Format.

8	4	4	112 bits

- FF01:0:0:0:0:0:0:77 would refer to all WWW servers on the same node as the sender.
- FF02:0:0:0:0:0:0:77 would refer to all WWW servers on the same link as the sender.
- FF05:0:0:0:0:0:0:77 would refer to all WWW servers at the same site as the sender.
- FF0E:0:0:0:0:0:0:77 would refer to all WWW servers in the Internet.

Finally, a number of well-known multicast addresses are predefined, including:

- **Reserved multicast addresses** These are reserved and are never assigned to any multicast group. These addresses have the form FF0x:0:0:0:0:0:0:0, where x is any hexadecimal digit.
- **All nodes' addresses** These identify the group of all IPv6 nodes within the given scope. These addresses are of the form FF0t:0:0:0:0:0:0:1, where t = 1 (node-local) or 2 (link-local).
- **All routers' addresses** These identify the group of all IPv6 routers within the given scope. These addresses are of the form FF0t:0:0:0:0:0:0:2, where t = 1 (node-local) or 2 (link-local).
- **The DHCP server/relay-agent address** This identifies the group of all IPv6 DHCP servers and relay agents with the link-local scope; this address is FF02:0:0:0:0:0:0:C.

IPV6 EXTENSION HEADERS AND OPTIONS

In IPv6, optional IP layer information is encoded in separate extension headers that are placed between the IPv6 basic header and the higher-layer protocol header. An IPv6 packet may carry zero, one, or more such extension headers, each identified by the next header field of the preceding header, and each containing an even multiple of 64 bits (see Exhibit 4.6). A fully compliant implementation of IPv6 includes support for the following extension headers and corresponding options:

- **The hop-by-hop options header** This header carries information that must be examined by every node along a packet's path. Three options are included in this category. The Pad1 option is used to insert a single octet of padding into the options area of a header for 64-bit alignment, whereas the PadN option is used to insert two or more octets of padding. The Jumbo Payload option is used to indicate the length of the packet when the payload portion is longer than 65,535 octets. This option is employed when the payload length field is set to zero.
- **The routing header** This header is used by an IPv6 source to list one or more intermediate nodes that must be visited as part of the packet's path to the destination; this option is functionally similar to IPv4's loose and strict source routing options. This header contains a list of addresses and an indication of whether each address is strict or loose.

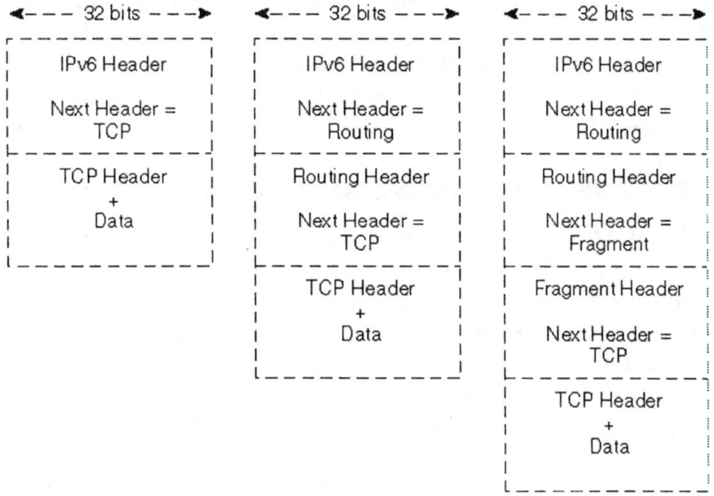

Note:
The left-hand column represents a TCP segment encapsulated in IP without additional options, the middle column represents a TCP segment following a routing header, and the right-hand column represents a TCP segment fragment following a fragment header following a routing header.

Exhibit 4.6 IPv6 Extension Header Examples.

If an address is marked strict, it means that this node must be a neighbor of the previously addressed node; if an address is marked loose, this node does not have to be a neighbor of the previous node.

- **The fragment header** This header is used by an IPv6 source to send packets that are larger than the MTU on the path to the destination. This header contains a packet identifier, fragment offset, and final fragment indicator. Unlike IPv4, where fragmentation information is carried in every packet header, IPv6 only carries fragmentation/reassembly information in those packets that are fragmented. In another departure from IPv4, fragmentation in IPv6 is performed only by the source and not by the routers along a packet's path. All IPv6 hosts and routers must support an MTU of 576 octets; it is recommended that path MTU discovery procedures (per RFC 1981) be invoked to discover and take advantage of those paths with a larger MTU.
- **The destination options header** This header carries optional information that has to be examined only by a packet's destination node(s). The only destination options defined so far are Pad1 and PadN, as described above.
- **The IP AH and IP ESP** These are IPv6 security mechanisms (a section on IPv6 security appears later in this chapter).

With the exception of the hop-by-hop option, extension headers are only examined or processed by the intended destination nodes. The contents of each extension header determine whether or not to proceed to the next header and, therefore, extension headers must be processed in the order that they appear in the packet.

IPV6 QUALITY-OF-SERVICE (QOS) PARAMETERS

The priority and flow label fields in the IPv6 header are used by a source to identify packets needing special handling by network routers. The concept of a flow in IP is a major departure from IPv4 and most other connectionless protocols; flows are sometimes referred to as a form of connectionless virtual circuit because all packets with the same flow label are treated similarly and the network views them as associated entities.

Special handling for nondefault quality of service is an important capability for supporting applications that require guaranteed throughput, end-to-end delay, and jitter, such as multimedia or real-time communication. These QOS parameters are an extension of IPv4's TOS capability.

The priority field allows the source to identify the desired priority of a packet. Values 0 through 7 are used for congestion-controlled traffic, or traffic that backs off in response to network congestion, such as TCP segments. For this type of traffic, the following priority values are recommended:

- Zero is recommended for uncharacterized traffic.
- One is recommended for "filler" traffic (e.g., Netnews).
- Two is recommended for unattended data transfer (e.g., E-mail).
- Three is recommended for reserved traffic.
- Four is recommended for attended bulk transfer, such as FTP or HTTP.
- Five is also recommended for reserved traffic.
- Six is recommended for interactive traffic (i.e., Telnet).
- Seven is recommended for Internet control traffic (i.e., routing protocols and SNMP).

Values 8 through 15 are defined for noncongestion-controlled traffic, or traffic that does not back off in response to network congestion, such as real-time packets being sent at a constant rate. For this type of traffic, the lowest priority value (8) should be used for packets that the sender is most willing to have discarded under congestion conditions (e.g., high-fidelity video traffic), and the highest value (15) should be used for those packets that the sender is least willing to have discarded (e.g., low-fidelity audio traffic).

The flow label is used by a source to identify packets needing nondefault QOS. The nature of the special handling might be conveyed to the network routers by a control protocol such as the RSVP, or by information within the flow packets themselves, such as a hop-by-hop option. There may be multiple active flows from a source to a destination, as well as traffic that is not associated with any flow (i.e., flow label = 0). A flow is uniquely identified by the combination of a source address and a nonzero flow label. This aspect of IPv6 is still in the experimental stage and future definition is expected.

IPv6 SECURITY

In the early days of TCP/IP, the ARPANET user community was small and close, and security mechanisms were not the primary concern. As the number of TCP/IP hosts grew, and the user community became one of strangers (some nefarious) rather than friends, security became more important. As critical and sensitive data travels on today's Internet, security is of paramount concern.

Although many of today's TCP/IP applications have their own devices, security should be implemented at the lowest possible protocol layer. IPv4 has few, if any, security mechanisms, and authentication and privacy at lower protocol layers is largely absent. IPv6 builds two security schemes into the basic protocol.

IP Authentication Header

The first mechanism is the IP authentication header (RFC 1826), an extension header that can provide integrity and authentication for IP packets. Although many different authentication techniques are supported, use of the keyed message Digest 5 (MD5, described in RFC 1321) algorithm is required to ensure interoperability. Use of this option can eliminate a large number of network attacks, such as IP address spoofing. This option is also valuable in overcoming some of the security weaknesses of IP source routing.

IPv4 provides no host authentication. It can only supply the sending host's address as advertised by the sending host in the IP datagram. Placing host authentication information at the Internet layer in IPv6 provides significant protection to higher-layer protocols and services that currently lack meaningful authentication processes.

IP Encapsulating Security Payload

The second mechanism is the IP encapsulating security payload (ESP, described in RFC 1827), an extension header that can provide integrity and confidentiality for IP packets. Although the ESP definition is algorithm-independent, the DES-CBC is specified as the standard encryption scheme to ensure interoperability. The ESP mechanism can be used to encrypt an entire IP packet (tunnel-mode ESP) or just the higher-layer portion of the payload (transport-mode ESP).

These features add to the secure nature of IP traffic while actually reducing the security effort; authentication performed on an end-to-end basis during session establishment provides more secure communications even in the absence of firewall routers.

ICMPv6

The ICMP provides error and information messages that are beyond the scope of IP. ICMPv6 is functionally similar to ICMPv4 and also uses a similar message format and forms an integral part of IPv6. ICMPv6 messages are carried in an IPv6 datagram with a next header field value of 58.

ICMPv6 error messages include:

- **Destination unreachable** This is sent when a packet cannot be delivered to its destination address for reasons other than congestion.
- **Packet too big** This is sent by a router when it has a packet that it cannot forward because the packet is larger than the MTU of the outgoing link.
- **Time exceeded** This is sent by a router when the packet's hop limit reaches zero, or if all fragments of a datagram are not received within the fragment reassembly time.
- **Parameter problem** This is sent by a node that finds some problem in a field in the packet header that results in an inability to process the header.

ICMPv6 informational messages are echo request and echo reply (used by IPv6 nodes for diagnostic purposes), as well as group membership query, group membership report, and group membership reduction (all used to convey information about multicast group membership from nodes to their neighboring routers).

MIGRATION TO IPv6

When IPv4 became the official ARPANET standard in 1983, use of previous protocols ceased and there was no planned interoperability between the old and the new. This is not the case with the introduction of IPv6.

Although IPv6 is currently being rolled out for the Internet backbone, there is no scheduled date of a flash cut from one to the other; coexistence of IPv4 and IPv6 is anticipated for many years to come. The sheer number of hosts using IPv4 today suggests that no other policy even begins to make sense. IPv6 will appear in the large ISP backbones sooner rather than later, and some smaller service providers and local network administrators will not make the conversion quickly unless they perceive some benefit from IPv6.

Exhibit 4.7 Common Short-Term Scenario where an IPv4 Network Interconnects IPv6 Networks.

The coexistence of IPv4 and IPv6 in the network means that different protocols and procedures need to be accommodated. In one common short-term scenario, IPv6 networks will be interconnected via an IPv4 backbone (see Exhibit 4.7). The boundary routers will be IPv4-compatible IPv6 nodes, and the routers' interfaces will be given IPv4-compatible IPv6 addresses. The IPv6 packet is transported over the IPv4 network by encapsulating the packet in an IPv4 header in a process called "tunneling." Tunneling can also be performed when an organization has converted a part of its subnet to IPv6. This process can be used on host-host, router-router, or host-router links.

Although the introduction of IPv6 is inevitable, many of the market pressures for its development have been rendered somewhat unnecessary because of parallel developments that enhance the capabilities of IPv4. The address limitations of IPv4, for example, are minimized by use of CIDR. Nomadic user address allocation can be managed by the DHCP servers and relay agents. QOS management can be handled by the RSVP protocol. And the IP authentication header and encapsulating security payload procedures can be applied to IPv4 as well as to IPv6.

This is not meant to suggest that IP vendors are waiting. IPv6 has already started to appear in many new products and production networks. Support for IPv6 on several versions of UNIX have been announced by such organizations as Digital Equipment Corp.; IBM Corp.; INRIA (or The French National Institute for Research in Computer Science and Control); Japan's WIDE Project; Sun Microsystems, Inc.; the SICS; and the U.S. Naval Research Laboratory.

Other companies have announced support for IPv6 in other operating environments, including Apple Computer, Inc.'s MacOS; FTP Software, Inc.'s DOS/Windows; Mentat's STREAMS; Novell, Inc.'s NetWare; and Siemens Nixdorf, Inc.'s BS2000. Major router vendors that have announced support for IPv6 include Bay Networks, Inc.; Cisco Systems, Inc.; Digital Equipment Corp.; Ipsilon Networks; Penril Datability Networks; and Telebit Corp.

6bone Trials

One of the important proving grounds of IPv6 is the 6bone, a testbed network spanning North America, Europe, and Japan, which began operating in 1996. The 6bone is a virtual network built on top of portions of today's IPv4-based Internet, designed specifically to route IPv6 packets. The goal of this collaborative trial is to test IPv6 implementations and to define early policies and procedures that will be necessary to support IPv6 in the future. In addition, it will demonstrate IPv6's new capabilities and will provide a basis for user confidence in the new protocol.

For most users, the transition from IPv4 to IPv6 will occur when the version of their host's operating system software is updated; in some cases, it means running dual-stacked systems with both versions of IP. For larger user networks, it may make sense to follow the model of the larger global Internet — in particular, to predesign the IPv6 network topology and addressing scheme, to build a testbed IPv6 network with routers and a DNS, and then slowly to migrate applications, users, and subnetworks to the new backbone. The lessons learned from the 6bone activity are useful for individual networks as well as for the Internet backbone.

CONCLUSION

The transition to IPv6 has already started, even though most Internet and TCP/IP users have not yet seen new software on their local systems or on local networks. Before IPv6 can be widely deployed, the network infrastructure must be upgraded to employ software that accommodates the new protocol.

In addition, the new address format must be accommodated by every TCP/IP protocol that uses addresses. The DNS, for example, has defined an AAAA resource record for IPv6 128-bit addresses (IPv4's 32-bit addresses use an A record) and the IP6.INT address domain (IPv4 uses the ARPA address domain). Other protocols that must be modified for IPv6 include DHCP, the ARP family, and IP routing protocols such as the RIP, OSPF protocol, and the BGP. Only after the routers and the backbones are upgraded will hosts start to transition to the new protocol and applications be modified to take advantage of IPv6's capabilities.

Chapter 5
Moving toward One Network: Practitioner's Steps to Consolidation of Data and Voice

Lisa Bobbitt

EVERYWHERE, EVERYONE IS LOOKING AT THE WEB, THE INTERNET, AND THE INFORMATION THAT THEY AVAIL TO US TO MAKE A DIFFERENCE IN THE WAY WE DO BUSINESS. With the changes required for the year 2000 as well as new applications enhancing the work process, TCP/IP has grown from an academic open standard protocol to the strategic application and network base of choice for many corporations. Everything from mainframe information access to intra-office phone service is expected to travel over the IP foundation, giving us the first opportunity to truly integrate all our communication needs over one network.

With one network as an end goal, how does one get from the current environment to this nirvana? Currently, many corporations have divided the communication mission into separate and distinct processes, staffs, and infrastructures to ensure that the network, data center/applications, and voice services meet their current business needs. To bring these services together onto a common base, a smooth migration and operation must be planned and implemented.

In working with customers whose goal is one network, the Cisco Systems design verification team has been exposed to the problems and solutions used to reach this goal. As customers verify their designs and prove their infrastructures' ability to meet their networking needs in a lab environment, some basic practicalities have been observed. Breaking this immense task into four major steps, a series of four articles will look at both the obstacles and solutions given the customers' current environments. The four steps are:

1. Ensure that the IP infrastructure can provide the access, availability, performance, and scalability required for corporatewide TCP/IP applications.
2. Determine the policies and levels of service needed to support time-sensitive applications such as SNA and voice over the IP network. If these levels of service are higher than those in Step 1, provide the additional infrastructure and functions as needed. They include quality of service in bandwidth and prioritization, security, and policies required to manage these services.
3. Verify the SNA over IP technology and make changes to the existing SNA network, end users/devices, and applications as needed to efficiently use the IP infrastructure.
4. Verify the Voice-over-IP technology and make changes to the existing voice network, end devices, and applications as needed to effectively use the IP infrastructure.

There are two other factors that have been observed in this author's lab while working with customers.

1. The selection of Layer 2 technology. IP runs on all of the switching technologies, most of the WAN protocols, and many of the end devices in use today.
2. The education required to consolidate the skills of the three staffs. Exchange of knowledge on voice, data center, and network is necessary for effective design, operations, and troubleshooting. However, given the ability of staffs to cross-teach and the wealth of documentation and classes available, this article will not delve further into this as the division of control and responsibility in each corporate environment is unique to that environment and its personnel.

While these factors will affect the overall convergence of the communications networks, they are outside the scope of this article and will not be discussed herein.

MAKING AN IP INFRASTRUCTURE CORPORATE READY

IP networks in their simplest form require devices to have an IP address (access) and a path between the two devices on which to exchange data (availability). If only these two devices are on the network, only the users of the two devices can use the information at the full speed of the media and end stations (performance). Building the corporate IP backbone — where thousands of end users need access, multiple paths are needed to ensure nondisruptive availability, and performance encompasses not all users — is a nontrivial task (see Exhibit 5.1).

Exhibit 5.1 Benefits and Concerns of IP Networks.

Benefits Added	Current Network Concerns
Competitive advantage of information to your business	Cost of overall system in capital, expenses, and staff
Access for employees, suppliers, customers to provide a competitive advantage for your company	Limited pool of IP addresses Statically defined end devices
24×7 Availability of information and applications for web sites and business operations	Devices that do not converge sufficiently fast to support availability requirements Multiple path costs versus the cost of unavailable
Scalability/performance to support existing and future business requirements	WAN costs (VCs, CIR, etc.) Number of Sites Requiring Change

One can look at some of the scenarios that Cisco Systems customers have verified when building their robust IP infrastructures. There are many technologies and functions available to provide the levels of service required for the corporate IP network foundation. The designers and implementers who come for design verification of IP functions look for assurance that they have chosen the best technology available to meet their companies' ongoing business needs. Exhibit 5.2 indicates some of these technologies and functions.

BUILDING THE IP FOUNDATION

When customers initially focus on access and availability, IP addressing is their first step.

Addressing

The ultimate goal for most customers is to make addressing as dynamic and painless as possible — controlling the policy instead of the static lists of addresses during all the moves, adds, and changes that occur in the network. DHCP and DNS are being used more and more to accomplish this. Migrating the devices to this dynamic environment takes time and effort because every end station must be configured to support the dynamic environment. Many make these changes as end stations are upgraded. Others make wholesale changes. Either way, the most important step is to plan the addressing scheme to support the availability and scalability requirements of the network. Addresses need to be assigned so summarization can occur.

The next addressing decision is determined by the amount of redundancy required.

Exhibit 5.2 Access, Availability, and Performance of Technologies and Functions

Technology/Function	Access	Availability	Scalability/ Performance
Fixed length subnet mask IP addressing	Basic	—	Basic
Variable length subnet mask IP addressing	Advanced	—	Advanced
Single path	Basic	Basic	Basic
Duplicate path	Advanced	Advanced	Advanced
RIP, IGRP routing protocols	—	Basic	Basic
EIGRP routing protocol	—	Advanced	Advanced
OSPF routing protocol	—	Advanced	Advanced
BGP routing protocol	—	Advanced	Advanced
STATIC IP routing	Basic	Basic	Basic
Fixed IP addresses for end users	Basic	Basic	—
Dynamic Host Configuration Protocol (DHCP)	Advanced	Basic	—
Fixed IP address for server	Basic	Basic	—
Domain Name Services (DNS)	Advanced	Basic	—
DNS with load balancing	Advanced	Advanced	—
Virtual addressing of servers	Advanced	Advanced	—
Hot standby router protocol	—	High	—
Policy routing	—	Advanced	Advanced
Fast switched policy routing	—	Advanced	Advanced
NAT	Advanced	—	—
Compression	—	—	Advanced
Disaster recovery	Advanced	Advanced	Advanced

Testing functions such as DHCP or DNS with load balancing determine the dynamics of access in their environments. Virtual addressing of servers combines access and availability by allowing servers to be added and taken out of service without any actions required from the end users' perspective. Using these dynamic ways of assigning a source address or obtaining the destination address allows less disruption during moves and changes in the network. Exhibit 5.3 displays a diagram of an environment using both virtual addressing and load balancing of functions. During the test, this design allowed servers to be added and taken out of service without affecting the availability of the information to the end users.

Customers' tests recommend the following:

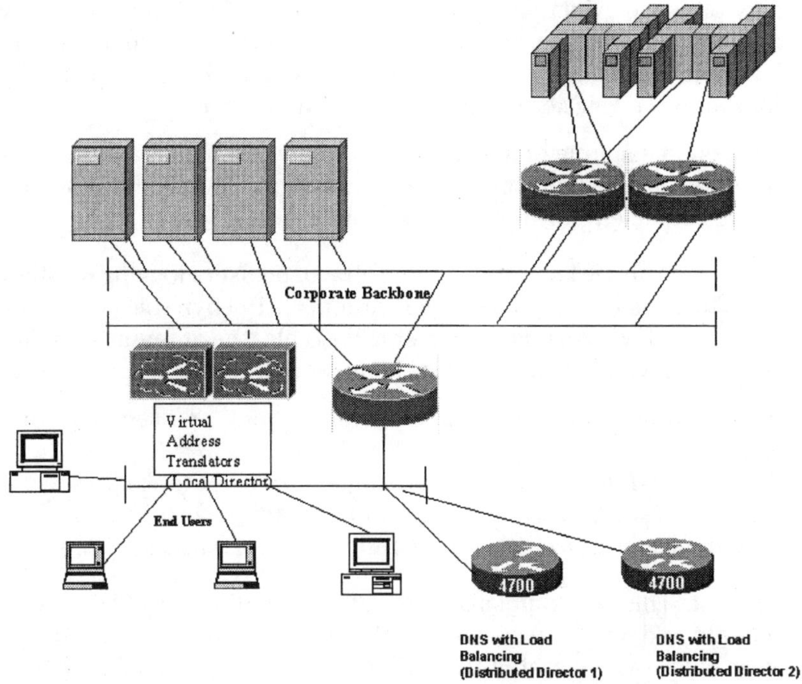

Exhibit 5.3 Environment Using Both Virtual Addressing and Load Balancing of Functions.

- Use Dynamic Host Configuration Protocol (DHCP) for any end-user device that supports it.
- Use Domain Name Service (DNS) to get server addresses when there is more than one physical device providing the service or backup to the service. In fact, whenever one adds new end users, DNS should be considered a standard way to access services to avoid having to change every end device at a later time.
- If load balancing will allow one to use the service platforms more effectively, the DNS service being used should provide this. Load-balancing via round-robin if all the servers are local or by distance if the servers are distributed are two of the more favored designs of Cisco customers. This is especially true for web sites that are supporting advertising and marketing endeavors for national or international mass markets.
- If the servers truly mirror the same services, load balancing using virtual addresses greatly adds to availability and scalability. With the ability to advertise one address but scale to meet the demands of the end users with multiple servers, the servers can be checked for not

only physical availability but for memory and CPU utilization, as well as determining if a service, such as FTP, is currently running. With this function, distributing the work across unequal resources is possible. This allows every server to be used most efficiently.

Following is an example of a test that has allowed a financial trading company to determine if its level of access and availability, as well as performance, was sufficient for its business requirements.

Logon Test. This test simulated the traffic of 60 users logging on simultaneously. This was accomplished by sending 30, 100-byte packets at a rate of 10PPS. An FTP was used to transfer a 24KB file, the amount of traffic the customer observed during an actual login, and this transfer. was timed.

- With no other traffic running, the transfer took 28.3 seconds (benchmark).
- With the 30, 100-byte packets running, the transfer took 28.43 seconds.
- The processor utilization on the central router was 0 percent.
- The processor utilization on the remote router was 15 percent.

Trade Test. This test simulated the traffic during trading. Data was sent to 59 clients and an FTP was used to simulate a single trade. In this case, multicast data was generated to all of the remote locations. The data was a 500-byte packet which was sent at the rate of 1 PPS. A successful test would allow for ten trades per minute. We would be able to complete over 100 from the results.

- With no other traffic running, the transfer took 0.55 seconds.
- With multicast traffic running, the transfer took 0.55 seconds.
- The processor utilization on the central router was 0 percent.
- The processor utilization on the remote router was 15 percent.

This test allowed the customer to roll out its network knowing that growth was also possible in that environment.

ROUTING PROTOCOLS

Once the end point addressing question is answered, the requirement for constantly available paths becomes the issue. As one of Cisco Systems' network design architects indicates in the following quote, it begins with an understanding of the critical paths in one's business and the level of service required.

> Network design is a process of connecting devices with the goal of achieving a highly available network. To achieve highly available networks, you need to stay focused on the critical path, that is, the relationship between the user and how information flows to the user. Therefore, on a highly available network there are failures, but those downed events don't generate phone calls. Let the number of users af-

fected on a particular path determine the investment you make. So for example, in the core of your campus network, you would invest in a more reliable design than at the access points of your network.

Physical paths and the determination of the logical paths for the data are the mainstay of availability.

1. Determination of the level of availability must be made. If a remote site does not need to be available all the time, a dial-in or casual connection can be made. If a remote site can handle delay in its service request being handled, a single path may be sufficient, knowing that there may be times when it will not be up. The service-level agreement for availability for each site is a policy decision.
2. If a resource requires 24×7 availability and the infrastructure is designed with no single point of failure in the path, it becomes the duty of the routing protocol to utilize the multiple paths. Working toward nondisruptive paths between end users and servers, tests of OSPF, and EIGRP are the bases of most of the tests performed in the Cisco lab.

Both OSPF and EIGRP are robust routing protocols for IP foundations. Both can converge in less than ten seconds. When determining which is the best choice for a particular environment, the following tests have been implemented:

- *Interoperability between multiple vendors.* OSPF as a standard is the choice if multiple vendors are being used.
- *IP address layout.* The addressing scheme required by each routing protocol differs. EIGRP works with any network addressing scheme, while OSPF requires a very structured network address layout. The cost of changing the addresses in the existing environment often leads customers to choose EIGRP.
- *OSPF Area 0 determination.* Is there a natural center of the network that can be used as Area 0 for OSPF? Is this center's infrastructure robust enough to handle all the traffic in the network? EIGRP does not require a designated core.
- *Routing protocol convergence.* How well does the routing process converge with redistributing the other routing processes? How quick is the routing table available after a network failure? Both EIGRP and OSPF are designed for fast convergence.

One customer wanted to determine the viability of EIGRP and OSPF in failure situations. The tests included:

- determining routing protocol path recovery in the event of a total network failure
- testing network recovery loads on mission-critical applications
- headquarter site disaster recovery

The goal was to determine network recovery capability and traffic rerouting in the event of a catastrophic failure at the headquarters I or II locations, defined as Atlanta or St. Louis. With the results of the tests, it was clear to all participants that the switches were capable of rerouting connections from failed trunks in subsecond response time with either of the two routing protocols. Fail-over of connections from one trunk to another was not the important issue. When a single truck failed, the entire network (22 sites switched from primary to disaster recovery switch) was backed up within 35 seconds. The choice of either EIGRP or OSPF then became a decision of other routing devices in the network as their addressing scheme supported the summarization required for OSPF and was useful in EIGRP.

The routing protocol of choice must also provide the third leg of support: performance. As seen in Exhibit 5.4, convergence performance when there are multiple paths is a major decision factor for availability. However, routing protocols alone cannot ensure high-performance IP networks. Performance is most directly affected by the amount of bandwidth available as determined by infrastructure selection and policies. Infrastructure designs minimizing routing hops and broadcast domains also enhance end-to-end performance.

Given a well-designed physical infrastructure, the routing protocol then again must have the ability to use all the paths as efficiently as possible. Load balancing allows the use of all the bandwidth purchased in a WAN or LAN environment. Where some routing protocols such as RIP or RIP V2 only allow a single path in use, OSPF and EIGRP both allow use of all equal-cost paths as determined by policy.

Performance is sometimes focused on the single device's ability to meet the network needs during a disaster scenario. In the test displayed in Exhibit 5.5, the performance of a single router was the focus. It needed to be able to handle a double load during a failure scenario. While under full loading of PVCs (600) and data traffic, the primary WAN router was failed via physical link failure. All 600 virtual circuits were moved to the secondary router, which was already running a full load. The performance of the router was found to meet the requirements of disaster recovery for the network.

The fourth leg of a strong IP foundation is scalability. The ability for the IP network to quickly expand and contract as required by the dynamics of the business has become a necessity in today's electronic business environment. Questions such as the ones below are the focus of many Cisco Systems scalability tests.

1. Is our web site able to support the information requests/business applications? Are all the end users able to access it when they want to?

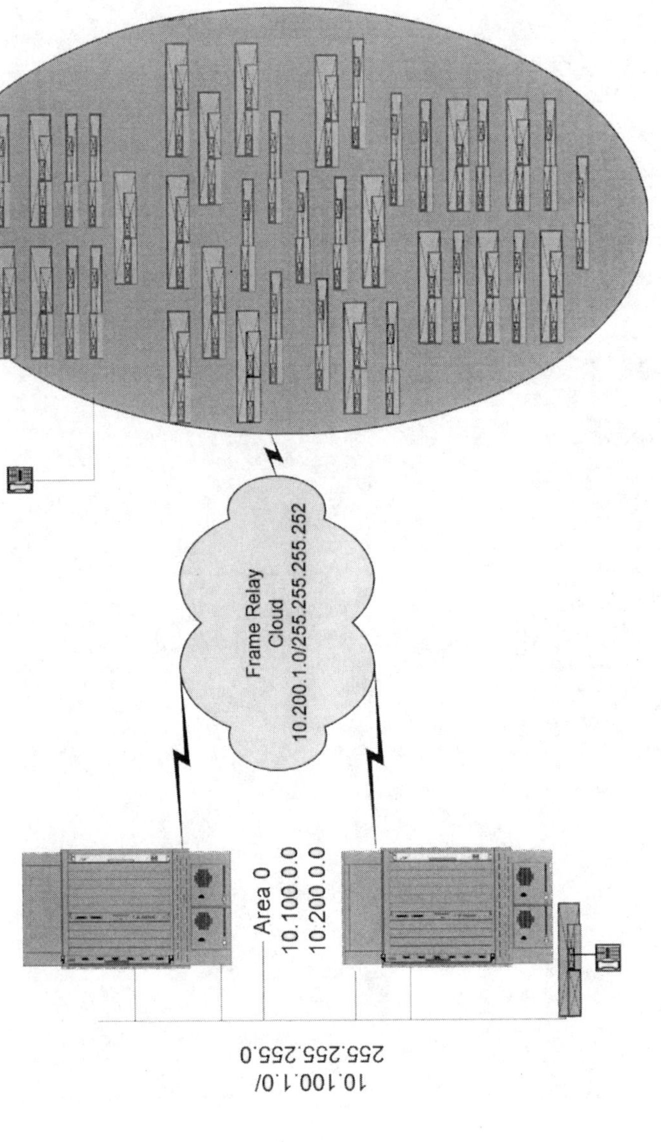

Exhibit 5.4 OSPF Design.

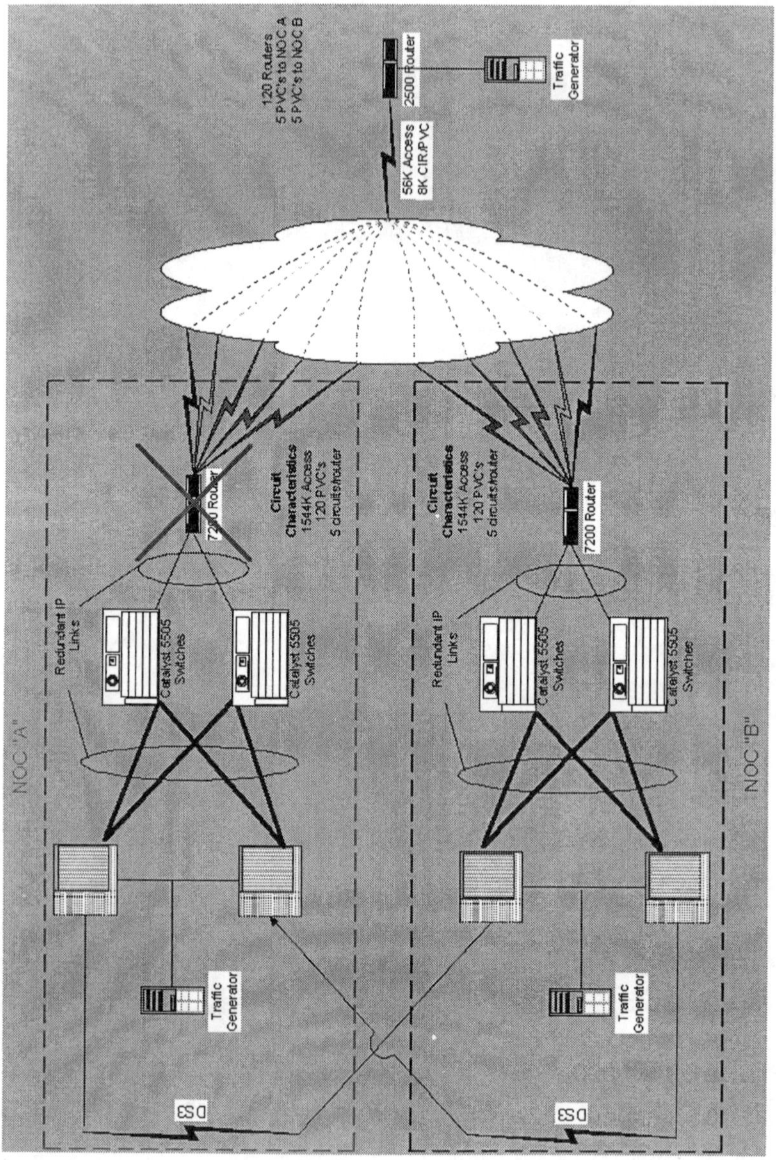

Exhibit 5.5 Test of Performance of Single Router.

2. How does one scale a routing protocol when the internal gateway protocol (EIGRP or OSPF) hits its limits?
3. Is the physical infrastructure sufficient for current and short-term data traffic?

The ultimate goal for the network is to be able to support any sales, marketing, manufacturing, advertising, or support issue that one's business requires. Scalability requirements range from the internal IP network to the external connections to partners, vendors, and customers. One such scalability test really brought home clearly the requirement of robust IP scalability for electronic business for Cisco lab engineers.

A leading technology manufacturer requested that the Cisco Systems lab determine the best routing technology to use as the gateway to the Internet. They wanted to test 50,000 BGP routes in their routing tables and traffic from three million different random client IP addresses. This volume was determined after a television advertisement highlighting their web site ran during the previous Super Bowl. The company had not been prepared for the overwhelming response at halftime, and some of its potential customers were not able to reach the customer's web information as the site was not scaled to meet the demand. They wanted to be prepared for the next time.

Web-based transactions were simulated by causing the web clients to send flows that consisted of five consecutive 64-byte request packets and having the web server send seven consecutive 384-byte response packets. Four 75xx routers with RSP4s, 256MB of memory, nine VIP2-40s each with a single PA-FE-TX interface, and one Catalyst 5500 with multiple 10/100 FE ports simulated the web site. With this test, the customer found the appropriate level of route processor and memory to handle its environment (see Exhibit 5.6).

In scaling a network across multiple sites or countries as shown in Exhibit 5.7, the limits of the internal gateway protocols such as EIGRP or OSPF start to be reached. For customers at this point, the Cisco Systems lab has verified designs using BGP to add another layer of hierarchy to the IP network.

The environment pictured in Exhibit 5.7 was verified for providing inherent stability to a large distributed network via BGP and strategic IP addressing/summarization. The previous network topology did not implement significant levels of redundancy or resilience in all cases. This test demonstrated the new routing structure and its ability to demonstrate the continued operation of the network under various failure scenarios within the environment.

The failures tested were representative of network failures that have impacted the correct operation of the network in the past. The purpose of the

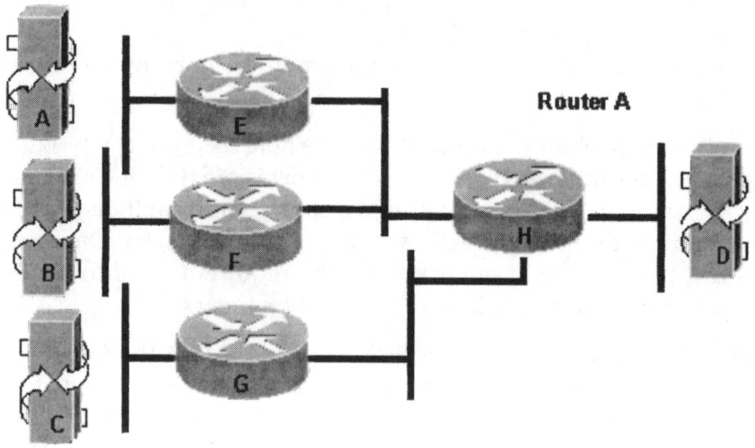

A, B, and C were the source of the "WEB requests"

B was the source of the "WEB replies"

E, F, and G were ISP Peers

H was the outer firewall

Exhibit 5.6 Lab Topology.

test was to identify any failures that could cause lasting or severe impact on network devices. The regions/networks (Asia, Germany, North and South America, Europe, and the core net) were connected to the core and interregional routing exchanges were via BGP. BGP was also be used between Austria and Europe, with IBGP peering between Vienna and Graz in the Europe region, and the Munich distribution routers via route reflectors. In these regions, EIGRP was the regional routing protocol. The tests were successful in determining alternate paths in 30 seconds or less (i.e., once the regional routing protocol EIGRP learned the new path, BGP added no extra latency in its learning process). The customers also verified that their migration order would work. The tests also focused on the need to ensure that the addressing scheme was structured such that summarization could minimize the routing tables. This summarization provided for even further growth in the future.

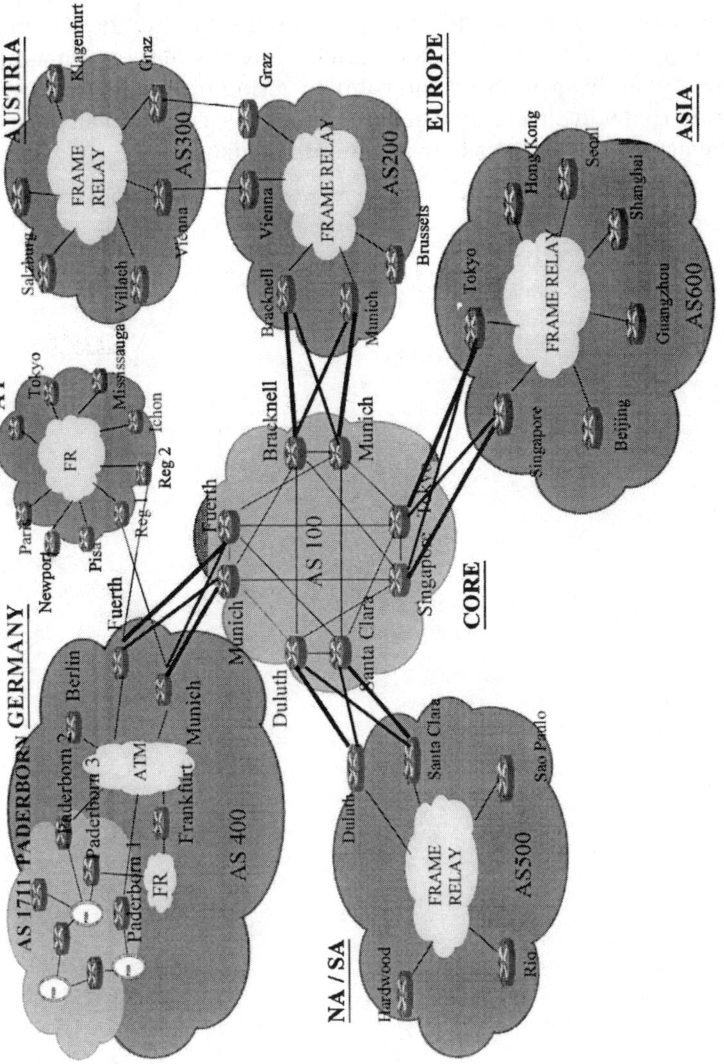

Exhibit 5.7 A Network Scaled Across Multiple Countries.

CONCLUSION

To make an IP infrastructure the networking foundation, the IP network must provide corporate-level services. With the addressing technologies and routing protocols available today, this is possible. Determining the level of access, availability, performance, and scalability appropriate for one's business requirements is the first step. With this knowledge in hand, ensure that IP network hardware and software vendors provide the functions needed to build an end-to-end IP foundation for internal and external business. Use the dynamic addressing and robust routing protocols to provide the foundation that will allow the IP network to support any and all requirements as the electronic business environment flourishes.

Chapter 6
Integrating Data Centers with Intranets

Anura Gurugé

NEARLY ALL ENTERPRISES THAT HAVE MAINFRAMES OR LARGE, NETWORKED AS/400S NOW HAVE AN INTRANET. Most, in addition, already have a presence on the Internet in the form of a home page, and many are actively exploring the possibilities of using the Internet for electronic commerce, customer support, and as an ultra cost-effective means of global remote access. In parallel, intranet-to-intranet communication via extranets is being viewed as the means of streamlining and expediting enterprise transactions. There is, however, a beguiling disconnect vis-à-vis these new strategic and burgeoning TCP/IP-centric networks and the traditional data center functions that continue to be imperative for the day-to-day operations of these enterprises. Very few enterprises at present have tightly integrated their intranets with their data centers. This is despite the fact that up to 70 percent of the vital data, and many of the mission-critical applications required by these enterprises, are still likely to reside on their mainframes or AS/400s. That is akin to baking an apple pie with no apple filling.

Integrating an intranet with a data center is not simply a matter of implementing TCP/IP on a mainframe or AS/400 along with a web server. Many of the host resident, mission-critical applications still required were developed, typically 15 years ago, such that they only work in Systems Network Architecture mode. The nearest that one can come to making theses applications TCP/IP compatible is to use them in conjunction with a host resident or "Off-Board" tn3270(E) (or tn5250 in the case of AS/400s) server which will perform standards-based SNA-to-TCP/IP protocol conversion. Otherwise, the applications will have to be rewritten to work in TCP/IP mode. This is not feasible since the cost and effort of doing so for the $20 trillion installed base of SNA mission-critical applications would make all the tribulations associated with the Y2K challenge appear trivial!

While some of the data center resident data could be accessed using an Open Database Connectivity type scheme, this is certainly not true for all of the data center resources. Some data, especially if stored on "flat files" or nonrelational databases (such as IBM's still widely used Information Management System), can only be accessed via SNA applications. In other instances, the data make sense only when combined with the "business logic" embedded within an SNA mission-critical application. In addition to these crucial SNA applications, there is inevitably a large installed base of SNA-only "legacy" devices such as IBM 4700 Financial Systems, automated teller machines, and control units that still need to be supported. Thus, there is a need for explicit SNA-related technologies in order to get the most from your host vis-à-vis your intranet.

The good news is that highly proven and stable technology from more than 40 credible vendors including IBM, Cisco, Attachmate, OpenConnect Systems, Wall Data, Eicon, Novell, WRQ, Farabi, Client/Server Technology, Sterling Software, Blue Lobster, etc. is now readily available to facilitate data-center-to-intranet integration in a seamless and synergistic manner. Enterprises around the world such as GM, FedEx, Ohio State University, Royal Jordanian Airlines, Nestles, The Chickering Group, National Van Lines, the State of Idaho, Al Rajhi Banking & Investment Corp. (Saudi Arabia's largest bank), and Gazprom (a $30 billion natural gas company in Russia) are already gainfully using this intranet-to-data center integration technology on a daily basis for business-critical production use. Al Rajhi Bank, for example, uses browser-based access to SNA to provide home banking, while GM, National Van Lines, Royal Jordanian Airlines, and The Chickering Group use it to permit agents to access applications or databases resident on mainframes or AS/400s over the Internet.

INTRANET TO DATA CENTER INTEGRATION TECHNOLOGIES

To be viable, integration technologies need to be able to accommodate an extremely broad and disparate population of client equipment and functionality including PCs, UNIX workstations, coax-attached 3270/5250 terminals, printers, minicomputers, SNA applications that communicate program-to-program using LU 6.2 or LU-LU Session Type 0-based protocols, SNA-only devices, SNA-LAN gateways (e.g. NetWare for SAA), and legacy control units. The PCs, workstations, and printers may work in either SNA or TCP/IP mode. Consequently, you will need SNA Access technologies to deal with TCP/IP clients, particularly PCs and workstations, and SNA Transport technologies to deal with SNA-only clients. The most pertinent technologies are:

- SNA Access technologies that permit non-SNA clients to gain access to SNA applications
  - *ip3270/ip5250*—the use of existing PC/workstation SNA emulators (e.g. Attachmate EXTRA!Personal Client) and existing SNA-LAN

gateways (e.g., Microsoft's SNA server) with proprietary encapsulation schemes for conveying a 3270/5250 data stream within TCP/IP

— *tn3270(E)/tn5250*—IETF standard that enables TCP/IP clients (e.g., Attachmate EXTRA!Personal Client) to access SNA applications via tn3270(E) (e.g. IBM 2216) or tn5250 servers

— *Browser-based Access with 3270/5250-to-HTML Conversion*—thin client solution where a server resident SNA-Web gateway performs 3270/5250 data stream to HTML conversion replete with some amount of user interface rejuvenation so that SNA applications can be accessed directly from a browser.

— *Browser-invoked Java or ActiveX applets*—dynamically downloadable applets, which can optionally be cached on a PC/workstation hard disk, that provide 3270/5250 emulation either directly or in conjunction with an intermediate SNA-Web gateway

— *Browser-invoked applets as "4" above,* but with user interface rejuvenation

— *Application-specific web-to-data-center gateways,* e.g., IBM's CICS Web Interface or Interlink's ActiveCICX

— *Programmatic (or Middleware) Servers,* e.g., IBM's MQSeries, Blue Stone's Sapphire/Web, or Blue Lobster's Stingray SDK

- SNA end-to-end Transport

— *Data Link Switching*—ubiquitous, standards-based encapsulation scheme performed by bridge/routers that permits any kind of SNA/APPN traffic, independent of session type, to be transported end-to-end across a TCP/IP WAN. Desktop DLSw (DDLSw) is also available where SNA traffic can be encapsulated within TPC/IP at the source PC

— *High Performance Routing-over-IP*—alternative to the DLSw championed by IBM, whereby SNA-oriented routing is performed across IP

— *AnyNet*—IBM protocol conversion technology, integrated within IBM server software including Comm. Server/NT and OS/390 as well as within some SNA/3270 emulation packages, that converts SNA message units into corresponding TCP/IP packets

The three transport technologies ensure that the still large installed base of SNA devices and control units are able to communicate with mainframe- or AS/400-resident SNA/APPN applications across an intranet using SNA on an end-to-end basis. Of the three, standards-based DLSw, which is available on nearly all major brands of bridge/routers, is by far the most widely used and the most strategic. AnyNet, in marked contrast, is not available on bridge/routers or within SNA devices such as 3174s, 4700s, etc. Consequently, it cannot be used easily as a universal scheme for supporting any and all SNA devices and control units as can DLSw. Thus, AnyNet is not as strategic or useful as DLSw. High Performance Routing (HPR) is IBM's follow-on architecture to APPN and SNA. HPR-over-IP, now available on IBM

2216 and CS/NT, has irrefutable advantages over DLSw: it can support native, data-center-to-data-center SNA/APPN routing over TCP/IP; SNA LU 6.2 Class-of-Service (COS)-based path selection; and traffic prioritization. If and when this technology is more readily available, corporations that require SNA/APPN routing to obtain optimum traffic routing in multidata center networks, or those that have LU 6.2-based applications that rely on COS, may want to consider HPR-over-IP as an alternative to DLSw.

DLSw's ability to support any and all types of SNA/APPN traffic effortlessly could be easily abused when trying to integrate intranets with data centers. DLSw could be used all by itself to realize the integration by grafting the existing SNA/APPN network, totally unchanged, onto the intranet through the extensive deployment of DLSw all around the periphery of the intranet. This brute force, "no SNA-reengineering whatsoever" approach has been used in the past to integrate SNA networks into TCP/IP networks. With this type of DLSw-only network you would find SNA-LAN gateways being used downstream of the intranet, and then DLSw being used to transport the SNA output of these gateways across the intranet. While such networks indubitably work, there are other strategic techniques such as a 3270-to-HTML and applet-based 3270/5250 emulation that should typically be used in conjunction with DLSw to achieve the necessary integration. Exhibit 6.1 summarizes how the various SNA Transport and SNA Access integration techniques can be gainfully synthesized to integrate data centers with intranets.

TOO MUCH CHOICE IS THE BIGGEST HURDLE

The availability of seven very different SNA access solutions is indeed one of major distractions in the intranet-to-data-center integration process. However, having such a wide range of solutions is also beneficial and should be gainfully exploited. With the solutions at hand you can tailor highly effective and pragmatic configurations where different access solutions are employed depending on both the requirements of the end user, and whether the access is across an intranet or the Internet. Exhibit 6.2 provides guidelines as to how the various access solutions may be best deployed.

The following is a comprehensive list of all the SNA access-related components that may be required to integrate an intranet to a data center, grouped, where appropriate, into "functional" classes:

- Web Server—to download applets or HTML-converted 3270 screens
- Web Browser—for browser-based access
- Client software other than the web browser:
  - —full-function, 3270/5250 emulator for ip3270 or ip5250
  - —tn3270(E) or tn5250 client

Exhibit 6.1 Synthesizing SNA Transport and SNA Access Integration Techniques.

PCs, Macs, workstations	• Browser-based • tn3270/tn5250 • ip3270/ip5250 • Programmatic	If a "full-stack" SNA/3270 emulator is currently installed: • DLSw • DDLSw • AnyNet • HPR-over-IP
Network Computers	• Browser-based • Programmatic	
SNA-Only Devices	• DLSw • HPR-over-IP	
Remote SNA-LAN Gateways	Eliminate the remote gateway and opt for ip3230/ip5250, tn3270/tn5250, or browser-based access	Prior to gateway elimination: • DLSw • HPR-over-IP
Devices that work in SNA or TCP/IP Mode	• Convert to IP and then use TCP/IP-based access schemes • tn5250/tn3270 • Programmatic	If the device cannot be converted to TCP/IP: • DLSw • AnyNet • HPR-over-IP

—terminal (and printer) emulation applet, with or without interface rejuvenation capability
—programmatic client which could be in the form of an applet
—optional, very small (e.g. 5K) "keyboard support" applet used by some 3270-to-HTML conversion solutions (e.g. Novell's HostPublisher) to overcome the problem that browsers currently do not support all the function keys found on a PC/workstation keyboard

- TCP/IP-to-SNA Gateway:
  —SNA-LAN gateway for ip3270/ip5250, such as Microsoft's SNA Server, located on a PC server or a channel-attached controller such as Bus-Tech's NetShuttle
  —tn3270(E) or tn5250 Server, located on a PC server, bridge/router, channel-attached bridge/router, channel-attached controller, or mainframe
- SNA-Web Gateway—for applet-based solutions either to augment the applet's functionality, provide security features, or both. Examples include Attachmate's HostView Server and OCS's OC://WebConnect Pro. Typically resident on an NT or UNIX server, though OCS has a version, it has an integrated TCP/IP-to-SNA gateway functionality that runs on a mainframe
- 3270-to-HTML or 5250-to-HTML Gateway

Exhibit 6.2 Deploying the Best Access Solution.

	intranet		**The Internet**	
	interim	**mid-term**	**interim**	**mid-term**
Employees:				
• Data Entry	tn3270(E)	Browser: tn3270(E) emulation		
• Power User (e.g., programmer)	ip3270	ip3270		
• Senior Management; queries, e-mail, calendar …	ip3270	Browser: with rejuvenation, or programmatic		
• Professional: < 2 hours/day mainframe access	tn3270(E)	Browser: some rejuvenation, or programmatic		
• Professional: > 2 hours/day mainframe access	ip3270	Browser: some rejuvenation, possibly programmatic		
• Telecommuter			Browser-based access	
• Mobile User			Browser-based access	
• "Agent" working for the company; e.g., dealership, travel agent			Browser: tn3270(E) emulation-cached applet	Browser: rejuvenation either with applet or 3270-to-HTML
Public:				
• Simple Query			Browser: 3270-to-HTML	
• Insecure, Multistep Query			Browser: 3270-to-HTML with some form of screen sequencing	
• Secure, Multistep Transaction			Browser: applet with rejuvenation	

- Application-Specific Web Gateway—for example, IBM's CICS Internet Gateway or CICS Gateway for Java
- Programmatic Server—an external, server-resident component required for certain programmatic access schemes

Of the seven access techniques, ip3270/ip5250 and tn3270(E)/tn5250 represent the "old guard", well-known and widely deployed schemes employed

Exhibit 6.3 Pros and Cons of Traditional Emulation Approaches.

	ip3270/ip5250	tn3270(E)/tn5250
Pros	• In most cases works with existing 3270 emulators and SNA-LAN gateway configuration, in which case a good tactical solution • Total support for all terminal emulation and workstation customization features • Extensive support for printing • Support for LU 6.2 • Support for other emulator-provided application APIs • Availability of channel-attached SNA-LAN gateways, e.g., Bus-Tech NetShuttle • Wall Data plans to have an applet that talks ip3270	• Widely adopted industry standard • Highly scalable tn3270(E) servers available on mainframes and channel-attached routers (e.g., Cisco 7xxx/CIP) • tn3270/tn5250 is the underlying "applet-to-gateway" protocol used by many browser-based access solutions including IBM's "Host-on-Demand" and WRQ's Reflection EnterView • Efficient, nonverbose protocol • Deemed by all to be a strategic technology for building SNA-Capable i•nets • Enjoys both vendor and customer mind-share
Cons	• Proprietary protocol • No mainframe-resident gateways • Not promoted by any vendors—even though this is probably an oversight because many forget the distinction between tn3270 and ip3270 • Client emulator may be more expensive than a tn3270 client • Most customers do not understand that this is an option open to them • No standard encryption schemes	• No support for LU 6.2 as yet • Printing options not as extensive or flexible as that potentially possible with ip3270/ip5250 • Cannot match ip3270 when it comes to the esoteric, power-user-oriented terminal emulation and customization features • Unlikely to support the application APIs supported by ip3270 emulators • Any end-to-end encryption available is going to be vendor specific or realized through the use of Virtual Private Networks (VPNs)

by over 20 million users. The best way to put these two techniques in context is to look at their pros and cons as shown in Exhibit 6.3.

Browser-Based Access via 3270-to-HTML Conversion

Browser-based access to SNA applications was initially made possible in late 1995 by 3270-to-HTML conversion. With HTML being the native language for creating web pages, converting 3270 data stream to HTML and vice versa was the obvious, logical, and most straightforward way to web-enable SNA applications. Alluding to the fact that web site and web page creation is often referred to as "Web Publishing," 3270-to-HTML conversion came to be known as "Host Publishing," with many product names such as Attachmate's HostPublishing System, Farabi's HostFront Publishing, Novell's HostPublisher, and IBM's Host Publisher feature in Ver. 6.0 of CS/NT amplifying this theme of how host (i.e., SNA) data can now be readily published

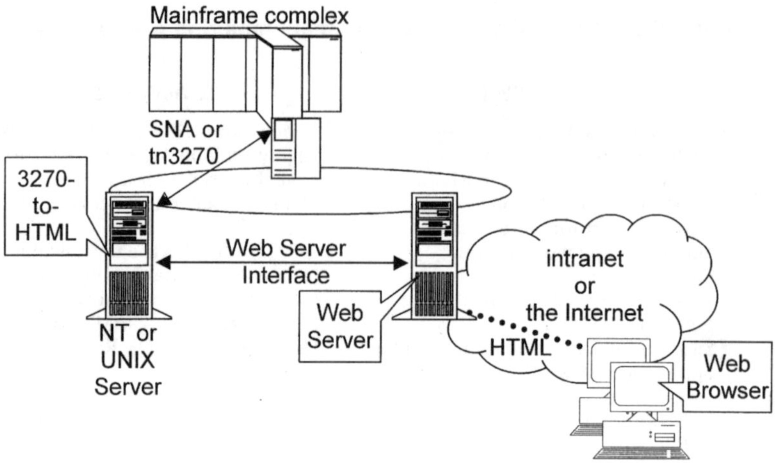

Exhibit 6.4 Typical Architecture of 3270-to-HTML Conversion.

on the Web. Exhibit 6.4 shows the general architecture of a 3270-to-HTML scheme.

3270-to-HTML has two incontrovertible advantages over that of an applet-based, "green-on-black" emulation approach:

1. *It only requires a browser at the client end.* This obviates any and all issues related to applet download time and the time taken to establish a persistent end-to-end connection between the applet and the pertinent data center component. Eliminating the need for an applet also makes 3270-to-HTML essentially browser-agnostic.
2. *It always delivers at least a "default" amount of user interface rejuvenation.* It is referred to as "Auto GUI," e.g., background and web page-like "trench" input fields. Extensive rejuvenation is possible, usually with the aid of visual programming tools such as JavaScript, Jscript, or Microsoft's Active Server Page technology.

3270-to-HTML is, thus, ideally suited for Internet-based SNA access, particularly to permit the public casual access (e.g., querying the delivery status of a package) to mainframe or AS/400 applications. Not having to download an applet will expedite the access process, while the rejuvenation capabilities will protect the innocent from the brutalities of the "green-on-black" screens. 3270-to-HTML used to get short shrift because the early implementations did not have good, cogent answers to issues such as session integrity, end-to-end persistence, file transfer, function key support, light pen support, etc. Some of these limitations were due to shortcomings of the browser or the HTTP protocol. The good news, however, is that some of today's implementations, for example Novell's intraNetWare Host-

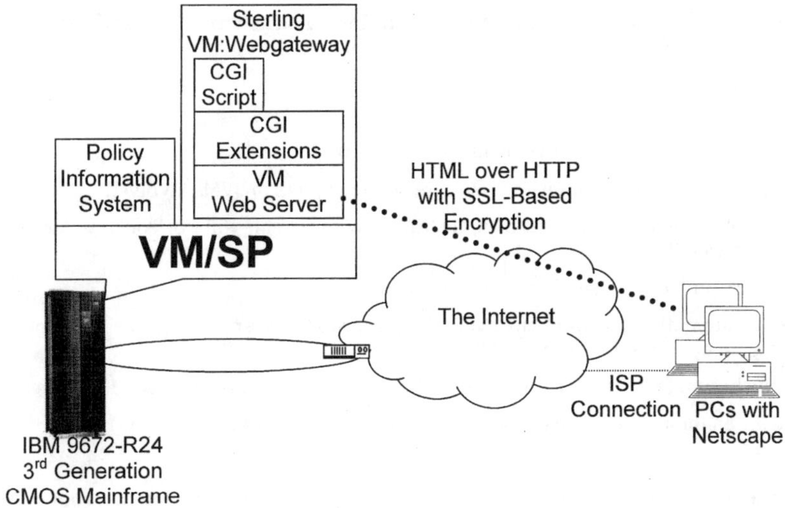

Exhibit 6.5 The Sterling Software VM:Webgateway-based architecture of the 3270-to-HTML Solution used by Lafayette Life Insurance.

Publisher and Eicon's Aviva Server, have gone to great lengths to come up with innovative but pragmatic solutions to nearly all of these previous limitations.

Take the sacrosanct session integrity and "persistence" issue. HostPublisher, through the use of session IDs, screen sequence numbers, timeouts, and unique "session name" validation (as opposed to IP-address-based correlation) can persuasively circumvent most, if not all, of the "horror" scenarios that could potentially occur given that 3270-to-HTML solutions do not maintain an end-to-end persistent connection à la applet-based solutions. Exhibit 6.5 shows the actual architecture of a mainframe-centric 3270-to-HTML solution actively used by Lafayette Life Insurance (Lafayette, IN) to provide 1,000+ field agents, as well as corporate staff in Indiana, real-time, online access to policy information via the Internet. Exhibit 6.6 weighs the pros and cons of such a solution.

Browser-Based Access Using Applets for 3270/5250 Emulation

Applet-based access schemes have the inalienable advantage that they provide a near complete "green-on-black" tn3270(E)/tn5250 emulation scheme, replete with TCP-based, end-to-end persistent connections in the form of a dynamically downloadable "thin-client." Distributing and, moreover, maintaining 3270/5250 emulation and tn3270(E)/tn5250 clients, especially given the incessant barrage of updates for such software, had become a costly and time-sapping endeavor for most SNA shops. Once

Exhibit 6.6 Pros and Cons of Lafayette Life Insurance 3270-to-HTML Solution.

Pros

- Only requires a browser at the client side
- Works with "any" browser given that there are no applets or JVMs involved
- A client platform-agnostic solution in that this access scheme will work on any platform that can run a web browser
- Facilitates and forces user interface rejuvenation
- Able to automatically exploit any standard, server-to-browser security schemes, including end-to-end encryption à la SSL V. 3.0
- Relatively simple, cost-effective solution that can be easily implemented on a PC server
- Ideal for enabling Internet users to gain casual access to SNA applications

Cons

- Most solutions do not support function keys and light pens since browsers do not currently recognize those items
- No SNA-specific printing or file transfer—with printing typically limited to screen printing functions provided by a browser
- Cannot deal with unsolicited screens from an application
- Schemes to ensure data integrity and session security in the absence of end-to-end persistent connections are implementation-specific, with some implementations being significantly better than others
- Rejuvenation schemes, though typically using off-the-shelf products such as JavaScript, are implementation-specific with no real commonality between the various products
- Corporate users, who require access to SNA applications and other web information concurrently, need to open multiple browser windows
- Some solutions may only support a few hundred concurrent SNA sessions per server

---

host printing, IND$FILE-based file transfer, and cut-and-paste between windows was in place by Spring 1998, applet-based emulation was a tempting proposition. It delivered tn3270(E)/tn5250 functionality, albeit by some esoteric and little used 3270 graphical capabilities, in the form of a browser-invoked "thin client." In addition, it promised the possibility of user interface rejuvenation via drag-and-drop tools, Java-based APIs, rules-based systems, or Visual Café-type applet programming.

Versions 4 or greater of Netscape and Microsoft Internet Explorer permit both Java and ActiveX applets to be cached on the hard disk of a PC/workstation. Caching eliminates the need for an applet to be downloaded from a web server each time it is invoked. Caching, however, does not compromise the version control and automatic software update advantages of downloaded applets. With cached applets, a web server is still automatically queried each time the applet is invoked to determine if there is a newer version of the applet. If there is, the user is given the option of dynamically downloading it. Cached applets, thus, provide users as well as Network Administrators with the best of all worlds—automatic version control without the continual wait for an applet to be downloaded.

A browser user invokes a terminal emulation applet by clicking on a "button" on a pre-designated web page in much the same way that 3270-to-HTML conversion is invoked. A key difference, however, is that there is no "script" per se, as is the case with 3270-to-HTML. Instead, the applet is embedded within the web page. Invoking the applet causes it to be activated. The applet may either be downloaded from the web server hosting the "mother" page that has the applet embedded, or be started up from a cached version following a quick version validation. Today, most of the applets perform a tn3270(E) or tn5250 client emulation, with Wall Data's Cyberprise Host Pro and Farabi being in the minority in that they work with SNA-LAN gateways rather than just tn3270(E) servers. Nearly all of the SNA access applets were Java-based at the beginning. Now, however, some vendors such as Attachmate, Wall Data, and Farabi are offering both ActiveX and Java applets.

Most of the applet solutions open up a separate emulation window alongside the browser window rather than having the emulation screen appearing as a "pane" inside the browser window. The advantage of keeping the applet emulation window separate is that it does not block the browser from being used for other purposes. Thus, the browser is not locked into an SNA session, as is the case with 3270-to-HTML conversion. Note, however, that regardless of whether the applet window runs alongside or within the browser, today's SNA access solutions invariably rely on a virtual machine provided by the browser, as opposed to the operating system, on which to run the applet. This means that the applet window will be abruptly and unceremoniously terminated if the browser hosting that applet is closed. An SNA access applet could, in theory, be converted and made to run as an application on a virtual machine provided by the operating system if one really wanted to eliminate dependence on the browser.

Quite a few of the applet-based emulation schemes require an intermediate SNA-Web gateway between the applet and the tn3270(E)/tn5250 server. Providing security, both in terms of authentication and encryption, is the overriding rationale for these intermediary SNA-Web gateways, so much so that, if an applet-based access scheme does not use an intermediary server component, you need to start thinking immediately of auxiliary security measures such as VPN. (See Exhibit 6.7) This will, however, change in the future when native end-to-end encryption is added to the tn3270(E) standard.

Exhibit 6.7 Pros and Cons of the Applet-Based Emulation Scheme.

Pros

- Powerful terminal emulation comparable to tn3270(E)/tn5250 clients
- "Thin-client" solution that minimizes software distribution and maintenance cost and effort—with the added attraction of cache-able applets
- Support for host printing à la tn3270(E), IND$FILE file-transfer, and cut-and-paste between windows
- Applet window can run alongside browser, thus leaving the browser window free for other interactions
- End-to-end persistent connections ensure data integrity and session security
- Encryption possible when an SNA-Web gateway is used
- Java applets facilitate cross-platform portability

Cons

- Rejuvenation only possible with certain solutions
- Encryption contingent on the presence of an SNA-Web gateway
- Potential delays when the applet is being downloaded, when caching is not used
- Most solutions currently only support Java applets
- May not work with "older," e.g., Ver. 3, browsers

The Bottom Line

Proven technologies now abound to facilitate intranet-to-data-center integration. Some of these technologies, such as DLSw and tn3270(E)/tn5250, are standards-based, mature, and very widely used. Even newer technologies such as browser-based access are now field tested and increasingly heavily deployed. Technology is not a barrier when it comes to intranet-to-data-center integration. The one distraction could be that there is too much choice, but that should not be used as an excuse; it is more an opportunity. You will require multiple access options and at least one transport scheme to satisfactorily meet all your needs. The technology and the products are here today to ensure that you have plenty of flexibility and variety in choosing that solution set.

Chapter 7
A Guide to Internet Telephony

Nathan J. Muller

AS MORE COMPANIES CONTINUE TO FLATTEN OUT AND DISTRIBUTE OPERATIONS ALL OVER THE GLOBE, AND MORE EMPLOYEES AWAKEN TO THE POSSIBILITIES OF TELECOMMUTING AND SELF-EMPLOYMENT, there is greater reliance on communications services for maintaining relationships and collaborating on projects. The cost of communications services can quickly escalate out of control.

As a result, companies and individuals are considering new media to expand their use of communications and contain costs. Internet telephony offers the opportunity to meet all of these objectives. The Internet is also being considered as a means of increasing the interactivity between customers and sales representatives in electronic commerce applications, or extending the reach of the help desk to employees worldwide. Even in these cases, Internet telephony can play a key role.

With special telephone software available from about 30 vendors — including IBM, Intel, Lucent Technologies, Microsoft, and Netscape Communications — users can call virtually anyone, anywhere, anytime, and talk for as long as they want without worrying about mounting phone charges. If the user pays a flat monthly fee for Internet access, the calls are free; if the user pays an hourly charge for Internet access, at least the phone calls will cost no more than the normal hourly rate to use the Web.

Telephony software, or phoneware, compresses the user's voice and packetizes it for transmission over the Internet. On the Internet, these IP (Internet Protocol) packets are treated as ordinary data, indistinguishable from any other kind of data. This means that voice is subject to the inherent variable delay of the Internet and, in extreme cases, dropped packets, which can result in clipped speech. Consequently, the quality of phone calls over the Internet is significantly lower than calls placed over the public switched telephone network (PSTN). At its best, the quality of a voice conversation over the Internet is only as good as that offered by today's cellular phones.

As the growth of the cellular industry demonstrates, however, people are willing to give up a significant level of quality in exchange for other benefits. In the case of cellular technology, the benefit is the ability to make a call from virtually anywhere. In the case of Internet telephony, the benefit is a vastly lower price, especially on international calls. Although Internet telephony users may have to sacrifice quality, they do not have to go without advanced call-handling features. Users can implement such features as call waiting, call screening, call transfer, call conferencing, and voice mail, again without having to pay extra monthly charges to a local telephone company or long-distance carrier. Most software also permits the user to adjust microphone and speaker volume during conversations. Some products allow the user to adjust the voice sampling rate and compression ratio to suit the modem speed and line conditions.

HOW INTERNET TELEPHONY WORKS

Placing calls over the Internet is as easy as clicking on a name in the public white pages or the user's private phone book. White pages are simply public directories that are located at one or more servers on the Internet, whereas a phone book is a private list of names and addresses that resides on the user's own PC. The public directories are maintained by various phoneware vendors and are organized by user name and topic of interest, making it easy for like-minded users to strike up conversations with others. Some directories are posted on vendor web pages. When the user opens the phoneware on his or her computer, he or she is automatically registered with a directory server (see Exhibit 7.1). People find each other by viewing the directory, which is periodically updated and sent to each computer, reflecting changes as they log on or off the server. Several phoneware products even allow the user to set up private directories (i.e., calling circles) on their server that can only be viewed by friends, family, or coworkers. Other phoneware vendors do not require that the user register with a server before using their product. Instead, the user can call someone merely by entering their E-mail or IP address. This has the added benefit of privacy — since the user's name and address do not show up on a public directory, he or she will not be bothered by unwanted calls.

APPLICATIONS

Internet telephony has passed the hobbyist stage and is seeing wider use in business applications such as electronic commerce. Virtual telephone companies that rely on the Internet for their backbone connections are springing up. In addition, new gateway products have become available that enable companies to put PBX-originated voice traffic onto their IP-based private network.

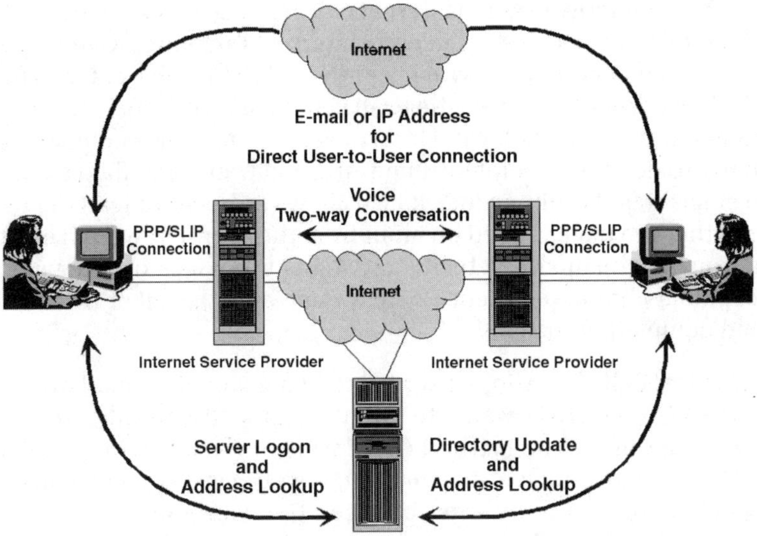

Exhibit 7.1 Establishing an Internet Telephony Conversation

Electronic Commerce

One of the fastest growing Internet applications is electronic commerce. Some industry estimates peg electronic commerce to grow to between $2 and $3 billion by the year 2000. By then, one-third of U.S. households are expected to be connected to the Web, propelling the electronic commerce market to even greater heights. Voice access to Internet-enabled call centers will be a key ingredient in the continued success of electronic commerce.

Several Internet call center products are available that give potential customers the ability to click on a company's web site icon and begin talking with a sales representative through their PC. This lets users ask questions, obtain pricing and availability information, and discuss options and customization. Speaking with a sales representative also provides users with a higher comfort level when making purchases over the Internet — something text alone cannot do. The setup enables the sales representative at the other end to view the same web page as the caller and push appropriate images and data over the Web to the customer's PC screen. Besides using Internet call centers for hard selling, retailers and others can use the technology to answer frequently asked questions, update customer profiles, and provide customers with periodic product updates.

Among the vendors offering Internet call centers are MCI, Sprint, and Lucent Technologies. Of these, Lucent's Internet Telephony Gateway is the most advanced. It combines with Lucent's DEFINITY Call Center to fully integrate the Internet into an existing call center, allowing businesses to take customer care to a higher level. The arrangement provides companies with the ability to deliver a complete, uninterrupted web voice/data experience over a single, regular phone line. It also allows an agent to lead the user on a tour of the web pages based on input from the user. The agent selects the pages from the business web site and loads the pages on the user's PC. This capability improves agent efficiency and aids the agent in closing the sale and completing the call.

In a typical call scenario, a user is browsing an online mail order catalog on the Web. The user wants to order shirts with a credit card and prefers not to use an online method (i.e., forms or E-mail). The user mouse-clicks a "Talk with Us Live" button on the web page. A voice connection is established over the Internet between the user and the Internet call center. The Internet call center determines the most appropriate agent to receive the call based on the user's experience by providing options while in queue — such as view new products or leave a message. When the call is answered, the agent's display presents the same web page as that seen by the user. The agent and the user can speak with each other, lead each other in browsing the site, and exchange data entered on web pages.

Virtual Phone Companies

Several companies see opportunities to offer alternative telephone services over IP networks at discounts of 20 percent to 40 percent lower than traditional long-distance services. Unlike Internet phones for home PC users, these voice-over IP services enable telephone-to-telephone communication over the Internet. Several companies offer or have announced plans to offer this type of service: IDT Corp. (Hackensack, NJ), Global Exchange Carrier Company, Inc. (Abingdon, VA), Vienna Systems Ltd. (Kanata, Canada), and AlphaNet Telecom, Inc. (Toronto, Canada).

IDT's Net2Phone service allows PCs to call other PCs or telephones. The service makes use of a debit card that users purchase from IDT. The ten-digit debit card number is manually entered whenever the user wants to place a call. Without this debit card, users can call only toll-free numbers. International calls to the U.S. cost ten cents a minute, regardless of country of origin, while the costs for calls between the U.S. and other countries varies by location. Thus, a one-minute call from Brazil to New York will cost ten cents, but a call from New York to Brazil will cost 63 cents a minute. Call rates from the U.S. to such high-volume locations as Japan, Australia, and Great Britain are 29 cents, 20 cents, and 18 cents a minute, respectively.

IDT's Net2Phone Direct, which began service in January 1997, dispenses with the need for a PC entirely and allows telephone-to-telephone communication through the Internet. Callers dial a local or toll-free access number, which connects the call to IDT's inbound switch server in New Jersey. The server switches the call from the public circuit-switched network to the Internet's packet-switched network, digitizing and packetizing voice along the way. By eliminating toll charges, users can call virtually anywhere in the world from their telephones at a substantial discount rate.

GXC offers service to all area codes in the U.S. For destinations within the U.S., GXC offers several calling plans tailored to individuals and corporations. For individuals, rates range from 8 cents to 25 cents per minute, regardless of the point-of-call origin. For corporations, reserved line accounts provide exclusive line use at scheduled recurring intervals at rates ranging from $9 to $12 per hour. GXC also provides corporations with private telephony gateways for intranets and WANs (wide area networks).

Vienna.Way consists of server software that runs on a UNIX or Windows NT Intel Corp.-based PC, client software, DSP (digital-signal processing) cards, PSTN interface cards supporting ISDN BRI (integrated services digital network basic rate interface), PRI (primary rate interface), or T1 telecommunications links. The server software establishes and manages all calls, provides telephony features such as call hold and call forwarding, and switches traffic between IP networks and, through its gateway module, between an IP network and the public telephone network. End users interact with the Vienna.Way server either through a traditional telephone or through client software installed on their PCs. The company offers users the choice of using its own client software, any standard TAPI-compliant (telephone application programming interface) client software, or H.323-compliant client software from such vendors as Microsoft and Intel.

AlphaNet introduced an IP-based international voice service in early 1997 that also relies on Sprint's GlobalOne international data service. The savings will come from compressing the voice traffic by as much as 15-to-1 to limit the amount of bandwidth used, and from using less costly data trunks that avoid conventional fees for landing voice traffic on foreign public phone systems. The service is intended for users who have no dedicated international links of their own.

Customers simply dial a toll-free number and a six-digit access code to connect with an AlphaNet server, then enter the number of the called party. The server compresses the voice call and puts the voice traffic into IP packets. Those packets are routed over frame relay links to a spring point of presence where the traffic is shipped over a Spring GlobalOne international data link. An AlphaNet server in the appropriate foreign country translates the IP packets into a digital voice signal so that the call can be completed on the local public-switched phone network.

Private Networks

Large corporations are the primary market for voice-over-IP gateways that can carry voice traffic over existing IP-based networks. Some of these networks are quite large. Hewlett-Packard alone, which is one of the world's largest users of the Internet, claims that five terabytes of information traverse its global IP-based intranet every month. Carrying voice over such networks is a matter of adding servers that digitize and packetize voice for transmission.

Among the providers of such servers is VocalTec. The company's Internet Phone Telephony Gateway provides connections between the Internet and local telephone lines, enabling callers to combine the low cost of Internet connections, the convenience of initiating calls from either PCs or telephones, and the ability to communicate with virtually anybody through the public-switched telephone network.

As voice over the Net becomes more popular, there will be a flood of new products seeking to leverage the free bandwidth available on private IP intranets and the greater Internet. The development of new products will be made easier with a Java application program interface (API) from Sun Microsystems called the Java Telephony Toolkit, or JavaTel. The new API will be used for linking telephones, computers, and so-called information appliances together over IP-based networks through Java-based applications. Among the initial supports of JavaTel are IBM, Intel, Lucent Technologies, and Northern Telecom.

HARDWARE REQUIREMENTS

To use phoneware, not just any computer will do. The user must have a computer with enough processing power to handle the voice compression/decompression algorithms provided by the phoneware vendors. In essence, these algorithms package samples of voice into packets suitable for transmission over the Internet. At the other end, the packets are opened and the sampled voice is rebuilt, preserving most of its unique characteristics. For this to work properly, a 486/33 DX or better computer is needed. With anything less — even a 486 SX — there will be too much processing delay, causing problems that range from lost syllables to total operational failure.

Memory

The user will also need enough memory, between 4 and 8MB, depending on the operating environment and how many applications are expected to be open at any given time. Although some phoneware vendors claim that only 4MB of RAM are needed to run their products, other applications will probably be running at the same time until a call comes in or one is established.

If the user is running Windows 95 and Netscape Navigator at the same time as the phoneware, at least 8MB of RAM are required. To provide an extra margin of safety, 16MB of RAM are recommended, especially with the propensity of some phoneware products to gobble up increasing amounts of RAM as they are continually opened and closed.

Phoneware products typically do not take up any more hard disk space than other graphically enriched applications. Depending on the product, only 1MB to 7MB of available disk space is required for a full installation; the higher amount is needed if the phoneware product comes with optional add-ons or is bundled with a web browser.

Although the amount of temporary disk space used during a conversation is negligible, if any, considerably more space may be needed to take advantage of such features as voice mail. If the hard disk is already running low on space, be aware that when voice mail is sent, the files will have to go to disk before they can be played. A single voice message can consume between 30KB and 60KB, depending on the length of the message and the vendor's audio format. For example, a 15-second voice mail message can be 30KB, whereas a 30-second voice mail can be 60KB. If allowed to accumulate, there must be enough disk space to store them. Allowing a hard disk to fill to capacity can render it useless, leaving no room to execute a delete command to free up space. In effect, the disk crashes. Users still getting by with old 240MB hard disks may need more of a cushion. With 1GB hard drives selling for under $200, the time may be right for an upgrade.

Modems

The correct modem is essential for any serious Internet user, and more so with phoneware. At a minimum, a 14.4Kbps modem is required to move packets of voice fast enough over the Internet with an acceptable delay of 4 to 5 seconds — similar to that experienced on international calls carried by satellite. A modem that operates at 28.8Kbps or higher can reduce the delay to 2 or 3 seconds, giving phone conversations a much smoother flow.

Sound Devices

To use phoneware, a Windows-compatible sound card, speakers (or headset), and a microphone are needed. If the sound card is being installed for the first time, it is recommended that it be tested for proper operation before installing the phoneware. That way, if there is a problem, the most likely source is the phoneware. Some sound cards have a setup utility that finds out what resources are already in use and automatically configures the board to avoid most conflicts. If the test feature detects a conflict, an appropriate notification is displayed so the user can select another setting. For example, if a conflict is reported with a COM port or IRQ (interrupt request), the user would check the computer's current configuration to see which COM ports and IRQs are free. The user would then enter these parameters and run the test again until no conflicts are reported. Once the sound card is properly installed and configured, the user can install the phoneware and test it for proper operation. If there is trouble at this point, it is most likely a problem with the phoneware.

The job of the sound card is to sample voice signals. The sampling rate must be at least 8KHz. Because virtually all Windows-compatible sound cards support an 8KHz sampling frequency, this is not a serious concern for buyers. Most sound boards provide a means of adjusting the sampling rate to higher levels for better quality audio. There may also be a means for adjusting the resolution. The higher the resolution, the more bits are used to represent each sample. Although higher sampling and resolution rates make for better quality audio, they also take up more memory and disk space (in the case of voice mail and greeting messages).

There are two types of sound cards: half-duplex and full-duplex. Half-duplex audio cards work like a citizen's band (CB) radio, where one person can talk at a time and says "over," indicating when he or she is finished talking. This method of communicating requires patience, and there is always the chance of one party dominating the conversation and preventing others from getting a chance to talk. With full-duplex audio cards, both parties can talk at once, just like an ordinary telephone call. However, if a full-duplex user is connected to a half-duplex user, the conversation defaults to the half-duplex mode.

A common problem with full-duplex conversation is acoustic feedback. However, this can be minimized using a headset to better isolate the received signal from the microphone. Alternatively, the speaker volume can be lowered, or the microphone and speaker can be spaced farther apart.

A half-duplex sound card, speakers, and microphone can be purchased separately or as a bundle for under $200. There are even sound cards that contain an on-board modem. Although more expensive, these combination boards are useful when the computer is running out of spare ISA card slots.

An economical alternative to computer speakers and microphone is a device like NET-FONE, a telephone handset that plugs directly into the sound card's 1/8-inch jacks. The NET-FONE, which costs less than $40, does not require any external power source, and there are no drivers to install. Alternatively, there are several earpieces available. JABRA Corp., for example, offers an earpiece that consists of a lightweight microphone and a speaker in a single tiny unit that fits into the user's ear. The device costs about $90.

Exhibit 7.2 Summary of Internet Telephony Hardware and Software Requirements

Hardware Requirements	Software Requirements
Computer (486/33MHz or better)	TCP/IP dialer (for Internet access)
4 to 16MB of RAM	Telephony software (includes GUI, voice compression, and call management features)
1 to 7MB of available disk space	Web browser (optional)
Modem (14.4Kbps or better)	
Sound card (half- or full-duplex)	
Microphone	
Speakers or headset	

Combined Audio- and Videoconferencing

Some phoneware products, such as White Pine Software's Enhanced CU-SeeMee and VDOnet's VDOPhone, combine audio- and videoconferencing. Users can transmit images using a camera that can be purchased for under $100, but a camera is not needed if the software will only be used for audio or text conferencing.

Video images over CU-SeeMe are transmitted in black and white to reduce demand on the network. One of the main advantages of CU-SeeMe for audioconferencing is that users can either connect directly to each other or they can enter a conference of up to 400 people at a reflector (a server-like component). A 14.4Kbps modem is adequate for audioconferencing, but to convey images while conversing, a 28.8Kbps or higher-speed modem is recommended.

Exhibit 7.2 summarizes the hardware requirements of Internet telephony, as well as the software requirements discussed in depth in the following section. A glossary of common terms associated with Internet telephony is provided in the Appendix.

SOFTWARE REQUIREMENTS

To use the Internet for phone calls, three software products are needed. Two of them are familiar to most Internet users: a TCP/IP (Transmission Control Protocol/Internet Protocol) dialer program and a web browser. The other is the phone software.

TCP/IP Dialer

The most basic piece of software users need is a TCP/IP dialer program, which enables packet mode operation so they can communicate over the Internet through the modem. If the user already has an Internet access account, he or she is probably using this software to log onto the Internet. Nevertheless, TCP/IP dialers are available as low-cost shareware. One of the most popular and easy to use third-party TCP/IP dialers is Trumpet, which can be downloaded for a 30-day evaluation from numerous Internet and BBS sites.

Some phoneware products include TCP/IP communications software for first-time Internet users. A good TCP/IP dialer comes with Windows 95 and multiple instances of the dialer can be easily set up, each with a different phone number. In addition, Windows 95 allows multiple users to share the same computer, each with a different user name and password. This allows the same machine to be used for multiple Internet access accounts.

Phoneware

The obvious software the user will need is the phoneware itself. Assuming that the user has a reasonably powerful PC that is equipped with the appropriate hardware, it is probably running Windows 3.1 or Windows 95. There are commercial phoneware products for each of these, in addition to products for the Macintosh. There is even phoneware for UNIX systems, and phoneware that works across multiple hardware platforms.

One of the oldest phoneware products is Vat, which traditionally has been a UNIX program. A new version has recently become available for Windows 95 and Windows NT. The latest version of Maven, which runs on Macintosh computers, is interoperable with Vat running on UNIX machines. Speak Freely offers cross-platform communication between Macintosh and PC users.

The phoneware provides algorithms that compress the recorded speech obtained from the sound card and employ optimization techniques to ensure its efficient delivery over the Internet. Phoneware vendors use a variety of compression algorithms to minimize bandwidth consumption over the Internet. For example, NetSpeak's WebPhone uses two audio compression algorithms — GSM and TrueSpeech. GSM is the Global System for Mobile communications and a worldwide standard for digital cellular communications. It provides close to a 5:1 compression of raw audio with an acceptable loss of audio quality on decompression.

TrueSpeech is a product of DSP Group, Inc. It can provide upwards of 18:1 compression of raw audio with an imperceptible loss of audio quality on decompression. WebPhone uses GSM compression when it is installed on a 486-based computer, and TrueSpeech when it is installed on a Pentium-based computer. Offering a high-compression ratio, TrueSpeech is more CPU-intensive than GSM, so it requires a faster processor to compress the same audio signal in real time.

VocalTec supports three compression schemes in its Internet phone product: TrueSpeech, GSM, and its very own VSC code, which squeezes raw audio data down to 7.7Kbps. Users have the option of adding VocalTec's VC Card to reduce bandwidth consumption further, to about 6.72Kbps, which provides the added benefit of offloading this chore from the computer's main processor. Although the VSC scheme is not as efficient as GSM, it is designed so that it can be used on a variety of hardware platforms, regardless of processor type.

It is advisable to try out various phoneware products before buying them. They differ in terms of installation ease, features, and performance. Most vendors have a web page or FTP (file transfer protocol) site from which trial versions of the software can be downloaded for evaluation. During installation, users have the opportunity to register their name with the directory server and specify any discussion groups to which they would like to belong. Typically, users will not have access to all the advanced features and calls will be limited to only a few minutes of conversation until they apply for a permanent account (i.e., pay for the software). In most cases, this can be done conveniently over the Internet through an order form posted on the vendor's web page, provided that the user wants to supply a credit card number.

In addition to the algorithms that compress/decompress sampled voice, phoneware products may include optimization capabilities to deal with the inherent delay of the Internet. The packets may take different paths to their destination and may not all arrive in time to be reassembled in the proper sequence. If this was ordinary data, late or bad packets would simply be dropped and the host's error-checking protocols would request a retransmission of these packets. However, this concept cannot be applied to packets containing compressed audio without causing major disruption to voice conversations, which are supposed to be conducted in real time. If only a small percentage of the packets are dropped, say 2 percent to 5 percent, the users at each end may not notice the gaps in their conversation. When packet loss approaches 20 percent, however, the quality of the conversation begins to deteriorate. Some products, such as VocalTec's Internet Phone, use predictive analysis techniques to reconstruct lost packets, thereby minimizing this problem.

Web Browser

The only other software needed is a web browser such as Mosaic or Netscape Navigator, or some other graphical web navigation software. Some phoneware products do not require that the browser software be opened to make calls over the Internet. If the browser is required, usually it must be opened up before an attempt is made to use the phoneware. Some products use the Web to convey directory updates. Users can also include a special link in their web page that enables others to call them at the click of a mouse. In the case of FreeTel Communications' FreeTel, such a link is formatted using standard HyperText Markup Language (HTML) tags as follows:

<A HREF="FreeTel:John Doe">Call John Doe via FreeTel</A>

From this example, it can be seen that HTML is a system of embedding codes, or tags, into text files to create multimedia documents that can be accessed on the Web. The tags are contained within angle brackets. In this case, the use of a hypertext reference tag <A HREF ... > and an anchor tag </A> cause the encapsulated text — Call John Doe via FreeTel — to be rendered by a Web browser as a hypertext link. Upon clicking on the link, a user will activate FreeTel and be able to converse with John Doe. When the user is running both Netscape Navigator and FreeTel, he or she can click on the link to place a FreeTel call to the owner of the web page. In this case, the web page owner just needs to change "John Doe" to his or her own name. Of course, to make a phone connection, both parties must be online at the same time and be running FreeTel software.

FreeTel Communications offers its phoneware at no cost. It is entirely supported by vendor advertising that appears at the bottom of the user interface. In addition to voice, FreeTel provides a split screen for text chatting, which is a handy feature to have if the Internet connection deteriorates during the voice conversation, or if a user wants to spell out a difficult name, word, or phrase.

Users can also open phoneware as a local file from their web browser and use it as a private phone book to access people they call often. In addition, such links can be added to call a topic group instead of a person. With VocalTec's Internet Phone, for example, a link to the topic, "surfing," is formatted in HTML as follows:

<A HREF="iphone:TOPIC/surfing">Internet Phone</A>

Another way to use such links is to create a communications center page, which not only contains links to the most frequently called people, but provides a convenient way of determining what phoneware they use. Because the phoneware of different vendors is not compatible, users can load several phoneware products into their computer to expand the possibilities for communication with others.

TELCO PARTICIPATION

Many U.S. and international telephone companies are considering ways to offer IP-based telephone service to their customers as a low-cost alternative to conventional calling services. Germany's Deutsche Telekom AG, for example, is planning worldwide rollout of Internet telephony services and has signed an agreement with VocalTec Communications, Ltd. to purchase more than $30 million of VocalTec products, services, and support. Subject to regulatory approval, Deutsche Telekom also intends to acquire a 21.1 percent stake in VocalTec.

Europe's largest telecommunications company and the third largest carrier in the world, Deutsche Telekom becomes the first major carrier to develop a portfolio of Internet telephony services to supplement its existing telephony services worldwide. Under the expanded partnership with VocalTec, Deutsche Telekom expects to quickly develop a high-quality product portfolio for Internet telephony. The two companies are already working on a pilot project in which Deutsche Telekom is giving 1,000 selected customers the ability to communicate through the Internet using a conventional or mobile telephone.

After providing a phone system that has not changed fundamentally in 30 years, VocalTec observes that the telecommunications industry is beginning to recognize the need for change. The company expects more and more major telcos to shift away from further massive investment in the traditional circuit-switched infrastructure in favor of developing a more flexible, cost-effective packet-switched IP network infrastructure on which to base services that include voice, fax, video, whiteboard, multipoint conferencing, and a host of other applications.

STANDARDS

A key issue being addressed by the vendor community is that of interoperability between the various phoneware products. Interoperability is especially important if voice communication and voice-enabled applications are to migrate into mainstream business use. After all, the conventional telephone's success has depended in large part on the interoperability of different vendors' phones, switches, and software — all working together and in conformance with a uniform set of standards. Until recently, the phoneware vendors have focused on promoting the concept of voice over the Internet and in building market share, with the result that most products are not interoperable or are only narrowly interoperable with those of other vendors.

In May 1997, the International Multimedia Teleconferencing Consortium (IMTC) approved G.723.1 as the low bit-rate default or baseline voice codec for Internet telephony applications based on the H.323 conferencing standard. The San Ramon, CA-based consortium includes more than 140 members and associates from 20 countries.

IMTC approval of the G.723.1 voice codec for H.323 is a significant step in completing the groundwork for interoperability in IP telephony and overall IP conferencing. In addition to greatly benefiting the industry and end users, IMTC approval provides a much clearer development path for manufacturers and third-party application programmers. The G.723.1 voice codec is based in large part on the DSP Group's TrueSpeech, a widely accepted audio technology that already has an installed base of several million users. TrueSpeech G.723.1, which combines low bit-rate compression with toll-quality communications at highly compressed rates of 6.3 and 5.3Kbps, was developed for video phone and videoconferencing applications over PSTN, and adopted by the International Telecommunications Union in 1995. Jointly developed by AudioCodes, DSP Group, France Telecom, and the University of Sherbrooke for the ITU H.324 POTS Video Telephony standard, G.723.1 was also included in the ITU H.323 specifications. DSP Group licenses the TrueSpeech G.723.1 technology on behalf of the four co-developers. To date, G.723.1 has been licensed by Microsoft, Intel, PictureTel, Dialogic, USRobotics, Creative Labs, NEC, Rockwell, VDONet, VIVO Software, and Smith Micro, among others.

FUTURE DIRECTIONS

With the new deregulated and competitive telecommunications industry ushered in by passage of the Telecommunications Act of 1996, the bandwidth, computer power, and general network infrastructure will evolve quickly. This, in turn, will promote the use of the Internet for voice communication and encourage more vendors to voice-enable their applications to enhance project collaboration and provide interactivity in customer service and electronic commerce applications.

One of the impediments to running voice over IP networks has been the problem of delay. A key disadvantage of the Internet Protocol is that it does not allocate a specific path or amount of bandwidth to a particular session. The resulting delay can vary wildly and unpredictably, wreaking havoc with real-time voice conversations. Several protocols are now available to deal with this problem by performing resource setup and management functions similar to those used on the public telephone network.

Among the most promising of these are RSVP (resource ReSerVation Protocol) and RPT (Realtime Transport Protocol) — both developed within the Internet Engineering Task Force (IETF) community. RSVP improves the quality of service by making enough bandwidth available on a priority basis end-to-end to support telephony or multimedia applications. Although RTP does not enhance the reliability of a transmission the way RSVP does, applications make use of its timing and sequencing information to enable audio and video streams to be played smoothly, despite occasional dropped packets.

Although the major router vendors support these and other protocols for transporting voice and multimedia applications, it will be several years before they are widely deployed throughout the Internet. However, organizations with private intranets have more control and can make full use of these protocols now by upgrading their network equipment with appropriate software. The quality and reliability of voice conversations in such closed environments will be vastly improved.

APPENDIX

Common Terms Associated with Internet Telephony

Term	Definition
Codec (coder/decoder)	An algorithm implemented in software that compresses digitized speech (i.e., coder) into a smaller number of bits, making it easier to send over the Internet. At the receiving end, the compressed speech is restored to its original form (decoder), so it can be understood by the other party.
COM (communications port)	This is the computer's communications port, usually numbered from 1 to 4. One is assigned to the modem.
IP (Internet Protocol)	IP is a key part of the TCP/IP suite of protocols. IP gets the raw data — the packets — from one place to another over the Internet, whereas TCP (Transmission Control Protocol) manages the flow and ensures that packets arrive correctly and, when received, are arranged in the correct order.
IRQ (Interrupt Request)	Interrupts are signals used by various devices, including COM ports, to get the CPU's attention. When two devices are configured to share the same IRQ, the CPU becomes confused and may shut down.
Phoneware	An informal term for software that implements phone calls over the Internet. Phoneware is available as shrink-wrapped commercial software, or may be downloaded from the Internet as shareware, freeware, or betaware.
PSTN (Public Switched Telephone Network)	This is a generic term for the circuit-switched public telephone network.
Voice sampling	A method for converting speech into digital form. The higher the sampling rate, the better the sound reproduction. However, a high sampling rate consumes more system resources and takes longer to process.

Chapter 8
Implementing and Supporting Extranets

Phillip Q. Maier

EXTRANETS HAVE BEEN AROUND AS LONG AS THE FIRST RUDIMENTARY LAN-TO-LAN NETWORKS BEGAN CONNECTING TWO DIFFERENT BUSINESS ENTITIES TOGETHER TO FORM WANS. In its basic form, an extranet is the interconnection of two previous separate LANs or WANs with origins from different business entities. This term emerged to differentiate between the previous definitions of external "Internet" connection and a company's internal intranet. Exhibit 8.1 depicts an extranet as a Venn diagram, where the intersection of two (or more) nets form the extranet. The network in this intersection was previously part of the "intranet" and has now been made accessible to external parties.

Under this design, one of the simplest definitions comes from R.H. Baker,[1] "An extranet is an intranet that is open to selective access by outside parties." The critical security concept of the extranet is the new network area that was previously excluded from external access now being made available to some external party or group. The critical security issue evolves from the potential vulnerability of allowing more than the intended party, or allowing more access than was intended originally for the extranet. These critical areas will be addressed in this article, from basic extranet setup to more complex methods and some of the ongoing support issues.

The rapid adoption of the extranet will change how a business looks at its security practices, as the old paradigm of a hard outer security shell for a business LAN environment has now been disassembled or breached with a hole to support the need for extranets. In many cases, the age-old firewall will remain in place, but it will have to be modified to allow this "hole" for the extranet to enable access to some degree to internal resources that have now been deemed part of the extranet.

Recognizing the growth of extranets as a common part of doing business today is important, and therefore the business enterprise must be ready with architectures, policy, and approaches to handle the introduction of

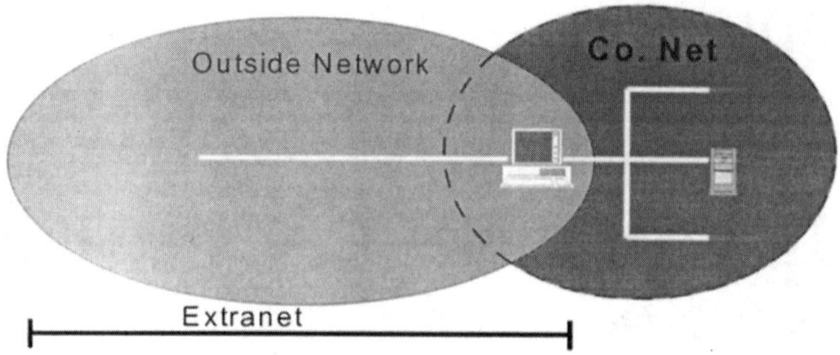

Exhibit 8.1 Extranet Venn Diagram.

extranets into its environment. A few of the considerations are the requirements-versus-security balance, policy considerations, risk assessments, and implementation and maintenance costs.

From a requirements-versus-security balance standpoint, the issue is the initial claim by business that extranets are an immediate need and absolutely must be established "if we are to remain competitive." But from a security standpoint, such a drastic change to the environment, which may not have had any form of an extranet in place, may well be throwing their financial data assets out the door with the first implementation of an extranet. Therefore, care must be taken from a security perspective and put in balance with the claimed business need for an extranet implementation.

One of the first areas of review and (possibly) update is the inner company's security policy. This policy most likely was not written with extranets in mind and thus may need modification if a common security philosophy is to be established regarding how a company can securely implement extranets. However, the policy review does not stop with one company's review of its own policy, but also includes connecting the company or companies on the outside. In the case of strategic business relationships that will be ongoing, it is important that both parties fully understand each other's responsibilities for the extranet, what traffic they will and will not pass over the joined link — what degree of access, and by whom, will occur over this link.

Part of any company's policy on extranets must include an initial requirement for a security risk assessment, the main question being: what additional levels of risk or network vulnerability will be introduced with the implementation of the proposed extranet? As well as vulnerability assessment, a performance assessment should be conducted to assist in the design of the extranet to ensure that the proposed architecture not only addresses the security risk, but that it also will meet performance expectations. Some of the questions to be asked in a combined security and performance assessment should be:

- Data classification/value of data
- Data location(s) in the network
- Internal users' access requirements to extranet components (internal access design)
- Data accessibility by time of day (for estimating support costs)
- Protocol, access services used to enter extranet (network design implications)
- Degree of exposure by transmission mechanism (Internet, private net, wireless transmission)
- End user environment (dial-up, Internet)
- Number of users, total/expectation for concurrent users access (line sizing)
- Growth rate of user base (for estimating administrative costs)
- CONUS (continental U.S.), international access (encryption implications)

The risk and performance assessment would, of course, be followed by a risk mitigation plan, which comes in the form of selecting an acceptable extranet architecture and identifying the costs. The cost aspect of this plan is, of course, one of the critical drivers in the business decision to implement an extranet. Is the cost of implementing and maintaining the extranet (in a secure manner) less than the benefit gained by putting the extranet in place? This cost must include the costs associated with implementing it securely; otherwise, the full costs will not be realistically reflected.

Finally, the company implementing the extranet must have a clear set of architectures that best mitigate the identified vulnerabilities, at the least cost, without introducing an unacceptable degree of risk into its computing environment. The following section reviews various extranet architectures, each with differing costs and degrees of risk to the environment.

EXTRANET ARCHITECTURES

Router-Based Extranet Architecture

The earliest extranet implementations were created with network routers that have the capability to be programmed with rudimentary "access control lists" or rules. These rules were implemented based solely on TCP/IP addresses. A rule could be written to allow External User A access to a given computer B, where B may have been previously unreachable due to some form of private enterprise network firewall (and in the early days, this firewall may have been a router also). Exhibit 8.2 depicts this very basic extranet. A more realistic rule can be written where all computers in an "outside network" are allowed to access computer B in a company network, thus forming an extranet. This is depicted in Exhibit 8.3.

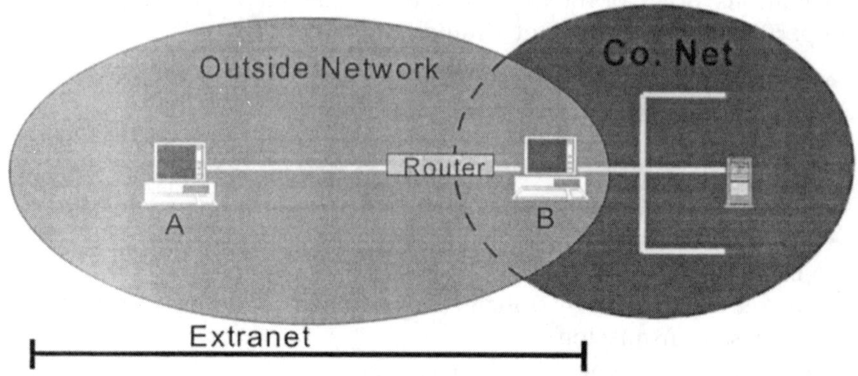

Exhibit 8.2 Basic Extranet with Router.

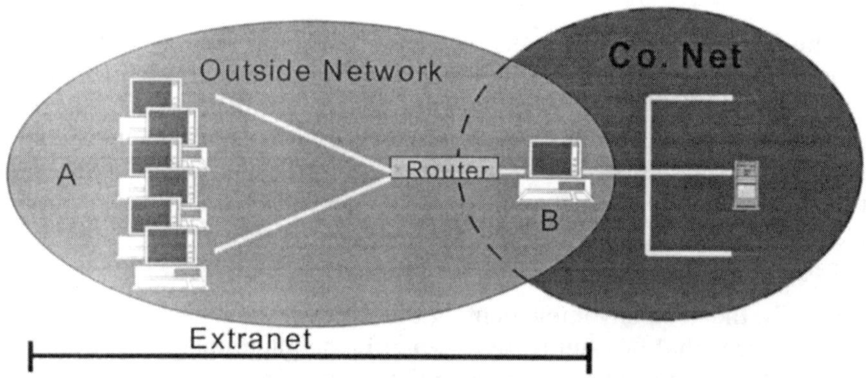

Exhibit 8.3 More Realistic Extranet.

As network security architectures matured, routers as the sole network access control device were replaced by more specific security mechanisms. Routers were originally intended as network devices — and not as security mechanisms — and lost functionality as more and more security rules were placed in them. Additionally, the security rules that were put into them were based on TCP/IP addresses, which were found to be subject to spoofing/masquerading and thus deemed ineffective in positively identifying the real external device being granted access. Therefore, routers alone do not provide an entirely secure extranet implementation; but when used in conjunction with one of the following extranet architectures, routers can be a component to add some degree of security, but only when used in conjunction with other network security devices.

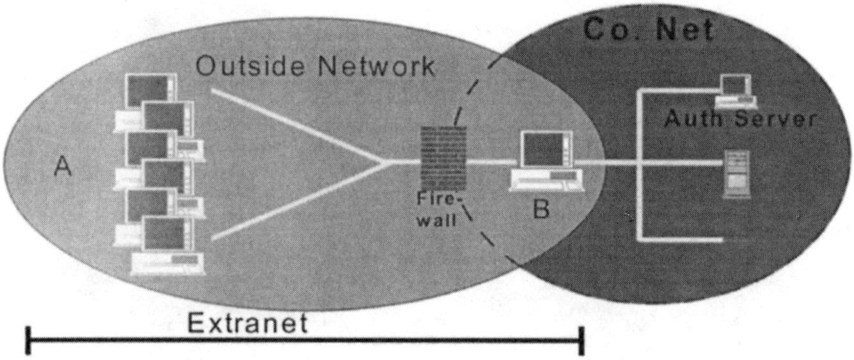

Exhibit 8.4 Extranet Using an Application Layer Gateway Firewall.

Application Gateway Firewalls

As network security architectures matured, the introduction of application layer gateway firewalls, software tools on dedicated machines, usually dual-homed (two network interfaces, one internal, one external), became the more accepted external protection tool. These software tools have the ability to not only perform router-type functions with access control rules, but also provide user authentication services on a per-user basis. This user authentication can take the form of an internal user authentication list, or an external authentication call to token-based authentication services, such as the ACE SecureID™ system. Exhibit 8.4 depicts this type of architecture setup to support an extranet using an a application layer gateway firewall to enable authenticated users inward access to an enterprise in a controlled manner.

In addition to supporting access control by IP address and user, some gateways have the further capability to restrict access by specific TCP/IP service port, such as Port 80, HTTP, so the extranet users can only access the internal resource on the specific application port and not expose the internal machine to any greater vulnerability than necessary.

Follow-on application layer gateway implementations have since emerged to provide varying additional degrees of extranet connectivity and security. One such method is the implementation of a proxy mechanism from an outside network to a portion of an internal company network. Normally, a proxy performs control and address translation for access from an intranet to the external Internet. These types of proxies normally reside on the firewall, and all user access to the Internet is directed through the proxy. The proxy has the ability to exert access control over who in the intranet is allowed external access, as well as where they can go on the Internet. The proxy also provides address translation such that the

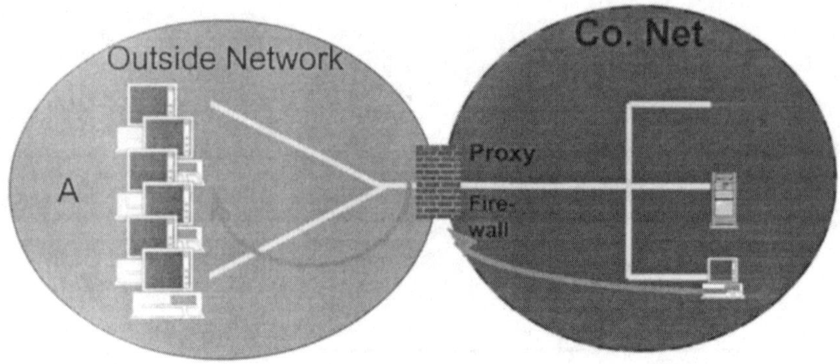

Exhibit 8.5 Outbound Proxy Architecture.

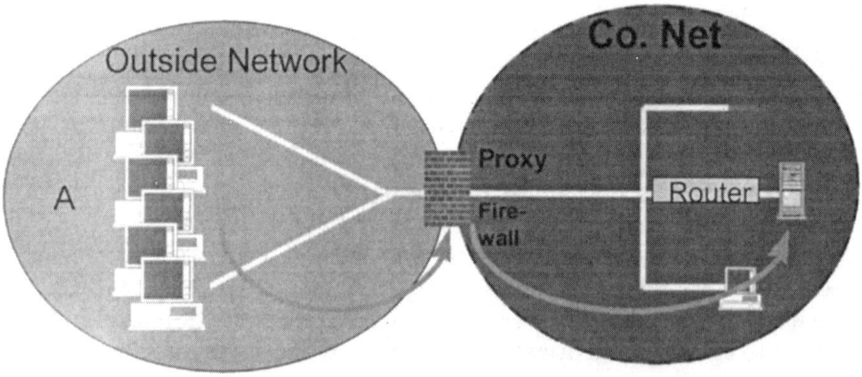

Exhibit 8.6 Reverse Proxy Extranet Architecture.

access packet going to the Internet is stripped of the user's original internal address, and only the external gateway address of the enterprise is seen on the packet as it traverses the Internet. Exhibit 8.5 depicts these proxy functions.

The proxy provides both security and network address functions, although the entire process can be used in its reverse to provide an extranet architecture because of its ability to provide access rules over who can use the proxy, where these proxy users are allowed to go, and what resources they can access. Exhibit 8.6 depicts a *reverse proxy* extranet architecture.

Today, most proxies are set up for HTTP or HTTP-S access, although application layer gateway proxies exist for most popular Internet access services (Telnet, ftp, sql, etc.). One of the major issues with proxy servers, however, is the amount of cycle time or machine overhead it takes to manage many concurrent proxy sessions through a single gateway. With highly scalable hardware and optimized proxy software, it can be carried to potentially handle high user demands, but the system architecture must be specifically designed for high loads to be able to meet user response expectations while still providing the security of an authenticated proxy architecture. On the *inward* proxy depicted in Exhibit 8.6, the proxy can be configured to only allow access to a single internal resource on a given TCP/IP port. Further protection can be added to this reverse proxy architecture by putting the target internal resource behind a router with specific access control rules, limiting the portion on the company intranet that inbound proxies can reach, which can ensure limited access on the intranet; should the internal machine ever be compromised, it cannot be used as a 'jumping off point' into the rest of company intranet.

A somewhat *hybrid* architecture extranet, where some firewall controls are put in place but the external user is not granted direct inward access to an enterprise's internal domain, has been evolving and put in place as a more popular extranet implementation. In this architecture, the external user is granted access to an external resource (something outside of the enterprise firewall), but still on the property of the enterprise. Then, this external resource is granted access to one or more internal resources through the enterprise firewall. This architecture is based on minimizing the full external access to the intranet, but still makes intranet-based data available to external users. The most popular implementation is to place an authenticating web server outside the firewall and program it to make the data queries to an internal resource on the enterprise intranet, over a specific port and via a specific firewall rule, allowing only that one external resource to have access to the one internal resource, thus reducing the external exposure of the intranet. Exhibit 8.7 depicts this type of extranet.

Issues with this type of architecture include reliance on a single user interface that can be safely placed outside the enterprise firewall, which makes it vulnerable to attack. Additionally, there is the issue of whether tight enough access rules can be placed on the access method between the external user interface resource (the web server in this example) and the internal resources that it needs access to on the protected enterprise intranet. If these two issues can be safely addressed, then this form of extranet can be very useful for an enterprise extranet with a high volume or varied user base and a large intranet-based data repository.

The user front end has been deployed as a web server, usually SSL-enabled to ensure data integrity and protection by encrypting the data as it passes over an external SSL link. Access to this external server is also

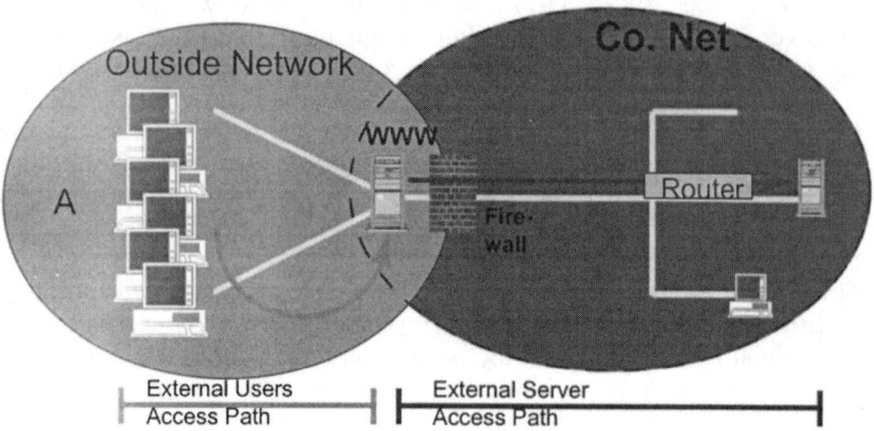

Exhibit 8.7 Extranet with Athenticating Web Server.

associated with some form of user authentication, either a static ID and password over the SSL link, and more recently with client digital certificates where each individual accessing the SSL-enabled site is issued his own unique digital certificate from an acknowledged certificate authority, thereby validating his identity. Each client maintains its own digital certificate, with the Web server having some record of the public-key portion of the client's digital certificate, either directly in the Web server internally, or accessible from a standalone directory server (usually LDAP reachable).

The most recent entrant in the extranet architecture arena is the Virtual Private Network (VPN). This architecture is based on a *software tunnel* established between some external entity, either client or external network, and a gateway VPN server. Exhibit 8.8 depicts both types of VPN architectures. External Network A has a VPN server at its border which encrypts all traffic targeted for Company Network C; this is a gateway-to-gateway VPN. Or, External Client B may have client VPN software on his workstation which would enable him to establish a single VPN tunnel from his workstation over the external network to Company C's VPN server.

Although both server-to-server VPN and client-to-server VPN architectures are offered in the industry today, it is this author's experience that the more popular extranet architect is the client-to-server VPN architecture, as it offers the most flexibility for the most diverse audience of external users. This flexibility does add to the complexity of the implementation, as it can involve a potentially large number of external desktops, all with differing configurations. The benefits of VPNs include the ability to safely traverse external public networks with some assurance of data integrity

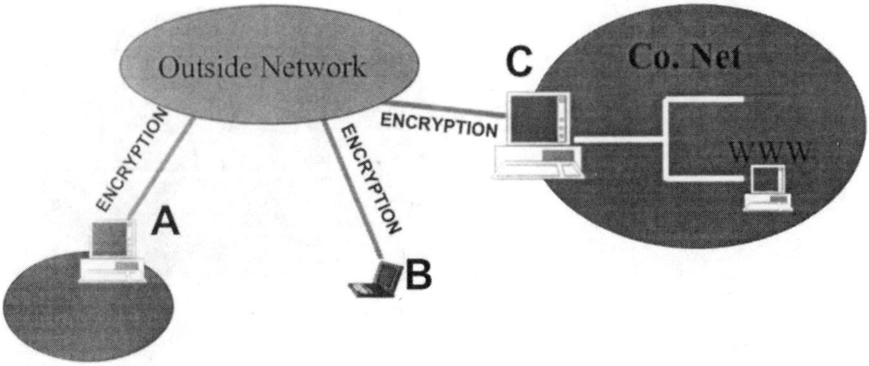

Exhibit 8.8 VPN Architectures.

and authentication as part of the VPN implementation. This architecture shows the most promise to meet the needs of extranets and cost savings for a world hungry for connectivity over public/external networks, although it still has some growing pains to go through to reach full product maturity.

An emerging standard for VPNs is coming out of the ITEF IPSec implementation, which draws a roadmap for the next-generation TCP/IP security protocol. Under this protocol, standards are being drafted that will enable differing devices to securely communicate under an agreed-upon security protocol, including key exchange for encryption and standardized authentication. Today, there are IPSec-compliant products on the market; however, the standard is still evolving and tests are being conducted to evaluate differing vendor compatibilities with each under the IPSec standard. One of the leading initiatives to evaluate this compliance is the Automotive Network Exchange (ANX) test, which is intended to establish a large extranet environment between the core automotive manufacturers and their vendors.

In the meantime, there are a wide variety of VPN product vendors on the market — some touting IPSec compliance and others, with proprietary implementations with IPSec in their future product roadmaps, choosing to wait until the standard stabilizes. The recommendation is to either select a vendor offering IPSec if it has some degree of maturity within its own product line, or one that is planning on adopting the standard; IPSec appears to be a viable standard once it fully matures.

Regardless of what VPN solution is being considered for implementing secure extranets, a few technical considerations must be understood and planned for before selecting and implementing a VPN extranet architecture.

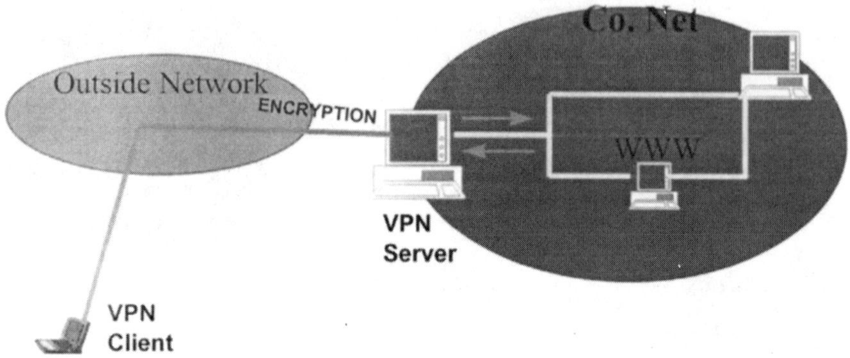

Exhibit 8.9 Traffic Patterns for Multi-homed Intranet with a Single VPN Gateway and an External VPN Client.

Scalability. Similar to proxy servers, VPN servers incur a fair amount of processing overhead that consumes processing resources as high levels of concurrent VPN sessions pass through a single server. It is important to attempt to estimate one's projected user base and current access to appropriately size a VPN server. Some servers are on established lower-level processors for smaller environments and should not be implemented where high concurrent access rates are expected, although there is some benefit to physical load balancing spreading the access among multiple servers. However, there is also concern about implementing too many servers to manage easily. A balance between installing a single large server and creating a single point of failure — versus implementing many smaller servers — creates an administrative nightmare.

Multihomed Intranets and Address Translation. In large intranet environments, many operate under a *split DNS* (domain naming structure) where intranet addresses are not "advertised" to the external networks, and external addresses are kept external so as not to flood the internal network. Additionally, many larger intranet environments have multiple gateways to external networks. If one of the gateways is established with a VPN gateway and an external client makes a connection to the internal intranet, it is important that the tunnel comes in through the appropriate VPN gateway, but also that the return traffic goes back out through that same gateway so that it gets re-encrypted and properly returned to the external VPN client. Exhibit 8.9 depicts the correct traffic patterns for a multihomed intranet with a single VPN gateway and an external VPN client.

VPN-Based Access Control. Many forms of gateway VPN servers offer the ability to restrict user access to a company intranet based on access groupings. This is especially important when intranets are being established for a diverse set of external users and it is important to minimize user access to the intranet. This type of access control is, of course, critical in establishing secure extranets, which further highlights the importance of understanding VPN access control capabilities.

User Authentication. Multiple options exist for user authentication, although the recommended option is to select a high-level authentication method (e.g., one-time passwords) or a time-synchronized password method. Under the IPSec standard, client-side digital certificates are evolving as a standard for high-level authentication. Unfortunately, initial implementations of client-side digital certificates for user authentication are entirely software based, eliminating the second-factor authentication, the "something the user physically has" in their possession. The return to true two-factor authentication under digital certificates will not really occur until physical smart cards become part of the authentication architecture. (Smart cards are credit card-type tokens that have a physically embedded chip which can be read electronically and written to, either with a portion of the client's digital certificate or the encryption algorithm used to unlock the digital certificate.)

IPSec Interoperability. Ultimately, the IPSec standard will stabilize, and all vendors following the established standard will allow different vendors' VPN products to interoperate. Under this environment, a company can implement a vendor's VPN server, and their acknowledged clients can purchase and use an IPSec-compliant client to gain access to the company intranet once they are authorized.

SUMMARY

Secure extranets are becoming the external network of choice in today's business world. There are multiple implementation options, as discussed in this article, each with varying degrees of risk and implementation complexity. Each implementation must be evaluated against a business case using the recommended risk and performance analysis outline. The basic router-controlled extranets are only recommended for the least valuable data environments, while the more sophisticated VPN extranet architectures appear to be the future for extranets, especially when the IPSec standard matures and gains industry adoption.

Chapter 9

Internet Site Development in a Bandwidth-Hungry World: An Enterprise Dilemma

Daniel Gonneau

EVERYONE LIKES TO SEE WEB SITES WITH LOTS OF MULTIMEDIA CONTENT — video clips, animations, audio, etc. — to make the online experience richer, more interesting, and more informative. Unfortunately, many users use modems to access the Internet, and such content — even with the latest streaming media technologies — can be excruciatingly slow. Even enterprise intranets can have these bandwidth problems, particularly with so many people using modems to dial into the network while telecommuting or on sales calls. The addition of megabyte video files to an overtaxed network can be a serious problem. Not all companies have the technological or financial wherewithal to upgrade their networks to meet these challenges.

Unfortunately, this situation is unlikely to change anytime soon. Few regular consumers have shown a willingness to spend real money on high-speed access, and even many Internet-savvy users have stayed with their 33.6 or 56KB/s modems. This is the case even though high-speed services are currently available in many of the largest markets, generally at $50 or more per month. Yes, everyone will all have inexpensive high-speed home access (cable modems, DSL, etc.) at some point in the future. But the question is: will these services be in wide enough use in the near future to affect the many site design and development decisions one needs to make today? For most sites, the appropriate answer is "no."

The situation at enterprises (both large and small) is different — but perhaps not as different as it might first appear. Networks are not free, after all, and network managers are under constant pressure to justify their budgets. The addition of multimegabyte multimedia content to an intranet will probably not be received with open arms. And, as mentioned above, what happens to the telecommuters and sales personnel who are dialing into the network using slowpoke modems? For these reasons, bandwidth will continue to be a major factor in site development for most enterprises.

LOW-BANDWIDTH, CONTENT-RICH INTERNET SITES

What, then, can be done to improve the functionality and attractiveness of our sites without imposing unrealistic bandwidth burdens on users and networks? This article reviews some of the techniques and technologies for adding brandwidth-friendly multimedia content to a site. It also looks at community and personalization technologies that add real interactivity to a site, as well as some of the browser compatibility issues that surround the use of these technologies. Whether one is responsible for an enterprise intranet, E-commerce effort, or any other type of web site, one will want to know more about these technologies.

Dynamic HTML

DHTML is a loosely knit group of web technologies that allows "dynamic" features without using applets or plug-ins. It is an extension to HTML that lets browsers refresh web pages without going back to the server for updates. This dynamic updating can take many forms, from purely graphical features like rollover buttons, to tables and forms that update themselves based on user input. DHTML-compliant browsers can do a surprising amount of processing without going back to the server for updates.

To implement DHTML, a few interlocking technologies are used:

Dynamic Positioning. DHTML allows unprecedented control over the positioning of elements on a web page. This positioning can be dynamically changed based on user input or other events.

Cascading Style Sheets (CSS). Style sheets can control many features of the text on a web page, from font size to indenting to font style. In addition, these features can be dynamically updated based on user inputs or other events.

Dynamic Fonts. Many headings, titles, and other text features on the Net are actually graphics since browsers can only use the fonts found on users' computers. Thus, Helvetica, Times Roman, and Arial are extensively (excessively?) used on the Net. With dynamic fonts, designers can specify custom fonts that download with the page.

Interactive Content. DHTML pages can capture user mouse and keyboard clicks and process the web page without going back to the server. A good example of this is a multicolumn table that lists, say, product manufacturer, price, and shipping cost. Without DHTML, the server would have to be contacted each time the user wanted to change the sorting of the table from, for example, price to manufacturer. With DHTML, the mouse click is captured and the table resorted without going back to the server. Another interesting capability is the ability to drag and drop elements on a web page. With DHTML, one can imagine a user dragging a product photo to a shopping cart rather than clicking a static button that says, "Click here to add to Shopping Cart."

Layers. Features can be layered, with only one layer visible at a time. The visible layer can be dynamically changed based on user input.

Using these technologies, simple animations, rollover graphics, and other graphical features — as well as interactive content — can be incorporated into web pages without relying on plug-ins or applets.

To get a flavor of the possibilities, take a look at http://www.dhtmlzone.com. This site is a DHTML demonstrator, so it probably includes more DHTML than anyone needs at a particular site, but it will give a good idea of the possibilities.

Vector Graphics

Regular bitmap graphics (.gifs, .jpgs, etc.) store graphical information as a series of pixels. Vector graphics, on the other hand, store a mathematical description of the graphic. This mathematical description is thus available for processing by the browser. Why is this important? One of the more dramatic capabilities of vector graphics is the ability to dynamically resize to the size of the user's browser window. Put another way, a large vector graphic is the same size as a smaller version; vector graphics are resolution independent. This is because vector graphics separate the mathematical description of the graphic from the rendering of the graphic. Thus, the size of the file is the same whether the browser renders it large or small. This capability thus eliminates the need for thumbnail graphics, coding, extra page sets, etc., for handling multiple resolutions.

A good example of vector graphics can be found at the home page of Macromedia, the developer of the Flash vector graphics standard. The Flash plug-in is one of the most popular on the Internet — and, as of this writing, it should be built into Netscape Communicator 5. Once the plug-in is downloaded — quite small, at about 125K — one sees that the page automatically resizes whenever the size of the browser window is changed. While this is a neat trick, the real advantage is that developers only need to create one file, which will be automatically resized for all users, regardless of monitor size and resolution. This takes care of one of the nagging problems of developing for users with widely varying monitor resolutions. A monitor set at 1280×1024 — typical for most 19" and many 17" monitors — has fully four times the area of a monitor at 640×480. This can be a real problem for developing an interesting interface for all users.

However, the ability to dynamically resize graphics is just the beginning. Vector graphics can also be used for low-bandwidth animations, as one can see on the Macromedia home page. Click on one of the left-hand links, hear a "click" sound, and the appropriate links will appear. Unlike animated gifs (or any animation scheme based on bitmap graphics), these animations can be full-screen with no increase in file size. Flash files can include rollovers, sounds, and other features. All these animations, sounds, site navigation features, text, rollovers, etc. are included in a single 50K file that automatically resizes to the user's screen. This is indeed an impressive technology.

Flash is by no means the only vector graphics technology available. While Macromedia has made the Flash standard available to the Internet community, there are others and it remains to be seen which one will "win." Among them are Adobe's Precision Graphics Markup Language (PGML) and Microsoft's Vector Markup Language (VML). PGML is an interesting standard because:

- all text is searchable (not so with Flash)
- it is built on Postscript and PDF (Adobe's Acrobat document model)
- it is ASCII-based (Flash is binary)

For more information, see the online resources on vector graphics listed in Exhibit 9.1.

XML

Both the familiar HTML and the not-so-familiar XML (eXtensible Markup Language) have their roots in SGML (Standard Generalized Markup Language), but they have very different purposes: HTML is the display language for the Web, while XML works behind the scenes to link web pages to data. While today's web servers can certainly handle huge volumes of static page hits, database access can be awkward and slow. The same is true of searches, which can generate lots of unnecessary hits (false positives) and take a long time to process. Thus, technologies that streamline and speed up how web pages link to databases are crucial to the further development of the Web as a medium for doing business.

XML allows developers to specify types of information. This can make connections to back-end databases much more efficient — and, thus, faster — as well as making searches much more precise.

Exhibit 9.1 Online Resources.

DHTML
- Netscape DHTML information (http://developer.netscape.com)
- Microsoft DHTML information (http://www.microsoft.com/workshop/author/)
- Inside DHTML (http://www.insidedhtml.com)
- Cross-browser DHTML (http://www.dansteinman.com/dynduo/)
- ZDNet HTMLUser (http://www.zdnet.com/products/htmluser.html)

XML
- XML.com (http://www.xml.com)
- w3.org's XML information (http://www.w3.org/XML/)

Vector Graphics
- Flash (http://www.macromedia.com/software/flash/)
- Quick Start Guide to Flash 3 (http://webreview.com/wr/pub/98/07/31/feature/index.html)
- PGML (http://www.w3.org/TR/1998/NOTE-PGML)
- VML (http://www.w3.org/TR/NOTE-VML)
- Vector Graphics and XML (http://www.xml.com/xml/pub/98/06/vector/intro.html)

Browsers
- Microsoft Internet Explorer (http://www.microsoft.com/windows/ie/)
- Netscape Communicator (http://www.netscape.com)

To implement these features, XML allows the creation of custom tags. This is both a blessing and a curse. It is a blessing since, for example, developers will finally be able to identify <PRODUCT_NAME> or <PRODUCT_PRICE> or any other type of information. It can be a curse because the meanings of these terms vary from company to company and from industry to industry. Anyone who has worked on EDI (electronic data interchange) is familiar with this problem. For example, the word "purchase order" has widely varying meanings in different industries.

This may not be a problem within an individual enterprise, but it seems that cross-enterprise exploitation of XML will rely on creating standard XML tags — even more importantly, standard meanings for XML tags — for different industries. An example of this type of effort is the Open Financial Exchange (http://www.ofx.net). While not an XML effort, OFX allows customers using Microsoft Money and Intuit Quicken to connect to financial institutions to pay bills, check account balances, etc. The cost savings are considerable because the institutions do not need to develop separate interfaces for every financial package. A similar process will happen with XML, with similar benefits: much better and faster web-based access to data and cost savings as standardized ways of interacting with that data are developed.

XML makes extensive use of style sheets also used with Dynamic HTML. Fortunately, there are both free and commercial tools available to help with XML development (see Exhibit 9.2).

Exhibit 9.2 Site Development and Maintenance Tools.

Site Management/Database Access
- FrontPage (http://www.microsoft.com/frontpage/)
- Cold Fusion (http://www.allaire.com/products/ColdFusion/)

Graphics
- Photoshop (http://www.adobe.com/prodindex/photoshop/)
- Fireworks (http://www.macromedia.com/software/fireworks/)
- Paint Shop Pro (http://www.jasc.com)
- ColorWorks (http://www.spg-net.com)

HTML/DHTML Editors
- Allaire HomeSite (http://www.allaire.com)
- BBEdit (Macintosh) (http://www.barebones.com/)
- Dreamweaver (http://www.macromedia.com)
- FrontPage (http://www.microsoft.com/frontpage/)

Java Development
- Visual Cafe (http://www.symantec.com/domain/cafe/vc4java.html)

Streaming Media
- NetShow (http://www.microsoft.com/NTServer/Basics/NetShowServices/)
- RealPlayer (http://www.real.com)

Chat/Mailing Lists
- iChat (http://www.ichat.com)
- ICQ (http:/www.icq.com)
- Microsoft Chat (http://www.microsoft.com/windows/ie/chat/)

Portable Documents
- Acrobat (http://www.adobe.com/prodindex/acrobat/)

XML
- XML Pro (http://www.vervet.com)
- List of free tools (http://www.stud.ifi.uio.no/~larsga/linker/XMLtools.html)

Streaming Media

Using various compression schemes, streaming video technology has made enormous strides over the last couple of years. Compressed video is even possible over 14.4Kbps modem links, albeit in a very jerky, low-resolution format. However, it looks like high-resolution Internet video will have to wait for true high-speed connections.

However, there are definitely applications — like Internet news broadcasts and customer-service applications — where low-resolution video can make a lot of sense. The two leading streaming media technologies are both proprietary:

- RealPlayer is an integrated environment, available as a multimegabyte plug-in, for video and audio over the Net. As the early leader, many companies use RealPlayer to deliver news clips (both video and audio), music clips for selling CDs, and other multimedia features.

Exhibit 9.3 Technologies for Enterprise Community Development.

	Private	Public
Asynchronous		
	E-mail	Mailing lists
	Newsgroups	
Synchronous		
	Chat (private)	Chat (public)
	Internet telephony	Videoconferencing
	Instant messaging	

- NetShow is a collection of technologies that perform many of the same functions as RealPlayer. Unlike RealPlayer, the NetShow server is free with Windows NT Server, while RealNetworks' server is not. It will be interesting to see how RealNetworks fares in direct competition with Microsoft.

Audio streaming technology is often excellent, even over modems. There are numerous potential applications here, from music clips for e-commerce music sites to audio customer service to audio instruction manuals for the visually impaired.

The TIME proposal from Microsoft, Macromedia, and Compaq offers some interesting possibilities. TIME (Timed Interactive Multimedia Extensions) is a submission to the World Wide Web Consortium to cover the delivery of time-based media (e.g., animations) in web browsers. Coming from Microsoft, one can only assume that it will be implemented in the next version of Internet Explorer. As of this writing, Netscape's participation in TIME is unknown.

Community Technologies

A low-bandwidth way to add considerable interest to a site is the addition of community technologies, that is, any technology that helps to make the user feel as if they are participating in a community of people with like interests rather than simply visiting a static web page. This can apply to both enterprise intranets — where "customers" might be users of a human resources site — and Internet sites for E-commerce or information delivery. There are two general types of community technologies (see Exhibit 9.3):

- Synchronous technologies allow realtime interaction between users and the owner of the web site, or between users. Examples are chat rooms, instant messaging, and Internet telephony.

- Asynchronous technologies allow non-realtime interaction between users. Examples include text-based chat rooms, e-mail, and newsgroups.

Community technologies can be further divided into two types:

- Public interactions are available for all to see; an example is a public newsgroup.
- Private interactions are user-to-user; an example is e-mail.

These distinctions can help one organize what types of technologies to consider for the various objectives of a site. For example, the customer-service pages of a commercial site might offer all these types of technologies because they tend to be used in different ways. Chat rooms and newsgroups tend to be used by people who have very specific questions. For more complex questions, there might never be a better solution than a customer service rep at the end of a phone line. But today's Internet telephony solutions allow this "phone line" to be integrated right into the online environment. A user struggling with a piece of software might thus be able to reach a customer service rep with one click on the screen, rather than looking up and calling a separate toll-free number. Even without Internet telephony, the ability to help customers with simple questions in an inexpensive, automated fashion can generate big cost savings. Cisco Systems estimates its annual savings in answering these types of questions without talking to a rep at $200,000.

Typically, community technologies are relatively inexpensive. For example, newsgroup servers are quite cheap to configure and operate. Newsgroup technology can help users help themselves. A good example is Macromedia's support newsgroup for their Dreamweaver Internet development product. This newsgroup, located at news://forums.macromedia.com, lets Dreamweaver users post questions about the product and how to use it. While it is monitored by Macromedia personnel, many times the questions are quickly resolved by the users themselves. This author has used it to post esoteric questions about site development and has been very satisfied with the responses. This contrasts markedly with many public newsgroups which are sometimes dominated by spam and flames (in plain English, junk mail and offensive messages). Admittedly, this example involves Internet-savvy users, but the same principle can be applied to other businesses.

Efforts to personalize a site can also greatly help make it as useful as possible. Typical examples are news and search services such as http://my.yahoo.com or http://my.excite.com. The advantage of building local news, weather, and sports right into one's own personalized home page is an obvious benefit for these sites. But these technologies may signal a permanent move toward one-on-one marketing over the Internet. For example, Office Depot offers personalized online catalogs to small-business buyers. And, for intranets, giving users the ability to self-configure web services can only help them be more efficient — with obvious benefits to the enterprise.

Like DHTML, personalization technology is really a cluster of interconnected technologies:

- Fast database access is a key technical issue because the server maintains information on each user.
- Collaborative filtering compares customers' purchase history, preferences, and other information to try to guess their overall product preferences, interests, etc. This technology is currently in its infancy, but has huge potential to offer users a valuable, personalized online experience.
- Dynamic page generation using systems like Microsoft's Active Server Pages (ASP) is also a key requirement, because some or all pages will be personalized for individual users.
- Secure transactions become even more important as potentially large amounts of information about individual consumers is processed.

Ultimately, these personalization technologies may lead to the mass customization of products and services, where users tell manufacturers exactly what they want before it is manufactured.

Some Basics for Speeding up Your Site

The above-mentioned technologies allow the development of vibrant low-bandwidth sites, but there are also some decidedly low-tech techniques for simply reducing bandwidth load:

- Specify height/width for all graphics. A classic, it allows the page to be presented to the user before all the graphics are downloaded.
- Do not put the whole page in a single table. If this is done, the whole page has to download before anything is presented to the user. To get around this, a common technique is to put the top banner in one table and the rest of the page in a second table. That way, the banner downloads quickly, giving the user at least something to look at while the rest of the page downloads. One needs to be careful with WYSIWYG HTML editors, which often generate single-table pages.
- Use advanced graphics tools to reduce graphics size. See Exhibit 9.2 for a listing of tools. Just about all modern web-centric graphics tools provide sophisticated ways of reducing the size of bitmap graphics. Some will even provide the download times with different download speeds (14.4, 28.8, etc.).
- Reduce the amount of copy. The amount of unnecessary copy on the Web can be astounding. Successful E-commerce sites such as

Exhibit 9.4 Internet Technologies: Microsoft Internet Explorer 4 versus Netscape Navigator/Communicator 4

	MSIE4	**Netscape4**
HTML 4	Yes*	Yes*
DHTML	Yes*	Yes*
CSS1	Yes*	Yes*
XML	Yes	No
VB/Jscript	Yes	No
JavaScript	No**	Yes
Java	Yes*	Yes*
ActiveX	Yes	Plug-in
Flash3	Plug-in	Plug-in
RealPlayer	Plug-in	Plug-in
Acrobat3	Plug-in	Plug-in
iChat	Plug-in	Plug-in

* The Microsoft and Netscape implementations are often incompatible.
** Microsoft's version of JavaScript is Jscript. The two versions are somewhat compatible. Fortunately, the development of ECMAscript may mean the end of the "script wars."

amazon.com reduce the amount of extraneous copy to a minimum. Simply reusing copy from print sources often generates a site with way too much copy for the way people actually use the Internet. Many users read very little of the introductory material on, for example, a company home page before trying to find the specific information they are seeking. Try to get users as quickly as possible to what they are looking for, putting in-depth editorial coverage into the product pages and other "meaty" pages of the site. One consequence of this strategy is smaller web pages and, thus, faster download times.

Browser Incompatibilities

No discussion of these techniques and technologies would be complete without some discussion of the serious problem of browser incompatibility. If one is fortunate enough to be developing an intranet, this may not be a problem. But anyone who develops for the wilds of the Internet needs to address the problem.

The two most popular browsers are, of course, Microsoft's Internet Explorer and Netscape's Navigator/Communicator. Each includes many industry-standard features (see the World Wide Web Consortium at http://www.w3.org) but also includes proprietary features to enhance their products. See Exhibit 9.4 for some of these differences. Each browser is a "stew" of technologies that often conflict. As for the technologies discussed in this article:

- Dynamic HTML. The versions in MSIE4 and NS4 are basically incompatible. While tools exist for cross-browser DHTML development (see Exhibit 9.2), this process can nonetheless be complex and time-consuming — and, thus, expensive.
- Vector graphics. Microsoft has proposed its own standard, VML, while, at this writing, Netscape had chosen to integrate Macromedia's Flash into Version 5 of its browser.
- Personalization/community technologies. These technologies are typically server based or rely on plug-ins. Thus, browser incompatibilities are less of an issue here.
- Streaming media. As of this writing, it is unclear which streaming media technology will dominate the market, or if a third standard will emerge. Microsoft's NetShow is built into their browser, but plug-ins like RealPlayer and QuickTime have large installed bases.

Developers who want to do anything beyond the basics are left with basically three choices:

1. selectively develop for one browser
2. develop advanced features for one browser, leaving people with the "other" browser to see basic features
3. develop multiple features/pages to handle both browsers

The frustration of developing advanced features for both browsers has forced many developers to select one of the first two options, but surely this is not an ideal solution. For example, the versions of DHTML implemented in the two browsers are almost entirely incompatible, so the choice is particularly stark in this case. One can only hope that the two browser giants, working in collaboration with standards bodies, will develop standardized feature sets.

CONCLUSION

With so many choices available, the world of Internet site development is certainly not getting any easier. However, there is no question that today's Web-based systems are far more powerful than those available just a year ago. The judicious use of Dynamic HTML, vector graphics, XML, communi-/personalization technologies, and streaming media should help in de-ping compelling, valuable content for a site *without* overwhelming the width of the enterprise network or user access speed.

Section III
LAN and Campus Networking

Chapter 10
High-Speed Token Ring

Martin Taylor

ALTHOUGH THE POPULATION OF 20 MILLION DESKTOPS CONNECTED TODAY BY TOKEN RING MAY BE DWARFED BY THE NUMBER OF ETHERNET CONNECTIONS that are out there, current users of Token Ring remain very loyal to the technology. The reasons are not hard to find — Token Ring's fault tolerance, manageability, performance, and efficient use of bandwidth get the job done. In large LAN installations where network availability is critical, one can depend on Token Ring to deliver.

The technology of Token Ring has been around for almost 15 years, and it is a testament to the robustness of the original design concept that many large networks today continue to make use of essentially the same network components that powered those first, early installations. But that is not to say that Token Ring technology has stood still. The demand for increasing network bandwidth drove the introduction in 1989 of 16M bps Token Ring to quadruple the speed of the original 4M bps offering. More recently, the development of Token Ring switching increased backbone capacity to hundreds of megabits per second and added the possibility of dedicated, full duplex links carrying Token Ring traffic at 16M bps in both directions at once.

By and large, this evolution of Token Ring technology has kept ahead of the growing demand for bandwidth in most LAN installations, but the anticipation of continued growth of traffic in the future has naturally made some large Token Ring users nervous. The availability of solutions based on ATM (Asynchronous Transfer Mode) that can switch Token Ring traffic at speeds of 155M bps and beyond have provided some comfort; but to date, Token Ring users have not enjoyed a choice of high-speed solutions — unlike Ethernet, where Fast Ethernet, and more recently Gigabit Ethernet, offer a simpler and lower-cost alternative to ATM.

High-speed Token Ring (HSTR) addresses this imbalance. Supporting 100M bps and 1G bps Token Ring connections for interswitch links and for switch-to-server connections, HSTR provides the first simple and cost-effective alternative to ATM for dealing with capacity issues in Token Ring backbones.

THE EMERGING NEED FOR HIGH-SPEED TOKEN RING

Continual growth in traffic loads is a seemingly inescapable fact of Local Area Networks. PCs get faster and deal in more complex and information-rich data objects; intranet web servers unlock corporate information resources for access by much wider communities of users; departmental server systems are displaced by corporate superservers. These and other trends contribute to a rising tide of data traffic that seems to devour all the bandwidth that can be thrown at it.

In Token Ring networks, switching is now the primary technology for tackling this growth in demand for LAN capacity. Shared 16M bps backbone rings that link multiple workgroup rings with corporate mainframe and server resources can now be replaced with switching systems that handle hundreds of megabits per second of Token Ring traffic. The effect on network performance can be dramatic because user access to central resources — which was once throttled through overloaded 16M bps backbone rings — now takes place at full wire speed through the fast backplane of a switch.

The 16M bps limitation of Token Ring does, however, limit the scalability of switched Token Ring networks. Connections between switches and connections from switches to servers represent traffic concentration points where 16M bps looks increasingly inadequate.

Some relief from overloaded interswitch connections can be obtained by taking advantage of the load-sharing capabilities that are inherent in Token Ring's source routing. Switches can be connected together with multiple parallel dedicated 16M bps Token Ring links, and the properties of source routing will ensure that network traffic is shared approximately equally among the multiple links.

But load sharing across multiple links provides only temporary relief from network overloads as traffic continues to grow. Each additional link adds only another 16M bps of capacity and occupies another port on the switches at each end of the link. Furthermore, the use of multiple links does not provide a satisfactory solution to the problem of server access bottlenecks.

TODAY'S HIGH-SPEED OPTIONS FOR TOKEN RING USERS

To solve the problem of overloaded interswitch links and switch-to-server connections, a higher speed networking technology is needed. Up until now, Token Ring users have effectively had a choice between FDDI (Fiber Distributed Data Interface) and ATM (Asynchronous Transfer Mode).

Fast Ethernet has not, so far, provided a satisfactory high-speed backbone solution for Token Ring users. Differences in frame format, address bit ordering, and maximum frame size make it very difficult to carry out a direct translation between Token Ring and Fast Ethernet without introducing unacceptable compromises. It is possible to use routers to connect the Token Ring environment to a Fast Ethernet backbone, but the software processing overhead of routers, and the necessity to carry out packet fragmentation to deal with differences in maximum frame size, result in severe performance limitations. This is also a costly solution.

FDDI offers a fault-tolerant, shared 100M bps medium and meets many of the backbone connectivity needs of Token Ring users. Its token-passing access mechanism, which is based on that of Token Ring, enables its full bandwidth potential to be realized, and its dual reconfiguring ring topology provides very effective protection against cable or transmission link faults.

However, FDDI suffers two disadvantages that are responsible for its fall from favor among Token Ring users over the last few years. First, the FDDI frame format is different from that of Token Ring, and Token Ring switches that support FDDI links must perform a number of complex packet translation functions to support connectivity. This makes FDDI a relatively expensive solution. Second, FDDI offers no scalability options beyond the existing 100M bps standard. Other technologies such as ATM and Fast Ethernet have next-generation standards that scale into the gigabit range, and FDDI appears far less future proof by comparison.

ATM looks like a much more promising candidate for Token Ring backbone and server connections. With its ability to scale to 2.4G bps and beyond, its inherent fault tolerance, and its support for native Token Ring frame formats (in LAN Emulation mode), ATM meets all of the key needs of large Token Ring backbone installations.

But ATM is not ideal for all situations. It is a far more complex technology than Token Ring; and although it can be relatively easy to install and configure, there is too much new information for network technicians to learn if they are to be able to troubleshoot ATM networks successfully. ATM may also be regarded by some as an overkill solution for many Token Ring needs in terms of both its raw power and its sophisticated features for bandwidth reservation and Quality of Service support.

DEFINING THE REQUIREMENTS FOR HIGH-SPEED TOKEN RING

The requirements for a High-Speed Token Ring solution have been the subject of discussion and agreement among members of the High-Speed Token Ring Alliance, a consortium of Token Ring equipment vendors that was formed in August 1997. The objective of the Alliance is to define a simple, cost-effective, and scalable solution for switched Token Ring networks that preserves as much of the existing investment in Token Ring as possible while providing a forward evolution path for Token Ring networks that meet all foreseeable applications needs.

In particular, the Alliance agreed to the following set of needs:

- Support for standard Token Ring frame formats
- Speeds of 100M bps and beyond
- Support for UTP, STP, and fiber cabling
- Support for existing maximum packet sizes as defined in the 802.5 Token Ring standard
- Support for all Token Ring bridging modes, including source routing, transparent, and SRT
- Compatibility with the 802.1Q standards effort for Virtual LAN tagging

The Alliance also agreed that the most urgent needs for High-Speed Token Ring were for interswitch and switch-to-server connections. Therefore, the focus of HSTR development would be on dedicated point-to-point full duplex links rather than shared media operation. This would reduce the time taken to arrive at a new standard and make it easier for vendors to bring products to market relatively quickly.

THE TECHNOLOGY AND TIMESCALES OF HIGH-SPEED TOKEN RING

The process of developing a new LAN standard can take years, but the need for HSTR is already here and demand is expected to grow rapidly during 1998. For this reason, the HSTR Alliance is cooperating very closely with the IEEE 802.5 committee on a "fast-track" resolution of technical issues to arrive at a completed standard as soon as possible.

Basing the standard as closely as possible on existing technologies is seen as the best way to get rapid agreement on the technical specification. The basis for HSTR is the existing 802.5r standard, which specifies full duplex, point-to-point, dedicated Token Ring operation at 4 and 16M bps. This will be adapted to operate first at 100M bps and then later at 1G bps.

The 100M bps speed was chosen in order to leverage the existing transmission technology on which Fast Ethernet is based. By making use of encoding schemes and transmission waveform specifications that have already been proven over both twisted pair and fiber optic cabling, the HSTR Alliance will minimize the amount of new technical thinking that needs to be done to arrive at a standard.

The first priority of the Alliance is to develop a specification for 100M bps HSTR over copper cabling, including both Category 5 UTP and IBM Type 1 STP. A specification for 100M bps HSTR over fiber cabling will follow soon after, based on the 802.5j Token Ring fiber standard adapted to run over the 100-Base-FX specification for Fast Ethernet on fiber.

Subsequent work will focus on Gigabit HSTR. As might be expected, this will leverage the transmission technologies that are being developed to support Gigabit Ethernet over both copper and fiber cabling.

At the time of this writing in October 1997, projections by the 802.5 committee suggested that the first HSTR standards would be completed by June 1998. The High-Speed Token Ring Alliance plans to demonstrate HSTR technology and multivendor interoperability at Networld+Interop in Las Vegas in May 1998, and vendors are predicting that HSTR products will begin shipping by the fourth quarter of 1998.

THE DEPLOYMENT OF HIGH-SPEED TOKEN RING

The first products supporting the new HSTR standards are likely to include 100M bps HSTR line cards for Token Ring switches, and 100M bps HSTR Network Interface Cards (NICs) for servers.

For most Token Ring users, the first application of HSTR will be for upgrading interswitch links. By installing HSTR line cards in existing Token Ring switches, interswitch links that are currently running at 16M bps can be very quickly upgraded to 100M bps.

HSTR provides full support for source routing, so backbone interswitch connections in fault tolerant networks with parallel load-sharing paths can be upgraded to 100M bps without compromising fault tolerance in any way. The HSTR connections between switches will behave in exactly the same way as existing 16M bps connections, except that they have more than six times the traffic capacity.

Some existing switched Token Ring networks make use of FDDI for interswitch connections. Because HSTR will offer both lower cost and better performance for interswitch connections, it may be desirable to use HSTR for future upgrades to these networks and freeze any further investment in FDDI. Some Token Ring switches will support both FDDI and HSTR line cards, and this makes it relatively easy to carry out a phased migration to HSTR in the backbone while preserving current investment in FDDI.

After upgrading interswitch links, the next area that may need attention is the connection to enterprise servers. If servers are already connected directly to Token Ring switches by full duplex 16M bps links, and these links are becoming overloaded, then the server connections may be upgraded by installing a 100M bps HSTR adapter card and connecting directly to a port on a 100M bps HSTR line card in a Token Ring switch.

Some vendors are likely to introduce server NIC products that support 4, 16, and 100M bps on the same card. This provides excellent investment protection, because the server may be connected initially to a 16M bps switch port, and then moved later to a 100M bps switch port if greater throughput is needed.

HSTR is not initially targeted at desktop connections, but there is nothing to prevent it from being deployed at the desktop if needed. HSTR will be supported over both Category 5 UTP and IBM Type 1 STP cabling, and 100M bps HSTR NIC cards could be installed directly in desktop systems for connection to 100M bps HSTR switch ports. However, it is likely that adding 100M bps HSTR to the desktop would only be justifiable for exceptionally demanding applications such as very high-resolution imaging.

HIGH-SPEED TOKEN RING AND VIRTUAL LANS

In some situations, it may be useful to divide the Token Ring LAN into a number of separate broadcast domains or "Virtual LANs" that are defined logically rather than physically, and to use routers to move traffic from one VLAN to another. Typically, VLANs are defined by assigning each switch port to a particular VLAN identity. Traffic arriving at that port from end stations is deemed to belong to the VLAN associated with that port.

If making use of VLANs, then in general it is useful to be able to define VLANs that span multiple switches. The key property of a VLAN is that a broadcast or multicast packet originating within a given VLAN must only be delivered to switch ports that belong to the same VLAN. When using High-Speed Token Ring connections between switches, it becomes necessary to mark each packet with its VLAN affiliation so that one knows which switch ports this packet may be delivered to.

Where source routing is used, the originating ring number in the source routing information field of each packet is sufficient to identify the packet's VLAN affiliation, and one does not need to do anything new or special. But packets that are being transparently switched do not contain any information from which VLAN information can be deduced. To solve this problem (for both Ethernet and Token Ring networks), the IEEE 802.1Q working group is defining a packet-tagging scheme that enables VLAN information to be associated with each packet.

The emerging 802.1Q standard is fully compatible with High-Speed Token Ring, and will enable VLAN-tagged packets to be carried between Token Ring switches. The 802.1Q specification is also expected to define a standard method for identifying Ethernet packets that are being carried over Token Ring links. The implication of this is that HSTR could provide a backbone solution that integrates Token Ring and Ethernet traffic across a common high-speed infrastructure.

HIGH-SPEED TOKEN RING AND QUALITY- OF- SERVICE

Until High-Speed Token Ring arrived on the scene, ATM was the first choice of Token Ring users looking for major upgrades to backbone switching capacity. This is because ATM can provide a high-capacity, scalable, fault-tolerant backbone that handles native Token Ring frame formats and source routing.

One additional capability of ATM has also been a factor in its adoption for Token Ring backbones: Quality-of-Service. ATM networks can provide not only "best effort" delivery of packet data like Ethernet and Token Ring LANs, but it can also deliver time-sensitive traffic streams like voice and video with a guaranteed upper limit on end-to-end delay.

This capability of ATM is being exploited today in a minority of ATM installations to provide backbone integration of voice or video trunks across campus LANs. However, Quality-of-Service cannot easily be exploited for packet-based LAN traffic because it requires much more sophisticated edge switches linking Ethernet and Token Ring to ATM than are currently available. This has not been an issue so far because there are few, if any, desktop applications in use today that can exploit Quality-of-Service.

However, there is growing interest in the deployment of desktop videoconferencing and voice telephony over the LAN, and network planners making backbone evolution decisions need to take into account the possible requirements of the enterprise looking several years forward. ATM has been seen as a safe choice for the large-scale LAN backbone because it is expected to have no problem accommodating future requirements for voice and video in the LAN.

So, how does HSTR stack up against ATM for handling integrated packet-based voice and video in the LAN? The answer is that although HSTR cannot provide the same level of guaranteed service for real-time traffic streams as ATM, it is likely to be good enough for most practical needs.

PRIORITY ACCESS FOR VOICE AND VIDEO TRAFFIC

The original designers of Token Ring had the foresight to build into the standard a feature that was intended to enable support for real-time traffic in the LAN. This capability, which allows prioritization of time-sensitive data packets, has been built into every Token Ring product since 1985 that complies with the 802.5 standard, but has remained effectively dormant for lack of applications that can exploit it.

The token-passing access mechanism of Token Ring is a deterministic protocol that guarantees a station will be able to access the medium to send a packet within some fixed time limit, depending on the number of stations on the ring and the speed of the ring. Under normal operation, when there are a number of stations on the ring waiting to send a packet, the token is passed from one station to the next around the ring, granting the right to transmit a packet. The priority mechanism allows a station that is waiting to send a higher priority packet to preempt all the other stations and grab the token next. If one thinks of all the stations waiting to transmit packets on the ring as being in an orderly queue to access the medium, then a station wanting to send a higher priority packet effectively jumps the queue and can start transmitting as soon as the packet currently being transmitted by another station has finished.

The Token Ring priority mechanism supports eight levels of priority, 0–7. The code of practice adopted by the industry assigns normal data packets to Priority 0, and ring management packets to Priority 7. Packets being transmitted by a bridge port or switch port should be sent with Priority 4. This is a recognition of the fact that when a ring is connected to a Token Ring switch, a high proportion of the traffic carried on the ring will come from the switch port. So, using a higher priority for transmissions from the switch port provides users with better overall performance.

Token priority levels 5 and 6 are intended for time-sensitive applications such as voice and video. Tests carried out on Token Ring networks with various populations of users connected to a ring have shown that the use of Priority 5 or 6 for voice and video packets can guarantee access to the medium within a few milliseconds, provided that the overall bandwidth consumed by voice or video packets is kept within a reasonable proportion of the total bandwidth on the ring.

So, the token priority mechanism provides a means to support real-time traffic in the context of a single ring. But what happens when the network consists of many rings interconnected by switches? Can the switches preserve the priority of voice and video packets end to end across the network? And can they also expedite the delivery of voice and video packets over normal data packets? The answers depend on how the switches have been designed.

PACKET PRIORITY AND TOKEN RING SWITCHES

When a switch receives a packet to be forwarded to another ring, it can establish what priority the packet was transmitted with by looking in the Logical Link Control (LLC) header that is part of the packet. It is therefore possible for the switch to preserve this priority across the switch fabric and ensure that the same priority level is used to transmit the packet onto the destination ring. This ability to preserve packet priority across the switch fabric, and to transmit the packet onto the ring at the output port of the switch with the appropriate priority level, is the first prerequisite of a Token Ring switch for handling voice and video traffic.

But just preserving packet priority across the switch fabric is not enough. Token Ring switches have internal queuing mechanisms that buffer packets being sent to a particular output port when that port is congested. If there is only a single queue for an output port, then it is possible for a high-priority packet to be trapped in a queue within the switch behind several lower priority packets. The result is likely to be that voice and video packets suffer unacceptable delay variations across the network at times of heavy traffic loading.

The solution to this is to implement multiple queues within the switch fabric so that high-priority packets are sent to a different output queue than lower-priority packets. The switch must give precedence to the higher priority queues when selecting which packet to send next on a congested switch output port. This way, voice and video packets have their own queues and do not get trapped behind data packets. Provided that the total bandwidth of voice and video packets being sent to a particular switch output port is limited to a reasonable fraction of the speed of that port, there will not be a buildup of voice or video packets in the high-priority queues, so end-to-end transmission delays will be kept within acceptable limits.

This, then, is the second prerequisite of a Token Ring switch for handling voice and video traffic: the switch must implement multiple queues to enable higher priority voice and video packets to be expedited over normal priority data packets at congested switch ports.

PACKET PRIORITY AND HIGH-SPEED TOKEN RING

As explained, the priority mechanisms embodied in Token Ring work in two distinct ways. First, high-priority packets enjoy privileged access to the shared medium when there are multiple stations connected to the ring. And second, an appropriately designed Token Ring switch can make use of the priority indication within the Token Ring packet to expedite delivery of high-priority packets over those with lower priority when switch ports become congested.

High-Speed Token Ring will not, initially, provide a shared media mode of operation and, therefore, the priority mechanism that provides privileged access to shared rings does not apply to HSTR.

However, priority information embedded in Token Ring packets is preserved across HSTR links, and Token Ring switches can be designed that support multiple queues on outgoing HSTR ports. Therefore, HSTR links for interswitch or switch-to-server connections can form part of an end-to-end transmission path that supports packet prioritization and expedited delivery of high-priority packets.

MEETING THE NEEDS OF VOICE AND VIDEO IN THE LAN

Real-time interactive voice and video in the LAN requires both adequate bandwidth for the delivery of constant streams of packets containing voice and video information, and consistent low transmission delay end-to-end across the network.

ATM provides the most sophisticated solution for the delivery of voice and video in the LAN. An application can request that bandwidth be reserved for a voice or video stream, and place an upper limit on transmission delay. The ATM switches that make up the network assess whether there are sufficient resources to meet the request from the application and either accept or refuse the request. Once a bandwidth reservation request has been accepted by the ATM network, resources are set aside on each link of the end-to-end transmission path to ensure that the voice or video stream is delivered within the requested delay bounds.

All this sophistication, of course, comes at a price. ATM is a substantially more complex and more costly technology than Token Ring. Because of this, ATM has made little impact at the desktop. And the popularity of ATM in the LAN backbone is based on other attributes of ATM besides Quality of Service: capacity, scalability, and fault tolerance.

Using end-to-end packet prioritization and expedited delivery of priority packets in the LAN is a much less sophisticated solution for voice and video than ATM can offer. However, it is also much simpler, and it leverages capabilities that are already built into the Token Ring network. The only circumstances under which it will not work so well are when bandwidth is very constrained and when voice and video traffic occupies a high proportion of the available bandwidth. But bandwidth in the LAN is relatively inexpensive, and it is generally cheaper and easier to over-provision capacity in the LAN to enable good voice and video performance with packet prioritization than it is to introduce a new and different technology for this purpose.

CONCLUSION

High-Speed Token Ring offers existing Token Ring users a simple incremental upgrade path for switched Token Ring backbones. Carrying native Token Ring frame formats with support for all Token Ring bridging schemes including source routing, HSTR is a logical extension of the Token Ring environment to 100M bps and 1G bps transmission speeds.

Both HSTR and ATM can meet all the practical backbone needs of large Token Ring installations in terms of capacity, scalability, and fault tolerance. HSTR will generally cost less than ATM for equivalent network capacity and involves less new knowledge to install and manage. ATM, on the other hand, can provide guaranteed quality of service, can be seamlessly extended over public and private WAN environments, and provides for a more peaceful coexistence with Ethernet in mixed-technology installations.

Token Ring users planning backbone upgrades now have the luxury of a choice. With HSTR and ATM to choose from, network planners are more likely to find a backbone solution that precisely fits their own particular needs.

Chapter 11
Should You Migrate Your Token Ring Network to Ethernet?

Randall Campbell

TOKEN RING NETWORK MANAGERS ARE FACED WITH THE DIFFICULT DECISION OF STAYING WITH TOKEN RING OR MIGRATING TO ETHERNET. Making matters worse is that analysts, consultants, and vendors do not always agree on the answer. Some recommend abandonment of Token Ring as soon as possible and a move to Ethernet. Others suggest that Token Ring is strategic and one should stay with it forever. Ultimately, one has to decide what to do. This article examines the technical and business factors that will help determine the future direction of the network.

First of all, take a look at a few of the more obvious facts about the Local Area Networking market. Regardless of how it is measured, the Ethernet market is much larger than the Token Ring market. This has resulted in increasing choices, decreasing prices, and rapid advancement of technology for the Ethernet user. Unfortunately, Token Ring technology has not advanced as quickly and prices remain much higher. Also, fewer vendors are offering Token Ring solutions, as many view this to be a declining market.

Recently there has been much press about new high-speed alternatives for Token Ring. Some argue that Token Ring customers have not needed higher speeds until recently, therefore these new technologies are arriving just in time. This is extolled as a revival of the Token Ring market and a reason for customers to stay with Token Ring. Others, however, say that this is too little and too late to stem the tide of customers migrating to Ethernet.

There is a business to run and one needs to make the best choice possible for the company, both in the short term and for the long term. Thus, one can examine the technical and business issues that should be considered. At the end, a checklist will be provided that will help weigh those factors that are most important in making this decision.

First, let's examine the technical issues. What are the differences between Token Ring and Ethernet? Will an Ethernet network support all applications?

DIFFERENCES BETWEEN TOKEN RING AND ETHERNET

With the current state of LAN switching technology, many of the important differences between Ethernet and Token Ring have disappeared. Both Token Ring and Ethernet can provide a full duplex, dedicated link to the desktop from a LAN switch. This gives each user his own LAN segment with dedicated 10/100 or 4/16Mbps of bandwidth. There are no collisions and no one with whom to share a token. In fact, for full duplex Token Ring there is no token and no ring. In this environment the only important differences are the speed, maximum frame size, source routing support, and IP multicast support.

Speed

The speed issue has recently become a topic of debate with the efforts of Token Ring vendors to establish a standard for 100Mbps and 1Gbps Token Ring. 10Mbps and 10/100Mbps Ethernet NICs, hubs, and switches are widely available today. Gigabit Ethernet is standardized and becoming widely available. Token Ring is widely available in 4 and 16Mbps versions.

The proposed High Speed Token Ring (HSTR) standards provide for 4/16/100Mbps Token Ring and 1Gbps in the future. There are no 100Mbps hubs planned for Token Ring because the proposed HSTR standards do not support shared media. At the time of this writing, 100Mbps HSTR NICs are just emerging and are priced at three to five times more than 10/100Mbps Ethernet NICs. Although there are stated goals to support both Token Ring and Ethernet over HSTR, initial offerings are Token Ring only. So, the only real difference in speed is 4/16Mbps vs. 10Mbps for client NICs. Server and backbone speeds for both Token Ring and Ethernet are the same.

Some vendors are providing support for Token Ring over Fast and Gigabit Ethernet links. This has the advantage of providing Token Ring frame support at or close to the price points of Ethernet. An important capability of these solutions is the ability to support both Token Ring and Ethernet over the same link at the same time. This makes migration easier by allowing installation of a single high-speed backbone for Token Ring and Ethernet clients and servers.

Maximum Frame Size

The other performance-related difference is the maximum frame size. All speeds of Ethernet are limited to 1500 bytes as the maximum frame size. The maximum frame size for Token Ring can be set as low as 516 bytes and as high as 17,749 bytes. The typical value is 4399 bytes. On the surface this appears to be a good thing. After all, the more data one can send per frame, the better the throughput, right?

As usual, things are not that simple. The benefit of large frames is most apparent on the server. Conventional wisdom states that the CPU in a server has to process each frame transmitted or received. The more bytes there are in each frame, the fewer frames are needed to send the same amount of data. This results in lower CPU utilization. This would be true if all NICs were created equal. CPU utilization depends a great deal on how efficiently data transfers occur across the bus. The NIC hardware, drivers, and operating system can impact performance and CPU utilization more significantly than frame size.

Well, if NICs are efficient, are large frames better? There is another tradeoff that needs to be considered. This has to do with the number of store and forward hops between the clients and servers. A store and forward hop can be a router, bridge, or switch in either the LAN or over the wide area network (WAN). Each store and forward hop must completely receive the frame before it forwards it on the next hop in the path. The larger the frame, the longer it takes to receive it, and the longer it takes to transmit it. This can result in increased end-to-end latency that can negatively impact performance.

Without going into great detail about queuing theory and flow control, the net result is that larger frames work best over short, fast paths. Smaller frames work best over longer, slower paths. Typically, the network is a mix of short and long paths. For example, the LAN has short, fast paths and the WAN has longer, slower paths. Based on the author's experience, a maximum frame size of around 2000 bytes provides good performance for both the LAN and WAN. By the way, cut-through switching (in which the bridge or switch forwards the frame on the output link as it is being received on the input link) can help, but cut-through only occurs if the input and output link are running at the same speed. If they are different speeds, even a cut-through switch will use store and forward.

Source Routing

Source routing is a technique for bridging LANs that is not available on Ethernet. Ethernet only supports transparent bridging. There are three advantages of source routing over transparent bridging: multiple parallel paths, duplicate MAC addresses, and network management.

Source routing allows traffic to be distributed over parallel bridge paths through the network. With Ethernet, parallel paths are allowed, but all traffic will take the same path. The other paths will be inactive until the active path fails. When the active path fails, the spanning tree protocol will automatically activate another path. There is an advantage to this approach. When a source routing path fails, the end systems have to rediscover a path. This may result in the failure of sessions (for example, SNA sessions). In Ethernet, a path change is transparent to the end system (this is why it is called transparent bridging). Therefore, sessions typically do not fail when the path changes in Ethernet. Pure transparent bridging can be used on Token Ring; however, this is not typical. Some networks use Source Route Transparent (SRT) bridging where some traffic will be bridged via source routing and other traffic, such as IPX, will be bridged using transparent bridging.

Duplicate MAC address support is a technique that was developed to get around a shortcoming of SNA. SNA did not implement an address resolution protocol on the LAN. As a result, SNA users had to configure the MAC address of their SNA gateway in their client. If the gateway failed, they would have to reconfigure a backup gateway or wait for the gateway to be repaired. Since source routing bridges or switches only look at the source routing information to determine the path, it is possible to put the same MAC address on two different ring numbers. This allowed the network manager to configure two identical SNA gateways with the same MAC address on two different Token Rings (with different ring numbers). Now, the SNA gateway would connect to the first gateway that responded to the client explorer frame (this is pretty much a random choice). If that gateway failed, the session would fail, but the client could immediately reconnect to the other gateway. This technique has been widely used in SNA to provide backup and load distribution between gateways.

Ethernet does not support source routing. All forwarding decisions in bridges and switches are made by looking only at the destination MAC address. As a result, a MAC address can only appear on a single Ethernet segment. Duplicate MAC addresses are simply not supported. One way to provide SNA gateway backup on Ethernet is to use a protocol in the gateway to activate a backup with the same MAC address if the primary fails, or a protocol in the client to reconfigure it to a different gateway. One can also use a single source route/translational (SR/TLB) bridge between the Ethernet network and the Token Ring attached SNA gateways as shown in Exhibit 11.1. Another alternative is to change from pure SNA clients to the tn3270 or tn5250 protocol. tn3270/5250 runs over TCP/IP and uses TCP/IP for alternate path support.

Source routing is helpful for network management in that each frame contains the bridging path from source to destination. This is useful when looking at traces since one can tell on which ring number a frame originated and where it is headed. On Ethernet, one must keep up with where MAC addresses are located. This can also be automated via a variety of tools and can include Layer 3 addresses such as IP, IPX, or NetBIOS.

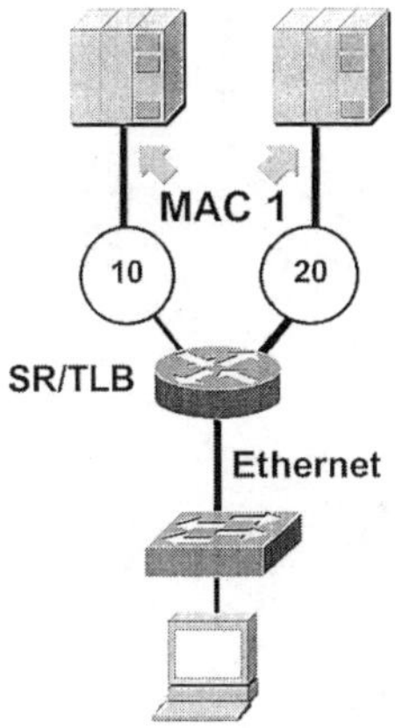

Exhibit 11.1 SNA Gateway Backup Using SR/TLB.

Source routing has its share of headaches as well. Since the procedures for discovering a source route were never well standardized, there are a variety of problems that occur with different implementation of protocol drivers and protocol stacks. Recently there was a situation where the client was configured to use source routing and the server was not using source routing. With this configuration, the server should not send or respond to source-routed frames. However, when the client and server are on the same ring, the server would actually respond to a source route frame from the client. When the client and server were on different rings it no longer worked. There are other cases where clients will send excessive explorers and flood the network when connections are lost to their servers. Ethernet is much simpler because it only supports transparent bridging. This is why some network managers have moved to transparent bridging on their Token Ring networks.

IP Multicast

IP multicast is increasingly being used for distribution of live or scheduled multimedia information over the network. A server can send a single multicast packet of voice, video, or data to the network that multiple clients can receive. The network will duplicate this multicast packet as needed to get a copy to each client registered to receive it. This technique greatly reduces the load on the server and the amount of traffic sent through the network.

Unfortunately, IP multicast is very difficult to do effectively on Token Ring networks. The reason has to do with the way different multicast streams are identified at the MAC layer. Think of a multicast stream as a TV

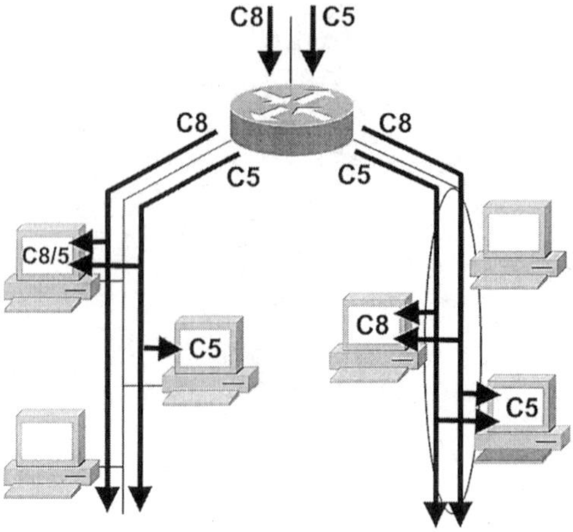

Exhibit 11.2 Token Ring IP Multicast Using Functional Address.

channel. On Ethernet, each channel uses a different MAC group address as the destination address of the frame. Ethernet NICs can support multiple group addresses at the same time. Therefore, an Ethernet NIC can tune in to Channels 8 and 5 and ignore all other channels. It does this by only copying frames destined to the MAC group address for Channels 8 and 5.

Unfortunately, most Token Ring NICs support only a single group address. Equally unfortunate is that most of them do not support the universally administered group addresses used by IP multicast. Therefore, on Token Ring, all IP multicast data is typically sent to the broadcast MAC address or a functional address. If the functional address is being used, a Token Ring NIC will either receive all of the channels or none of them, as illustrated in Exhibit 11.2. If the broadcast MAC address is being used, all Token Ring NICs on the LAN will receive all of the channels if any one is tuned in to a single channel. These received multicast frames are sent to the CPU in the client or server regardless of whether there is any application running to process them. This can result in significant unnecessary use of the CPU.

WILL AN ETHERNET NETWORK SUPPORT ALL OF MY APPLICATIONS?

The short answer is yes. The author cannot think of any applications that only run over Token Ring. For years SNA only ran over Token Ring, but SNA over Ethernet abounds today. One might have devices that only support Token Ring interfaces, such as 3174 control units. Source route translational bridging can be used to connect these Token Ring devices to an Ethernet network.

The only function Ethernet cannot provide is duplicate MAC address support. If one relies on this feature of source routing on Token Ring, it will be necessary to either migrate the SNA to TN3270/5250 or automate the gateway backup via other means. The other main difference between Token Ring and Ethernet is the maximum frame size. This may provide some performance advantage for Token Ring over Ethernet, but with the increasing performance of clients and servers these differences will be less noticeable.

If one has decided Ethernet can support the applications, what are the migration issues that need to be considered?

Migration Issues

The first migration issue is cabling. If one already has unshielded twisted pair (UTP) cabling for the Token Ring network, one is in good shape. For Category 5 UTP cable, one can run both 10Mbps and 100Mbps Ethernet; for Category 3 cable one can run 10Mbps Ethernet. What if one is currently using the IBM cabling system with Type 1 or 2 cable? This is a shielded twisted pair (STP) cable with 150-ohm impedance instead of the 100-ohm impedance for UTP. One cannot directly connect an Ethernet NIC or port to the IBM Type 1 or 2 STP cable.

There are a few choices in this situation. For IBM Type 2 cable, one can use the UTP Category 3 telephone pairs for Ethernet. It will be necessary to wire these cables to an RJ45 jack in the office. For IBM Type 1 cable and no UTP cable, one can either rewire the building (no fun) or use baluns to convert the impedance from 150 ohms to 100-ohm. These baluns have an RJ45 female connector and an IBM cabling system connector. One will need one of these baluns at each end of the cable.

The second migration issue is interoperability during the migration. The length of the migration will determine how best to provide this. There are two basic approaches. One is to migrate as quickly as possible by replacing the current Token Ring NICs and installing an Ethernet switched network. The other is to move to Ethernet as devices are replaced; when a new PC comes in the door it is installed on the Ethernet network. The existing PCs remain on the Token Ring network until they are replaced.

To migrate quickly, simply leave the current Token Ring backbone in place and install the new Ethernet backbone. Typically, one would use existing routers to interconnect the Token Ring and Ethernet backbones. As clients and servers are upgraded with new NICs, they are connected to the Ethernet backbone. This will likely require changing the client's or server's network address, which can be done automatically with DHCP for TCP/IP. For IPX, the clients will automatically find their nearest server, but servers will have to be reconfigured manually.

To migrate gradually, a single high-speed backbone that can support both the Token Ring and Ethernet users is desirable. The choices for a high-speed backbone are ATM, Fast/Gigabit Ethernet, or HSTR at some point in the future. When using ATM as a backbone, it is easy to support both Token Ring and Ethernet. If Fast/Gigabit Ethernet is used for the Ethernet network, then look for a vendor that can support full Token Ring capability over this backbone. It will be a nonstandard implementation, but it can provide an easy way to migrate to Ethernet without building separate Ethernet and Token Ring backbones. HSTR may be an option in the future, but at the time of this writing it only supports Token Ring frames at 100Mbps. Even if it does support Ethernet frames and 1Gbps in the future, it will likely be much more expensive than a Fast/Gigabit Ethernet solution.

Next, one needs a way to interconnect the Token Ring and Ethernet network. The best solution is to use routing, either in a router or in a switch. Protocols that are routable such as IP and IPX should be routed, with the Ethernet and Token Ring networks configured as different subnets or networks. Nonroutable protocols such as SNA and NetBIOS can be translationally (SR/TLB) bridged. Typically, routers will support both routing and SR/TLB bridging between Token Ring and Ethernet. Routing will block broadcast and explorer traffic between the LANs and thereby reduce the impact of any broadcast storms.

Many customers are eliminating nonroutable protocols. NetBIOS can be run over TCP/IP very effectively and provides superior scaling to the native version. SNA traffic can be converted to tn3270/5250 and therefore also run over TCP/IP. This eliminates the need to have SNA session routing in the network. Even Novell is migrating Netware to run over TCP/IP. Migration to TCP/IP as the only networking Layer 3 protocol is a good goal to set because it eliminates the complexity of managing multiple Layer 3 protocols and allows the use of high-speed, Layer 3 switching technologies that tend to be IP only.

For the routable protocols, parallel routers can be used between subnets and networks. The routing protocols will provide for automatic path selection and backup. For any nonroutable protocols, parallel paths can only be used if there is the capability to run spanning tree end-to-end across both the Ethernet and Token Ring network. This can be a tricky function to implement, so it is usually best to provide a single path for bridging between the Token Ring and Ethernet network. Network management can be used to activate backup paths in the event of a failure.

One of the issues sometimes mentioned by Token Ring vendors about mixing Token Ring and Ethernet is the maximum frame size differences. Some will claim that this can cause a severe problem with either dropped frames or degraded performance. It is true that there could be some problems in this area; however, there are effective ways to handle this issue.

For TCP/IP, there are two mechanisms available to solve mismatches in frame sizes. Modern TCP protocol implementations will negotiate the frame size down to the smallest supported on either end of the connection. If a Token Ring user is connecting to a server on Ethernet, the maximum frame size will be set to 1500 bytes and all will be fine. If TCP is not being used and the Token Ring user sends a frame larger than 1500 bytes, the router can perform IP fragmentation and convert the large frame into multiple frames that are no larger than 1500 bytes.

For nonroutable protocols, the SR/TLB function can use the routing information field (RIF), which is part of source routing, to bring the frame size down to 1500 bytes. This will result in the Token Ring client only sending 1500 byte frames to an Ethernet server while still being able to send larger frames to a Token Ring server. In most cases, these mechanisms will effectively resolve frame size differences.

To summarize, the main issues for migration are to choose a backbone technology that can support both Token Ring and Ethernet and the use of routing and translational bridging to support interoperability between the LANs. One should always be aware of possible issues with the different frame sizes, but in most cases this will happen automatically. If the application does not automatically adjust, one will need to configure the maximum frame size for the application on Token Ring to 1500 bytes.

BUSINESS ISSUES

Thus far, the technical differences between Token Ring and Ethernet and the issues involved in migration have been discussed. Understanding these issues will help in dealing with the business issues that usually revolve around satisfying a wide variety of demands from users and corporate management. How many times has one heard the following?

- The network is too slow, we are losing productivity.
- The network is critical; we lose lots of money every minute the network is down.
- We have a new application that we are rolling out. Can the network handle it?
- We keep getting a lot of dropouts when we play audio and video streams over the network.
- My new PC came with an Ethernet interface; why can't I use it?
- The CFO says we are spending too much on the network, so he has cut our budget.

It always seems that everyone wants the network manager to do more with less, but the network manager has problems to deal with as well. It will be necessary to manage the network, define and deliver service levels, train employees so that they are effective in dealing with problems, and make the most cost-effective use of the budget available. When faced with all of these requirements and pressures, the typical reaction is to try to tweak the current network architecture to meet the needs with the least amount of change.

There is no doubt that simply upgrading the existing Token Ring network will provide the least amount of change. Migration to a new technology is difficult, time consuming, and will likely result in some problems during the migration. The most conservative approach is simply to enhance the existing Token Ring network.

On the other hand, the network manager may not be able to meet all of the requirements placed on him by simply upgrading the Token Ring network. For example, if there are definite needs for Ethernet connectivity, heavy requirements for IP multicast, or requirements for high-speed backbones and Layer 3 switching today, one will at least need to provide an integrated Token Ring and Ethernet network. Most of the short-term requirements can likely be met by upgrading the Token Ring network or providing the integration of some Ethernet technology.

The real issue is the long-term viability of Token Ring. It is doubtful that Token Ring will ever increase in popularity or catch up with Ethernet in terms of technology or price. HSTR is unfortunately too late to attract the interest of many vendors other than traditional Token Ring NIC vendors. Since the number of Token Ring vendors is not likely to grow significantly, Token Ring customers will always have fewer choices than Ethernet customers. Over the long term, an Ethernet network clearly provides more choices, both in terms of vendors and technologies.

If one cannot, or does not want to migrate to Ethernet, then the best choice is to build the switched network so that one can easily support both Token Ring and Ethernet users. The short-term solution may be to only support Token Ring users. If one builds an infrastructure that can support both, one can add Ethernet users in the future if the requirements change. Therefore, choose vendors that can provide a single backbone for both Token Ring and Ethernet, whether it be ATM, Fast/Gigabit Ethernet, or even HSTR. Ideally, you should not have to pay more for a backbone infrastructure that can support both Ethernet and Token Ring than you would pay for one that only supports Ethernet.

SUMMARY

It is possible to migrate to Ethernet from Token Ring, and there are short-term and long-term advantages in doing so. One can build higher capacity LANs at lower cost with Ethernet than with Token Ring. There are a few features and functions that one has to work around (e.g., duplicate MAC address support), but one gains better support for IP multicast. So, should one migrate?

Exhibit 11.3 Determining Factors for Your Network.

Rank	Criteria	Advantage
___	New equipment cost must be kept as low as possible	Ethernet
___	CPU utilization must be kept as low as possible	Token Ring
___	Non-ATM, high-speed backbone and server connections are desired	Ethernet
___	Duplicate MAC address support is needed	Token Ring
___	IP multicast applications required	Ethernet
___	Active parallel switching path is needed	Token Ring
___	Layer 3 switching required	Ethernet
___	Minimize change to the network	Token Ring
___	Need to exploit latest technology for business advantage	Ethernet
___	Must use IBM type 1 or 2 cable	Token Ring

This is a decision that only the network manager can make. If one needs to keep up with the latest networking capabilities and new applications, migration to Ethernet is a wise choice. If one has limited resources and no major needs to change, enhancing the existing Token Ring network may be best. If one chooses to stay with Token Ring in the short-term, make sure that the backbone solution can support Ethernet users in the future. This will minimize headaches if one decides to migrate to Ethernet in the future.

Exhibit 11.3 summarizes a series of factors that can be used to determine which issues are the most important in your network. After ranking the items, sort them from high to low and see which technology best satisfies the most important items. Enter an importance factor for each item (1 is low; 5 is medium; 10 is high).

Chapter 12
Introduction to 100BASE-T: Fast (and Faster) Ethernet

Colin Mick

FAST ETHERNET (100BASE-T), an extension to the IEEE802.3 Ethernet standard to support service at 100M bps, is virtually identical to 10BASE-T, in that it uses the same media access control (MAC) layer, frame format, and carrier sense multiple access with collision detection (CSMA/CD) protocol. Network managers can use 100BASE-T to improve bandwidth and still maximize investments in equipment, management tools, applications, and network support personnel.

INTRODUCTION

Fast Ethernet (100BASE-T) is an extension to the IEEE802.3 Ethernet standard to support service at 100M bps. It is virtually identical to 10BASE-T in that it uses the same media access control layer, frame format, and carrier sense multiple access with collision detection (CSMA/CD) protocol. This means that network managers can use 100BASE-T to improve bandwidth and still make maximum use of investments in equipment, management tools, applications, and network support personnel.

100BASE-T is designed to work transparently with 10BASE-T systems. Switches (high-speed, multiport bridges) are used to connect existing 10BASE-T networks to 100BASE-T technology. By building networks with 100BASE-T and 10BASE-T linked with switches and repeating hubs, network designers can build networks that provide four levels of service:

- Shared 10M-bps service
- Dedicated (switched) 10M-bps service
- Shared 100M-bps service
- Dedicated 100M-bps service

Operating at higher speeds with the same frame size and the CSMA/CD protocol requires that 100BASE-T collision domain diameters be smaller — typically about 200 meters. In 100BASE-T, larger networks are built by combining collision domains by way of switches. Fiber (100BASE-FX) links are used to support long (i.e., 412 meters in half duplex, 2 kilometers in full duplex) cable runs. Within a single collision domain, port density is increased by using modular or stacking hubs.

The 100BASE-T standard (IEEE802.3u, 1995) currently defines four physical layer signaling systems:

- 100BASE-TX supports operation over two pairs of Category 5 unshielded twisted pair (UTP) or shielded twisted pair (STP) cables.
- 100BASE-T4 supports operation over four pairs of Category 3, Category 4, or Category 5 UTP or STP cables.
- 100BASE-T2 supports operation over two pairs of Category 3, Category 4, or Category 5 UTP or STP cables.
- 100BASE-FX supports operation over two 62.5-micron multimode fibers.

Products for 100BASE-TX and 100BASE-FX are available from a wide range of manufacturers. Products for 100BASE-T4 are supported by a smaller group of manufacturers, and no products have yet been offered for 100BASE-T2.

In addition, the Fast Ethernet standard has recently added support for full-duplex operation and flow control. Full-duplex operation is broadly available in current 10BASE-T and 100BASE-T products.

HOW IT WORKS: AN ISO VIEW

Exhibit 12.1 depicts an ISO seven-layer diagram comparing 10BASE-T and 100BASE-T. Both 10BASE-T and 100BASE-T defined operations at the lower half of the data link layer (known as the Media Access or MAC layer) and the physical layer. Extension of the Ethernet standard to 100M-bps operation required one small change to the MAC layer operation specified in the IEEE802.3 standard. Originally, timing was defined in absolute terms (i.e., an external reference clock). As a result, timing specifications were defined in milliseconds, nanoseconds, and picoseconds. To support 100M-bps operation, timing was respecified relative to the internal clock of the MAC. This meant that specifications were defined in bit times.

Several changes were made at the physical layer. In 10BASE-T, coding (i.e., conversion of data bits to symbols) is done in the PLS layer, directly below the MAC. A mechanical interface called the attachment unit interface (AUI) is situated directly below the PLS. Below the AUI is the PMA layer, which converts the digital symbols into analog symbols that can be sent

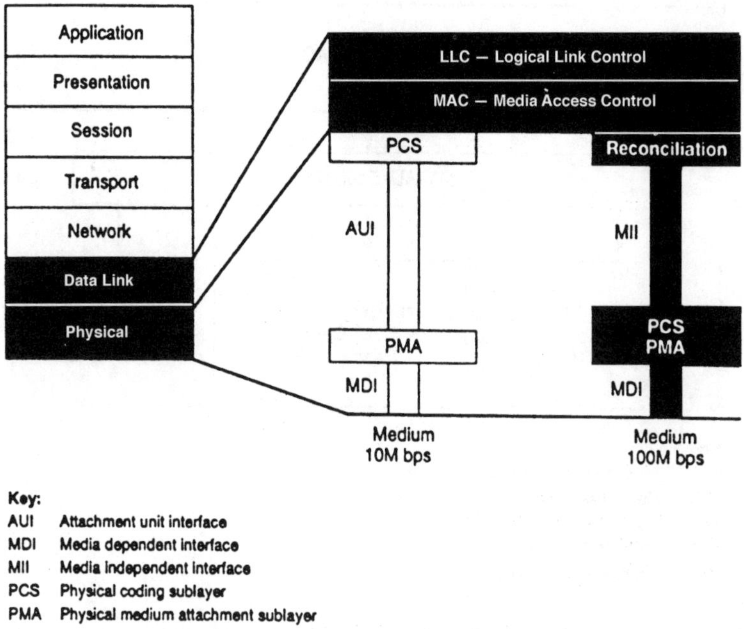

Exhibit 12.1 100M bps Standards Model.

across the wire, and a media-dependent interface (MDI) — a socket for connecting the cable.

100BASE-T puts the coding, called the physical coding sublayer (PCS), below the mechanical interface. This is done to make it possible to offer a variety of coding systems that can be packaged in a transceiver along with the analog/digital circuitry for connection via the mechanical interface. The mechanical interface used for 100BASE-T is called the Media Independent Interface (MII). It is similar to the AUI, but offers a larger data path and the ability to move management information between the PHY and the MAC. A simple mapping function, called the Reconciliation Sublayer, handles linking the MII to the MAC. As noted previously, 100BASE-T currently supports four signaling systems (see Exhibit 12.2): 100BASE-TX, 100BASE-T4, 100BASE-T2, and 100BASE-FX.

Two 100BASE-T signaling systems — 100BASE-TX and 100BASE-FX — are based on the transport protocol/physical medium dependent (TP/PMD) specification developed by the ANSI X3T12 committee to support sending fiber distributed data interface (FDDI) signals over copper wire (see Exhibit 12.3). TP/PMD uses continuous signaling, unlike the discrete signaling used with 10BASE-T. In 10BASE-T, when a station is finished

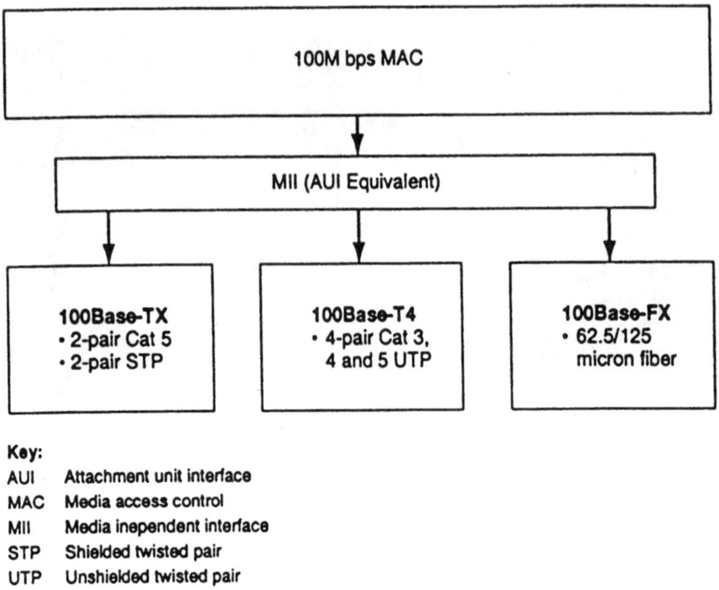

Exhibit 12.2 100BASE-T Physical Layers.

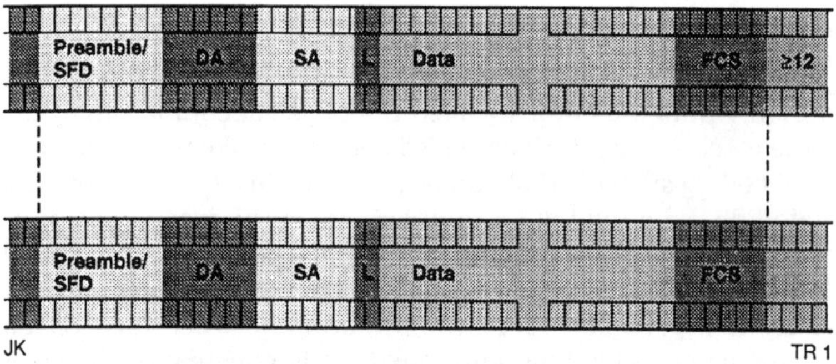

Exhibit 12.3 100BASE-T (TX and FX) Frames.

sending a frame, it sends a few idle signals and then goes quiet except for a link pulse which is sent every 16 ms to indicate that the link is still good.

In TP/PMD, a continuous stream of idle symbols is sent when data is not being transmitted. To ease the transition between data and idle signals, a JK symbol sequence is added to the front of a data frame and a TR symbol sequence is added to the end of the frame before transmission of idle symbols

Exhibit 12.4 100BASE-T-FX.

- Uses 2-strand, 62.5/125 micron fiber
- Connector: MIC, ST, SC (converters available)
- Uses FDDI TP/PMD specification
  - — Continuous signaling scheme
  - — 4B5B coding scheme
- Transmits over 1-fiber and receives over 1-fiber
- 100M bps data rate
- Full and half duplex
- Detects and signals far end faults

begins. The JK, TR, and idle transmission patterns must be added to Ethernet frames when they are transmitted via the TP/PMD specification.

Both 100BASE-TX and 100BASE-FX use 4B5B coding. This means it takes 5 baud (signal transitions on the wire) to transmit 4 bits of information. This is vastly more efficient than the Manchester coding used for 10BASE-T, which requires 2 baud to send each bit across the wire.

Exhibit 12.4 summarizes the attributes of 100BASE-FX. It uses two strands of 63.5-micron fiber. All standard connectors are listed in the specification — different manufacturers support different types of connectors. 100BASE-FX uses the FDDI TP/PMD specification with continuous signaling and 4B5B coding. The data clock runs at 125 MHz, providing a signaling rate of 100M bps with the 80% efficiency of 4B5B coding. One fiber is used for transmitting data, the other for receiving data. It can support both half-duplex and full-duplex operation and has automatic link detection.

100BASE-TX

Exhibit 12.5 summarizes the attributes of 100BASE-TX. It operates over two pairs of Category 5 UTP or STP, and uses Category 5 certified RJ-45 connectors. It uses the 125-MHz data clock, continuous signaling, and 4B5B coding of 100BASE-FX, but adds signal scrambling and MLT-3 conditioning to deal with noise problems associated with sending high-frequency signals over copper. 100BASE-TX uses exactly the same connector pinouts as 10BASE-T. It transmits over one pair and receives over the other. It supports half-duplex and full-duplex operation.

100BASE-T4

100BASE-T4 (see Exhibit 12.6) is a more complex signaling system because it must support a 100M-bps data rate over cable certified for operation at 16 MHz. This is accomplished by increasing the number of cable pairs used

- Uses 2-pair Category 5 twisted pair cable
- Connector: Category 5 certified RJ-45 (IEC 603-7), or DB-9
- Uses FDDI TP/PMD specification
  - Continuous signaling scheme
  - 4B5B coding scheme
  - Scrambled symbols
  - MLT-3
- Transmits over 1-pair and receives over 1-pair
- 100M bps data rate
- Full and half duplex modes
- Identical connector pin-out

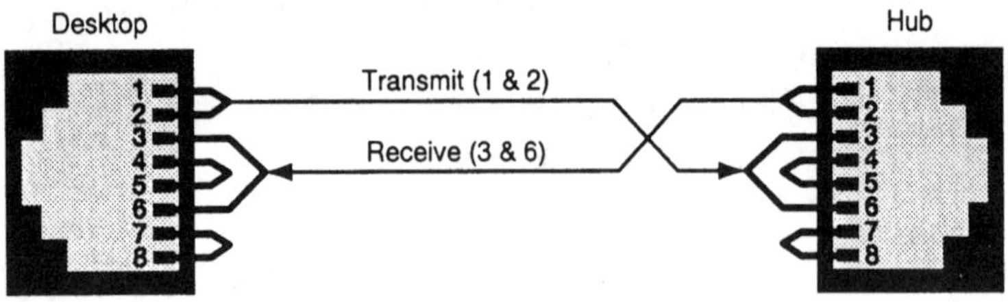

Exhibit 12.5 100BASE-TX.

- Uses 4-pair Cat 3, 4, or 5 cable
- Transmits over 3-pair
- Connector: Standard RJ-45 (IEC 603-7)
- Uses an 8B6T coding scheme
- Signaling: 100M bps = 3-pair x 25 Mhz x (133% for 8B6T encoding)
- Half Duplex

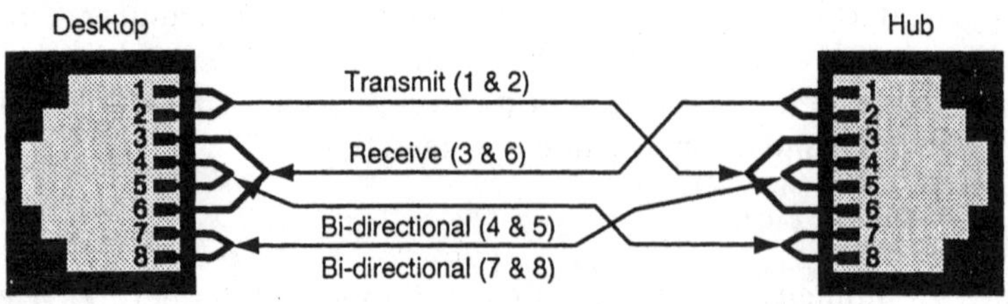

Exhibit 12.6. 100BASE-T4.

for data transmission and using a more sophisticated coding system. 100BASE-T4 starts with the two pairs used for 10BASE-T — one for transmit and one for receive — and adds two additional pairs that are used bidirectionally. This means that when transmitting, 100BASE-T4 always transmits over three pairs (one dedicated and two bidirectional) while listening for collisions on the remaining pair. It uses a much more sophisticated coding system called 8B6T.

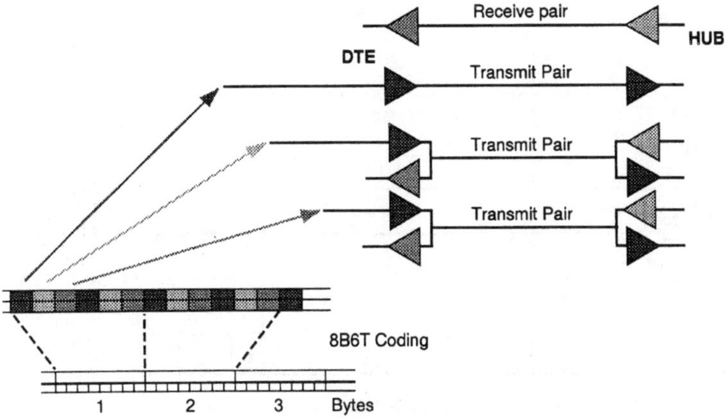

Exhibit 12.7 100BASE-T4 Signaling.

Unlike other coding systems that use binary (0, 1) codes, 100BASE-T4 uses ternary (+1, 0, –1) codes which enable it to pack 8 bits of data into 6 ternary symbols. By using 8B6T coding and three wire pairs for transmission, 100BASE-T4 provides a 100M-bps data transmission rate with a clock speed of only 25 MHz (8 bits transmitted as 6 ternary symbols over three wire pairs at 25 MHz). This process is diagrammed in Exhibit 12.7: 1 byte (8 bits) of data is encoded into 6 ternary symbols, which are transmitted sequentially across three wire pairs. Unlike 100BASE-TX and 100BASE-FX, 100BASE-T4 does not support full-duplex operation.

100BASE-T2

100BASE-T2 provides a more robust and noise-resistant signaling system capable of operating over two pairs of Category 3, Category 4, or Category 5 UTP, or over STP links and supporting both half-duplex and full duplex operation. It uses an extremely sophisticated coding system called PAM5X5 which employs quinary (five-level — +2, +1, 0, –1, –2) signaling. In addition, it uses hybrid circuitry to enable simultaneous bidirectional transmission of 50M-bps data streams over each of the two wire pairs (see Exhibit 12.8).

Because of its robust encoding, 100BASE-T2 emits less noise during use and is less susceptible to noise from external sources. When used with four-pair Category 5 cable bundles, it can coexist with other signaling systems. A single four-pair bundle can carry two 100BASE-T2 links, one 100BASE-T2 link, and one 10BASE-T link, or one 100BASE-T link and one voice (telephone) link.

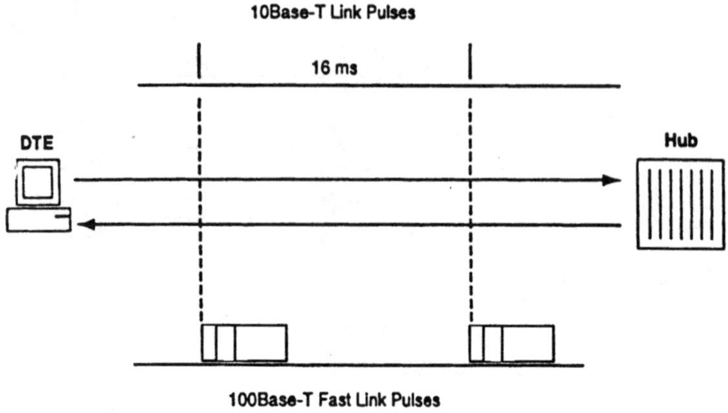

Exhibit 12.8 Media-Independent Interface (MII).

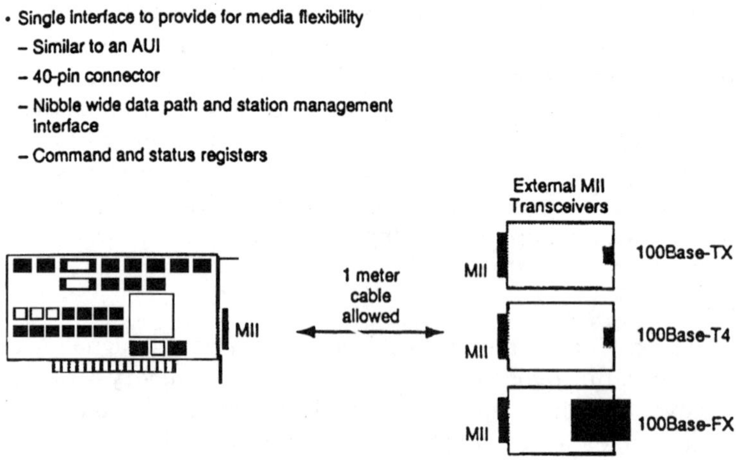

Exhibit 12.9 100BASE-T Auto Negotiation (2).

Media-Independent Interface (MII)

The Media-Independent Interface is a mechanical interface to the Ethernet MAC, similar to the AUI, which is used to connect transceivers (see Exhibit 12.9). The MII supports a nibble-wide data path, a station management interface, and command and status registers. It uses a 40-pin connector, similar in appearance to mini-small computer systems interface (mini-SCSI) connectors.

Auto-Negotiation

Auto-Negotiation provides automatic link testing and configuration for UTP signaling systems. All 100BASE-T systems using UTP or STP go through Auto-Negotiation prior to establishing a link. During this startup process, 100BASE-T systems on each side of a link:

- Check the link.
- Exchange coded information defining the abilities of each link partner (e.g., 10BASE-T half duplex operation, 10BASE-T full-duplex operation, 100BASE-TX half-duplex operation, 100BASE-TX full-duplex operation, 100BASE-T2 half-duplex operation, 1000BASE-T2 full-duplex operation, or 100BASE-T4 operation).
- Go to an internal lookup table to determine the highest common operation mode.
- Configure themselves as per the table.
- Turn off Auto-Negotiation.
- Open the link.

If one end of the link is a 10BASE-T system that does not support Auto-Negotiation, the partner is automatically configured for 10BASE-T half-duplex operation (default mode). When confronted with another networking technology that uses the RJ-45 connector (e.g., Token Ring), Auto-negotiation will automatically fail the link.

Auto-Negotiation is based on the link pulse used in 10BASE-T. For Auto-Negotiation, the link pulse is divided into 33 fast link pulses that are used to carry pages of coded information between link partners.

Full Duplex Operation

Full-duplex operation supports simultaneous signaling in both directions over dedicated links by turning off the CSMA/CD collision detection circuitry. It provides some increase in bandwidth over links that have a high proportion of bidirectional traffic, such as switch-switch and switch-server links. In addition, full-duplex operation increases the maximum length of fiber links. Whereas a half-duplex link is limited to 412 meters by the need to detect collisions, full-duplex operation supports links of up to 2 kilometers because no collision detection is required. This increased link length is only useful for fiber links, signal attenuation limits, and copper link length to 100 meters for both half- and full-duplex operation.

Flow Control

Flow control provides a method for controlling traffic flows between intermediate devices (primarily switches and routers), and between intermediate devices and servers to avoid dropping packets. Currently, two-speed (10/100 or 100/1000) operation requires large buffers to reduce the

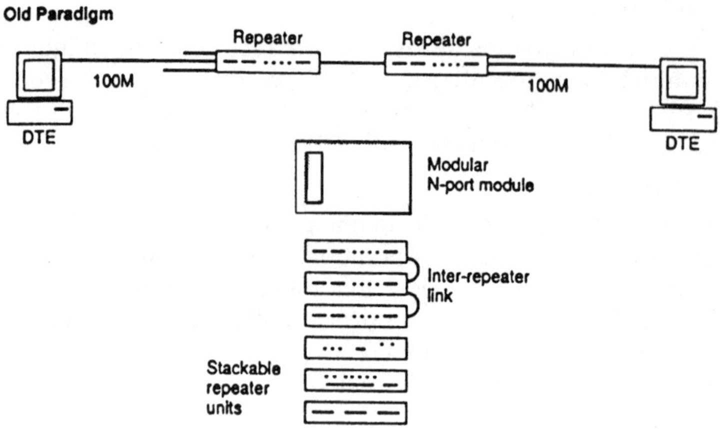

Exhibit 12.10 Repeater Connection Styles.

probability of dropping packets when a continuous stream of packets is sent from a high-speed to a low-speed device (e.g., 100M bps to 10M bps, or 1000M bps to 100M bps). In such a scenario, when the buffers fill, the intermediate device drops the unbuffered packets.

Flow control provides a management alternative to having large buffers. When a buffer approaches full, the receiving device can send a flow control packet back to the sending device to stop the incoming packet stream. When the buffers of the receiving device empty, packet transmission starts again. This eliminates dropped packets and allows manufacturers to build switches with smaller buffers, which reduces costs.

Repeaters and Repeater Connections

Repeaters provide for shared media operation in 10BASE-T and 100BASE-T via the CSMA/CD protocol. 10BASE-T networks have a collision domain diameter of 1000 meters. This permits building large, single-collision domain networks using hierarchical, cascaded repeating hubs to increase port density. 100BASE-T does not permit hierarchical cascading of hubs because the maximum collision domain for UTP is slightly more than 200 meters (see Exhibit 12.10).

Two techniques can be used to build large, single-collision domain networks (i.e., increase port density). One technique is to use modular hubs where ports can be added by inserting additional multiport cards into the hub chassis. A second is to use stackable hubs — standalone repeaters that can be connected via high-bandwidth stacking ports that do not impact the collision domain.

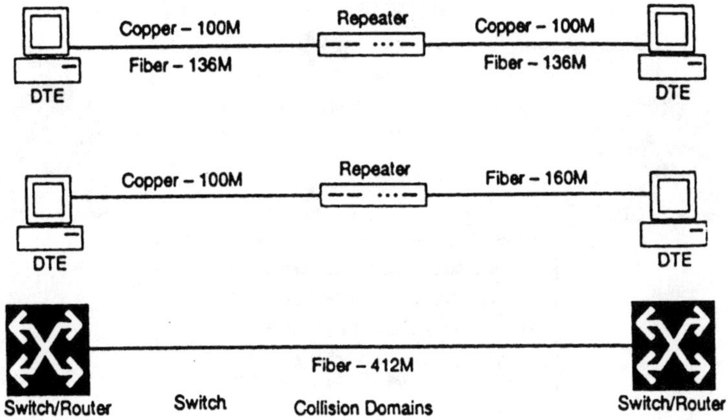

Exhibit 12.11 100BASE-T Topologies.

Topology Rules

Topology rules for half-duplex 100BASE-T networks are shown in Exhibit 12.11. Copper links are limited to 100 meters by the U.S. cabling standard EIA/TIA-568-A. A collision domain containing two copper links can contain one Class I repeater and two 100-meter copper links; or two Class II repeaters, two 100-meter copper links, and a 5-meter copper inter-repeater link. A collision domain containing a Class I repeater with two fiber links can support two fiber links of 136 meters, for a collision domain diameter of 272 meters. A collision domain containing a Class I repeater can also support one copper link of 100 meters and a single fiber link of 160 meters.

A fiber DTE-DTE half-duplex collision domain (e.g., switch-to-switch or switch-to-server) can support a 412-meter fiber link. Links of up to 2 kilometers can be supported over fiber by operating in full-duplex mode, which turns off the CSMA/CD portion of the protocol and requires a dedicated link (see Exhibit 12.12).

Gigabit Ethernet

Work to extend the Ethernet family to 1000M-bps (gigabit) operation is well underway. The first products using the new technology were demonstrated at Networld + Interop Las Vegas in May 1997, and the first products started shipping during the summer of 1997. Initial products will support operation over 62.5-micron multimode fiber (1000BASE-SX), 50-micron single-mode fiber (1000BASE-LX), or short lengths (to 25 meters) of coaxial cable (1000BASE-CX). The operation of these products is being defined in a supplement to the IEEE 802.3 standard entitled 802.3z

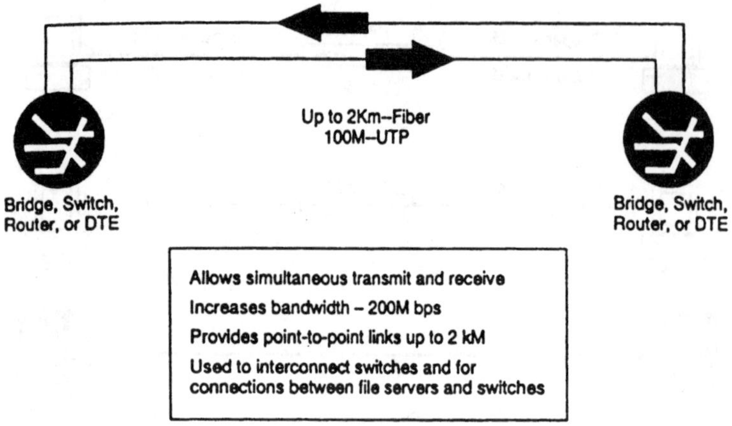

Exhibit 12.12 Full Duplex.

It is scheduled for completion in 1998. A second supplement, entitled 802.3ab, will define gigabit Ethernet over 100-meter, four-pair Category 5 copper links (1000BASE-T). It is scheduled for completion in late 1998.

IMPACTS ON NETWORK DESIGN

Fast Ethernet is a family of 100M-bps signaling systems for use with the standard Ethernet MAC layer. The family consists of four signaling systems (100BASE-TX, 100BASE-T4, 100BASE-T2, and 100BASE-FX) and technologies that support automatic startup (Auto-Negotiation), shared media operation (Repeaters), full-duplex operation, and flow control to manage traffic flow.

Fast Ethernet devices work seamlessly with legacy Ethernet systems: they have the same MAC layer, the same frame format, and the same CSMA/CD protocol for shared media operation. Auto-Negotiation ensures that all 100BASE-T devices operating over copper links automatically configure themselves to operate with link partners. This makes 100BASE-T a very economical technology for adding high-bandwidth links to legacy systems.

Higher-speed operation reduces the diameter of 100BASE-T shared media collision domains to approximately 200 meters for copper. Collision domains can be extended through the use of fiber and connected via switches to build large, complex networks. Full-duplex operation improves bandwidth for bidirectional links and increases the maximum length of fiber links to kilometers. Port density within a single collision domain is expanded through the use of modular and stackable hubs.

100M bps is not the endpoint for Ethernet. 1000M-bps (gigabit) devices were demonstrated in the spring of 1997 and began shipping in the summer of 1997. Targets for gigabit Ethernet operation are 700 meters for full-duplex single mode fiber links, 25 meters for short-haul copper coax links, and 100 meters for Category 5 copper links.

Chapter 13
Understanding Gigabit Ethernet Performance

Gilbert Held

GIGABIT ETHERNET represents the latest Institute of Electrical and Electronic Engineers (IEEE) Carrier Sense Multiple Access with Collision Detection (CSMA/CD) 802.3 standard. In developing the Gigabit Ethernet standard, the IEEE decided that it was important to retain the CSMA/CD access protocol and frame format to provide a truly scalable technology, ranging from Ethernet's 10Mbps operating rate to the 100Mbps operating rate of Fast Ethernet, and to the 1Gbps operating rate of Gigabit Ethernet. Retaining the same media access and frame format simplifies equipment design as the difference between each version of Ethernet becomes primarily speed adaptation on clocking. Unfortunately, retaining the frame structure of Ethernet and Fast Ethernet introduced a new problem in the form of a collision domain.

Both Ethernet and Fast Ethernet provide sufficient time between the next-to-last bit in a frame being placed on a network and the receipt of a propagated collision signal to enable a cabling length of 100 meters from a workstation to a hub port. In a Gigabit Ethernet environment, the higher operating rate restricts the diameter of a network to approximately ten meters, as cabling beyond that distance closes the collision window and makes it possible for a station to transmit every bit in a frame when a collision has occurred on some prior bits and the collision signal is still propagating toward the transmitting station. Because a ten-meter diameter would be overly restrictive, the IEEE standardized a technique referred to as "carrier extension technology" that adds special symbols to the end of relatively short frames on a Gigabit network. While the resulting technique literally opens the collision window and permits a much more practical 200-meter network diameter, the carrier symbols occupy bandwidth and adversely affect the ability of Gigabit Ethernet to transport information per unit of time under certain conditions. Thus, the trade-off was between obtaining a reasonable network cabling diameter and lowering the ability of the network to transport information under certain network situations.

It should be noted that carrier extension technology is only applicable to half-duplex, shared-media Gigabit Ethernet. This is because full-duplex, non-shared media Gigabit Ethernet uses separate paths for the transmission and reception of data, precluding the possibility of collisions. However, because shared-media Gigabit Ethernet is copper based and costs significantly less per port than full-duplex, non-shared media, it is reasonable to expect that many network managers and LAN administrators will focus their attention on the less expensive technology. After all, 1Gbps of shared media bandwidth represents a considerable improvement over 100Mbps Fast Ethernet. Or does it? To answer this question, one must first examine the potential frame flow on different types of Ethernet networks.

ETHERNET FRAME FLOW

Exhibit 13.1 illustrates the composition of an Ethernet frame. Note that the "Data" field must be a minimum of 46 bytes in length and is filled with pad characters if the number of bytes to be transported is less than 46. Thus, the minimum length of an Ethernet frame is 72 bytes.

In some literature one may have read that the minimum length of an Ethernet frame is 64 bytes. That metric references the IEEE standard that defines the frame prior to its placement on the network at the physical layer. Because network adapter cards automatically add the "Preamble" and "Start of Frame Delimiter" fields that add eight bytes to the frame for synchronization, the minimum length of the frame as it flows on the network is 72 bytes.

Returning to the Ethernet frame's "Data" field, its maximum value is 1500 bytes. When one considers the 26 overhead bytes in each frame, this means that the maximum frame length on the network is 1526 bytes. Many publications reference a maximum frame length of 1518 bytes. Once again, such publications are referencing the length of the frame prior to its placement on the network and do not consider the additional eight bytes added by the network adapter card.

Fields	Preamble	Start of Frame Delimiter	Destination Address	Source Address	Type/ Length	Data	Frame Check Sequence
Bytes	7	1	6	6	2	46 to 1500	

Exhibit 13.1 The Ethernet Frame Format.

By understanding that the minimum length of an Ethernet frame is 72 bytes and its maximum length is 1526 bytes, one can compute the maximum number of minimum length and maximum length frames that can flow on an Ethernet network. Because many readers may be from the "Show Me" state of Missouri, the computations are given here.

At a 10Mbps operating rate, the Ethernet standard requires a "dead time" of 9.6 μs between frames, while the bit duration is 100 ns. Thus, the time required to transmit a 72-byte minimum length Ethernet frame becomes:

$$9.6\ \mu s + 72\ \text{bytes} \times 8\ \text{bits/byte} \times 100\ \text{ns/bit}$$

or 67.2×10^{-6} s. Thus, in one second, there can be a maximum of $1/67.2 \times 10^{-6}$, or 14,880, minimum size 72-byte frames flowing on an Ethernet 10Mbps network.

Similar to the manner by which the maximum number of minimum length frames that can flow on a 10Mbps Ethernet network was calculated, one can also compute the maximum number of maximum length frames that can flow on that network. To do so, first compute the time per frame as follows:

$$9.6\ \mu s + 1526\ \text{bytes} \times 8\ \text{bits/byte} \times 100\ \text{ns/bit}$$

or 1.23 ns. Then, in one second, there can be a maximum of 1/1.23 ns, or 812, maximum length frames that can flow on a 10Mbps Ethernet network.

In examining Fast Ethernet technology, one notes that the dead times between frames and bit duration are one tenth of 10Mbps and that both technologies use the same frame format with the same minimum frame length of 72 bytes and maximum frame length of 1526 bytes. Due to this, the maximum frame rate obtainable on Fast Ethernet is ten times the frame rate of 10Mbps Ethernet.

In addition to determining the maximum frame rate on Ethernet and Fast Ethernet, one can also compute their information transfer capability. To do so, multiply their maximum and minimum frame rates by the number of bytes capable of being transported in maximum length and minimum length frames. For example, consider the maximum frame rate when minimum length frames are transported on an Ethernet network. Since a minimum length 72-byte frame is capable of transporting 46 bytes and 14,880 minimum length frames per second can flow on an Ethernet network, the data transfer rate becomes:

$$14{,}880\ \text{frames/s} \times 46\ \text{bytes/frame} \times 8\ \text{bits/byte} = 5.476\text{Mbps}$$

Similarly, one can also compute the information transfer capability of 10Mbps Ethernet for maximum length 1526-byte frames. Since a maximum

Exhibit 13.2 Ethernet Frame and Information Transfer Capability.

	Average Frame Size (Bytes)	Maximum Frame Rate	Maximum Information Transfer
Ethernet			
	72	14880	5.476 Mbps
	1526	812	9.744 Mbps
Fast Ethernet			
	72	14880	54.76 Mbps
	1526	8120	97.44 Mbps

of 812 frames per second can flow on that network and each maximum length field can transport 1500 bytes, then the maximum data transfer rate for maximum length frames becomes:

$$812 \text{ frames/s} \times 1500 \text{ bytes/frame} \times 8 \text{ bits/byte} = 9.744\text{Mbps}$$

Exhibit 13.2 summarizes the maximum frame rates and information transfer capability for minimum length and maximum length frames flowing on Ethernet and Fast Ethernet networks. Because Fast Ethernet uses the same frame format and only differs from Ethernet by placing bits on the network faster by an order of ten, its transfer rate is ten times that of Ethernet. Thus, although one could compute the information transfer of Fast Ethernet in the same manner as for the information transfer for Ethernet, this will not be done. Instead, note that because the maximum frame rate on Fast Ethernet is ten times that of Ethernet, this results in the information transfer capability also being ten times that of Ethernet.

Returning to the examination of the entries in Exhibit 13.2, one can now focus on the maximum information transfer capability of Ethernet and Fast Ethernet. Note that when the minimum frame length occurs that corresponds to the transmission of frames carrying interactive traffic, the maximum information transfer on Ethernet and Fast Ethernet is slightly more than half of the operating rate of each network. At the opposite end of the range of frame lengths, the transmission of maximum length frames that corresponds to file transfers results in the information transfer rate approaching the operating rate of each network. Thus, the average frame length flowing on a network has a significant effect on the information transfer rate obtainable on an Ethernet network. With this in mind, one can focus on Gigabit Ethernet by first examining the extension of the Ethernet frame via the use of carrier extension technology.

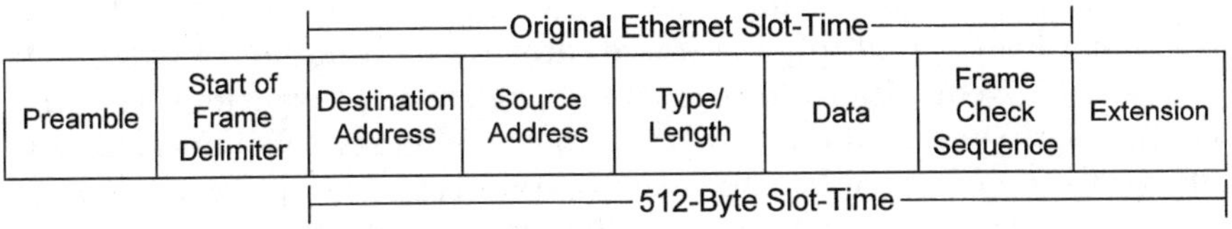

Exhibit 13.3 Half-Duplex Gigabit Ethernet Using Carrier Extension to Extend Timing.

THE GIGABIT ETHERNET FRAME

As previously mentioned, the higher operating rate of Gigabit Ethernet resulted in the IEEE extending its frame through the use of carrier extension technology for shared-media, half-duplex operations. This frame extension occurs by the addition of carrier extension symbols to ensure all frames have a minimum length of 512 bytes in the adapter and 520 bytes when placed onto the network. Through the use of carrier extension technology, a 200-meter network diameter can be supported. Otherwise, the need to ensure that a station could hear a collision propagate on the network between transmitting the next-to-last bit in a short frame and the last bit in the frame would have significantly reduced the allowable cabling distance to 10 meters (or approximately 33 feet).

Exhibit 13.3 illustrates the Gigabit Ethernet frame format. Note that the carrier extension extends the frame timing to guarantee at least a 512-byte slot time (520 bytes on the network) for half-duplex, shared-media operations. The increase in the minimum length frame does not change the frame size and only alters the time the frame is on the network. This enables compatibility to be maintained between Ethernet, Fast Ethernet, and Gigabit Ethernet.

Although the carrier extension scheme permits the support of a more reasonable and practical network diameter, that extension is not without a price. That price is one of additional network bandwidth utilization by carrier extension symbols, which reduces the information transfer capability of Gigabit Ethernet when carrier extension symbols are required.

One can determine the frame rate for Gigabit Ethernet in the same manner as for Ethernet; that is, first compute the transmission time required for placing a Gigabit Ethernet frame on the network. Once this is accomplished, divide that time into one second to determine the number of frames per second that can flow on the network. Because Gigabit Ethernet also has minimum length and maximum length frames, perform two computations to determine the maximum number of minimum length frames and maximum length frames that can flow on a Gigabit network. Once this is accomplished, multiply the frame rates by the number of bytes in the "Data" field of each frame to compute the information transfer rate for minimum length and maximum length frames. Obtaining this information will illustrate the effect of transporting certain types of data on the performance of Gigabit Ethernet, which will in turn illustrate the importance of throughput above the operating rate of a network.

To compute the frame rate on Gigabit Ethernet, note that the minimum length frame is now 520 bytes when placed on the network. Because carrier extension symbols are added to the end of an Ethernet frame, the actual composition of the frame is not altered. This means that the "Data" field will vary between 46 and 1500 bytes. Thus, the minimum number of bytes that can be transported in a minimum length Gigabit Ethernet frame is 46, which is the same as Ethernet and Fast Ethernet. However, instead of being 72 bytes in length, the minimum frame is 520 bytes on the network to include 448 carrier extension symbols.

To compute the maximum number of minimum length frames that can flow on a Gigabit Ethernet network, use a dead time between frames of 0.096 μs, a frame length of 520 bytes, and a bit duration of 1 ns. Thus, the time per minimum length frame becomes:

$$0.096 \times 10^{-6} + 520 \text{ bytes} \times 8 \text{ bits/byte} \times 1 \text{ ns/bit} = 4.256 \times 10^{-6} \text{ s}$$

Then, in one second, there can be a maximum of $1/4.256 \times 10^{-6}$, or 234,962, frames per second. Because each frame can transport up to 46 bytes of data, the maximum information transfer capability for minimum length Gigabit Ethernet frames becomes:

$$234{,}962 \text{ frames/s} \times 46 \text{ bytes/frame} \times 8 \text{ bits/byte} = 86.46\text{Mbps}$$

In comparing the value just computed to the entries in Exhibit 13.2, note that the ratio of Gigabit Ethernet's Information transfer to that of Fast Ethernet for minimum length frames is 86.46/54.76, or approximately 1.58 to 1. Thus, although Gigabit Ethernet's operating rate is ten times the operating rate of Fast Ethernet, when used to transport relatively short interactive queries its information transfer capability is only slightly better than Fast Ethernet.

A similar computation for the maximum length Gigabit frame shows that Gigabit Ethernet indeed provides ten times the information transfer capability of Fast Ethernet. This results from the fact that as the "Data" field increases, the number of carrier extensions decreases until a "Data" field transporting 494 or more bytes of data requires no carrier extension symbols and provides exactly ten times the information transfer capability of Fast Ethernet. Thus, for all frames with "Data" fields between 494 and 1500 bytes, half-duplex, shared-media Gigabit Ethernet provides what one would normally expect from the technology — an information transfer rate ten times that of Fast Ethernet and 100 times that of Ethernet.

RECOMMENDED COURSE OF ACTION

While Gigabit Ethernet provides an operating rate ten times that of Fast Ethernet, its information transfer rate can be much lower. This means it is important to examine the type of data being transmitted on a network prior to considering the use of a shared-media, half-duplex Gigabit Ethernet backbone. If the organization's average frame length is relatively low, one may need to consider installing a more expensive, full-duplex Gigabit Ethernet backbone, or consider an alternate technology such as ATM.

Chapter 14
Asynchronous Transfer Mode (ATM) Implementation

John R. Vacca

ATM TECHNOLOGY is being placed in public and private networks in a cost-effective manner. It is an evolving technology designed for the high speed transfer of voice, video, and data. ATM applies very large-scale integration (VLSI) technology to the transfer of data within public networks. It is based on the efforts of Study Group XVIII of the International Telecommunication Union[1] Telecommunication Standardization Sector (ITU-T, formerly the Consultative Committee for International Telegraph and Telephone [CCITT] and the American National Standards Institute [ANSI]).[2] The ATM layer of the Broadband Integrated Services Digital Network (BISDN) model is officially defined by CCITT I.361. Founded in 1991 by Cisco Systems,[3] NET/ADAPTIVE,[4] Northern Telecom,[5] and Sprint,[6] the ATM Forum's[7] current efforts are to bring ATM technology to private networks and to guarantee interoperability between private and public networks.

ROLE OF ATM TECHNOLOGY IN INTERNETWORKS

Despite a common misconception that ATM is a complex and expensive technology, it continues to be embraced worldwide at a growth rate of 24 percent annually to 2004, according to multiple sources (IDC, Cahners In-Stat). In fact, the current estimate for ATM market size in 2004 is $7.9 billion. So why are enterprises continuing to select and implement ATM technology?

ATM introduces several new technical capabilities into existing enterprise networks. But, perhaps the single most important tangible benefit that ATM offers is its ability to consolidate disparate services into a common transmission network for wide-area deployment. Unlike comparable alternatives, ATM does not suffer from the bandwidth and distance constraints of many traditional, narrow-band wide area services. For this reason alone there is over a 95 percent chance that when you use the Internet, your traffic will at some point be transported by ATM.

Enterprises that have selected a LAN based on Fast Ethernet or Gigabit Ethernet switching can use ATM to scale that bandwidth out of the LAN for interconnection of private networks, or for connection to public services such as the Internet. At the same time, ATM allows those enterprises to integrate disparate services such as video, voice, and data into a common network infrastructure, reducing costs while dynamically optimizing the network for the different traffic trends and application requirements.

The bottom line is, ATM can protect your existing investment while providing for future growth in terms of bandwidth or applications and services deployment. So, what are the benefits of ATM?

ATM Benefits

Because ATM can be deployed for local area, campus, metropolitan, or wide area services, the technology has many unique advantages, depending on where you choose to implement it. ATM offers capabilities such as:

- True Quality of Service with support for Class of Service
- Broadband Integration
- Virtual Networking

True Quality of Service with Support for Class of Service. As the Internet moves from traditional *best effort* services to a Class of Service (CoS) model, it will be necessary for enterprises to select network equipment and technologies that can take advantage of such service classes offered by their ISP or regional provider. More importantly, ATM can offer Quality of Service (QoS) in addition to CoS support. This approach means that enterprises that deploy ATM can offer value-added services tailored to the unique requirements of their customers, and can cater to specific application demands over and above a simple allocated Class of Service.

Broadband Integration. ATM has the unique ability to combine data, voice, and video traffic on the same physical link. Data traffic connects to the ATM enterprise via LAN-to-ATM internetworking devices such as routers and switches. In addition, using ATM interfaces within the PBX, external codecs, or Circuit Emulation Services (CES), voice and video traffic can be multiplexed on the same broadband backbone. With this union, ATM connectivity can truly be ubiquitous on the enterprise network.

Virtual Networking. Virtual Networking is accomplished through a combination of hardware and software that allows a network administrator to manage, secure, and account for the amount of bandwidth utilized by each user on the ATM network. When it becomes necessary to set up temporary subgroups of workers, often for projects involving multiple departments, a network administrator can functionally limit broadcast and multicast traffic by creating virtual workgroups.

These workgroups can be created on the fly and dismantled just as quickly and easily. This also allows the network administrator to create secure environments by limiting users' access to other users and to network resources, and can eliminate the significant costs associated with adds, moves, and changes. Finally, the administrator can monitor the amount of data exchanged in virtual workgroups and generate a network usage bill, much like a telephone bill.

These benefits result in greater network flexibility allowing for increased network management automation which, in turn, lowers operating costs. As a result, ATM technology is a leading method for reducing the costs of ownership in dynamic environments where change is a necessity to survive.

The Multiservice Networks

One of the technologies for integrating LANs and WANs has been through the emergence of ATM. According to Cisco Systems, Inc., ATM can support any traffic type in separate or mixed streams, delay-sensitive traffic, and nondelay-sensitive traffic as shown in Exhibit 14.1.[8]

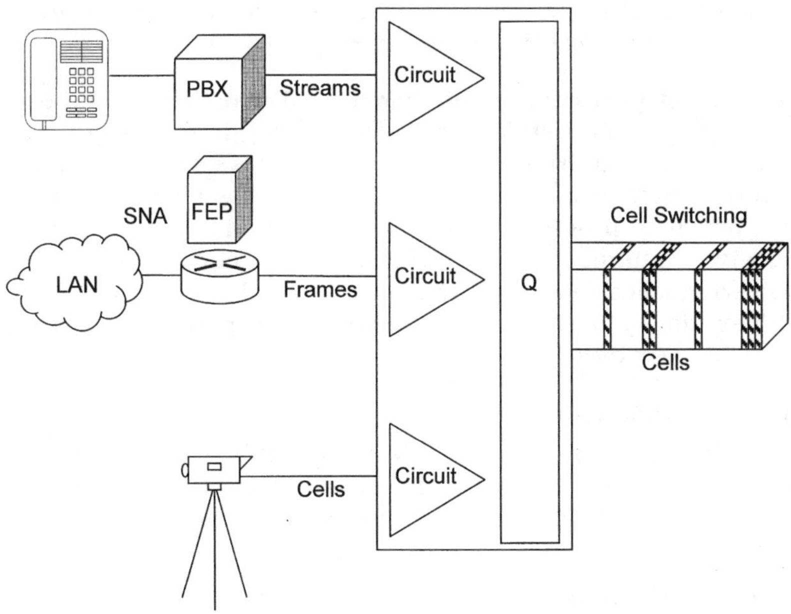

Exhibit 14.1 ATM Support of Various Traffic Types.

ATM can also scale from low to high speeds; from LAN to private branch exchange (PBX), it has been adopted by all the industry's equipment vendors. Network designers can now integrate LANs and WANs, support emerging applications with economy in the enterprise, and support legacy protocols with added efficiency.

ATM Cell Switching and Internetworking Modules. ATM cell switching (CS) and ATM interworking unit (IWU) modules provide high-performance ATM services for multiservice WAN switches at DS3/E3 and OC-3/STM-1 speeds. According to Ascend Communications, these high-performance ATM modules provide four functions:

- ATM OPTimum trunking
- Frame Relay to ATM Network Interworking
- Frame Relay to ATM Service Interworking
- Native ATM cell switching[9]

The modules represent a major expansion in trunking capacity, native ATM services, and Frame Relay to ATM interworking capabilities for the multiservice networking platform.

By selecting the ATM Cell Switching and Interworking Unit modules, service providers can position their networks for the expanding data communications opportunities that lie ahead. Consequently, network capacity and performance can keep pace with users' growing demands. Service providers can meet the challenge by quickly and easily implementing interworking between today's Frame Relay network and the ATM network of the future while protecting existing investments.

ATM Technology Ensures Smooth Migration to Future Growth. Because networks built with multiservice networks can grow to meet increasing demands without requiring extensive upgrades, service providers can smoothly migrate their existing networks to ATM technology using multiservice networks' proven interworking solutions. Multiservice networking along with multiprocessing architecture delivers unsurpassed service-ready performance, long-term investment protection through compatibility with existing systems, and high scalability to support a virtually unlimited number of end users.

Migration of TDM Network

Network designers are deploying ATM technology to migrate from time-division multiplexing (TDM) networks. This is in addition to using ATM to combine multiple networks into one multiservice network for the following reasons:

- To reduce WAN bandwidth cost
- To improve performance
- To reduce downtime

WAN Bandwidth Cost Reduction. ATM switches provide additional bandwidth through the use of voice compression, silence compression, repetitive pattern suppression, and dynamic bandwidth allocation. The implementation of ATM combines the strengths of TDM (whose fixed time slots are used by telephone companies to deliver voice without distortion) with the strengths of packet-switching data networks (whose variable size data units are used by computer networks such as the Internet to deliver data efficiently).

ATM avoids the weaknesses of TDM (which wastes bandwidth by transmitting the fixed time slots even when no one is speaking) while building on the strengths of TDM. Because PSDNs are designed for transmitting bursty data, this includes packet-switched data networks (PSDNs) which cannot accommodate time-sensitive traffic (such as voice and video). ATM combines the isochronicity of TDM with the efficiency of PSDN by using fixed-size cells.

Performance Improvement. ATM offers improved performance through performance guarantees and robust WAN traffic management that support the following capabilities:

- Feedback congestion notification
- Large buffers that guarantee Quality of Service (QoS) for bursty data traffic and demanding multimedia applications
- Per-virtual circuit (VC) queuing and rate scheduling

Downtime Reduction. By reducing downtime, ATM offers high reliability. Because of the following ATM capabilities, this high reliability is available: the capability to rapidly reroute around failed trunks, and the capability to support redundant processors, port and trunk interfaces, and power supplies.

INTEGRATED SOLUTIONS

Without creating multiple networks or writing off existing data communications investments, the trend in internetworking is to provide network designers greater flexibility in solving multiple internetworking problems. Acting as a barrier against inadvertent broadcast storms in the local networks, routers can provide a reliable, secure network. According to Cisco, switches, which can be divided into two main categories (LAN switches and WAN switches) can be deployed at the workgroup, campus backbone, or WAN level as shown in Exhibit 14.2.[10]

Cabletron Systems' integrated solution, on the other hand, is tied to a multitiered product approach.[11] In other words, ATM switches are developed and then positioned where they are best suited to meet an enterprise's needs. Clearly the needs of an enterprise connecting multimedia services over a wide area public network are very different from that of a customer

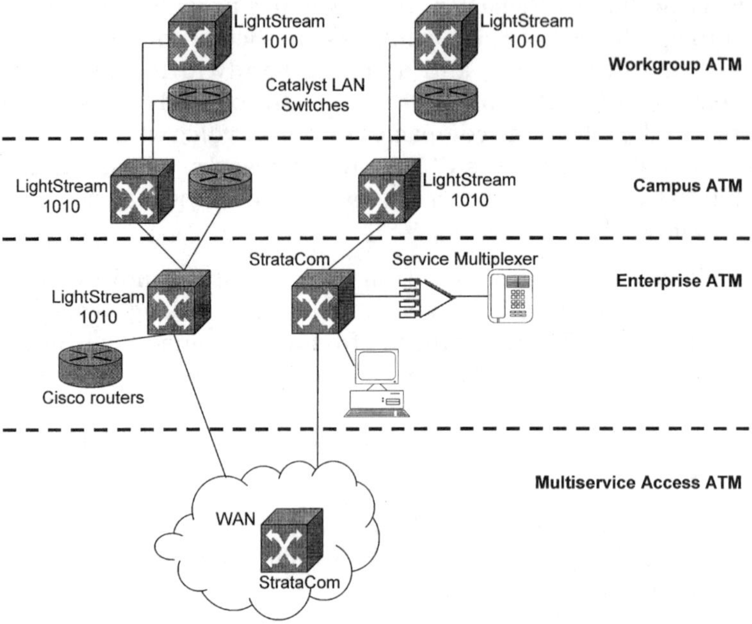

Exhibit 14.2 The Role of ATM Switches in an Internetwork.

requiring a campus LAN backbone switch. Also, tight interaction among the various development groups allows common elements of the signaling software and hardware to be shared between groups on various projects. This ensures that plug-and-play operation is implemented wherever possible to ease the initial implementation and expansion of the ATM network.

Nevertheless, FORE Systems' ATM-based integrated solution incorporates all the characteristics of an intelligent infrastructure.[12] By its very nature, FORE's ATM technology provides a high-capacity, high-performance network infrastructure. Communication paths are established in a connection-oriented way. This means that a connection is made as required and all the appropriate traffic is passed along this path and switched through the actual node elements. Switching takes place at extremely high speed using hardware-based cell-switching techniques.

All of this is done regardless of the nature of the traffic. Whether it is nonreal-time traffic such as conventional data, or real-time traffic such as voice and video, the intelligent infrastructure adjusts or tunes the connection according to the bandwidth requirements. This approach is often referred to as "smart bandwidth allocation" since each connection, regardless of traffic type, is given the appropriate quality of service. To achieve this, the intelligent infrastructure uses a mechanism called Private Network-Network (or Node) Interface (PNNI). PNNI is a highly-efficient and optimized routing and signaling system that offers QoS-based routing and, where appropriate, functions such as resilience, load-balancing, and routing authentication.

DIFFERENT TYPES OF ATM SWITCHES

Even though all ATM switches perform cell relay and embrace fast packet technology, ATM switches differ markedly in the following ways:

- Depth of ATM internetworking software
- Redundancy
- Sophistication of traffic management mechanism
- Variety of interfaces and services that are supported

ATM Integrated Switching Technology

ATM is a newer high speed (45-600Mbps) wide area (WAN) and local area (LAN) multiuser integrated networking technology. Enterprises can connect their remote-site LANs over public ATM services to headquarters, along with all their remote-site voice, fax, video, and data traffic. ATM will play an important part in the information highway, especially for large enterprises that require high-speed connections.

The strong appeal of ATM comes from its simple networking protocol. This protocol is very efficient, able to dynamically allocate bandwidth to a user only when the user is active, and only in the direction of activity. The protocol is based on the ATM cell containing 48 bytes of data and 5 bytes of network routing information, 53 bytes (424 bits) altogether. ATM's cell basis qualifies it as a cell relay technology.

In wide area networking, WAN costs are proportional to bandwidth consumed. The more bandwidth you need, the more you pay. ATM is no exception. ATM at 45Mbps is about 8 times more costly than a T1/E1 (1.5-2.0Mbps) connection and about 50 times more than 56/64Kbps. The majority (90 percent +) of connections operate at 56/64Kbps or below, creating a strong case to offer ATM at speeds much slower than 45Mbps.

The disappointing news, however, is that ATM does not scale down to speeds slower than T1/E1. ATM carries an inherent 12 percent overhead in its cell structure, and also requires certain control information that further reduces its efficiency to around 71 percent at T1/E1 speeds. Beyond this, ATM's cell size, designed for 45Mbps speeds, is much too large to guarantee the timely delivery of delay-sensitive voice and video at speeds much lower than T1/E1. In fact, the combination of these two factors prohibits ATM from running at speeds lower than T1/E1.

To address the requirements of users of connections running at sub-T1/E1 speeds, new ATM integrated switching technology has been developed. For example, MICOM's MicroBand ATM is a combination of Fast Packet and Cell Relay technologies that is designed to efficiently integrate remote-office voice and data traffic. According to MICOM, MicroBand ATM (M-ATM) runs on connections as low as 8.5Kbps (hence, MicroBand), with network efficiencies of 91 percent to 96 percent. Cell sizes range from 30 bits (for 8.5Kbps connections) to 60 bits (for 256Kbps connections), providing optimal efficiency at varying speeds.[13]

Types of Switches

ATM switches can be segmented into the following four distinct types that reflect the needs of particular applications and markets (just as there are routers and LAN switches available at various price/performance points with different levels of functionality):

- Workgroup ATM switches
- Campus ATM switches
- Enterprise ATM switches
- Multiservice access switches

Workgroup and Campus ATM Switches. To connect to a campus ATM switch, workgroup ATM switches are characterized by having Ethernet switch ports and an ATM uplink. The Cisco Catalyst 5000 is an example of a workgroup ATM switch. High-performance switching between workstations, servers, switches, and routers in wiring closet, workgroup, and campus backbone environments is provided by the Catalyst 5500 switch. The Catalyst 5500 LAN is a 13-slot switch. Slot 1 is reserved for the supervisor engine module which provides switching, local and remote management, and dual Fast Ethernet uplinks. Slot 2 is available for a second, redundant supervisor engine or any of the other supported modules. Slots 3-12 support any of the supported modules.

A 3.6-Gbps media-independent switch fabric and a 5-Gbps cell-switch fabric are contained in the Catalyst 5500. The connection between power supplies, supervisor engine, interface modules, and backbone module is provided by the backplane. The 3.6-Gbps media-independent fabric supports Ethernet, Fast Ethernet, Fiber Distributed Data Interface/Copper Distributed Data Interface (FDDI/CDDI), ATM LAN Emulation, and Route Switch Module (RSM) modules.

Enterprise ATM Switches. Enterprise ATM switches are sophisticated multiservice devices. They are designed to form the core backbones of large enterprise networks. They are also intended to complement the role played by today's high-end multiprotocol routers. Campus ATM switches are interconnected by enterprise ATM switches. Enterprise-class switches, however, can act not only as ATM backbones but can serve as the single point of integration for all of the disparate services and technology found in enterprise backbones today. By integrating all of these services onto a common platform and a common ATM transport infrastructure, network designers can gain greater manageability and eliminate the need for multiple overlay networks.

Multiservice Access Switches. ATM platforms will also be widely deployed by service providers, both as customer premises equipment (CPE) and within public networks and beyond private networks. On a common ATM infrastructure, such equipment will be used to support multiple MAN and WAN services (for instance, Frame Relay switching, LAN interconnect, or public ATM services). Enterprise ATM switches will often be used in these public network applications because of their emphasis on high availability and redundancy, their support of multiple interfaces, and their capability to integrate voice and data.

ATM OVERVIEW

The following ATM concepts are discussed in this part of the article:

- Structure of an ATM Network
- ATM Functional Layers
- ATM Addressing
- ATM Media
- ATM Data Exchange Interface

ATM Network Structure

By means of intermediate switches, ATM is based on the concept of two end-point devices communicating with each other. As Exhibit 14.3 shows, an ATM network is made up of a series of switches and end-point devices.[8] According to Cisco, the end-point devices can be ATM-attached end stations, ATM-attached servers, or ATM-attached routers.

There are two types of interfaces in an ATM network, User-to-Network Interface (UNI) and Network-to-Network Interface (NNI), as shown in Exhibit 14.3. An end-point device and a private or public ATM switch make up the UNI connection. The NNI is the connection between two ATM switches. The UNI and NNI connections can be carried by different physical connections.

The ATM Forum has defined a set of LAN Emulation (LANE) standards and a Private Network to Network Interface (PNNI) Phase 0 protocol in addition to the UNI and NNI protocols. Network designers use Ethernet and Token Ring with ATM-attached devices with LANE technology to internetwork legacy LANs. Most LANE networks consist of multiple ATM switches and typically employ the PNNI protocol.

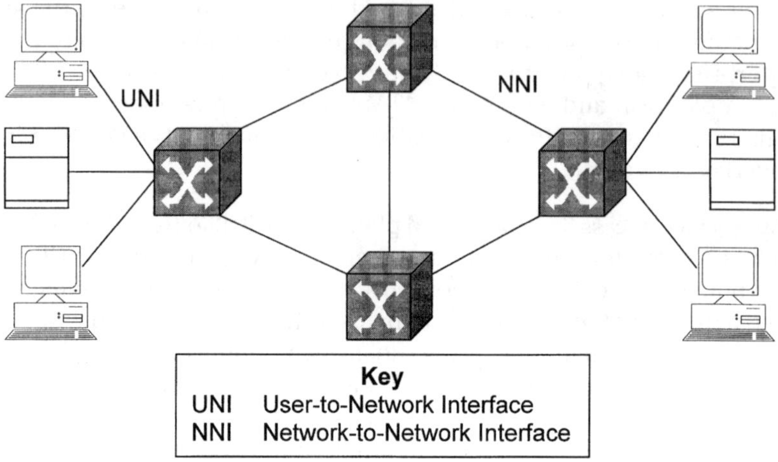

Exhibit 14.3 Components of an ATM Network.

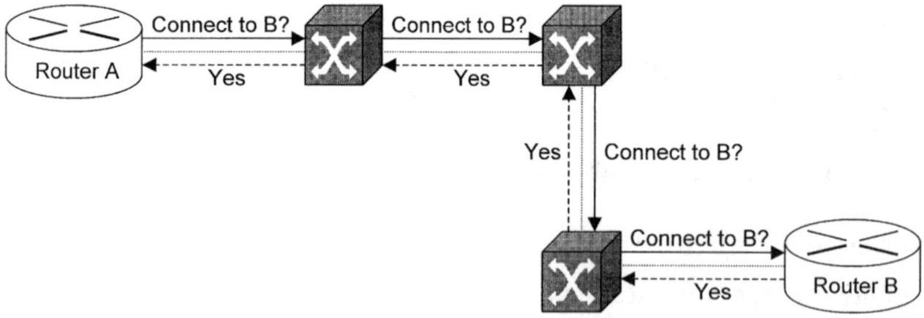

Exhibit 14.4 Establishing a Connection in an ATM Network.

ATM Network General Operation

A connection must be established between two end points before any data transfer can occur because ATM is connection-oriented. According to Cisco, this connection is accomplished through a signaling protocol as shown in Exhibit 14.4.[8]

For Router A to connect to Router B, the following must occur: Router A sends a signaling request packet to its directly-connected ATM switch (ATM Switch 1). This request contains the ATM address of the Router B. It also contains any QoS parameters required for the connection. ATM Switch 1 then reassembles the signaling packet from Router A and examines it. If ATM Switch 1 has an entry for Router B's ATM address in its switch table and it can accommodate the QoS requested for the connection, it sets up the virtual connection and forwards the request to the next switch (ATM Switch 2) along the path.

Next, every switch along the path to Router B reassembles and examines the signaling packet and forwards it to the next switch if the QoS parameters can be supported. Each switch also sets up the virtual connection as the signaling packet is forwarded. If any switch along the path cannot accommodate the requested QoS parameters, the request is rejected and a rejection message is sent back to Router A.

When the signaling packet arrives at Router B, Router B reassembles it and evaluates it. If Router B can support the requested QoS, it responds with an accept message. As the accept message is propagated back to Router A, the switches set up a virtual circuit.[14] Finally, Router A receives the accept message from its directly-connected ATM switch (ATM Switch 1), as well as the Virtual path identifier (VPI) and Virtual channel identifier (VCI) values that it should use for cells sent to Router B.[15]

ATM Functional Layers

The ATM protocol model describes how two end systems communicate through ATM switches, just as the Open System Interconnection (OSI) reference model describes how two computers communicate over a network. According to Cisco, the ATM protocol model consists of the following three functional layers:

- ATM physical layer
- ATM layer
- ATM adaptation layer

These layers correspond roughly to Layer 1 and parts of Layer 2 (such as error control and data framing) of the OSI reference model as shown in Exhibit 14.5.[8]

Physical Layer. Transmission and receipt of bits on the physical medium are controlled by the ATM physical layer. For the physical medium being used, it also keeps track of ATM cell boundaries and packages cells into the appropriate type of frame. The ATM physical layer is divided into two parts, physical medium sublayer and transmission convergence sublayer.

Physical Medium Sublayer. To synchronize transmission and reception, the physical medium sublayer is responsible for sending and receiving a continuous flow of bits with associated timing information. Its specification depends on the physical medium used because it includes only physical-medium-dependent functions. ATM can use any physical medium capable of carrying ATM cells. Some existing standards that can carry ATM cells are Synchronous Optical Network/Synchronous Digital Hierarchy

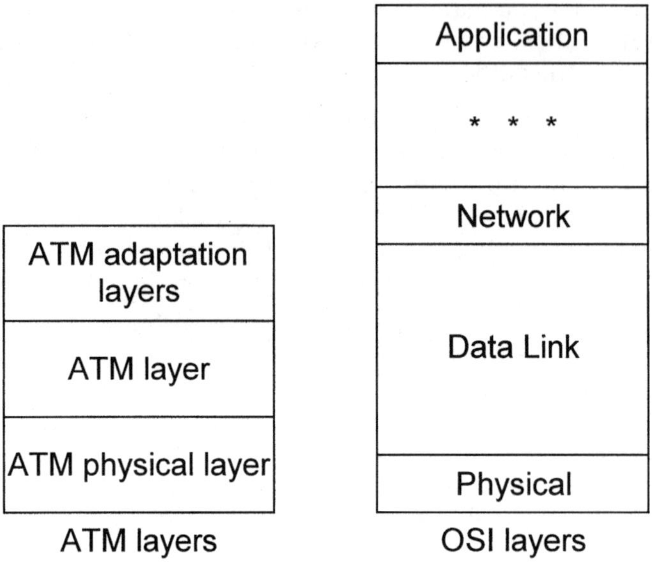

Exhibit 14.5 Relationship of ATM Functional Layers to the OSI Reference Model.

(SONET/SDH), Digital Signal Level 3/European3 lines (DS-3/E3), 100-Mbps local fiber (Fiber Distributed Data Interface [FDDI] physical layer), and 155-Mbps local fiber (Fiber Channel physical layer). Various proposals for use over twisted-pair wire are also under consideration.

Transmission Convergence Sublayer. The following are the responsibility of the transmission convergence sublayer:

- *Cell delineation* maintains ATM cell boundaries.
- *Cell rate decoupling* inserts or suppresses idle (unassigned) ATM cells to adapt the rate of valid ATM cells to the payload capacity of the transmission system.
- *Header error control sequence generation and verification* generates and checks the header error control code to ensure valid data.
- *Transmission frame adaptation* packages ATM cells into frames acceptable to the particular physical layer implementation.
- *Transmission frame generation and recovery* generates and maintains the appropriate physical layer frame structure.

ATM Layer. Virtual connections are established by the ATM layer. It also passes ATM cells through the ATM network. It uses the information contained in the header of each ATM cell to do this. The ATM layer is responsible for performing the following four basic functions:

Exhibit 14.6 ATM Adaption Layers.

Characteristics	AAL1	AAL3/4	AAL4	AAL5
Requires timing between source and destination	Yes	No	No	No
Data rate	Constant	Variable	Variable	Variable
Connection mode	Connection-oriented	Connection-oriented	Connectionless	Connection-oriented
Traffic types	Voice and circuit emulation	Data	Data	Data

- Extracting and inserting the header before or after the cell is delivered to or from the higher ATM adaptation layer
- Handling the implementation of a flow control mechanism at the UNI
- Multiplexing and demultiplexing the cells of different virtual connections; these connections are identified by their VCI and VPI values
- Translating the values of the VCI and VPI at the ATM switches or cross connects

ATM Adaptation Layer (AAL). The AAL translates between the larger service data units (SDUs) (for instance, video streams and data packets) of ATM cells and upper-layer processes. The AAL specifically receives packets from upper-level protocols (such as AppleTalk, Internet Protocols [IP], and NetWare). It also breaks them into the 48-byte segments that form the payload field of an ATM cell. Several ATM adaptation layers are currently specified. According to Cisco, Exhibit 14.6 summarizes the characteristics of each AAL.[8]

AAL1. A cell for transmission is prepared by AAL1. A synchronous sample is contained within the payload data. For instance, this could be one byte of data generated at a sampling rate of 125 microseconds. The sequence number field (SN) and sequence number protection (SNP) field provide the information that the receiving AAL1 needs to verify that it has received the cells in the correct order. The rest of the payload field is filled with enough single bytes to equal 48 bytes.

AAL3/4. AAL3/4 is closely aligned with Switched Multimegabit Data Service (SMDS) and designed for network service providers. It is also used to transmit SMDS packets over an ATM network. According to Cisco, the convergence sublayer (CS) creates a protocol data unit (PDU) by prepending a Beginning/End Tag header to the frame and appending a length field as a trailer as shown in Exhibit 14.7.[8]

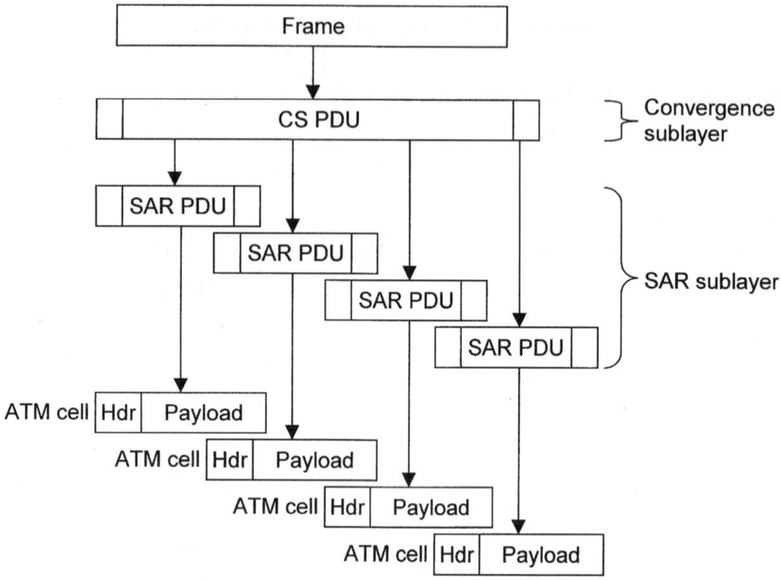

Exhibit 14.7 AAL3/4 Cell Preparation.

The PDU is fragmented by the segmentation and reassembly (SAR) sublayer. Consisting of the following fields, it prepends to each PDU fragment a header:

- *Multiplexing identifier* identifies cells from different traffic sources interleaved on the same virtual circuit connection (VCC) so that the correct cells are reassembled at the destination.
- *Sequence number* identifies the order in which cells should be reassembled.
- *Type* identifies whether the cell is the beginning of a message, continuation of a message, or end of a message.

AAL5. According to Cisco, AAL5 prepares a cell for transmission as shown in Exhibit 14.8.[8] First, the convergence sublayer of AAL5 appends an 8-byte trailer to a frame and a variable-length pad. The pad is long enough to ensure that the resulting PDU falls on the 48-byte boundary of the ATM cell. The trailer includes a 32-bit CRC computed across the entire PDU and the length of the frame. Bit errors and lost cells, or cells that are out of sequence, are detected by the AAL5 at the destination.

Next, the convergence sublayer (CS) PDU into 48-byte blocks is segmented by the segmentation and reassembly. Each block is then placed by the ATM layer into the payload field of an ATM cell. For all cells except the last cell, a bit in the payload type (PT) field is set to zero to indicate that the cell

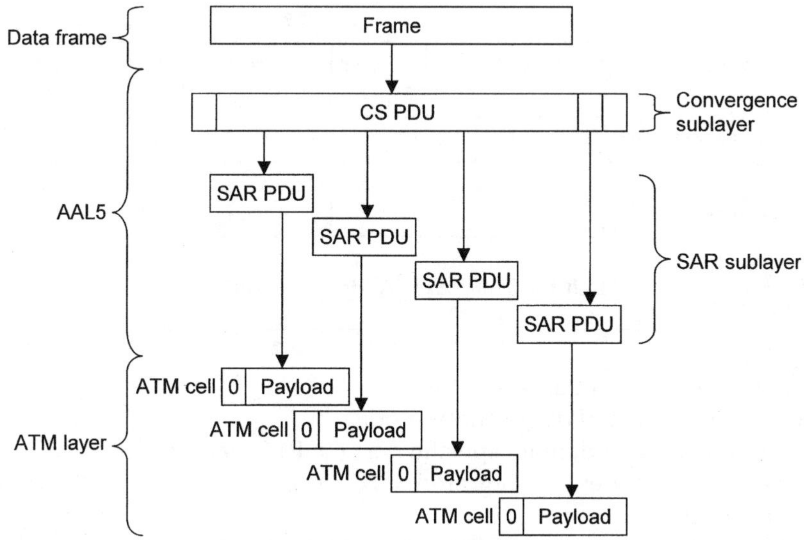

Exhibit 14.8 AAL5 Cell Preparation.

is not the last cell in a series that represents a single frame. For the last cell, the bit in the PT field is set to one. When the cell arrives at its destination, the ATM layer extracts the payload field from the cell, the SAR sublayer reassembles the CS PDU, and the CS uses the CRC and the length field to verify that the frame has been transmitted and reassembled correctly. AAL5 is the adaptation layer used to transfer most non-SMDS data such as classical IP over ATM and local area network (LAN) emulation.

ATM Addressing

The responsibility for mapping network layer addresses to ATM addresses is through the ATM layer. This is because the ATM Forum has adapted the subnetwork model of addressing. According to Cisco, several ATM address formats have been developed. Public ATM networks typically use E.164 numbers, which are also used by Narrowband ISDN (N-ISDN) networks.

Exhibit 14.9 shows the format of private network ATM addresses.[8] The three formats are Data Country Code (DCC), International Code Designator (ICD), and Network Service Access Point (NSAP) encapsulated E.164 addresses.

Fields of an ATM Address. The following are the fields of an ATM address:

- *AA*: three bytes of administrative authority
- *AFI*: one byte of authority and format identifier; the AFI field identifies the type of address. The defined values are 45, 47, and 39 for E.164, ICD, and DCC addresses, respectively

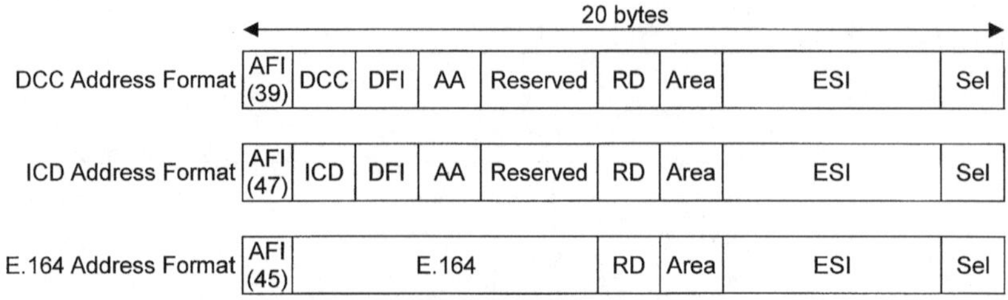

Exhibit 14.9 ATM Address Formats.

- *Area*: two bytes of area identifier
- *DCC*: two bytes of data country code
- *DFI*: one byte of domain specific part (DSP) format identifier
- *E.164*: eight bytes of Integrated Services Digital Network (ISDN) telephone number
- *ESI*: six bytes of end system identifier, which is an IEEE 802 Media Access Control (MAC) address
- *ICD*: two bytes of international code designator
- *RD*: two bytes of routing domain
- *Sel*: one byte of Network Service Access Point (NSAP) selector

The ATM address formats are modeled on International Organization for Standardization (ISO) NSAP addresses. But, they identify subnetwork point of attachment (SNPA) addresses. So, it's easy to map ATM addresses into existing LANs by incorporating the Media Access Control (MAC) address into the ATM address.

ATM Media

Multiple standards have been defined by the ATM Forum for encoding ATM over various types of media. Exhibit 14.10 lists the framing type and data rates for the various media, including unshielded twisted-pair (UTP) and shielded twisted-pair (STP) cable.[8]

The ATM Forum encouraged initial ATM development efforts by endorsing TAXI 4B/5B as one of the first ATM media encoding standards because the FDDI chipset standard (Transparent Asynchronous Transmitter/Receiver Interface 4-byte/5-byte [TAXI 4B/5B]) was readily available. However, the Synchronous Transport Signal Level 3, concatenated/Synchronous Transport Module (STS3c/STM) is the most common fiber interface today.

ATM Data Exchange Interface

The ATM Forum developed a standard known as the ATM Data Exchange Interface (DXI) in order to make ATM functionality available as soon as

Exhibit 14.10 ATM Physical Rates.

Framing	Data Rate (Mbps)	Multimode Fiber	Single Mode Fiber	Coaxial Cable	UTP-3	UTP-5	STP
DS-1	1.544			D			
E1	2.048			D			
DS-3	45			D			
E3	34			D			
STS-1	51				D		
SONET STS3c SDH STM1	155	D	D	D		D	
SONET STS12c SDH STM4	622	D	D				
TAXI 4B/5B	100	D					
8B/10B (Fiber Channel)	155	D					D

possible. Network designers can use DXI to provide UNI support between Cisco routers and ATM networks as shown in Exhibit 14.11.[8]

Over a High-Speed Serial Interface (HSSI), the ATM Data Service Unit (ADSU) receives data from the router in ATM DXI format. Over a DS-3/E3 line, the DSU converts the data into ATM cells and transfers them to the ATM network. ATM DXI is available in several modes:

- *Mode 1a* supports AAL5 only, a 9232 octet maximum, and a 16-bit frame check sequence (FCS), and provides 1023 virtual circuits.
- *Mode 1b* supports AAL3/4 and AAL5, a 9224 octet maximum, and a 16-bit FCS. AAL5 support is the same as Mode 1a. AAL3/4 is supported on one virtual circuit.

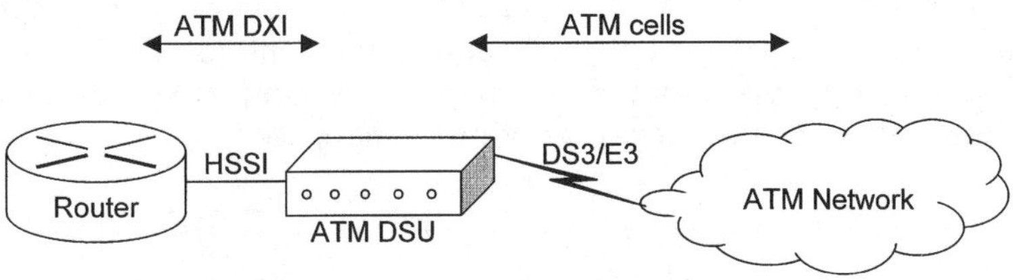

Exhibit 14.11 ATM DXI Topology.

	Flag	Header	SDU	FCS	Flag
Field size in octets	1	2	0-9232	2	1

Exhibit 14.12 ATM DXI Frame Format.

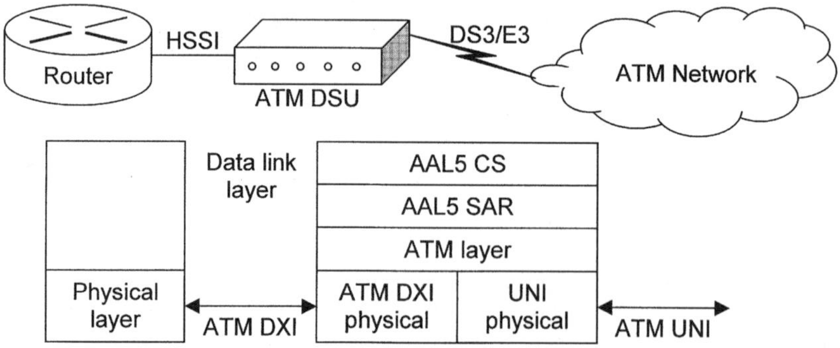

Exhibit 14.13ATM DXI Mode 1a and Mode 1b Protocol Architecture for AAL5.

- *Mode 2* supports AAL3/4 and AAL5 with 16,777,215 virtual circuits, a 65535 octet maximum, and 32-bit FCS.

On the router, according to Cisco, data from upper-layer protocols is encapsulated into ATM DXI frame format. Exhibit 14.12 shows the format of a Mode 1a ATM DXI frame.[8]

In Exhibit 14.13, a router configured as a data terminal equipment (DTE) device is connected to an ATM data service unit (ADSU).[8] According to Cisco, the ADSU is configured as a data communications equipment (DCE) device. By processing ATM DXI frames through the AAL5 CS and the SAR sublayer, the router sends the frames to the ADSU, which converts the frames to ATM cells. The ATM layer attaches the header. The cells are then sent out the ATM UNI interface.

According to Cisco, ATM DXI addressing consists of a DFA, which is equivalent to a Frame Relay data link connection identifier (DLCI). The DSU maps the DFA into appropriate VPI and VCI values in the ATM cell. Exhibit 14.14 shows how the DSU performs address mapping.[8]

MIGRATION: INTEGRATION WITH CURRENT ATM TECHNOLOGY

ATM technology is extremely mature in terms of standards compliance and deployment. Many of these standards have been developed to allow the incremental migration of present networks to an ATM intelligent infrastructure.

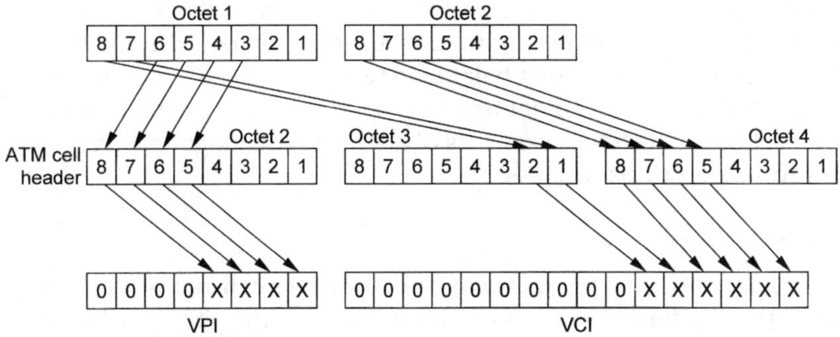

Exhibit 14.14 ATM DXI Address Mapping.

These standards include the highly successful ATM Forum LAN Emulation (LANE) standard which allows an ATM network to emulate an Ethernet or Token Ring LAN segment. This capability protects your investment in conventional LAN applications during the migration to ATM. It also means that this migration can be made at your own pace because LANE allows seamless interworking between ATM-attached and LAN-attached devices.

Another key ATM Forum standard is Multi-Protocol Over ATM (MPOA). Every network manager today appreciates the benefits that routing provides in their networks. Without routing, enterprises could never hope to build large, hierarchical, secure enterprise networks using conventional LAN protocols.

MPOA is focused on delivering a routed networking service over the ATM intelligent infrastructure without performance compromises. Emerging ASIC-based routers (marketed as Layer 3 switches, wire-speed routers, or routing switches) promise to address the performance bottlenecks of conventional routers, but they do not improve the operational costs of conventional routers. By distributing router functionality across the ATM intelligent infrastructure, MPOA reduces the number of routers in the network and, consequently, reduces the operational cost.

All in all, the various standardized techniques for using ATM allow you to change elements of your network as the need arises. The implementation cost is balanced against the dramatic, ongoing savings that result from an ATM intelligent infrastructure.

Reliability

PNNI builds heavily upon the tried-and-tested techniques of routing systems including Open Shortest Path First (OSPF), Intermediate System-Intermediate System (IS-IS), and Border Gateway Protocol (BGP). Inevitably, part of your ATM intelligent infrastructure may develop a fault. For example, a transmission link may fail, or a switch may be taken offline by mistake. By using PNNI, the ATM intelligent infrastructure minimizes the effect of these failures through resilient load-balancing techniques. Failed connections are reconnected rapidly, even before the users know anything has happened. The smart bandwidth requirements of all the applications are taken into account throughout this process.

Security

A secure network ensures that only specific users can access certain information. Due to the connection-oriented nature of the ATM intelligent infrastructure, policy definitions within PNNI ensure that connections meet your security requirements. In addition, standards such as LANE and MPOA contain the concept of a standard Policy Server that acts as a single point of configuration for all security requirements. This feature is not available in a conventional infrastructure.

The network must also protect itself from unintended abuse from the applications themselves. Application *bandwidth greed* is a typical feature on today's data-only networks, but is unacceptable on a mixed-media backbone. The ATM intelligent infrastructure supports traffic connection types that can be used to protect time-critical information streams from the unruly behavior of data traffic. Through facilities known as Call Admission Control (CAC) and Usage Parameter Control (UPC), the ATM intelligent infrastructure ensures that users cannot compromise network integrity either by establishing new connections or by putting too much traffic into the network.

Scalability

The ATM intelligent infrastructure is extremely scalable. By moving routing responsibility and policy decisions to the edge of the network, the ATM intelligent infrastructure accommodates new capacity whenever necessary. Additional links or ATM switches can be added without disruption. No complex planning or configuration is required. You simply install the new link or switch and PNNI automatically reconfigures the network to use the new path to its optimum capability.

Future-Proof Technology

Through the use of the advanced capabilities of PNNI, LANE, and MPOA, the ATM intelligent infrastructure supports any new user application without reconfiguration. Whether time-sensitive or otherwise, you simply add the new application to your edge system, request the necessary connections from the ATM intelligent infrastructure, and the appropriate smart bandwidth is allocated.

The network infrastructure should not dictate how you run your business. By using the ATM intelligent infrastructure, the physical topology of the underlying network will match your communications needs, whatever they might be.

SUMMARY

ATM is an efficient technology to integrate LANs and WANs, as well as to combine multiple networks into one multiservice network. This article has described the current ATM technologies that network designers can use in their networks. It also made recommendations for designing non-ATM networks.

Notes

1. International Telecommunication Union, Place des Nations, 1211 Geneva 20, Switzerland.

2. American National Standards Institute, Inc. (ANSI), 11 West 42nd, Street, New York, NY 10036.

3. Cisco Systems, Inc., 170 West Tasman Drive, San Jose, CA, 95134-1706.

4. NET/Adaptive, Europe Office, Av. De Tervueren 402, 1150 Brussels, Belgium.

5. Northern Telecom, West Road, Harlow, Essex, England, CM20 2SH.

6. Sprint World Headquarters, 2330 Shawnee Mission Parkway, Westwood, KS 66205.

7. ATM Forum, Worldwide Headquarters, 2570 West El Camino Real, Suite 304, Mountain View, CA 94040-1313.

8. Cisco Systems, Inc. COPYRIGHT©1999 Cisco Systems, Inc. ALL RIGHTS RESERVED, 170 West Tasman Drive, San Jose, CA 95134-1706

9. Ascend Communications, Inc., One Ascend Plaza, 1701 Harbor Bay Parkway, Alameda, CA 94502-3002, 1999.

10. Cisco Systems, Inc.

11. Cabletron Systems, Inc., 35 Industrial Way, Rochester, NH 03867, 1999.

12. FORE Systems, Corporate Headquarters, 1000 FORE Drive, Pittsburgh, PA 15086-7502, 1999.

13. MICOM Communications Corp., 4100 Guardian Street, Simi Valley, CA 93063, 1999.

14. A virtual channel is equivalent to a virtual circuit, that is, both terms describe a logical connection between the two ends of a communications connection. A virtual path is a logical grouping of virtual circuits that allows an ATM switch to perform operations on groups of virtual circuits.

15. ATM cells consist of five bytes of header information and 48 bytes of payload data. The VPI and VCI fields in the ATM header are used to route cells through ATM networks. The VPI and VCI fields of the cell header identify the next network segment that a cell needs to transmit on its way to its final destination.

Chapter 15
Virtual LANs

Gilbert Held

THIS ARTICLE DISCUSSES THE RATIONALE FOR VIRTUAL LOCAL AREA NETWORKS (VLANS), THEIR CONSTRUCTION METHODS, AND EMERGING STANDARDS. Since it may be late 1997 or early 1998 when VLAN standards are agreed upon, readers may prefer to satisfy their virtual networking requirements using proprietary equipment. In doing so, this article will serve as a guide as it covers basic construction techniques to include compatibility issues associated with attempting to construct a network using equipment from different vendors.

DEFINITION

A virtual LAN (VLAN) represents a logical broadcast domain established on a physical network topology. To illustrate the effect of this definition, the focus will be on an Ethernet switch.

Exhibit 15.1 illustrates an eight-port Ethernet switch used to connect six workstations to two servers. In this example, the eight ports are labeled 0 through 7 and represent the eight physical ports built into the switch.

Assume workstations connected to Ethernet switch ports 0 through 2 and 6 represent computers and a server that belong to the engineering department. Also assume that workstations connected to Ethernet switch ports 3 through 5 and 7 represent computers and a server that belong to the sales department. If the Ethernet switch is capable of supporting the creation of VLANs based on the physical port that devices are connected to, two virtual LANs could be constructed. The engineering department VLAN could be established to form a broadcast domain consisting of switch ports 0 through 2 and 6.

In comparison, the sales department could have a VLAN established by logically grouping switch ports 3 through 5 and 7 into a broadcast domain. Exhibit 15.2 illustrates the logical establishment of the two virtual LANs on the physical topology of the eight-port Ethernet switch.

In examining the two VLANs established in Exhibit 15.2, note that each represents a broadcast domain separate from the other. This separation

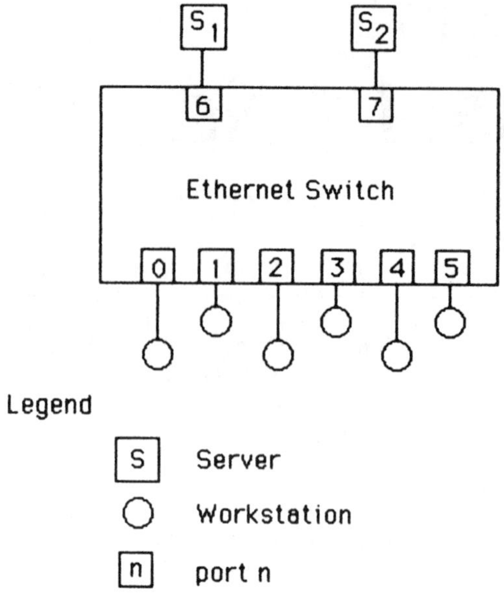

Exhibit 15.1 Using An Ethernet Switch.

can represent a problem if, for example, a workstation on one VLAN requires access to a server assigned to a different VLAN. Later in this article, interoperability issues will be examined when the different VLAN construction techniques are described.

Rationale

There are several key reasons that form the basis for organizations building virtual LANs. Those reasons include supporting the virtual networking requirements of an organization; facilitating adds, moves, and changes; enhancing network performance; and enhancing network security. Each of these reasons will be discussed briefly to obtain an appreciation for how the use of VLAN technology may be able to enhance an organization's networking operations.

Support for Virtual Networking Requirements

A virtual LAN represents a logically created broadcast domain. As such, it is not restricted to following the exact topology of a LAN. Instead, it must reside within the physical topology of a LAN. This means a VLAN can be constructed, modified, and deleted in tandem with the assignment of personnel to a project that requires communications support. In addition, the construction and modification of stations into a broadcast domain can correspond to the computers attached to LANs and used by employees

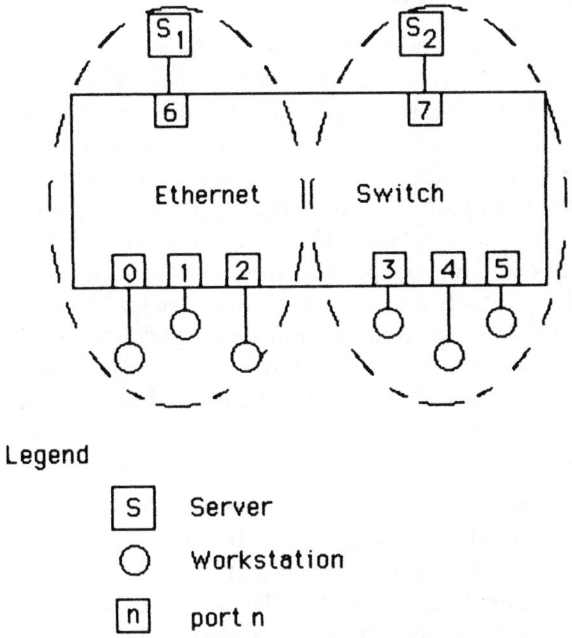

Exhibit 15.2 Establishing VLANs Based on Switch Port Connection.

---

assigned to a particular project. Thus, a VLAN can be used to support the virtual networking requirements of an organization.

Facilitating Adds, Moves, and Changes

A VLAN associates network nodes into a broadcast domain based on a predefined networking characteristic, such as a switch port. By adding, deleting, or modifying the characteristics used to define network nodes to a broadcast domain, one obtains the capability to add, modify, or delete members in the VLAN. Thus, a VLAN facilitates adds, moves, and changes.

Enhancing Network Performance

Many upper-layer protocols use broadcast packets to advertise the presence of certain network devices. In doing so, such packets can flood a network as they are repeated by bridges onto other network segments, even if no device on the new segment requires the service of the device. For example, in a Novell NetWare environment, servers transmit a Service Advertising Protocol (SAP) Packet every 60 seconds. Since broadcast packets or frames originating on one VLAN do not normally flow onto other VLANs, the use of VLANs can improve performance as broadcast packets can be restricted to specific domains requiring knowledge of server offerings.

Enhancing Network Security

Although a VLAN is not a security device and does not provide such services as encryption, authentication, and verification, it can provide enhanced network security. The reason it can do so is due to the fact that transmissions are normally restricted to predefined broadcast domains. This means that only a member of the broadcast domain will normally receive a packet originated by another member of the VLAN domain. Conversely, this also can result in a network user who is not a member of the domain having an inability to receive packets sent on the domain. Thus, a common method used to breach security by the connection of a LAN monitor to read all network packets would not normally work if the user attempting to use the monitor was not a member of the VLAN. An exception to the use of VLANs to enhance network security is discussed later in this article.

Construction Methods

There are two basic methods used to construct VLANs — implicit and explicit tagging. Implicit tagging results in the use of a characteristic or feature of the node connected to a LAN, such as the port on a switch through which a node connects to a network. Through the use of implicit tagging, the LAN frame remains unmodified. In comparison, explicit tagging results in the modification of a LAN frame as it flows through a switch or router, during which a field is added that defines the VLAN associated with a network node.

Currently, most work in the area of explicit tagging is being performed by the Institute of Electrical and Electronic Engineers (IEEE) 802.1Q committee. That committee prepared a Project Authorization Request (PAR) in March 1996 that defined the general scope of the VLAN standardization effort. That effort is expected to be finalized in late 1997 or early 1998. In the interim, one can acquire equipment that uses implicit tagging or proprietary explicit tagging. Both techniques are examined in the following paragraphs.

Implicit Tagging

There are three main methods of implicit tagging used to create VLANs, with each method corresponding to one of the three lower layers of the International Standards Organization (ISO) Open System Interconnection Reference Model. Those methods include the use of ports, MAC addresses, and protocols that correspond to the Physical Layer, Data Link Layer, and Network Layer, respectively. Inasmuch as the creation of VLANs via port assignments was previously examined, VLANs can also be created using MAC addresses and protocols.

The use of MAC addresses to create VLANs provides more flexibility than the use of ports. This is because a segment-based switch would otherwise

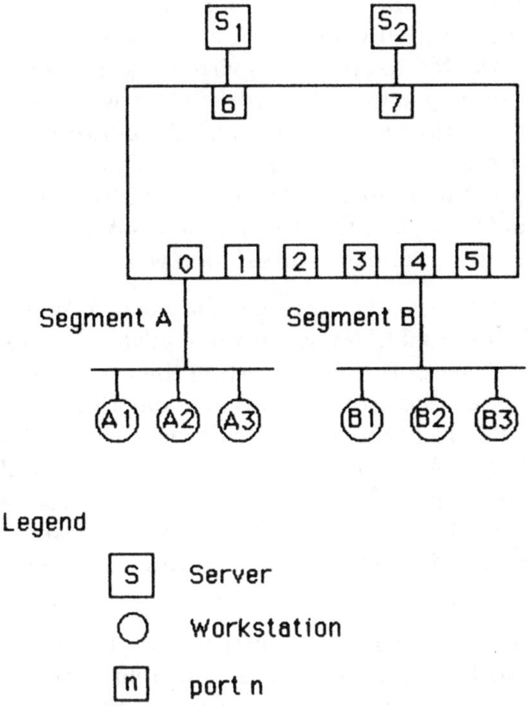

Exhibit 15.3 VLAN Creation Using MAC Addresses.

have to associate all nodes on a segment with a predefined VLAN when VLAN assignments occur based on associating a port connection to a virtual LAN.

Exhibit 15.3 illustrates an eight-port switch that supports both VLAN creation via MAC addresses and segment switching. For simplicity, MAC addresses are shown as two hex digits instead of their IEEE 6-byte or 48-bit address. In examining Exhibit 15.3, note that only two segments are shown populated — segments A and B. Each of those segments has three workstations with their MAC addresses shown as A1, A2, A3, and B1, B2, B3, for each segment.

When a VLAN is created based on port use, all nodes on a segment are required to be assigned to the same VLAN. For example, if Ports 0 and 6 were used to form a VLAN, then nodes with MAC addresses A1, A2, and A3, and Server S1 would be assigned to the same broadcast domain.

Suppose workstations with MAC addresses A1 and A2 were used by members of the engineering department, while workstations with MAC addresses A3, B1, B2, and B3 were used by members of the sales department.

If the creation of VLANs was based on port use, one would have to physically move the workstation whose MAC address is A3 from segment A to segment B if one wanted to create two virtual LANs, each based upon the department the employee using the workstation was assigned to. In comparison, if VLAN creation is based upon a MAC address, one could assign workstations with MAC addresses A1 and A2 to one virtual LAN, while workstations with MAC addresses A3, B1, B2, and B3 could be assigned to the second virtual LAN. Note that this method of VLAN assignment alleviates the necessity to recable the workstation whose MAC address is A3.

Although this is clearly advantageous, the use of MAC addresses on segments to establish virtual LANs negates one advantage of virtual LANs — enhanced network security. By supporting the creation of VLANs based upon MAC addresses, the broadcast domain must be extended to the entire segment on which a particular MAC address resides. Thus, a packet associated with the VLAN formed by workstations A1 and A2 would be broadcast onto segment A. This means that, with appropriate equipment, the user at the workstation whose MAC address is A3 would be capable of reading packets on a different virtual LAN from the one he or she is assigned to.

The third major method used to construct VLANs is based on the protocol used by network nodes. This is a very flexible approach to VLAN creation as it enables workstations and servers to belong to multiple VLANs. For example, returning to Exhibit 15.3, assume workstations whose MAC addresses are A1 and A2 use a TCP/IP protocol stack, while workstations whose MAC addresses are B1, B2, and B3 use a IPX/SPX protocol stack. Further assume that the workstation whose MAC address is A3 has dual TCP/IP and IPX/SPX protocol stacks. Then, if VLAN creation is based upon the protocol used, the workstation whose MAC address is A3 can belong to both virtual LANs.

Explicit Tagging

Explicit tagging references the physical modification of a LAN frame as it flows through a networking device. That modification adds two or more fields into a LAN frame. Currently, the IEEE 802.1Q committee has proposed two models of VLAN tagging, referred to as a one-level model and a two-level model. Exhibit 15.4 illustrates the formats of the one-level and two-level models. Both models are shown for Ethernet, and the proposed tagging formats differ for FDDI and Token Ring LANs, which are not illustrated.

In the one-level model, the destination and source addresses from the original, untagged Ethernet frame are split apart from the remainder of the frame. Two new fields, labeled Ethertype and VLAN-ID, are inserted. The Ethertype field is used to identify the frame as a tagged packet, while the VLAN-ID field is used to identify the VLAN to which the packet is assigned.

A. One-level model

Destination Address	Source Address	Ethertype	VLAN-ID	Remainder of frame	FCS

B. Two-level model

Virtual Destination	Virtual Source	Ethertype	VLAN-ID	CRC	Remainder of frame with FCS

Exhibit 15.4 Potential IEEE VLAN Tagging Models.

For the two-level model, the Virtual Destination and Virtual Source fields represent new addresses added to the frame that represent virtual addresses. The Ethertype field and VLAN-ID fields retain the same meanings as the one-level model; however, through the use of the two-level model, frames can be addressed explicitly.

Currently, a large degree of effort remains in developing VLAN standards. For example, such questions as how explicit and implicit tagging methods can co-exist, or will even co-exist, remain to be determined. Another area that remains to be resolved is the effect of extended frames upon many communications products such as bridges, repeaters, routers, and gateways. For example, the addition of a VLAN-ID field can result in an Ethernet frame carrying a maximum 1500-byte information field, having its total length exceed the maximum allowable Ethernet frame length. When this situation occurs, some communications devices that check frame lengths will treat such frames as errors. This can result in such frames being dropped and could result in session timeouts as the frames are repeatedly retransmitted and dropped prior to their arrival at their intended destinations. Thus, a considerable amount of effort remains prior to explicit tagging becoming a viable VLAN creation method.

Interoperability Issues

Today, one can acquire VLAN-capable switches and routers from more than 30 vendors. Unfortunately, the ability of a product from one vendor to interoperate with a product from another vendor with respect to their VLAN capability cannot be guaranteed. In fact, due to the lack of standards, one can only configure dissimilar equipment to operate together when implicit tagging is used. When doing so, one must restrict the method of implicit tagging to a common denominator supported by all products, and more than likely manually configure all equipment, as vendor equipment management capability may not be interoperable. Due to this, many organizations commonly acquire VLAN equipment from the same vendor.

CONCLUSION

Due to explicit tagging being several years away from possible standardization, most virtual LANs will continue to be developed using implicit tagging. This means organizations with immediate requirements to establish virtual LANs should focus their attention on equipment that has an implicit tagging capability. However, since organizations are always looking for methods by which they can economize on the purchase of equipment, network managers and administrators should ensure that the VLAN-capable switches and routers they purchase are software upgradeable. This means that microprocessor-based switches and routers should normally be considered, since such equipment is commonly upgradeable via a ROM change or PROM upgrade.

In comparison, lower-cost products based upon the use of Application Specific Integrated Circuits (ASICs) cannot be upgraded. Thus, if an organization anticipates a requirement to obtain the support of explicit tagging once this method of VLAN creation is standardized, it should also consider acquiring upgradeable network devices, even if such devices are more costly initially.

Chapter 16
Multicast Networking

C. Kenneth Miller

IN A BUSINESS ENVIRONMENT WHERE EFFICIENT INFORMATION DELIVERY IS CRITICAL TO STAYING COMPETITIVE, MULTICAST NETWORK TECHNOLOGY OFFERS A NEW METHOD OF ELECTRONIC DISTRIBUTION. Multicast networking enables group-oriented applications such as videoconferencing, dataconferencing, electronic software distribution, and database updates.

INTRODUCTION

Multicast transmission is the sending of one message to many, but not all, receivers. Multicast network infrastructures are becoming available in all kinds of data networks, including wide area, satellite, and wireless. This new infrastructure is being used for group-oriented data networks.

Broadcast transmission is the sending of one message to all receivers and has been used extensively in LAN environments. Broadcast traffic over WANs should be avoided, however, because it can flood the WAN with unwanted traffic or broadcast storms.

Multicast provides the mechanism for one-to-many transmission over a WAN without creating broadcast storms. Multicast network infrastructures can be created at Layer 2 (i.e., the link layer) or at Layer 3 (i.e., the network layer). The primary Layer 3 multicast transport technique is multicast IP, which many router vendors support. Layer 3 multicast transport is independent of the underlying network architecture.

Different physical and link layer architectures support multicast and broadcast services. For example, satellite data transmission is a broadcast architecture that easily supports multicast services. Other network infrastructures are multicast LANs, multicast frame relay, and multicast SMDS.

MULTICAST LANS

Every station on the LAN listens to all transmissions. Nodes on a LAN have MAC addresses which are sometimes called physical addresses because they designate a physical node on the network. MAC addresses are global, which means that each one is unique.

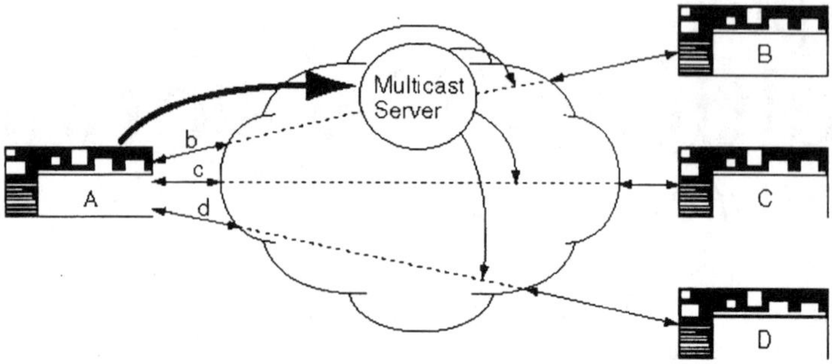

Exhibit 16.1 One-way Multicast.

MAC addresses come in three varieties: individual (unicast), multicast, and broadcast. Unicast addresses identify end points in point-to-point communications. The MAC broadcast address is the all address. MAC multicast addresses are primarily used for mapping to upper-layer multicast addresses.

MAC broadcast frames are usually filtered by bridges and routers, confining them to the LAN and preventing broadcast storms.

Multicast Frame Relay

Frame Relay is a Layer 2 protocol designed for use over a WAN. Frame Relay is a connection-oriented protocol, which means that it emulates actual physical links with PVCs and SVCs. PVCs and SVCs represent point-to-point connections with DLCIs and do not usually have the facility for one-to-many connections. However, the Frame Relay Forum recently released specifications for one-way, two-way, and N-way multicast services over Frame Relay.

One-Way Multicast

One-way multicast is suitable for electronic distribution of information from an information service provider. (See Exhibit 16.1.) A multicast server in the network maps the multicast data link connection identifiers to the individual data link connection identifiers. Individual DLCIs are also present from the members of the group to the multicast transmitter.

Two-Way Multicast

Two-way multicast enables migration of old IBM SDLC multidrop configurations to a Frame Relay environment. (See Exhibit 16.2.)

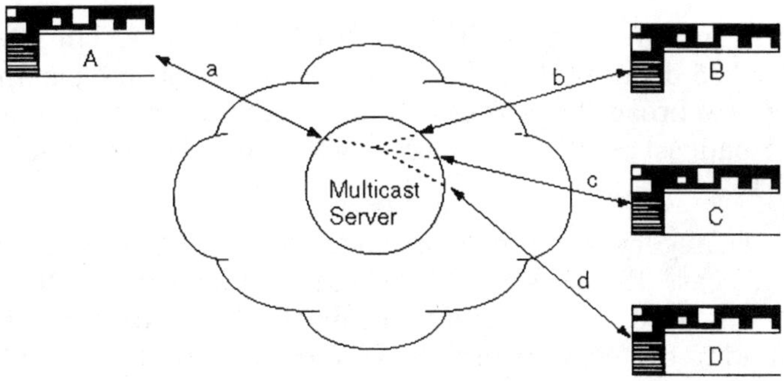

Exhibit 16.2 Two-way Multicast.

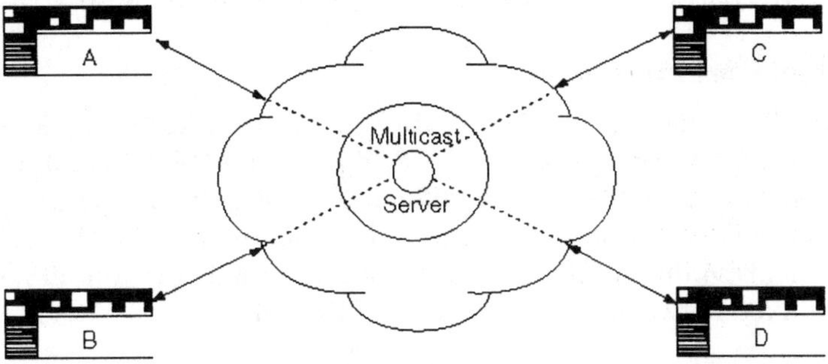

Exhibit 16.3 N-way Multicast.

N-Way Multicast.

N-way multicast allows any node to be either a transmitter to the group or a receiver. N-way multicast (see Exhibit 16.3) was designed for teleconferencing applications.

Multicast Switched Multimegabit Data Service

SMDS was developed by Bellcore for LEC to provide high-speed connectivity between LANs in a metropolitan area. SMDS is a subset of the IEEE 802.6 MAN standard and is offered as a service by several carriers.

SMDS uses a connectionless network architecture in which all nodes can communicate to all other nodes on the network, similar to LANs. All packets sent on the network are available for reception by all nodes, which means that SMDS can support multicast, unicast, and broadcast services.

COMPARING BROADCAST AND MULTICAST IP

IP supports broadcast as well as multicast services. The IP address 255.255.255.255 defines the global broadcast address and is mapped into the MAC global broadcast address on the LAN to which it is attached. The IP global broadcast address is not usually forwarded by routers out of a local network.

A second category of IP broadcast address is the network broadcast address. In this case, the network portion of the Class A, B, or C IP address is sent to the desired network destination; the host portion is the all 1s broadcast address. Network broadcast addresses are forwarded by routers to the designated network where they are mapped into that network's MAC broadcast address. For example, the Class C IP address 206.42.32.255 is the network broadcast address for network 206.42.32.

Network broadcast addresses can provide so-called directed broadcast services to a particular network or subnetwork over an internetwork.

The IP Multicast Protocol

Multicast IP is a new technology that provides network layer routing of IP Class D group address packets in TCP/IP networks. Multicast IP operates over any network architecture, and the multicast groups can be set up and torn down within seconds. Multicast IP is destined to be the dominant means for providing multicast services in data networks of all kinds, including mixed network environments or intranets.

Class D Addressing

Multicast IP uses Class D IP addresses, as shown in Exhibit 16.4. IP Class A, B, and C addresses are used for point-to-point (unicast) communications and consist of network and host components. Class D addresses, by contrast, have only one component that identifies the multicast group.

Class D addresses occupy the range from 224.0.0.0 to 239.255.255.255 and can be assigned semipermanently or temporarily for the length of time a group is in place. Groups may be set up and torn down in seconds.

Hosts that belong to a group need to support RFC 1112, which is the standard for host extensions for IP multicasting. RFC 1112 specifies the IGMP that is used by members of a multicast group to inform the nearest router supporting multicast routing of their presence in a particular group. The router is responsible for updating routing tables so that multicast packets with the Class D address associated with that group are forwarded to the subnetwork that includes members of the group.

The IGMP dialog is shown in Exhibit 16.5. Two messages are provided in the dialog: the IGMP query and the IGMP response. Queries are sent to the

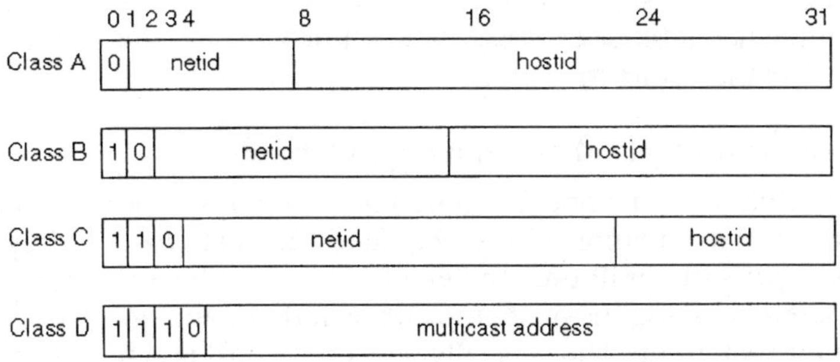

Exhibit 16.4 IP Address Types.

```
SYSCON 3.82                    Monday, November 8, 1993, 9:34 a.m.
                User CXHELD on File Server MDPC-1

                      System Login Script
map ins s1; =sys;public
if member of "WP40" then
MAP ins S16;=SYS;WPPO OFDOS40
map ins s16;=sys;wppo ofwin40
end
#capture L=2 noti ti=40 q=hplaserjetii-209 nt nti nb
display public welcome.txt
IF MEMBER OF 'APS' THEN
#capture L=3 noti ti=40 q=hplj3-254 nt ntf ab
else
IF MEMBER OF "ADM" THEN
#capture L=2 noti ti=40 q=HPLJ3-ADM nt ntf nb
else
END
```

Exhibit 16.5 IGMP Dialog.

all hosts' multicast group address — 224.0.0.1 — and carry an IP time-to-live of 1. Hosts respond to queries with host membership reports for each host group to which they belong.

There are two techniques for reducing the number of concurrent reports and the total number of reports transmitted. A host can start a report delay timer, or it can send a report that other members of the same group on the network can overhear.

Starting a Report Delay Timer

When a host receives a query, it can start a report delay timer rather than send reports of the query immediately for each of its group members on the network interface. Timers are set to different, randomly chosen values between zero and D seconds. A report is generated for the corresponding host group when a timer expires. Thus, reports occur over a D-second interval rather than concurrently.

Sending a Report that Other Members Overhear

To ensure that other members of the same group overhear the report, it is sent with an IP destination address equal to the host group address being reported, and with an IP time-to-live of 1. If a host hears a report for its group on the network, the host stops its own timer and does not generate a report for that group. Thus, usually only one report is generated for each group on the network by the member host whose delay timer expires first. The multicast routers receive all IP multicast datagrams; therefore, they need not be addressed explicitly. The routers need not know which hosts belong to a group, only that at least one host belongs to a group on a particular network.

Multicast routers send queries periodically to refresh their knowledge about memberships on a particular network. Queries are usually sent at intervals of approximately one minute to minimize overhead traffic on the network. However, when a multicast router starts up, it may often issue several closely spaced queries to obtain knowledge of local group membership quickly.

Similarly, when a host joins a new group, it issues a report immediately without waiting for a query in case it is the first member of the group on the network.

New members notify the nearest router of their presence in a group almost instantly, whereas notification of leaving a group depends on query timeout, which can be a minute or more. A new version of IGMP is being proposed that provides for explicit host leave group reports to speed up the notification to routers of hosts leaving a group.

The Multicast Backbone

The Mbone of the Internet is based on multicast IP and covers 3,000 subnetworks. It is predicted that the Internet will be fully multicast IP-enabled soon.

The Mbone uses computers to provide the multicast routing capability, and multicast packets are tunneled through unicast links to tie together the islands of multicast networks. It is used mostly by Internet researchers for videoconferencing and dataconferencing.

Two applications used on the Mbone are based on the ALF lightweight session model originated by Internet researchers. ALF dictates that the best way to meet diverse multicast application requirements is to leave as much flexibility as possible to the application in the application layer. This means that connectivity is viewed as unreliable, using the UDP connectionless transport layer.

Two prominent ALF model applications come from Lawrence B. Livermore Laboratory, which provides the Visual/Audio Tool in multimedia videoconferencing. In addition, the Mbone enables a whiteboard application that lets users of a multicast group electronically write on a whiteboard. All members of the group can see what each member has written. Whiteboarding is essentially an electronic dataconference that allows group members to brainstorm visually.

Other videoconferencing tools that provide use over the Mbone are being used in some universities; however, they are not yet commercially available. The Mbone is still in the research stage.

Multicast Routing Protocols

Multicast routing protocols are required to perform optimal routing through router networks, just as unicast routing protocols such as IGMP, OSPF, and RIP are needed to perform optimal unicast routing in router networks. Multicast routing protocols should efficiently minimize the necessary traffic for routing multicast data in the network or internetwork.

There are three predominant multicast routing protocols: DVMRP, Multicast OSPF, and PIM.

DVMRP. DVMRP is the oldest multicast routing protocol and uses a technique known as "reverse path forwarding." When a router receives a multicast packet (i.e., one with a Class D destination address), it floods the packet out of all paths except the one that leads back to the packet's source, as shown in Exhibit 16.6. This allows the packet to reach all subnetworks, possibly multiple times. If a router is attached to a set of subnetworks that do not want to receive packets destined to a particular multicast group, the router can send a prune message back up the distribution tree to stop subsequent multicast packets from being forwarded to destinations where there are no members.

DVMRP periodically refloods to reach any new hosts that want to receive a particular group. There is a direct relationship between the time it takes for a new receiver to get the data stream and the frequency of flooding.

DVMRP implements its own unicast routing protocol to determine which interface leads back to the source of the packets for a particular group. This unicast routing protocol is similar to Routing Information Protocol and is based purely on hop count. As a result, the path that the multicast traffic follows may not be the same as the path the unicast traffic follows.

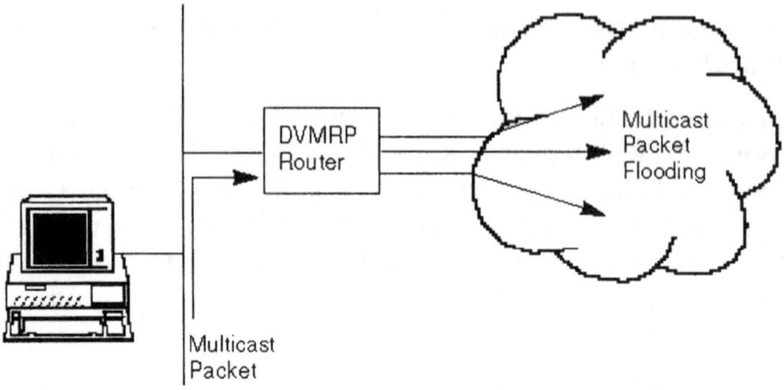

Exhibit 16.6 DVMRP Flooding.

DVMRP is an inefficient multicast routing protocol because of scaling problems. The periodic flooding with broadcast traffic generates significant overhead in the network, and early versions of the protocol did not implement pruning.

Nevertheless, DVMRP has been used to build the Mbone and has been implemented by some router companies. Bay Networks, Inc. (Santa Clara, CA) has chosen DVMRP as its first offering to support multicast routing, and 3Com Corp. (Santa Clara, CA) is planning to implement it in the next few months.

Multicast Open Shortest Path First (MOSPF). MOSFP is an extension of the OSPF unicast routing protocol. OSPF is a link state routing protocol, which means that each router in a network understands all of the available links in the network. Each OSPF router calculates the routes from itself to all possible destinations.

MOSPF includes multicast information in OSPF link state advertisements. MOSPF routers learn which multicast groups are active on which subnetworks.

MOSPF builds a distribution tree for each source/group pair and computes a tree for active sources sending to the group. The tree states are stored, and trees must be recomputed when a link state change occurs or when the timer for the store of the link expires.

The main disadvantage of Multicast OSPF is that it works only in networks supporting OSPF. Multicast OSPF, authored by Proteon, Inc. (Westborough, MA) has been implemented in routers available from Proteon and Xyplex, Inc. (Littleton, MA).

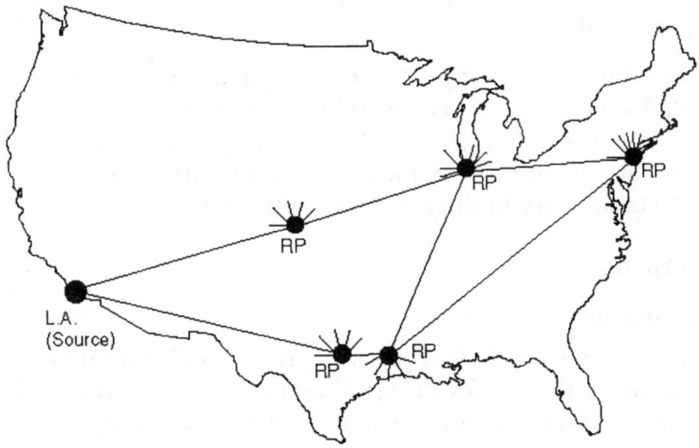

Exhibit 16.7 Sparse Mode Topology Showing Rendezvous Points.

Protocol Independent Multicast (PIM). PIM works with all existing unicast routing protocols and supports two types of traffic distribution topologies: dense and sparse. Dense mode is most useful when:

- Senders and receivers are in close proximity to one another.
- There are few senders and many receivers.
- The volume of multicast traffic is high.
- The multicast groups do not change very often.

Dense mode PIM uses reverse path forwarding and is similar to DVMRP. Sparse mode PIM is most useful in network topologies scattered over a large geographic area when:

- There are a only a few receivers in a group.
- Senders and receivers are geographically separated by WAN links.
- The groups change often.

Sparse mode PIM is designed for environments where there are many data streams destined for group members, as shown in Exhibit 16.7. Each stream goes to a relatively small number of group members. For this topology, reverse path forwarding techniques waste bandwidth.

Sparse mode PIM defines an RP. Receiving members of the group join to a particular RP to create a shared distribution tree. There can be many RPs in a single group. Sources must register with all RPs in a group. Once the data stream begins to flow from sender to RP to receiver, the routers optimize the path automatically to remove any unnecessary hops. Sparse mode PIM assumes that no hosts want the multicast traffic unless they specifically request it.

PIM simultaneously supports dense and sparse mode for different groups. PIM was authored by Cisco Systems, Inc. (San Jose CA) and researchers from the University of Southern California, Lawrence B. Livermore Laboratory, and Xerox Parc. Work is continuing in the Internet community to improve multicast routing protocols.

APPLICATIONS

Dynamic Groups

Groups defined by Class D addresses can be created and torn down in seconds. (Groups that are easily created and torn down are referred to as dynamic.) Changes to the group can also be performed in seconds, which is useful in many applications.

Videoconferencing

Videoconferencing is a temporary group setup that may exist for the duration of a meeting of any length. It is essentially an electronic meeting.

Dataconferencing

Dataconferencing is the same as videoconferencing, but without the video. Dataconferencing often involves a whiteboard, such as the program used by the academic community in the Mbone.

Real-Time Distribution of Multimedia Information

There are several new services that provide real-time business news to the desktop. Some of the information provided includes audio, video, and data. Subscribers to these services are constantly entering and leaving the group, therefore the ability to change the size and composition of the group dynamically is very important for this application.

Multicast File Transfer. There are several multicast file transfer applications that involve one-to-many transmission of information in the form of files.

Subscription-based Information Delivery. Subscription-based information delivery is a service that distributes information electronically using a multicast network and a multicast file transfer program. The information might be, for example, news in text and image form, or financial trend analysis in audio form.

Information is in the form of files that need guaranteed delivery. Dynamic groups are important in this application because subscribers are constantly changing.

Electronic subscription delivery of information could become a popular service once the Internet is fully multicast-enabled. Today, much time is wasted searching for information on the Internet, even with the advent of sophisticated browsing tools. Consumers have newspapers and magazines delivered as print subscriptions; they should be able to receive the equivalent electronically.

Electronic Software Distribution. Software is usually distributed at the time of its release. As the number of hosts on networks grows, software updates become more of a burden, which has led to great interest in the electronic distribution of software using a multicast file transfer program that is guaranteed.

Because distribution of software is not a daily or even a weekly event, it is desirable to terminate the group right after the transfer.

Distributed Publishing. National newspapers and other print publications use regional printing facilities to reduce shipping costs. They send information using a series of point-to-point transfers. Multicast file transfer is a more efficient and less costly solution. In this application, the membership group is relatively static.

Kiosk Applications. Kiosks are being planned to provide a number of new services, most of which need to be updated with information common for all of the kiosks. For example, the U.S. Department of Transportation is sponsoring a project to implement kiosks at various urban and suburban locations. Information about traffic will be periodically downloaded to the kiosks so that travelers can obtain information about traffic patterns. These kiosks would also be used to disperse information of general public interest, such as facts about major sporting events such as the Olympic games.

Kiosks may also be placed in retail stores as a means of showcasing products using multimedia presentations. The latest fashions and sales promotions can be downloaded into the kiosks using multicast networks and a file transfer program.

Database Updates. Linked databases in remote locations can be updated easily using a multicast file transfer program on a multicast network infrastructure.

VENDOR OFFERINGS

There are not many network infrastructures on a wide area basis that support multicast. There are, however, a number of companies providing low-cost videoconferencing products, and many of them support multicast IP or plan to support it. Intel Corp. (Santa Clara, CA), Connectix Corp. (San Mateo, CA), PictureTel Corp. (Danvers, MA), Insoft, Inc. (Mechanicsburg, PA), and Xing Technology Corp. (Arroyo Grande, CA) all have PC-based videoconferencing products.

Multicast file transfer products are available from StarBurst Communications Corp. (Concord, MA) and Legent, a division of Computer Associates International, Inc. (Islandia, NY).

CONCLUSION

Multicast network technology and the associated applications are poised to dramatically change the use of networked PCs. Videoconferencing and dataconferencing over IP data networks will become common as network speeds increase and multicast IP router networks become pervasive. Multicast file transfer over multicast IP networks makes electronic information delivery on a subscription basis a reality, and electronic delivery of information more like the model for printed information delivery.

Multicast file transfer also facilitates software maintenance by providing the ability to easily distribute software updates. Other distributed applications such as groupware and E-mail directory updates can also benefit, proving that multicast technology will be a major component of information distribution in the not-too-distant future.

Chapter 17
Voice and Video on the LAN

Martin Taylor

VOICE AND DATA CONVERGENCE IN THE LAN is about to become a hot topic in the industry, thanks to advances in switching and processors, as well as the H.323 standard. This chapter first looks at the business reasons for considering the deployment of voice and video over the LAN, and then discusses the technical issues and requirements.

INTRODUCTION

Most desktops in enterprises today are equipped with two network connections: a LAN connection to the PC or workstation for data communications, and a phone connection to the PBX for voice communications. The LAN and the PBX exist as two separate networks with little or no connectivity between them. Each has evolved to meet the very specific and differing needs of data and voice communications, respectively.

Despite much talk in the industry about the convergence of computers and communications, LANs and PBXs have not really moved any closer together during the last decade. In the mid-1980s, some PBX vendors sought to bring data services to the desktop via ISDN technology, but the advent of PCs requiring far more than 64K-bps communications bandwidth favored the emerging LAN standards of Ethernet and Token Ring. So far, most LAN vendors have not attempted to support voice communications on the LAN. But all this is about to change.

There are three key factors at work today that suggest that voice and data convergence in the LAN is about to become a hot topic in the industry:

- The widespread acceptance of advanced LAN switching technologies, including ATM, which makes it possible for the first time to deliver reliable, high-quality, low-delay voice transmissions over the LAN
- The emergence of the first standard for LAN-based videoconferencing and voice telephony, H.323, which removes objections about the use of proprietary protocols for voice and video over the LAN

- The deployment of the latest generation of Intel processors, featuring MMX technology, which makes high-quality, software-based, real-time voice and video processing feasible for the first time, and the new PC hardware architectures with Universal Serial Bus that permit voice and video peripherals to be attached without additional hardware inside the PC

THE VALUE OF VOICE AND VIDEO ON THE LAN

There are essentially two main kinds of motivation for considering voice and/or video on the LAN: the need to support new types of applications that involve real-time communications, and the desire to improve the overall cost-effectiveness of the local communications infrastructure.

New Types of Applications

Desktop videoconferencing, real-time multimedia collaboration, and video-based training are all examples of new kinds of applications that can benefit from the delivery of voice and video over the LAN.

The uptake of desktop videoconferencing has been held back by a combination of high costs and the difficulty of delivering appropriate network services to the desktop. Standards-based H.320 desktop videoconferencing systems require costly video compression and ISDN interface hardware, as well as the provision of new ISDN connections at the desktop alongside the LAN and the phone system. New systems based on the H.323 standard and designed to run over the LAN will leverage the processing power of the latest PCs and the existing switched LAN infrastructure to lower cost and simplify deployment dramatically.

Desktop videoconferencing may be used either to support internal meetings and discussions between groups located at remote sites, or to support direct interaction with customers and clients. For example, some enterprises in the mortgage lending business use videoconferencing to conduct mortgage approval interviews with potential borrowers so as to greatly reduce the overall time to complete a mortgage sale.

Real-time collaboration applications, involving any mix of video and voice with data conferencing to support application sharing and interactive whiteboarding, provide a new way for individuals and small groups to collaborate and work together remotely in real-time. This emerging class of applications, typified by Microsoft NetMeeting, is being evaluated by many enterprises, particularly for help desk applications.

By contrast, video-based training is already widely used in enterprise LANs. By delivering self-paced video learning materials to the desktop, training needs can be met in a more timely and less disruptive fashion than traditional classroom methods.

The growing popularity of these kinds of applications should be noted by network planners and designers. A preplanned strategy for local LAN upgrades to support voice and video will reduce the lead time for the deployment of these applications, and enable the enterprise to move swiftly, when the application need has been identified, to obtain the business benefits with the least possible delay.

Infrastructure Efficiencies

A single local communications infrastructure based on a LAN that handles data, voice, and video has the potential of costing less to own and operate than separate PBX and data-only LAN infrastructures.

The average capital cost of a fully featured PBX for large enterprises is between $700 and $750 per user, according to a leading U.S. telecommunications consultant, TEQConsult Group. Furthermore, this is expected to rise slightly over the next few years as users demand more sophisticated features from their phone systems. It is not difficult to see how a switched LAN that has been enhanced to handle voice could provide a solution for telephony at a fraction of this cost.

Most large PBX installations are equipped with additional facilities such as voice mail and Interactive Voice Response systems for auto-attendant operation. These systems are typically connected directly to the PBX via proprietary interfaces, and they, too, represent major capital investments. With voice on the LAN, such voice processing applications could be based on open server platforms and leverage the low-cost processing power and disk storage that is a feature of today's PC server market, thereby lowering the system's capital cost further.

Separate PBX and LAN infrastructures each incur their own management and operational costs. For example, moves, adds, and changes require separate actions to patch physical LAN and voice connections, and to update LAN logon and voice directories. With telephony provided over a voice-enabled LAN supporting combined directory services, the management effort required to administer moves and changes would be substantially reduced.

These cost-of-ownership benefits come with a raft of usability improvements for telephony. The PC (with phone handset attached) becomes the communications terminal for making and receiving phone calls, and the processing power and graphical user interface of the PC can be leveraged to provide point-and-click call launch and manipulation. Features of PBXs such as call transfer, divert, and hold, which are hard to invoke from a phone keypad, become very easy to use from a Windows interface.

Incoming callers can be identified on the PC display by matching Calling Line Identifier with directory entries. And with voice mail and E-mail supported on a unified messaging platform such as Microsoft Exchange or Lotus Notes, all messages are accessible and manageable via a single user interface.

These usability benefits for voice telephony over the LAN extend also to videoconferencing — a single consistent user interface may be applied to both video and voice-only calls.

LAN TECHNOLOGIES FOR INTEGRATED VOICE AND VIDEO

The LAN technologies in widespread use today — Ethernet, Fast Ethernet, FDDI, and Token Ring — were not designed with the needs of real-time voice and video in mind. These LAN technologies provide "best-effort" delivery of data packets, but offer no guarantees about how long delivery will take. Interactive real-time voice and video communications over the LAN require the delivery of a steady stream of packets with very low end-to-end delay, and this cannot generally be achieved with the current LAN technologies as they stand.

Asynchronous Transfer Mode (ATM)

At one time, there was a belief that ATM networking to the desktop would be embraced by LAN users to solve this problem. ATM is a networking technology that was designed specifically to handle a combination of the low-delay steady stream characteristics of voice and video, and the bursty, intermittent characteristics of data communications.

The ATM Forum, the industry body responsible for publishing ATM specifications, has developed a number of standards that enable desktops connected directly to ATM networks to support existing LAN data applications as well as voice telephony and videoconferencing. The ATM Forum standards for the support of voice and video over ATM to the desktop typically avoid the use of traditional LAN protocols such as IP, and instead place the voice or video streams directly over the ATM protocols.

While it is clear that ATM to the desktop provides an elegant and effective solution for combining voice, video, and data over the LAN, this approach implies a "forklift" to the LAN infrastructure and the end station connection. The cost and disruptive impact of such an upgrade tend to limit its appeal and, as a result, desktop ATM is not expected to be widely adopted.

However, the ability of ATM to provide Quality of Service — that is, to deliver real-time voice or video streams with a guaranteed upper bound on delay — makes ATM an excellent choice for the LAN backbone where voice and video over the LAN is needed.

Shared and Switched LANs

It is generally accepted that shared LANs are unsuitable for handling real-time voice and video because of the widely varying delays observed when multiple stations are contending for access to the transmission medium. The CSMA/CD access method used in shared Ethernet is particularly poor in this respect. Token Ring, on the other hand, is based on a token-passing access method with multiple levels of priority. Stations waiting to send data packets can be preempted by other stations on the ring with higher priority voice or video packets to send. As a result, Token Ring has excellent potential to handle real-time voice and video traffic, although this potential has yet to be realized in currently available networking products.

LAN switching does much to overcome the limitations of shared LANs, although today's products are still a long way from providing an answer for voice and video over the LAN. It is now cost-effective to provide users with dedicated 10M-bps Ethernet connections to the desktop, and 100M-bps Fast Ethernet uplinks from the wiring closet to the backbone.

However, despite the vast increase in bandwidth provision per user that this represents over and above a shared LAN scenario, there is still contention in the network leading to unacceptable delay characteristics. For example, multiple users connected to the switch may demand file transfers from several servers connected via 100M-bps Fast Ethernet to the backbone. Each server may send a burst of packets that temporarily overwhelms the Fast Ethernet uplink to the wiring closet. A queue will form in the backbone switch that is driving this link, and any voice or video packets being sent to the same wiring closet will have to wait their turn behind the data packets in this queue. The resultant delays will compromise the perceived quality of the voice or video transmission.

The only way to overcome this problem is to find a way of treating real-time voice and video packets differently from data packets in the network, and to give them preferential treatment when transient data overloads cause queues to form on busy network links. In practice, this means that LAN packets must be tagged with some kind of priority information that enables switches to identify which packets need to jump the queue.

The IEEE 802, which oversees standards for LAN technologies, has initiated a project identified as 802.1p, which is concerned with Traffic Class Expediting in LAN switches.

The principal problem faced by 802.1p is that there is no spare information field in the standard Ethernet packet format that could carry the required priority tag. As a result, it has been necessary to propose a new Ethernet packet format with an additional four bytes of information in the packet header that can contain a 3-bit priority tag field (offering eight levels of priority), together with some other information concerned with Virtual LANs.

With the new Ethernet packet format containing a priority tag, end-station applications can identify real-time voice or video packets by assigning them a high priority value in the tag. LAN switches that have been enhanced to process the priority tags can separate high- and low-priority traffic in the switching fabric and place them in separate queues at outgoing switch ports. The LAN switches need to implement a queue scheduling algorithm that gives preference to the higher-priority queues on outgoing ports, and by this means it is hoped that real-time voice and video can be carried over the LAN without incurring unacceptable delays during periods of heavy data traffic.

As of July 1997, the 802.1p standard was still in draft form and the standard is not expected to be completed until 1998. Ethernet switches that support the 802.1p priority tags with multiple internal queuing structures will require a new generation of switching silicon, and the earliest we could expect to see products that conform to the standard would be late 1998 or into 1999. Surprisingly, we may see Token Ring switches that handle multiple priority levels before that time, leveraging the capabilities of the existing Token Ring standard that supports eight levels of priority.

Hybrid ATM Networks

The discussion of ATM described how it offers guaranteed Quality of Service for real-time voice and video streams. Today, ATM is increasingly used as a LAN backbone for pure data applications because it offers greater scalability and fault tolerance than other LAN technologies. Ethernet and Token Ring LANs are connected to ATM via "edge switches" equipped with ATM uplinks typically supporting the ATM Forum standard for carrying LAN traffic over ATM and known as LAN Emulation.

It is possible to enhance ATM edge switches to enable desktops connected via Ethernet or Token Ring to enjoy the benefits of ATM Quality of Service across the LAN backbone. Two techniques have been proposed to achieve this.

The first technique, known as "Cell-in-Frame," extends the native ATM signaling protocols over dedicated Ethernet connections from the edge switch to the end station. The voice or video application in the end station places the voice or video stream in ATM cells using the ATM Forum standards for native ATM transport, and then encapsulates the ATM cells in Ethernet packets for transport to the edge switch for onward transmission onto the ATM network. Effectively, this is ATM to the desktop, but using physical Ethernet with standard Ethernet adapter cards as a kind of physical transport layer for ATM traffic.

The second technique makes use of an emerging standard protocol for end stations to request Quality of Service for IP-based voice or video applications known as the Resource Reservation Protocol, or RSVP. The enhanced edge switch intercepts RSVP requests originated by end stations and converts them into ATM signaling to request the setup of connections across the ATM backbone with the appropriate Quality of Service. The edge switch then distinguishes between IP packets containing data and those containing voice or video, using the information provided by RSVP, and steers voice and video packets onto ATM connections that have Quality of Service.

At the time of this writing, the technique described here for RSVP-to-ATM mapping enjoys somewhat broader industry support than Cell-in-Frame, perhaps because of its relationship with Internet technology.

Until LAN switches supporting 802.1p priority tagging have proven themselves capable of meeting the very stringent end-to-end delay requirements for real-time voice and video communications, hybrid approaches based on ATM in the backbone and switched Ethernet or Token Ring to the desktop are likely to find acceptance as the solution of choice for voice and video over the LAN.

Standards for LAN-based Voice and Video Applications

Standards for voice and video over the LAN fall into two categories: those designed for native ATM protocols, and those intended for general-purpose LAN protocols, particularly IP.

Standards for native ATM protocols, such as the ATM Forum's Voice Telephony over ATM (VTOA) are appropriate only for ATM-connected desktops, or desktops running Cell-in-Frame over Ethernet.

Standards for applications that run over IP are applicable both to ATM-connected desktops and desktops in general Ethernet or Token Ring environments. The most important standard in this space is H.323, which was developed by the International Telecommunications Union. While H.323 is designed to be independent of the underlying networking protocol, it will most often be deployed running over IP.

H.323 references other existing standards for the digital encoding and compression of voice and video signals, and describes how audio and video streams are carried in the payload of IP packets with the aid of the Real Time Protocol (RTP), which provides timing and synchronization information. H.323 also covers the handling of data streams for application sharing, shared whiteboarding, and real-time file transfer (referencing the T.120 standard), and includes signaling based on ISDN messaging protocols for call setup and teardown.

The H.323 standard is flexible and accommodates any combination of real-time voice, video, and data as part of a single point-to-point or multipoint conference call. It may be used with a voice stream alone as the basis of a LAN telephony solution. H.323 enjoys the broadest support in the industry as a proposed standard for Internet telephony.

Additional Components: Gateways and Gatekeepers

Creating a LAN infrastructure that can consistently deliver voice and video streams with sufficiently low delay is an absolute prerequisite for integrating voice and video on the LAN, but it is by no means the complete answer to the problem. There are two other key components of a complete voice and video solution, which in H.323 parlance are known as the *gateway* and the *gatekeeper.*

An H.323 gateway provides interconnection between voice and video services on the LAN, and external voice and video services typically provided over circuit-switched networks such as ISDN and the public telephone network. The gateway terminates the IP and RTP protocols carrying the voice and video streams, and converts them to appropriate formats for external networks. For videoconferencing, the conversion is most likely to be to H.320, another ITU standard that specifies how voice and video are carried over ISDN connections. For voice-only connections, the conversion will be to the G.711 standard for digital telephony. This allows voice interworking with any phone on a public network or connected to a PBX.

An H.323 gatekeeper is a pure software function that provides central call control services. While it is possible to run H.323 voice and video communications over the LAN without a gatekeeper, in practice this function is extremely useful. At the most basic level, the gatekeeper provides directory services and policy-based controls applied to the use of voice and video communications. For example, the gatekeeper can bar stations from accessing certain types of external phone numbers at certain times of day. The gatekeeper can be thought of as the "server" in a client/server model of LAN-based telephony and videoconferencing.

At a more sophisticated level, the gatekeeper may be able to support supplementary services, including call transfer, hold and divert, hunt groups, pick-up groups, attendant operation, etc. — features that are typically found in high-end PBXs for controlling and managing voice calls. While the H.323 standard does not explicitly describe how supplementary call control features may be supported, the standard does provide a framework for the addition of these advanced capabilities.

CONCLUSION

This chapter has explained the value of voice and video integration on the LAN in terms of both application-driven needs and the desire for infrastructure efficiencies. It has looked at the technology issues surrounding the transport of real-time voice and video streams over LAN infrastructures, and concluded that ATM backbones provide a solution in the near term, with the possibility later of a solution based entirely on switched Ethernet or Token Ring.

Finally, some additional functional elements have been described, such as gateways and gatekeepers, that are essential parts of a complete solution for voice and video over the LAN. Over the last decade, the open standards-based environment typified by PCs and LANs has revolutionized the way data is handled and processed in enterprise environments. Now, this open and standards-based approach is set to tackle the challenge of voice and video, formerly the exclusive domain of the PBX. The history of LAN evolution is set to repeat itself, and one can expect the traditional proprietary mainframe PBX to diminish in importance to the enterprise, giving way to client/server telephony and videoconferencing just as the mainframe computer has been pushed into the background by client/server techniques for data processing.

Chapter 18
Voice and Data Network Integration

Larry Schessel

VOICE OVER INTERNET PROTOCOL (VOIP) IS THE ENABLING TECHNOLOGY FOR THIS SERVICE INTEGRATION. In short, 64Kbps voice is converted into data packets and transferred over the data backbone. The savings of voice and data over a single corporate network can be considerable and the technology exists. But where does a company begin? This article reviews the enabling technologies and presents a viable private network evolution strategy that takes advantage of current technologies and standards. Savings are realized almost immediately.

VOICE AND DATA NETWORKS

Companies with private voice networks own one or more private branch exchanges (PBX)s to deliver the service. A PBX is the switching element that links two users together in a voice connection. PBXs can be broken done into the following basic functional components:

- Wiring: PBXs require dedicated wiring to each telephone in the company. This allows employees to call other employees connected to the PBX. To gain access to the public service telephone network (PSTN), PBXs require outside lines purchased from the local public telephone company.
- Hardware: PBX hardware includes line cards terminating the local wiring, a switch network that makes the connection between two telephones, and servers for PBX software.
- Software: PBX software controls the call setup and features such as call forwarding, call transfer, and call hold; software also provides operation and maintenance support as well as per-call statistics.

Companies with multiple locations require dedicated PBXs at each site and public leased lines connecting the multiple PBXs into a single logical network.

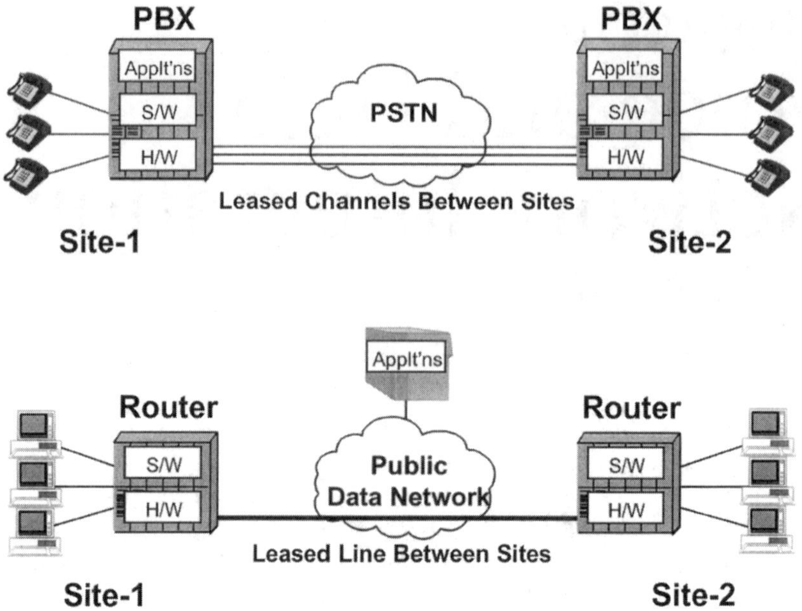

Exhibit 18.1 Comparison Between Private Voice and Data Networks.

Private data networks are delivered using equipment known as routers and hubs. Similar to voice networks, data networks also require dedicated wiring between the different network elements and to each desktop. PCs, servers, printers, and scanners are some of the data devices that can be connected to the network to deliver specific services. Once networked, these devices communicate with each other, passing information in the form of data packets. Exhibit 18.1 compares the voice and data network topologies.

VOICE OVER INTERNET PROTOCOL

When a user picks up and dials the telephone, the PBX provides dial tone, collects digits, and routes the call based on entered digits. In routing the call, the PBX sets up a dedicated connection between the two users. As the users speak into their phones, the PBX converts the analog voice into digital signals and sends the digitized voice across the connection at a rate of 64Kbps. The PBX converts the analog signal to digital format by sampling the analog voice every 125 microseconds and converting the voice sample to an 8-bit digital representation. Digitized voice is converted back to an analog signal at the terminating end of the connection.

IP data networks, on the other hand, handle device communication by packaging information into logical envelopes known as packets, addressing

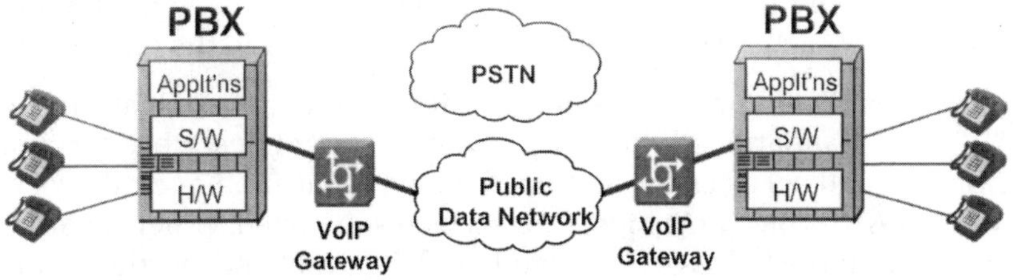

Step-1 Introduce VoIP Gateways for Inter-PBX Links

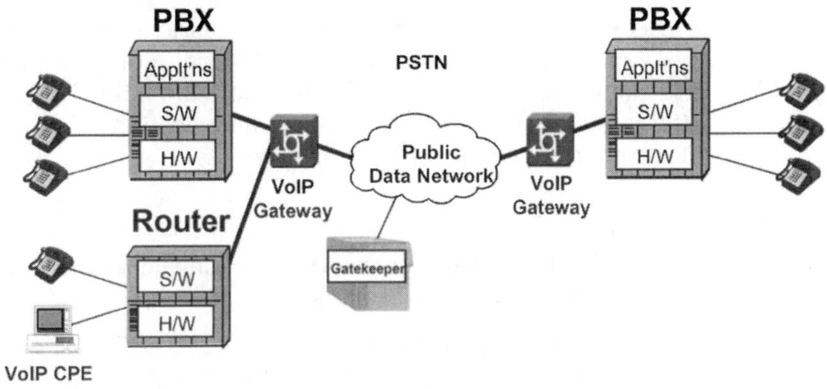

Step-2 Introduce VoIP Gatekeeper and CPE

Exhibit 18.1 *(Continued)* Comparison Between Private Voice and Data Networks.

the packets with destination information, and routing the packets through the network. Each network node looks at the addressing information and passes the packet onto the next network node until the packet arrives at its final destination. This routing method is connectionless because there is no dedicated connection between origination and termination as there is with a voice connection. A side benefit of connectionless communication is the fact that bandwidth is only required when there is information to send.

VoIP merges voice and data technologies into one ubiquitous voice service. Analog voice signals are sampled and digitized the same as the steps performed by the voice network. However, rather than send the 64Kbps digitized signals over dedicated voice channels, the digitized voice is packetized and sent over the IP data network. Furthermore, bandwidth efficiencies can be realized if the voice signal is compressed prior to packetization. Voice compression can lower bandwidth demands as far as 4Kbps while maintaining near toll-quality voice. This means that a single

64Kbps channel that normally carries a single voice conversation in the voice network can carry up to 16 voice conversations in the data network.

To support call control, the International Telecommunication Union (ITU-T) has standardized the H.323 protocol. H.323 describes terminals, equipment, and services for multimedia communication over a local area network (LAN), voice being just one service supported. Other recommendations within the H.323-Series include H.225 packet and synchronization; H.245 control; H.261 and H.263 video codecs; G.711, G.722, G.728, G.729, and G.723 audio codecs; and the T.120-series of multimedia communications protocols. Taken as a whole, these recommendations provide the standards to which many backbone, access, and customer premise equipment vendors are developing VoIP components and guaranteeing interoperability. Cisco, for example, has integrated H.323 protocols and voice capabilities into each of its newest routers. Microsoft and Intel also have H.323-compatible products for voice and data communications. The next section provides an overview of product categories.

VoIP PRODUCTS

H.323 VoIP products can be broken out into the following general product categories that loosely map to network layers:

- **Customer premise equipment (CPE):** Originally, H.323 CPE was mostly developed for multimedia communication. Products such as Microsoft Netmeeting, Intel Proshare®, and PictureTel LiveLAN™ were marketed for their desktop video and data sharing capabilities. More recently, however, CPE products have been developed that are more focused on IP-voice communication. The Selsius Ethernet phone, for example, resembles a standard phone but connects to an Ethernet port rather than a PBX port. Symbol Technologies has developed a wireless H.323 telephone that also connects directly into an Ethernet port.
- **Network infrastructure:** Network equipment includes standard routers, hubs, and switches. Voice, however, is very susceptible to network delays and packet loss. Router features such as random early drop/detect, weighed fair queuing, reservation protocol (RSVP), IP precedence, compressed real-time protocol, and multiclass multilink PPP have been developed to address these issues and are prescribed in a converged network. Asynchronous transfer mode (ATM) network equipment also supports quality of service parameters and enables the administrator to set up dedicated private virtual connections (PVCs) between sites.
- **Servers:** A major VoIP benefit is the fact that it builds on the Internet model where there is a clear separation between network infrastructure and network applications. While network infrastructure provides

the packet transport, servers support the applications. The H.323 gatekeeper server, for example, supports call control functions; an accounting, authorization, and administration (AAA) server provides billing and accounting; and network management typically requires a simple network management protocol (SNMP) server.

- **Gateways:** It will still be a long time before one sees the data network handling all voice communication. As an interim step, gateways will be necessary to link the new IP-based voice services with existing public and private voice networks. Gateways also provide a good evolutionary step for companies to selectively choose how and where they will apply VoIP for low risk and high returns.

VOICE AND DATA NETWORK INTEGRATION

The technology, standards, and products exist to migrate voice onto the corporate data network. But where does a company begin in order to minimize network risk and optimize savings? There are two major areas for network cost savings in a voice network: PBX private service (single location and multilocation) and public service (local and long distance). There is also another category representing value-added services that are enabled due to the voice, data, and video integration. These value-added services offer potential for companies to reengineer existing services such as automatic call distribution (ACD) and build new services such as click-to-dial where an Internet user can access a company's operator through the Web.

The following sections discuss each category in further detail. Included is a proposed network integration plan, as well as future value-added services to consider when building the next-generation converged communication network. Refer to Exhibit 18.2 for a summary of the proposed private voice network evolution.

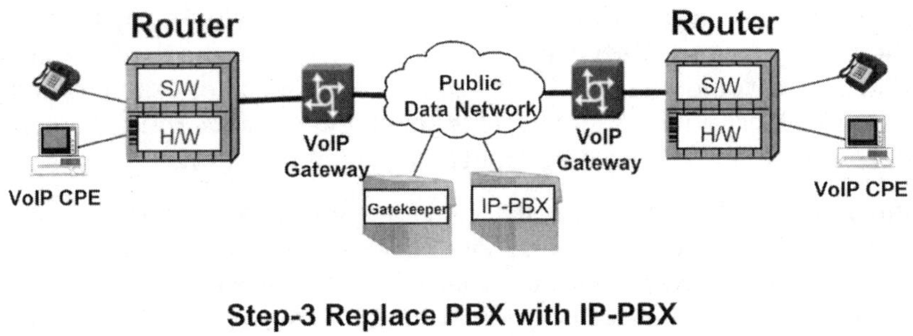

Exhibit 18.2 Stepwise Private Voice Network Evolution.

Private Voice Service

Purchasing a PBX is a big decision for a company. For one thing, it is expensive; and once purchased, the company is locked into the specific PBX vendor for hardware/software upgrades and maintenance. Companies are also limited to the call control features supported by the PBX so that little differentiation is possible between themselves and their competition.

In an IP world, networks are built in layers. Companies first choose the underlying infrastructure, for example, routers, switches, or ATM equipment. Then, companies choose the application protocol — IP for VoIP. Finally, they decide the specific services, servers, and CPE that they wish to run over the network. Taken as a whole, the decision is just as important as the PBX decision. But taken separately, each decision is much lower in cost and risk; new technologies can be introduced with little or no impact to other network layers.

VoIP assumes that there is already a data infrastructure supporting IP services; IP data infrastructure is a mature market with many third- and fourth-generation products. VoIP, however, is in its infancy. The most mature of the products are VoIP gateways providing interworking between the voice and data networks. Gateways terminate voice interfaces from a PBX or public voice switch and convert the signaling and voice to IP protocol. Gateways then send the information across the IP network to either another gateway or a VoIP CPE.

Gateway-to-gateway communication is a good first step for a company introducing VoIP capabilities. Back-to-back gateways provide an alternative to dedicated inter-PBX trunks, typically leased lines, which are very expensive if purchased from a public carrier. Instead, gateway-to-gateway service allows the network administrator to redirect leased-line inter-PBX traffic over the internal corporate IP network. Gateway costs can be recovered in months, depending on the number of locations and distance.

Gateway-to-gateway communication can also be used to link a remote office to the corporate office PBX. The remote office would appear as one or more extensions of the corporate office PBX, and traffic to and from the remote office would be routed through the PBX. Cisco, for example, has integrated VoIP capabilities into both its small remote office routers and larger corporate office routers to support such an application.

Introducing H.323 CPE to the network is a good second step in the corporate network VoIP evolution. This enables IP users to originate and terminate calls directly from their PC, Ethernet phone, or wireless IP-phone. IP devices add a level of network complexity due to the potentially large number of users, security issues, and varying IP device types. The H.323 gatekeeper provides the needed user authorization and routing capabilities.

Many vendors have introduced gatekeepers, both integrated into the router and contained on separate servers. Both solutions offer viable alternatives.

With the combination of gateway, gatekeeper, and IP device, companies can begin to introduce business solutions to work-at-home and CTI applications. Basically, the IP device can be logged into the corporate data network at the same time as calls are originated and terminated from the device. Help Desk operators working from home, for example, would require only a single link into the corporate data network to answer customer questions and access corporate information.

The third major step in evolving the private voice network is fully replacing the PBX with a next-generation IP-PBX. As discussed earlier, PBXs rely on wiring to the desktop and provide hardware and software for call control, accounting, etc. In a VoIP network, Ethernet wiring provides the desktop wiring, the data backbone provides the transport, and servers provide the PBX call control features. The wiring and data backbone exist, while companies such as Selsius and NBX are introducing the server and CPE components. It will not be long before IP-PBX features match and surpass capabilities of current PBXs, and the IP-PBX becomes a viable alternative.

Value-Added Services

One area where IP-PBX features will surpass PBX capabilities will be value-added services. These are services that take advantage of voice, data, and video integration as well as IP network and device capabilities to deliver services not possible on a PBX. Some of these services offer significant cost savings, while others offer revenue-generating opportunities. Following are several value-added service examples to consider when implementing a VoIP network. Note that many other services are possible by adding new servers to the network, similar to adding new web sites to an IP network.

- **Unified messaging.** Most corporate managers receive voicemail, e-mail, and faxes on a daily basis. While timely information is one of the keys to a successful business, it is also difficult to manage the information, especially when it arrives in different forms and locations. Unified messaging consolidates the different information into a single mailbox accessible from a telephone, web site, or pager. The unified messaging server supports media conversion so that e-mails can be read to phone users, faxes can be redirected to local fax machines, and voicemail can be converted to audio files and played back on a computer. Future enhancements may include videomail and interactive content.
- **IP call center/click-to-dial.** The number of Internet users and the amount of Internet commerce are growing faster than most estimates. Intel, for example, just announced that it is generating over $1 billion

in revenue per month directly from the Internet. Cisco also generates over 50 percent of its revenue from direct Internet sales. VoIP offers further opportunity to reach customers through the Internet by enabling direct and immediate interactions with customer service representatives. The click-to-dial application enables users to click a web site button to automatically set up a VoIP call between user and operator. IP call centers extend the model so that the call completes between IP user and IP attendant, thereby lowering 800 service costs.

- **Active directory services.** There is much electronic information within a corporation and contained on public servers. Active directories enable users to interact with the directory, potentially clicking an employee name and connecting to the user through the Internet. This service integrates the data server with VoIP capabilities and strengthens the argument for integrated voice and data services.
- **Virtual second line.** Eventually, VoIP devices will be in widespread use within the corporation and residence. Much of their success will depend on how they are addressed. Dialing an IP address to reach these devices is cumbersome and doomed to failure because it changes the way people are accustomed to making a call. The possibility exists, however, to address the IP device with a standard telephone number. This strategy merges seamlessly with today's telephones and cellular phones, and provides a good evolutionary path to the next-generation communication network.
- **Follow-me/find-me.** Once the IP device has an assigned telephone number, many standard telephone features can be carried over from the PSTN. Call forwarding, call transfer, and three-way calling are examples. Follow-me service allows users to predefine a search list of telephone numbers at which they may be contacted (e.g., work number, home number, and cellular number). When an incoming call arrives, the service rings each number on the list and, once located, transfers the incoming call to one's current location. It is an intuitive step to add the IP device telephone number to the list and direct incoming calls to the IP device when active.
- **Desktop video conferencing.** VoIP specifically refers to handling voice service over the IP network. The benefits of VoIP, however, extend beyond voice. If the devices at either end of a VoIP connection support voice, video, and data, then the underlying signaling will set up a multimedia connection. No user interface change is required; the user still dials a standard telephone number. For example, a company that builds an internal VoIP network can take advantage of full multimedia communication between employees. Gateways can provide the interoperability with older telephone devices and links to the public voice network. Not only are costs reduced because voice services are moved onto the private data network, but internal services are enhanced with video and data as well.

Public Voice Services

Public voice carriers have also recognized the potential cost savings and new revenue from VoIP solutions. Most carriers have ongoing trials to better define market segments and decide where best to apply internal resources. With the convergence of voice, data, and video, additional carriers are entering the fray. Internet service providers (ISPs), competitive local exchange carriers (CLECs), inter-LATA exchange carriers (IXCs), and cable companies have all begun investing in VoIP solutions. Some are looking to enhance existing services, while others are looking to compete with incumbent carriers.

This is all very good news to companies looking for competitive public service rates because increased competition is expected to drive down local and long-distance service rates. Already, VoIP-based companies such as ITXC have begun to offer cut-rate international and long-distance calling card services.

It is also good news for companies evolving their networks to VoIP solutions. In the short term, they can be assured that vendor priority is kept on standards and interoperability. Longer term, as carriers begin offering VoIP public services, company voice traffic can seamlessly transfer from private to public networks without the need for gateways, moving end-to-end via VoIP.

CONCLUSION

The communication paradigm is on the verge of change, signaling a new generation of lower-cost voice, video, and data services. Companies have the opportunity to take advantage of this paradigm shift in a low-risk and cost-effective manner. VoIP is the enabling technology to begin integrating voice, video, and data services onto a single data backbone network. VoIP to PSTN gateways are good first steps to begin the convergence because they are established products with many competitors. Gatekeepers and IP CPE are close behind, with the potential to introduce new value-added services. Finally, there is the opportunity to rebuild the infrastructure with IP-PBXs and public VoIP offerings. This enables ubiquitous voice, video, and data services over a data backbone, and offers services one can only begin to imagine.

Chapter 19
Securing Your Router

Gilbert Held

A ROUTER REPRESENTS AN ENTRY POINT INTO MOST NETWORKS AS WELL AS THE PRIMARY COMMUNICATIONS DEVICE USED TO MOVE DATA BETWEEN NETWORKS. As such, it represents a very strategic communications networking device because the inadvertent or intentional change in its configuration could have a major bearing on the ability of an organization to maintain its network in a desired state of operation. In addition, if routing tables or other parameters are altered, it becomes possible for organizational data to be directed to locations where such information could be recorded and read by third parties. Thus, it is important to understand how one can access and take control of a router, and steps one can employ to secure this communications networking device.

This article examines and discusses methods of router access in both general and specific terms. The discussion of router access in general terms will be applicable to products manufactured by different vendors. However, for specific methods of access and methods one can use to secure access to a router, specific details applicable to routers manufactured by Cisco Systems will be used to supplement the generalizations. The selection of the use of examples specific to routers manufactured by Cisco Systems is based on this vendor currently having approximately a 75 percent share of the market for routers. Although specific examples of methods to protect access to routers in this article will be oriented toward Cisco routers, most routers manufactured by other vendors include similar capabilities. If an organization uses routers manufactured by another vendor, one should check the vendor's router manual for the access security functionality of that router. One should also check the specific commands supported by the router to facilitate one or more access security features for enabling, disabling, and protecting access to the device.

NEED FOR ACCESS SECURITY

When considering router security, most people automatically think of router access lists. Router access lists are used to establish restrictions on the transfer of data through router ports, and are considered by many to represent the first line of defense of a network. Although router access lists are an extremely important aspect of network security, this author considers them to actually represent the second line of defense. This is because the ability to access and configure a router represents the first line of defense. If other than designated personnel obtain the ability to access and change the configuration of organizational routers, this means that any previously developed access lists can be altered or removed — in effect, stripping away any previously developed network protection. Similar to the farmer who constructs a solid henhouse but inadvertently goes home at the end of the day and leaves the door ajar, failure to secure access to organizational routers permits predators of the two-legged variety to gain access to valuable resources.

Probing deeper into router access, one will note that there are several methods to bar the proverbial door to this communications device. In fact, one method to be discussed involves the use of an access list as a mechanism to control access to the router to certain predefined IP addresses. However, prior to doing so, one must lock the door, which should be accomplished prior to the use of the access list capability of the router. Thus, the use of access lists should be viewed as a second line of defense.

ROUTER ACCESS

For the purposes of this article, the term "router access" represents the ability of a person to connect to a router and gain access to its operating system. Most routers include one or more serial ports built into the device that permit terminals or personal computers operating a specific type of terminal emulator to gain access to the router. This terminal access can occur directly via a direct cable connection, or remotely via a communications path that results in a modem or DSU connected to a router's serial port. Although the use of a serial port connection is the primary method used by most organizations to provide access to a router's operating system to enable the device to be configured, it is not the only method of access. Additional methods supported by many routers include Telnet access and the use of the Trival File Transfer Protocol (TFTP) to store and transmit system images and configurations files to and from routers and workstations.

Telnet Access

Telnet provides the ability to access a remote device to include a router as if the terminal device operating a Telnet client program was directly connected to the remote device. Telnet access to a router can occur from in front or behind a router. The term "in front" is used to reference access to a router via a wide area network connection from a station located on another network not directly connected to a specific router, while the term "behind" references a station located on a network directly connected to a router's local area network port. This means that Telnet access to a router can occur from a device located on a local organizational network or, if the router is connected to the Internet, from virtually any terminal device in the world that has Internet access. This also means that regardless of the location of the client operating the Telnet program, the operator of the program only needs to know the IP address of the network interface of the router to attempt to initiate a Telnet session to the router and gain access. If the operator of the Telnet client makes a connection to the router, the operator will receive a prompt, such as "routername" (where routername represents the name an organization assigned to the router).

It should be noted that many organizations have an IP addressing policy where they assign low addresses to router interfaces. For example, if an organization's Class C IP network address was 205.123.124.0, it might assign 205.123.456.1 as the address of the interface from the 205 network to the router. Due to this common scheme of addressing used by many organizations, it is often easy to determine the address of a router for subsequent Telnetting attempts. At this point in time, a Telnet client operator may be able to directly access all of the router's configuration capabilities and, in effect, take over control of the router. As an alternative, the Telnet client operator may be prompted to enter a password to gain access to the router. Concerning the latter, many routers are configured at the factory to have a default password for Telnet access. Unfortunately for many organizations that should know better, one should never use a default password. This is because such passwords are listed in the vendor's router manual, which may be available for purchase for $29.95, or available for access via the World Wide Web for free. This means that a virtually unlimited number of persons have the ability to discover the default password needed to access a router via a Telnet connection.

If the router administrator fails to change the default Telnet password or does not place any additional restrictions on Telnet access, anyone with knowledge of the IP address of the router interface can gain access to the device. As noted later in this article, if the router administrator overlooks another aspect of device access control, a security hole big enough to literally move a truck full of hackers will be opened, enabling literally millions of remote users to take control of an organization's router.

TFTP Access

Most routers have two types of memory: conventional random access memory (RAM) and nonvolatile memory. Unlike conventional RAM, wherein contents are erased upon the removal of power, the contents of nonvolatile memory remain in place. When configuring a router, nonvolatile memory is commonly used to store an image of router memory as well as backup or alternative router configurations. Because routers do not contain diskettes, nor do they have hard drives, their ability to store more than one or perhaps a few alternative configurations is severely limited.

This means that administrators who require the ability to store backup or alternative router configurations beyond the capacity of the limited amount of router nonvolatile memory typically do so on a workstation and use TFTP to load and save router system images and configuration files. This also means that if TFTP access is enabled, depending on how the router supports TFTP access, it may be possible for unauthorized persons to create configuration data that, when used by the router, results in a breach of security or an unintended operational environment.

Now that one has an appreciation for the main methods that can be used to gain access to a router, one can focus on the methods that can be used to either protect such access or literally lock the door on the access method, making it extremely difficult for nonauthorized persons to gain access to a router and obtain the ability to view and possibly change the configuration of the device. In doing so, this article will also, when applicable, discuss certain Cisco Systems router commands.

SECURING CONSOLE AND VIRTUAL TERMINALS

After unpacking a router and initiating its installation process, it is extremely important to consider the manner by which access to configuring the device will occur. If one only plans to enable configuration changes to occur from a directly connected terminal device, then one should ensure that Telnet and TFTP access are disabled. In a Cisco router environment, one can configure access from the console and virtual terminals via the use of the *line* command. That command has the format

```
line [type-keyword] first-line [last-line]
```

where information in brackets represent options. The type-keyword entry can be either *console, aux,* or *vty.*

The *console* entry is used to represent a console terminal line, a device directly cabled to a port on the router. In comparison, *aux* is used to indicate an auxiliary line and allows one to specify access via a port on the router connected to a CSU, DSU, or modem, permitting serial communications from afar. The third option, *vty,* represents a virtual terminal connection for remote console access. Note that when entering the line command, the first and last lines represent a number of contiguous entries that are applicable to a specific device and can be represented and associated with a line number.

When configuring access through the use of the *line* command, it is also important to consider associating a password with the device that one enables for access. Even if one only plans to allow access to a router via a directly cabled terminal device located in a secure technical control center, every once in awhile a situation can occur that would justify password protection. In one event this author is familiar with, a tour of the technical control center of a government network by a group of Boy Scouts resulted in one extremely inquisitive individual inadvertently causing a bit of havoc. As the rest of the group moved to an area of the technical control center to view a graphical display of the status of the network, this individual started playing with a terminal that was directly cabled to a router and that functioned as the router console. Not knowing what to enter, the Boy Scout entered a question mark (?), which resulted in the display of router commands. Within a short period of time, this Boy Scout managed to misconfigure the router while the rest of the group was on the other side of the center listening to a briefing by the manager of the center. Needless to say, if a password was previously associated with terminal access, the unintentional misconfiguration of the router and the resulting havoc it created would not have been possible.

In a Cisco router environment, one can associate a password with a remote access method. To do so, one would use the password command. For example, the line console password "bugs4bny" would block console access until the console operator responded with the password "bugs4bny" to a prompt generated by the router for a password.

The password associated with the Cisco password command can be up to 80 characters in length. The password is case sensitive and can contain any combination of alphanumeric characters, including spaces. While this capability provides the router administrator with the ability to be innovative, it also provides the ability to make it extremely difficult for authorized users to gain access. This is because selecting a password based on a large number of varying upper- and lower-case letters mixed with numerics makes it subject to erroneous entry. While this type of password will certainly be difficult to guess and should avoid the possibility of a successful dictionary attack, it is also easy for an authorized router administrator to enter incorrectly. If incorrectly entered three times, a Cisco router will return the terminal attempting access to the idle state of operation. Thus, when selecting a password, it is important to remember several password principles. First, use a mixture of alphabetic and numeric characters to alleviate the potential of a dictionary attack being successful. Second, when structuring a password, remember that as one extends the length of the password, one also increases the possibility of password entry error. In general, passwords that are between six and eight characters in length should be sufficient if they are structured to join a few abbreviated words with a sequence of numerics.

FILE TRANSFER

As previously noted, TFTP is commonly supported by routers as a mechanism to permit system image and configuration files to be stored on workstations. In a Cisco router environment, to enable the loading of network configuration files at router reboot time, one must specify the *service config* command. Fortunately, the default is the disabling of this capability. If this capability is enabled, the router will broadcast a TFTP read request message and the first station to respond will have the file with a specific name based on the router's configuration loaded into the router across the network. Because a standardized file-naming scheme is used, this author believes it is best to consider leaving this feature in its disabled state instead of opening the ability for inquisitive employees with a bit of knowledge to "see what would happen" if they create a configuration file.

INTERNAL ROUTER SECURITY

Once access is gained into a router, the operating system of the device can provide a further level of protection capability that one can use for additional router access security. In a Cisco router environment, the command interpreter included in the operating system is referred to as the EXEC. The EXEC has two levels of access: user and privileged.

The user level of access allows a person to use a small subset of all router commands, such as commands that enable the listing of open router connections, commands for providing a name to a logical connection, and displaying certain statistic concerning router operations. In comparison, the privileged level of access includes all user access commands as well as commands that govern the operation of the router, such as the "configure" command that allows a router administrator to configure the router, the "reload" command that halts the operation of the device and reloads its configuration, and similar commands that have an active effect on the operational state of the device.

Due to the ability of a person gaining access to the privileged mode of operation of a Cisco router to obtain the ability to directly control the operation of the router, this level of access can also be password-protected. Thus, when installing a Cisco router, it is important to use the *enable-password* configuration command to protect access to the privileged level of router access. For example, to assign the password "power4you" for the privileged command level, one would use the *enable password* command as follows:

```
enable password power4you
```

Similar to the password associated with a serial terminal line, the password assigned to the privileged command level is case sensitive, can contain any mixture of alphanumeric characters to include spaces, and can consist of up to 80 characters. Thus, by placing a password on the serial port or on any allowed virtual terminal connections, as well as on the privileged command level of the router, one protects both access into the router and the use of privileged commands once access is obtained.

ADDITIONAL PROTECTIVE MEASURES

If one needs to provide one or more persons on a network with the ability to configure one or more routers, one can add an additional layer of protection beyond passwords. To do so, one can program one or more router access lists. Although the use of router access lists will be the subject of a subsequent article by this author, one notes that they represent a sequential collection of permit and deny conditions that can be applied to Internet addresses. This means that if one can determine the IP addresses of stations that will require the ability to have operators configure one or more routers via a network connection, then one can use the access list capability of each router to restrict Telnet access to each router to one or more specific IP addresses. This means that not only does the terminal operator need to know the correct passwords to gain access to an appropriate router, but in addition, the operator can only perform such access from predefined locations. By combining password protection into a router with password protection to its privileged mode of operation, and restricting configuration access to predefined locations via the use of one or more access lists, one can, in effect, close the proverbial door to the router.

Section IV
Remote Access

Chapter 20
Modem Technology Update

Gilbert Held

IN MID-1996, MODEM DESIGNERS REALIZED THAT BY ELIMINATING ONE OF A NORMAL PAIR OF DIGITAL-TO-ANALOG CONVERSIONS OCCURRING ON THE SERVER SIDE OF A CLIENT/SERVER SESSION, a considerable amount of line noise associated with conventional transmission over the switched telephone network could be eliminated. This, in turn, enabled modem designers to achieve a higher data rate in one direction without violating Shannon's Law because the reduction in line noise allows a higher operating rate to be achieved. Unfortunately, two competing modem technologies that do not interoperate with one another were developed.

To add an additional problem to the many that the IS department has to contend with, the asymmetrical transmission capability of this recently developed series of modems advertised as 56K-bps devices rarely provides anywhere near that capability. Thus, it is important for IS support staff to obtain a firm understanding of the operational capabilities and limitations of the emerging series of 56K-bps modems and the effort to standardize this new communications product. After reading this article, IS staffers will be better prepared to evaluate the potential use of 56K-bps modems in their network infrastructure, as well as understand the capabilities and limitations of this new communications device.

AN OVERVIEW OF 56K-BPS MODEM CAPABILITIES

Until the announcement of 56K-bps modems, the highest operating rate obtainable on the switched telephone network was limited to the 33.6K-bps rate associated with V.34 modems. The ability of 56K-bps modems to obtain a significant improvement over the operating rate of V.34 modems is due to the increasing digital infrastructure of the telephone network. Today, almost 100 percent of the communications infrastructure that links one telephone company office to another has been converted to a digital transmission facility. Thus, a call made between two telephone subscribers will be carried as an analog transmission on the local loop between the call originator and the telephone company central office serving the subscriber.

At that location, the call is digitized via an analog-to-digital (A/D) conversion process and routed over the digital infrastructure that interconnects telephone company offices. After being routed to the telephone company office that serves the destination or called party, a digital-to-analog (D/A) conversion occurs, and the signal is transferred to the destination over another local loop as an analog signal.

In the preceding example, a call made between two subscribers over the public switched telephone network requires two signal conversions. One conversion occurs when an analog signal is digitized for transmission over the telephone company digital infrastructure that interconnects their offices, while the second conversion occurs when the digital signal is converted back into an analog signal. Because the telephone company infrastructure is digital, modem designers noted that Internet Service Providers (ISPs), or a corporation's remote access server, could be directly connected as a digital data stream to a special server-side modem capable of accepting digital data directly at data rates up to 56K bps.

This server side modem would then pass the digital data stream through the telephone company's digital infrastructure to the central office serving the subscriber. At that location, the data would be converted into an analog signal for transmission onto the local loop connecting the subscriber to the central office. In the reverse direction, data transmitted by a client-side modem would be modulated for transmission over the local loop to the telephone company's central office where it is then digitized for transmission over the digital infrastructure until it reaches the central office serving the ISP or a corporation's remote access server. At that location, the signal is demodulated by the server-side modem and passed in a digital format to the ISP or corporate remote access server.

Exhibit 20.1 illustrates the operation of 56K-bps modems. These devices have the capability, theoretically, to operate at data rates of up to 56K bps downstream as a digital signal passes through the modem, in a format compatible with the standard Pulse Code Modulation (PCM) signal used to digitize voice conversations, to the central office serving the client subscriber where it is then converted into an analog signal. The downstream link results in only one A/D conversion. Thus, some noise is eliminated that is normally associated with dual A/D and D/A conversions, enabling a higher operating rate to be achieved.

In the upstream direction from the client to the server, data is first modulated based on the modem modulation method being used. Although the analog signal is digitized at the serving telephone company central office for transmission over the digital infrastructure of the telephone network, it cannot be directly understood in its digital form by ISP equipment or by a remote access server. Thus, upon reaching the destination central office,

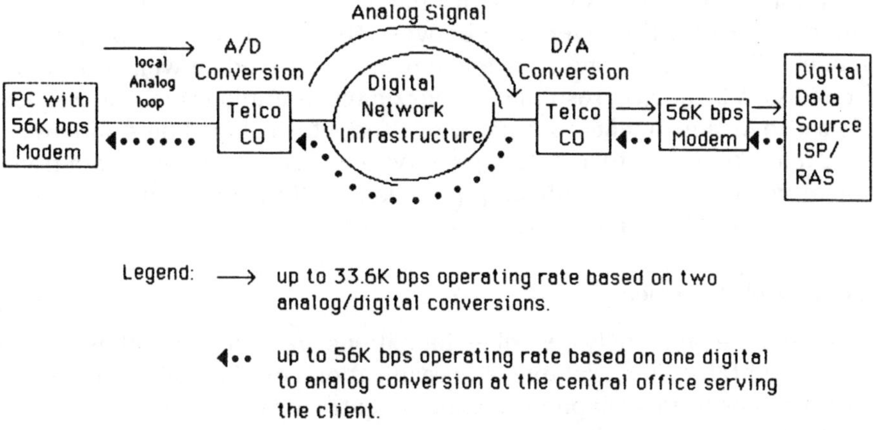

Exhibit 20.1 56K-bps Modem Operation.

the signal must be converted back into an analog format and demodulated by a compatible modulation scheme so it can be converted into an appropriate digital signal understood by digital equipment. This second conversion limits the data transmission rate in the upstream direction to the V.34 operating rate of 33.6K bps, which represents the theoretical limit based on Shannon's Law.

LIMITATIONS OF 56K-bps MODEMS

The use of 56K-bps modems involves an asymmetrical transmission method in which downstream data may be able to obtain a higher data transfer rate than data sent in the opposite direction. The actual ability to obtain a 56K-bps downstream data transmission capability depends on a number of factors, including current FCC regulations, the direct connection of ISPs to the digital side of 56K-bps modems, and a limitation of one conversion of the digital signal between the originator and recipient.

The FCC has a rule that restricts the output power of signals placed onto the public switched telephone network. The essence of this rule is to limit the upper boundary of the signal-to-noise ratio, which results in the lowering of the outbound or downstream 56K-bps modem's data rate to 53K bps.

The second key limitation associated with the use of 56K-bps modems is the fact that the ability to obtain a data transmission rate above 33.6K bps depends on the direct connection of ISP or a corporate remote access server via the installation of a digital leased line to the digital modem. Thus, the user cannot just dial any telephone number. Instead, the user must connect to a digital modem that is directly connected to a digital data source.

A third limitation associated with the use of 56K-bps modems involves the number of A/D conversions between the transmitting and receiving modem. If more than one conversion occurs, the modem will, at best, operate as a V.34 device, resulting in a maximum operating rate of 33.6K bps. This also means that 56K-bps modems cannot be installed on standard analog lines for a remote access capability to a LAN and obtain a data transmission rate above 33.6K bps, because dialing from home or when traveling would result in two A/D conversions.

Compatibility Issues

Although the previously described limitations are of concern, they are not insurmountable. For example, an organization could install a T1 or fractional T1 line from a telephone company central office to a corporate office and connect each 64K-bps channel at 56K bps to a 56K-bps digital modem. That modem in turn would be connected to a remote access server port. Then, remote users using 56K-bps modems could access the corporate LAN and download data at 53 K bps (due to current FCC restrictions). However, the preceding example assumes that the organization purchased all of its 56K-bps modems from the same vendor.

Competitive Technologies

Today, there are two competitive 56K-bps modem technologies. One technology, referred to as X2, was developed by U.S. Robotics. A second technology for achieving a 56K-bps downstream data rate was jointly developed by Lucent Technologies and Rockwell Semiconductor, Inc. Referred to as K56 Flex, this method is not compatible with the U.S. Robotics X2 technology. Although both modems support standardized V.34 operations at data rates up to 33.6K bps, their use of incompatible digital technologies means that the use of one modem to access the Internet or a remote LAN that is connected to a modem using a different method of achieving a downstream digital data link will force the downstream operation into a conventional modem modulation method.

Thus, the use of 56K-bps modems to obtain data rates beyond V.34's 33.6K-bps operating rate has two basic restrictions. First, if the pair of modems use the same technology, the maximum data rate obtainable will be restricted to 53K bps downstream to the client, while the upstream rate will be restricted to the maximum analog transmission rate, typically the 33.6K bps of V.34 modems or their fallback rates of 28.8K bps, or 26.0K bps when line noise adversely affects modem operations. Second, if the modems use different digital signaling technology, the maximum downstream and upstream data transmission rates will be fixed at the V.34 maximum data rate.

Evolving Standards

In mid-1997, U.S. Robotics controlled approximately 30 percent of the market for modems used on the public switched telephone network. In comparison, Rockwell Semiconductor modem chips were used in approximately 60 percent of all modems, with about 10 percent of modems manufactured using other vendor-designed chips. This means that for the foreseeable future, there will be two dominant manufacturers of incompatible 56K-bps modems — those manufactured by Rockwell Semiconductor as chips using Lucent technology, and products developed by U.S. Robotics or using U.S. Robotics technology.

A 56K-bps modem should be selected based upon the type of modem used at the dialed location. That is, if an organization uses 56K-bps modems to enhance its use of the Internet, the modem should be selected based on the type of modem used by the organization's Internet Service Provider (ISP). In this situation, the organization must either use the same modem technology as that used by the ISP, or be willing to change the ISP. If an organization uses 56K-bps modems as a mechanism to provide remote access to the network, IS staffers have an additional degree of control that allows them to select a particular 56K-bps modem technology. In this situation, IS can specify both the client and server modems to be used, which allows greater control over compatibility.

TIA Standards

Recognizing the problems associated with two competing 56K-bps modem technologies, the Telecommunications Industry Association (TIA) standards committee, TR-30.1, initiated the development of an interim standard for 56K-bps modems. Recognizing that such modems operate downstream by passing PCM data, the standard is called "TIA Interim Standard" for PCM Modems.

The standardization effort being performed by the TIA is designed to produce a faster development process that will enable international standards organizations, such as the International Telecommunications Union (ITU), to develop an international recommendation before the three-year life of an interim standard expires. The ITU formed a study group that has already used some of the work accomplished by the TIA. An interim TIA standard concerning the operation of PCM modems might be agreed to by the end of 1997, with the ITU issuing a recommendation for the operation of PCM modems sometime during 1998.

RECOMMENDED COURSE OF ACTION

The use of 56K-bps modem technology includes some risks as well as potential rewards. The key risk is the fact that the standards that evolve could result in any investment made today soon becoming obsolete. To minimize this risk, IS should consider purchasing upgradeable modems whose operating characteristics can be modified through the use of flash RAM, replacement of ROM chips, or the download of a file into the modem's memory. Other risks, such as the inability to achieve a true 56K-bps operating rate downstream and the limitation of one A/D conversion process, are not as serious.

The ability to download data to clients at 53K bps instead of 56K bps is still a considerable improvement over the 33.6K-bps maximum operating rate of a V.34 modem. IS should carefully review the potential use of 56K-bps modem technology. Once the ability to obtain an approximate 50 percent improvement in downstream data transfer capability is determined to justify the use of this new technology, modems that are upgradeable should be acquired and users should be briefed on their operational capability. Doing so will allow users to obtain the benefits of 56K-bps modem technology, prepare for an upgrade to an evolving standard, and become acquainted with the capabilities of the technology.

Chapter 21
Choosing Asymmetrical Digital Subscriber Lines

Gilbert Held

DIGITAL SUBSCRIBER LINES (DSLs) ENABLE HIGH-SPEED TRANSMISSIONS — including interLAN communications, videoconferencing, and mainframe access — over existing copper twisted-pair local loops. By determining the availability, operating rate of the service, and the capability of the carrier's internal broadband network, users can decide whether DSL service is a cost-effective transport replacement for their current service.

INTRODUCTION

Although all long-distance communications in North America are essentially now transported via fiberoptic cable, the majority of what is referred to as "the last cable mile" — the 18,000-foot local loop from the telephone company central office (CO) to subscribers — continues to be twisted-pair copper cable. Because of the high cost of installing fiber cable from telephone company central offices to individual residential and business subscribers, communications carriers are exploring methods that enable high-speed transmission to occur over their existing copper twisted-pair local loops. The result is the development of a series of digital subscriber lines.

Although several DSL products, including high data-rate digital subscriber lines (HDSL), single-line digital subscriber lines (SDSL), and very high data-rate digital subscriber lines (VDSL) are discussed in this chapter, the focus is on asymmetrical digital subscriber line (ADSL) technology because it is the most practical transmission facility. In addition, ADSL is being used in a series of field trials conducted by telephone companies around the world.

ASYMMETRICAL DIGITAL SUBSCRIBER LINES (ADSL)

ADSL is a technology for use over telephone twisted-pair loops in which data rates between 1.544 and 6M bps, and in some cases up to 8M bps, can be obtained from the carrier's central office to the subscriber (referred to as downstream transmission), while a transmission rate of 640K bps to 800K bps is obtained in the opposite direction (referred to as upstream transmission). Because the operating rate differs in each direction, this technique is referred to as asymmetrical transmission.

ADSL modems use either discrete multitone (DMT) or carrierless amplitude/phase (CAP) modulation. DMT was standardized by the American National Standards Institute (ANSI) as standard T1.413. CAP was developed by Paradyne, formerly a unit of AT&T and now an independent company. Although DMT is standardized, CAP is presently deployed in a number of field trials throughout the world for high-speed Internet access and has received high marks for its performance. It is quite possible that communications carriers will select both of the competitive methods to obtain an ADSL capability.

Discrete Multitone (DMT) Operation

DMT permits the transmission of high-speed data over conventional copper twisted-pair wire without adversely affecting voice communications. Through the use of frequency division multiplexing (FDM), a copper twisted-pair subscriber line is subdivided into three parts by frequency. One channel is assigned for downstream data transmission and a second is assigned for upstream data transmission. The first two channels have a variable amount of bandwidth based on the results of the ADSL modems communicating with one another. The third channel, which uses a fixed spectrum of bandwidth from 0 to 4 KHz, is used for normal telephone operations. Thus, an ADSL line simultaneously supports the bidirectional transmission of high-speed data and conventional voice connection.

In Exhibit 21.1, the ADSL modem is shown with two digital connections, one to a LAN and another to a mainframe. An ADSL modem can support virtually unlimited digital connections; however, most modems used in field trials have been limited to two to four connectors. The ADSL modem supports multiple digital devices through the use of time-division multiplexing (TDM), which can be used to subdivide the upstream and downstream data channels by time into independent digital connections.

Under the DMT standard, the frequency spectrum above 10 KHz is subdivided into a large number of independent subchannels. Because the amount of attenuation at high frequencies depends on the length of the subscriber line and its wire gauge, an ADSL modem based on DMT technology installed at a central office must first determine which subchannels are

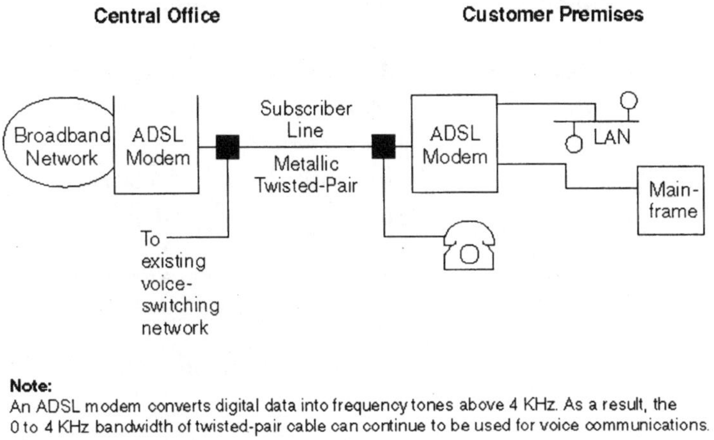

Exhibit 21.1 Using an Asymmetrical Digital Subscriber Line.

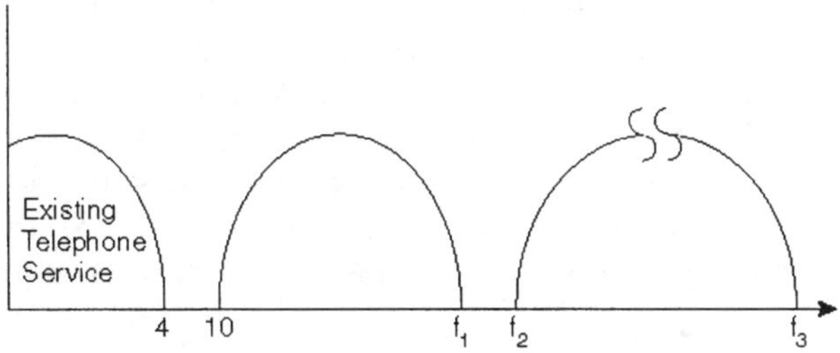

Exhibit 21.2 ADSL Modem Use of the Copper Twisted-Pair Frequency Spectrum.

usable. The modem transmits tones to the remote modem installed at the customer's premises, where the tones are analyzed. The remote modem responds during this initiation process at a relatively low data rate, which significantly reduces the possibility of the signal analysis performed by the remote modem being misinterpreted.

On receipt of the returned signal analysis, the ADSL modem located in the telephone company's central office will use up to 256 4-KHz-wide subchannels for downstream transmissions. Through a reverse learning process, the remote modem will use up to 32 4-KHz-wide subchannels for upstream transmission. Exhibit 21.2 illustrates the use of the copper twisted-pair frequency spectrum by an ADSL modem. In Exhibit 21.2, f1, f2, and f3 are variables based on the results of the ADSL modem's discrete multitone handshaking process; however, the actual limit for f3 is 1.1 MHz,

Exhibit 21.3 Generalized ADSL Performance.

Wire Gauge	Subscriber Line Distance	Operating Rate
24 AWG	18,000 feet	1.5/2.0M bps
26 AWG	15,000 feet	1.5/2.0M bps
24 AWG	12,000 feet	6.1M bps
26 AWG	9,000 feet	6.1M bps

Note: AWG is the American Wire Gauge standard for measuring connectors.

which represents the maximum usable bandwidth on a copper twisted-pair cable.

Benefits of DMT. One of the main benefits of DMT is its ability to take advantage of the characteristics of twisted-pair wire, which can vary from one local loop to another. This makes DMT modulation well suited for obtaining a higher data throughput than is possible through the use of a single carrier transmission technique. Because the transmission rate varies depending on the wire gauge and cable length as well as the physical characteristics of the cable, ADSL performance can vary considerably from one local loop to another. Ignoring bridge taps, which represent sections of unterminated twisted-pair cable connected in parallel across the twisted-pair cable under consideration, various tests of ADSL lines provide a general indication of their operating rate capability, which is summarized in Exhibit 21.3.

Carrierless Amplitude/Phase (CAP) Modulation

CAP modulation is a derivative of Quadrature Amplitude Modulation (QAM) which was developed by Paradyne when it was a subsidiary of AT&T, and which is used in most modems today. Unlike DMT, which subdivides the bandwidth of the twisted-pair wire into 4-KHz segments, CAP uses the entire bandwidth in upstream and downstream channels.

Under CAP, serial data is encoded by mapping a group of bits into a signal constellation point using a two-dimensional, eight-state trellis coding with Reed-Solomon forward error correction. Reed-Solomon automatically protects transmitted data against impairments due to crosstalk, impulse noise, and background noise.

In the CAP modulation process, once a group of bits is mapped to a predefined point in the signal constellation, the in-phase and quadrature filters implement the positioning in the signal constellation. Because this technique simply adjusts the amplitude and phase without requiring a constant carrier, the technique is referred to as "carrierless."

The ADSL modem developed by Paradyne uses a 256-point signal constellation for downstream operations. This constellation pattern is in the 120 KHz to 1,224 KHz frequency range and produces a composite signaling rate of 960K baud, with 7 bits packed into each signal change. Although the resulting downstream operating rate is 6.72M bps, the use of the Reed-Solomon forward error correction code reduces the actual payload to 6.312M bps, plus a 64K-bps control channel. In the upstream direction, the Paradyne ADSL modem uses a CAP-16 line code in the 35-KHz to 72-KHz frequency band to obtain a composite signaling rate of 24K baud across 16 subchannels. By packing three bits per signal change, an upstream line rate of 72K bps is obtained, of which 64K bps is available for data. Similar to an ADSL DMT modem, an ADSL CAP modem can use time-division multiplexing to derive multiple channels on the upstream and downstream channels.

ADSL Developments and Field Tests

Although the ANSI DMT standard defines an operating rate of 6M bps on local loops up to 12,000 feet in length, some vendors now offer rates approaching 8M bps downstream and up to 800K bps upstream. This additional data rate is accomplished through an automatic rate adaptation capability that correlates the connected data rate to distance and line noise, enabling additional bandwidth to be gained — which translates into a higher operating rate.

In the US, GTE Telephone Operations (Stamford, CT) began testing the use of ADSL technology as part of a public data trial in the Dallas-Fort Worth area during 1996. This trial tests downstream data transmission at operating rates of up to 4M bps, and upstream transmission at data rates up to 500K bps. The testing involves the Irving Public Library system, several bookstores, and area GTE employees. ADSL modems used during this trial are provided by Western Technologies of Oswego, IL, and Aware, Inc. of Bedford, MA.

On the West Coast, U.S. WEST (Olympia, WA) is testing ADSL-based services through its ENTERPRISE business unit, and will use equipment from several vendors before deciding on which equipment to deploy commercially. Other communications carriers are expected to announce field trials as well as commercial deployment of ADSL technology soon.

ADDITIONAL DSL TECHNOLOGIES

In addition to ADSL, three related technologies warrant a degree of discussion:

- High data-rate digital subscriber line (HDSL).
- Single-line digital subscriber line (SDSL).
- Very high data-rate digital subscriber line (VDSL).

HDSL modems are used on either end of one or more pairs of twisted-pair wires to obtain a T1 or E1 operating rate. Currently, a T1 operating rate requires two lines, and E1 requires three. HDSL is primarily used on college campuses to provide a T1 or E1 link to interconnect LANs.

SDSL represents HDSL over a single telephone line and is used where multiple lines do not exist or would be expensive to install.

VDSL modems for twisted-pair access operate at data rates from 12.9 to 52.8M bps on 24-gauge cable. Because of the higher operating rate, the maximum transmission distance supported by VDSL is limited to under 4,000 feet. Higher rates over shorter distances may limit VDSL's ability to support communications carrier customers.

VDSL standards may not be finalized for several years. There are four VDSL line coding methods — discrete multitone, discrete wavelet multitone, carrierless amplitude/phase modulation, and simple line coding. DMT for VDSL is similar to the technology used in the ADSL standard, with more carriers (supporting ADSL) being the primary difference between the two. Discrete wavelet multitone coding represents a multicarrier system that uses wavelet transformers for individual carriers. The carrierless amplitude/phase modulation method under consideration is similar to Paradyne's CAP, representing a version of suppressed carrier quadrature amplitude modulation. The fourth coding technique, simple line coding, represents a baseband signaling and filtering system.

Based on the higher transmission capacity of VDSL, it will probably be deployed by communications carriers to businesses as a mechanism for asynchronous transfer mode (ATM) connectivity. The ATM Forum, which produced a 51.84M-bps standard for a private user-to-network interface, is currently examining the potential delivery of ATM via VDSL transmission.

CONCLUSION

From the viewpoint of a corporate network manager, ADSL is probably the only digital subscriber line technology that will be offered by enough carriers in the next few years to deserve consideration for interconnecting geographically separated locations. Although two different types of ADSL technology, based on either DMT or CAP modulation, are being considered by communications carriers, from the perspective of the end user, the most important factors are:

- Availability of service
- Operating rate of the service
- Ability of the carrier's internal broadband network to provide connectivity

The availability of ADSL will govern whether or not the technology can be used to interconnect LANs, provide remote access to mainframes, support videoconferencing between locations, or perform other communications functions. Although it may be possible, for example, to use an ADSL connection in one location and a conventional T1 connection in another location for communications, carriers still have not addressed the conversion of data between different transmission facilities.

The second area of concern is the operating rate of ADSL service. Because the transmission rate varies — depending on line quality, the distance of the user's location from a carrier's central office, and the wire gauge of the wire pair — users cannot fully appreciate its transmission capability until an ADSL line becomes operational. Vendor specifications should be used as a guide and ADSL should be considered for applications that require a high data transmission capability but that can also be effectively performed if only a fraction of that capability is actually achieved.

Thus, videoconferencing or LAN access to the Internet are two applications for which ADSL may prove effective, even if the user achieves half or even a quarter of the theoretical ADSL rate due to a bad subscriber loop, long subscriber wire distance from the carrier's central office, or a similar factor.

A third area that should be examined before considering ADSL service is the scope of the carrier's internal broadband network. The user should determine whether the network interconnects only the ADSL offerings of that carrier, if it provides connectivity to other carriers, and if it provides connectivity to other data services offered by that carrier. By doing so, the user can determine if it will be possible to use ADSL services as a transport mechanism for such data services as X.25 and frame relay.

By contacting their communications carriers, users can ascertain when ADSL may be deployed in their areas and its projected cost. By asking the carrier key questions concerning the availability of the service in different areas where an organization may have offices, the operating rate of the service, and the capability of the carrier's internal broadband network, users can determine if ADSL service will be suitable for trial. If so, a test plan should be developed to compare ADSL with traditional T1 service to determine if it can be used as a cost-effective transport replacement.

Chapter 22
Preparing for Cable Modems

Gilbert Held

AS THE INTERNET GAINS IN POPULAR USAGE WITHIN CORPORATE AMERICA, organizations are trying to establish ways to connect to the Information Superhighway at speeds that make economic sense. This chapter describes the operation of cable modems, devices that provide high-speed connectivity and enable the organization to take advantage of transmission rates of up to tens of millions of bits per second. Cabling infrastructure — the cabling within buildings — is another key component defined and discussed to introduce the data center operations manager to a rapidly evolving new technology that can help keep the organization competitive.

PROBLEMS ADDRESSED

During 1995, the use of the Internet expanded considerably, with tens of thousands of corporations, universities, government agencies, and individuals creating home pages on servers, while tens of millions of users surfed the World Wide Web. As corporations began to recognize the value of the Internet for building software applications, promoting products and services, and locating and disseminating information, the addition of graphics to World Wide Web home pages literally slowed web surfing operations to a crawl, adversely affecting user productivity. Whereas the replacement of 14.4 Kbps modems by state-of-the-art 28.8 Kbps devices has assisted many users in speeding up their Internet search operations, even at that operating rate the display of a typical web page containing one or two graphic images can result in a delay of 10 to 15 seconds as the picture is "painted" on a monitor.

Recognizing the operating limitations associated with transmissions via the public switched telephone network, as well as looking for an additional source of revenue, several cable television (CATV) companies initiated broadband access trials to the Internet during 1995. Each of these trials involves the use of cable modems, which enables personal computers (PCs) to access the Internet via a common CATV coaxial cable at operating rates up to tens of millions of bits per second. Although cable modems are in their infancy, both independent market research organizations and many cable operators predict that the installed base of this new type of communications device will rapidly grow to over 10 million modems within a few years.

Due to the advantages associated with obtaining high-speed Internet access, as well as the potential economics associated with the use of cable modems to obtain such access, data center managers should consider preparing their facilities for the infrastructure required to use cable modems.

This chapter discusses the nature of cable modems and describes their operation. The scope of the discussion also includes the cabling infrastructure being developed to provide a megabit transmission facility to residences and businesses. The chapter outlines the cabling requirements for installation within buildings, and the requirements to access this new high-speed information highway via the use of cable modems. The data center manager should have a background of knowledge concerning a rapidly evolving new technology and be able to support its use when corporate policy begins to include Internet issues.

MODEM FUNDAMENTALS

The ability to appreciate why cable modems are able to provide a transmission capability that is an order of magnitude or more than conventional modems used for transmission on the switched telephone network requires knowledge of certain transmission concepts, including the Nyquist Theorem. This section concentrates on the operation of conventional analog modems that are used on the switched telephone network. This can provide the data center manager with an understanding of why analog modems' operating rates are limited, and how they may be able to overcome that operating rate limitation.

A conventional analog modem commonly used to transmit information over the switched telephone network is limited to a maximum operating rate of between 28.8 Kbps and 33.6 Kbps, with the rate achievable dependent upon the quality of the connection and according to the modulation technique employed. In theory, the maximum operating rate of an analog modem designed for use on the switched telephone network is limited by the 4 KHz bandwidth provided by the communications carrier for a switched telephone channel.

In 1924 Nyquist proved, in what is now referred to as the Nyquist Theorem, that the maximum signaling rate of a device is limited to twice the available bandwidth; beyond that rate, intersymbol interference starts to occur and adversely affects the transmission. As an example, for the 4 KHz telephone channel, this means the maximum signaling rate of a modem used to transmit on that medium is limited to 8,000 baud. "Baud" is a term used to indicate signal changes per second.

THE QUADRATURE AMPLITUDE MODULATION TECHNIQUE

The most commonly used modem modulation technique, Quadrature Amplitude Modulation (QAM), uses a combination of phase and amplitude to convey the settings of a group of bits in one signal change, enabling four bits to be represented by one baud change. This in turn enables an 8,000 baud signaling rate to transport data at a rate of 32 Kbps when QAM is used for modulation.

Due to the 4 Khz telephone channel limitation, however, data transmission rates are limited to approximately 32 Kbps, with a slightly higher rate of 33.6 Kbps recently achieved by a few modem vendors using a modified QAM technique. Although the incorporation of data compression into modems provides a potential doubling to quadrupling of modem throughput, to between 67.2 Kbps and 134.4 Kbps, the ability of a modem to compress data depends upon the susceptibility of data to the compression algorithm being used. Because that susceptibility varies considerably as a modem user performs different operations, the end result is a variable compression rate; even though it is not noticeable during file transfer operations, that variable rate becomes extremely noticeable during interactive operations. In addition, even with the ability to compress data at a high rate, the resulting information transfer rate of 134.4 Kbps pales by comparison to the operating rate obtainable through the use of cable modems. It is clear, however, that advances in modem and cabling technology are limited with respect to increasing the performance of modems used to communicate via the switched telephone network.

CABLE MODEMS

The key difference between an analog modem designed for use on the public switched telephone network and a cable modem is in the bandwidth of the channels they are designed to use. Cable TV uses RG-11 cable for the main CATV trunk, and RG-59 cable from trunk distribution points into and through residences and offices. Both types of coaxial cable have 75 ohms impedance and support broadband transmission, which means that two or more channels separated by frequency can be simultaneously transported on the cable.

From Unidirectional to Bidirectional Systems

A cable TV broadcasting infrastructure uses 6 MHz channels within the bandwidth of RG-11 and RG-59 cable to transmit a TV channel. Most CATV systems are currently unidirectional, which means that TV signals are broadcast from the CATV system operator without any provision for receiving a return signal. This transmission limitation is gradually being overcome as CATV operators begin to add bidirectional amplifiers to their networks that, when they are installed, will support transmission from subscribers in the reverse direction to conventional TV signal broadcasts. This will enable CATV systems to support the standardized transmit frequency range of 5 Mhz to 42 Mhz, and receive a frequency range of 54 Mhz to 550 Mhz, with 6 Mhz cable TV channels.

By using one or more 6 Mhz cable TV channels, a cable modem obtains the use of a bandwidth that is 1,500 times greater (6 Mhz/4 Khz) than that provided by a voice channel on the switched telephone network. This means that the modem can support a signaling rate of twice the bandwidth, or 12 Mbaud, on one TV channel based upon the Nyquist Theorem, before the occurrence of intersymbol interference.

The primary difference between cable modems currently in field trials are in their use of one or more 6 Mhz TV channels within the band of channels carried by a coaxial cable, and their methods of attachment to the CATV network. One cable modem manufactured by Zenith Network Systems, a subsidiary of Zenith Electronics of Glenview, IL, operates on 6 Mhz channels at 4 M bps to the subscriber, using a special filtering technique to prevent data channels from interfering with adjacent information, which can be in the form of either data or video, that would coexist with the data transmission provided by the cable modem. The uplink or return data rate occurs at 500 Kbps. Modem modulation is biphase shift key (BPSK), which means that two bits (bi) are encoded in each phase change, and the modem's phase changes are shifted in phase from one to another. This modem is also frequency-agile, which means it can be set to operate on any standardized channel on a broadband CATV system.

The Zenith cable modem is actually one portion of a series of components required for a PC to use the modem. A complete transmission system requires the use of a Zenith cable modem, Ethernet 10BASE-T adapter card with a 15-conductor pin connector, and a 15-conductor shielded cable to connect the cable modem to the adapter. Exhibit 22.1 illustrates the cabling required to connect a PC to a CATV network via the use of a Zenith Network Systems cable modem.

When the adapter card is installed in the PC, it in effect turns the computer into a client workstation. Because the adapter is an Ethernet 10BASE-T card, the channel being used by the cable modem operates as one long CSMA/CD LAN, with each PC user competing with other PC users for access to the channel. This means that the CATV operator should segment its cable distribution system to limit the number of cable modems attached to any segment, similar to the manner in which conventional LANs are limited with respect to the maximum number of workstations that can be connected.

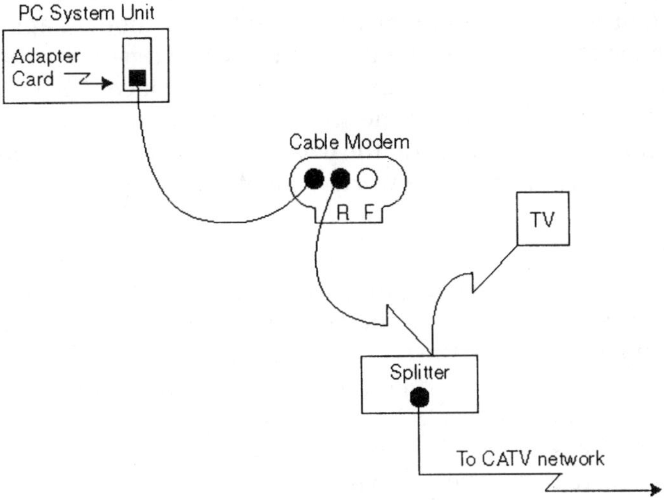

Exhibit 22.1 Cabling for a Zenith Network Cable Modem System.

The connector labeled "R" on the rear of the cable modem is a reverse cable connector designed for networks that use a single coaxial cable. The second connector, labeled "F," represents a forward cable connector that would be used if the modem were connected to a cable system that uses two cables. In such a system, one cable is dedicated to conventional CATV broadcasting through one-way amplifiers, which precludes reverse transmission on the same cable. This type of system also requires the use of a second cable to obtain a transmission capability in the reverse direction.

A High-Speed Cable Modem Architecture

In addition to the previously described cable modem based upon the exclusive use of RF technology and biphase shift key modulation, Zenith Electronics Corporation announced a high-speed cable modem architecture. This architecture is based on the use of 16-VSB (vestigial sideband), a technique developed by Zenith as part of the organization's high-definition research, as well as the 256 quadrature amplitude modulation technology. Through the use of more complex modulation techniques for which more data bits can be represented by one signal change, the Zenith modem architecture can support data rates up to 40M bps on a 6 Mhz cable channel.

Recognizing the fact that many cable TV systems will be limited to one-way transmission for the foreseeable future because of the time and cost associated with upgrading the CATV infrastructure, Zenith plans to support a range of options and speeds for upstream data transmission. According to Zenith, both telephone (analog modulation) and RF return path transmission capabilities will be incorporated into different versions of this new family of cable modems. For many cable modem applications, such as Internet operations, the use of the switched network for a return path should provide an acceptable level of performance. The rationale for this is best noted by examining the communications interaction between a potential cable modem user and the cable network as a user searches and accesses various points on the World Wide Web.

On the Web

When users access a web page, they transmit a universal resource locator (URL) address that represents the document they wish to view. This address is transported using the HTTP within a packet. The HTTP consists of an address that totals fewer than 100 characters, which are used to frame the address the message is being transported to and the address of the originator of the request. The destination web server uses the document address to locate the requested page on the server, retrieves it from disk, and forms a packet using the source address from the incoming packet as the destination address for the outgoing packets. If the requested document contains a full screen of text, the packet contain close to 2,000 characters because a full screen of text consists of 80 columns by 24 rows of data (i.e., 1,920 characters). However, because a typical web page contains one or more graphics, the total amount of data transmitted from the server to the user will actually be substantially more than 2,000 characters. For example, it is assumed that the web page in question includes a 3 × 3-inch photograph, drawing, or schematic diagram that has been scanned using a resolution of 300 dots per inch. Regardless of the color of the image, each square inch of the image requires 11,250 bytes of storage. If the image was scanned using a 256-color resolution, each pixel requires a byte to represent its color, resulting in 90,000 bytes of storage per square inch. Thus, a 3 × 3-inch color image requires 270,000 bytes of storage.

Because HTTP breaks large files into small packets for transmission, the image might be carried by a sequence of approximately 100 packets, each roughly 2,700 bytes in length, to include packet overhead. Thus, the short, 100-character transmission from a user can result in a response of 280,000 bytes. Because a user connected to the Web typically clicks on hotlinks that represent document addresses to view other documents, most web operations represent asymmetrical transmission, that is, more transmissions return to the user than the user actually originates. Thus, a high speed cable channel with a low speed reverse path occurring over the switch telephone network may actually be sufficient for most data transmission applications.

The previously described asymmetrical transmission operation of users was also recognized by Intel Corporation, which took it into consideration when designing its CablePort cable modem system. That cable modem is designed to provide an average downstream data rate of 27M bps and a 96 Kbps upstream rate. One interesting difference between Zenith and Intel concerning their cable modem systems is in the type of adapter card required in the PC. Because Intel provides a higher downstream operating rate than what is usable by a 10BASE-T adapter card, the user must install a Fast Ethernet (100M bps) adapter card in the PC to be able to use the Intel cable modem. Although no commercial costs were provided by Zenith or Intel for field trial operations, it is worth noting that a Fast Ethernet adapter has a retail cost of approximately $250, whereas a 10BASE-T adapter can be obtained for less than $50.

A second difference between the Zenith and Intel modems concerns their upstream capability. Although Zenith's new architecture permits support of the switched telephone network for locations where CATV operators cannot provide reverse direction transmission, the Intel system did not offer this capability when this chapter was researched.

RECOMMENDED COURSE OF ACTION

Although the technology of cable modems is in its infancy, the data center manager can still plan for their use. Whereas the type of cable modem offered will depend upon the CATV operators' cable infrastructure (i.e., either unidirectional or bidirectional), as well as the cable modem vendor the data center manager selects, each cable modem requires the use of RG-11 coaxial cable. Thus, if the manager has previously installed such cabling as part of a legacy terminal-to-mainframe or legacy LAN wiring system and is considering its removal, he or she may wish to leave the cabling in place. If RG-11 cabling has not been installed, the manager may wish to consider contacting the local CATV operator to determine when cable modems will be supported and the type the operator intends to use. If it intends to use bidirectional transmission via RF modulation, the data center manager can develop a wiring plan that requires only the use of RG-11 cable. If the CATV operator says it intends to provide reverse transmission via the public switched telephone network, the wiring plan must be modified to ensure that each cable modem user will have an available telephone jack. By understanding how cable modems operate and planning the organization's wiring infrastructure to use this evolving technology, the data center manager will be prepared for its future use.

Chapter 23
Virtual Private Networks: Secure Remote Access over the Internet

John R. Vacca

THE COMPONENTS AND RESOURCES OF ONE NETWORK OVER ANOTHER NETWORK ARE CONNECTED VIA A VIRTUAL PRIVATE NETWORK (VPN). As shown in Exhibit 23.1, VPNs accomplish this by allowing the user to tunnel through the Internet or another public network in a manner that lets the tunnel participants enjoy the same security and features formerly available only in private networks.

Using the routing infrastructure provided by a public internetwork (such as the Internet), VPNs allow telecommuters, remote employees like salespeople, or even branch offices to connect in a secure fashion to an enterprise server located at the edge of the enterprise local area network (LAN). The VPN is a point-to-point connection between the user's computer and an enterprise server from the user's perspective. It also appears as if the data is being sent over a dedicated private link because the nature of the intermediate internetwork is irrelevant to the user. As previously mentioned, while maintaining secure communications, VPN technology also allows an enterprise to connect to branch offices or to other enterprises (extranets) over a public internetwork (such as the Internet). The VPN connection across the Internet logically operates as a wide area network (WAN) link between the sites. In both cases, the secure connection across the internetwork appears to the user as a private network communication (despite the fact that this communication occurs over a public internetwork); hence the name Virtual Private Network.

VPN technology is designed to address issues surrounding the current enterprise trend toward increased telecommuting, widely distributed global

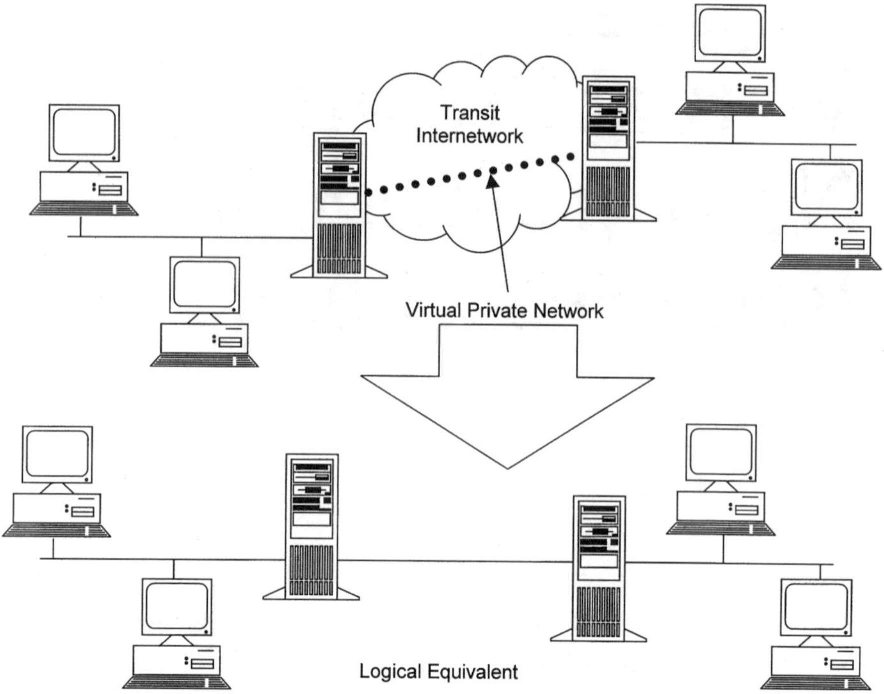

Exhibit 23.1 Virtual Private Network.

operations, and highly interdependent partner operations. Here, workers must be able to connect to central resources and communicate with each other. And, enterprises need to efficiently manage inventories for just-in-time production.

An enterprise must deploy a reliable and scalable remote access solution to provide employees with the ability to connect to enterprise computing resources regardless of their location. Enterprises typically choose one of the following:

- an IT department-driven solution, where an internal information systems department is charged with buying, installing, and maintaining enterprise modem pools and a private network infrastructure
- value-added network (VAN) solutions, where an enterprise pays an outsourced enterprise to buy, install, and maintain modem pools and a telco infrastructure

The optimum solution in terms of cost, reliability, scalability, flexible administration and management, and demand for connections is provided by neither of these traditional solutions. Therefore, it makes sense to find a

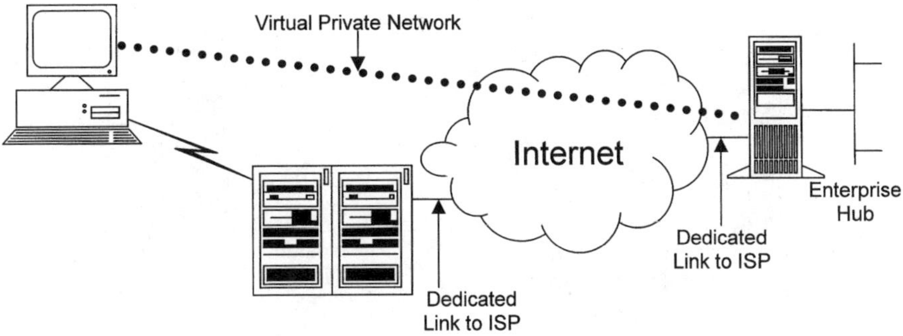

Exhibit 23.2 Using a VPN to Connect a Remote Client to a Private LAN.

middle ground where the enterprise either supplements or replaces its current investments in modem pools and its private network infrastructure with a less-expensive solution based on Internet technology. In this manner, the enterprise can focus on its core competencies with the assurance that accessibility will never be compromised, and that the most economical solution will be deployed. The availability of an Internet solution enables a few Internet connections (via Internet service providers, or ISPs) and deployment of several edge-of-network VPN server computers to serve the remote networking needs of thousands or even tens of thousands of remote clients and branch offices.

VPN Common Uses

The next few subsections of this chapter describe common VPN situations in more detail.

Secure Remote User Access over the Internet. While maintaining privacy of information, VPNs provide remote access to enterprise resources over the public Internet. A VPN that is used to connect a remote user to an enterprise intranet is shown in Exhibit 23.2. The user first calls a local ISP Network Access Server (NAS) phone number, rather than making a leased-line, long-distance (or 1-800) call to an enterprise or outsourced NAS. The VPN software creates a virtual private network between the dial-up user and the enterprise VPN server across the Internet using the local connection to the ISP.

Connecting Networks over the Internet. To connect local area networks at remote sites, there exist two methods for using VPNs: using dedicated lines to connect a branch office to an enterprise LAN, or using a dial-up line.

Using Dedicated Lines to Connect a Branch Office to an Enterprise LAN. Both the branch office and the enterprise hub routers can use a local dedicated

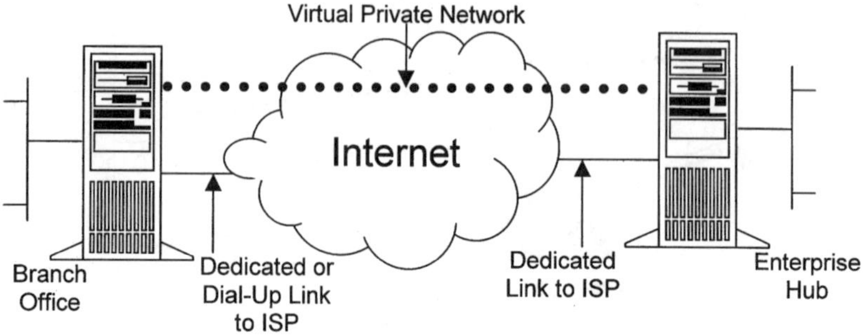

Exhibit 23.3 Using a VPN to Connect Two Remote Sites.

circuit and local ISP to connect to the Internet, rather than using an expensive long-haul dedicated circuit between the branch office and the enterprise hub. The local ISP connections and the public Internet are used by the VPN software to create a virtual private network between the branch office router and the enterprise hub router.

Using a Dial-Up Line to Connect a Branch Office to an Enterprise LAN. The router at the branch office can call the local ISP rather than having a router at the branch office make a leased-line, long-distance call to an enterprise or outsourced NAS. Also, in order to create a VPN between the branch office router and the enterprise hub router across the Internet, the VPN software uses the connection to the local ISP as shown in Exhibit 23.3.

The facilities that connect the branch office and enterprise offices to the Internet are local in both cases. To make a connection, both client/server, and server/server VPN cost savings are largely predicated on the use of a local access phone number. It is recommended that the enterprise hub router that acts as a VPN server be connected to a local ISP with a dedicated line. This VPN server must be listening 24 hours a day for incoming VPN traffic.

Connecting Computers over an Intranet

The departmental data is so sensitive that the department's LAN is physically disconnected from the rest of the enterprise internetwork in some enterprise internetworks. All of this creates information accessibility problems for those users not physically connected to the separate LAN, although the department's confidential information is protected.

VPNs allow the department's LAN to be separated by a VPN server (see Exhibit 23.4), but physically connected to the enterprise internetwork. One should note that the VPN server is not acting as a router between the enterprise internetwork and the department LAN. A router would interconnect

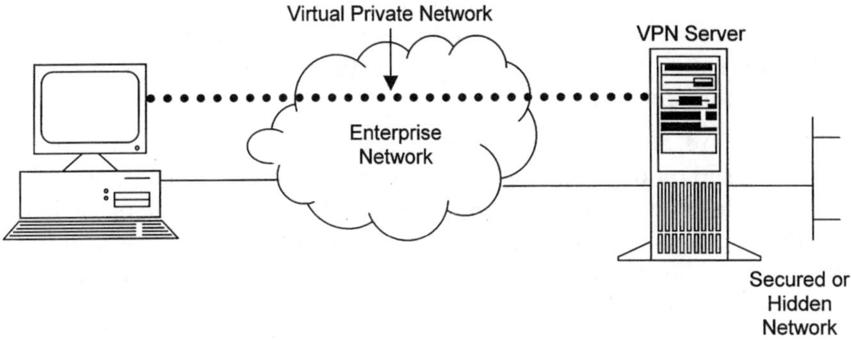

Exhibit 23.4 Using a VPN to Connect to Two Computers on the Same LAN.

the two networks, thus allowing everyone access to the sensitive LAN. The network administrator can ensure that only those users on the enterprise internetwork who have appropriate credentials (based on a need-to-know policy within the enterprise) can establish a VPN with the VPN server and gain access to the protected resources of the department. Additionally, all communication across the VPN can be encrypted for data confidentiality. Thus, the department LAN cannot be viewed by those users who do not have the proper credentials.

BASIC VPN REQUIREMENTS

Normally, an enterprise desires to facilitate controlled access to enterprise resources and information when deploying a remote networking solution. In order to easily connect to enterprise local area network (LAN) resources, the solution must allow freedom for authorized remote clients. And, in order to share resources and information (LAN-to-LAN connections), the solution must also allow remote offices to connect to each other. Finally, as the data traverses the public Internet, the solution must ensure the privacy and integrity of data. Also, in the case of sensitive data traversing an enterprise internetwork, the same concerns apply. A VPN solution should therefore provide all of the following at a minimum:

- *address management:* The solution must assign a client's address on the private net, and must ensure that private addresses are kept private.
- *data encryption:* Data carried on the public network must be rendered unreadable to unauthorized clients on the network.
- *key management:* The solution must generate and refresh encryption keys for the client and server.
- *multiprotocol support:* The solution must be able to handle common protocols used in the public network; these include Internet Protocol (IP), Internet Packet Exchange (IPX), etc.

- *user authentication:* The solution must verify a user's identity and restrict VPN access to authorized users; in addition, the solution must provide audit and accounting records to show who accessed what information and when.

Furthermore, all of these basic requirements are met by an Internet VPN solution based on the Point-to-Point Tunneling Protocol (PPTP) or Layer 2 Tunneling Protocol (L2TP). The solution also takes advantage of the broad availability of the worldwide Internet. Other solutions meet some of these requirements, but remain useful for specific situations, including the new IP Security Protocol (IPSec).

Point-to-Point Tunneling Protocol (PPTP)

PPTP is a Layer 2 protocol that encapsulates PPP frames in IP datagrams for transmission over an IP internetwork, such as the Internet. PPTP can also be used in private LAN-to-LAN networking.

PPTP is documented in the draft RFC, "Point-to-Point Tunneling Protocol."[1] This draft was submitted to the IETF in June 1996 by the member enterprises of the PPTP Forum, including Microsoft Corporation, Ascend Communications, 3Com/Primary Access, ECI Telematics, and U.S. Robotics (now 3Com).

The Point-to-Point Tunneling Protocol (PPTP) uses Generic Routing Encapsulation (GRE), encapsulated Point-to-Point Protocol (PPP) frames for tunneled data, and a TCP connection for tunnel maintenance. The payloads of the encapsulated PPP frames can be compressed as well as encrypted. How a PPTP packet is assembled prior to transmission is shown in Exhibit 23.5. The illustration shows a dial-up client creating a tunnel across an internetwork. The encapsulation for a dial-up client (PPP device driver) is shown in the final frame layout.

Layer 2 Forwarding (L2F)

L2F (a technology proposed by Cisco Systems, Inc.) is a transmission protocol that allows dial-up access servers to frame dial-up traffic in PPP and transmit it over WAN links to an L2F server (a router). The L2F server then unwraps the packets and injects them into the network. Unlike PPTP and L2TP, L2F has no defined client.[2]

Layer 2 Tunneling Protocol (L2TP)

A combination of PPTP and L2F make up L2TP. In other words, the best features of PPTP and L2F are incorporated into L2TP.

L2TP is a network protocol that encapsulates PPP frames to be sent over Asynchronous Transfer Mode (ATM), IP, X.25, or Frame Relay networks. L2TP can be used as a tunneling protocol over the Internet when configured to use IP as its datagram transport. Without an IP transport layer, L2TP can also be used directly over various WAN media (such as Frame

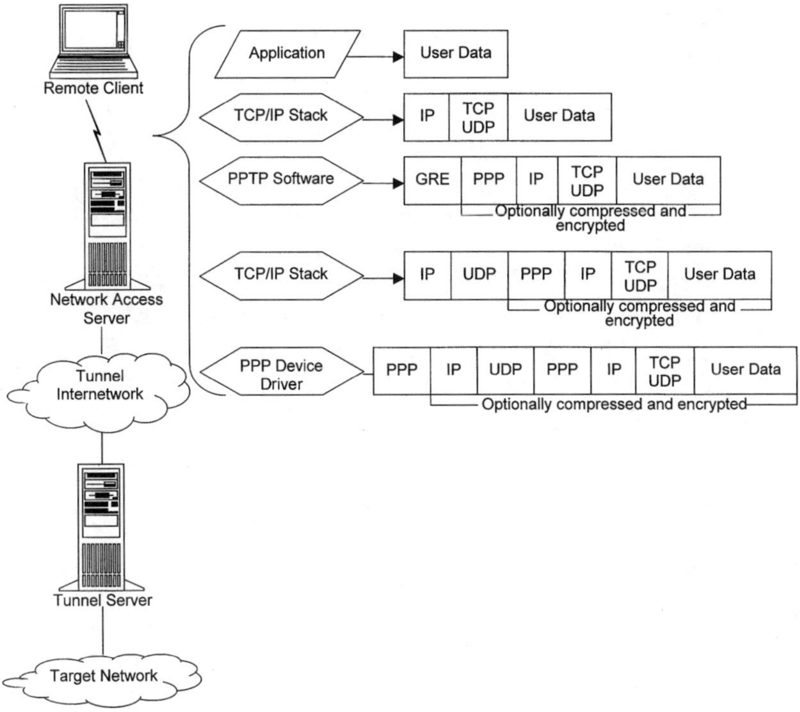

Exhibit 23.5 Construction of a PPTP Packet.

Relay). L2TP is documented in the draft RFC, Layer 2 Tunneling Protocol "L2TP." This document was submitted to the IETF in January 1998.

For tunnel maintenance, L2TP over IP internetworks uses UDP and a series of L2TP messages. As the tunneled data, L2TP also uses UDP to send L2TP-encapsulated PPP frames. The payloads of encapsulated PPP frames can be compressed as well as encrypted. How an L2TP packet is assembled prior to transmission is shown in Exhibit 23.6. A dial-up client creating a tunnel across an internetwork is shown in the exhibit. The encapsulation for a dial-up client (PPP device driver) is shown in the final frame layout. L2TP over IP is assumed in the encapsulation.

L2TP Compared to PPTP. PPP is used to provide an initial envelope for the data for both PPTP and L2TP. Then, it appends additional headers for transport through the internetwork. The two protocols are very similar. There are differences between PPTP and L2TP, however. For example,

- L2TP provides for header compression. When header compression is enabled, L2TP operates with four bytes of overhead, as compared to six bytes for PPTP.

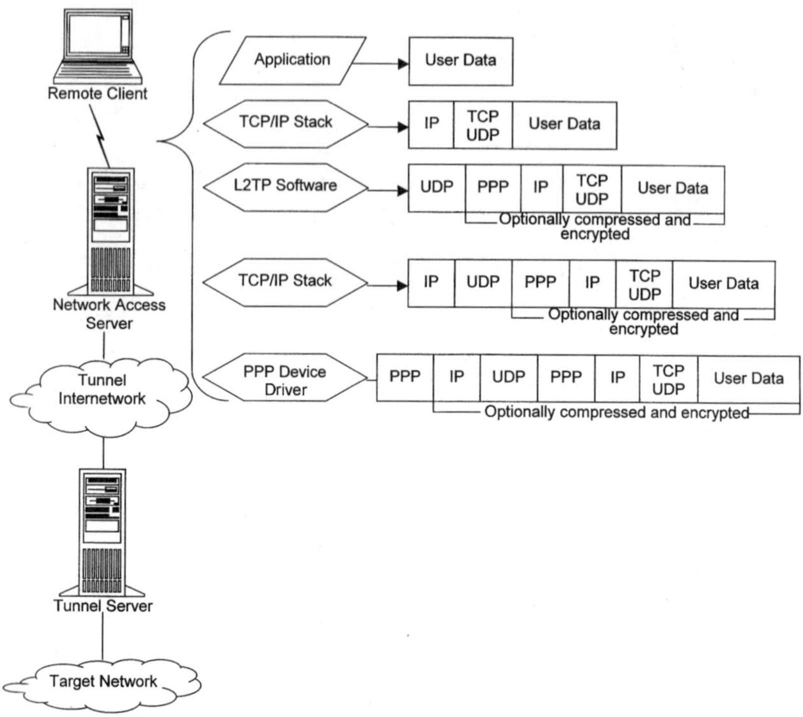

Exhibit 23.6 Construction of a L2TP Packet.

- L2TP provides for tunnel authentication, while PPTP does not. However, when either protocol is used over IPSec, tunnel authentication is provided by IPSec so that Layer 2 tunnel authentication is not necessary.
- PPTP can only support a single tunnel between endpoints. L2TP allows for the use of multiple tunnels between endpoints. With L2TP, one can create different tunnels for different qualities of service.
- PPTP requires that the internetwork be an IP internetwork. L2TP requires only that the tunnel media provide packet-oriented point-to-point connectivity. L2TP can be used over IP (using UDP), Frame Relay permanent virtual circuits (PVCs), X.25 virtual circuits (VCs), or ATM VCs.

Internet Protocol Security (IPSec) Tunnel Mode

The secured transfer of information across an IP internetwork is supported by IPSec (a Layer 3 protocol standard). Nevertheless, in the context of tunneling protocols, one aspect of IPSec is discussed here. IPSec defines the packet format for an IP over an IP tunnel mode (generally referred to as IPSec Tunnel Mode), in addition to its definition of encryption mechanisms for IP traffic. An IPSec tunnel consists of a tunnel server and tunnel client. These are both configured to use a negotiated encryption mechanism and IPSec tunneling.

For secure transfer across a private or public IP internetwork, IPSec Tunnel Mode uses the negotiated security method (if any) to encapsulate and encrypt entire IP packets. The encrypted payload is then encapsulated again with a plaintext IP header. It is then sent on the internetwork for delivery to the tunnel server. The tunnel server processes and discards the plaintext IP header, and then decrypts its contents to retrieve the original payload IP packet. Upon receipt of this datagram, the payload IP packet is then processed normally and routed to its destination on the target network. The following features and limitations are contained within the IPSec Tunnel Mode:

- It is controlled by a security policy: a set of filter-matching rules. This security policy establishes the encryption and tunneling mechanisms available in order of preference and the authentication methods available, also in order of preference. As soon as there is traffic, the two machines perform mutual authentication, and then negotiate the encryption methods to be used. Thereafter, all traffic is encrypted using the negotiated encryption mechanism and then wrapped in a tunnel header.
- It functions at the bottom of the IP stack; therefore, applications and higher-level protocols inherit its behavior.
- It supports IP traffic only.

The remainder of this article discusses VPNs and the use of these technologies by enterprises to do secure remote access (e.g., by traveling employees and sales reps) over the Internet in greater detail.

EASY TO MANAGE AND USE

While squeezing the maximum possible from budget and support staffs, today's enterprises are asking their Information Technology groups (ITGs) to deliver an increasing array of communication and networking services. It appears that the situation is no different at Microsoft Corporation (Redmond, WA). The Microsoft ITG needed to provide secure, Internet-based remote access for its more than 35,000 mobile sales personnel, telecommuters, and consultants around the world.

Microsoft's ITG is currently using and deploying a custom Windows-based remote dial-up and virtual private networking (VPN) solution by using Windows-based clients and enhanced Windows 2000® RAS (Remote Access Server) technology available in the Windows 2000 Option Pack (formerly named Windows NT 5.0). Users are given quick, easy, and low-cost network access. Additional user services are provided with new Windows-based network services from UUnet Technologies, Inc.[3]

Integrated RAS-VPN Clients

According to Microsoft, its ITG has learned that the widespread adoption and use of technology largely depends on how easy and transparent the experience is for the end user. Likewise, Microsoft's ITG has learned not to deploy technologies for which complexity results in an increased support burden on its limited support staff. Microsoft's ITG provided a single client interface with central management to simultaneously make the remote access solution easy to use and manage.

Single Client. A single client is used for both the direct dial-up and virtual private network connections. Users utilize the same client interface for secure transparent access, whether dialing directly to the enterprise network or connecting via a VPN, by using Windows integrated dial-up networking technology (DUN) and Microsoft Connection Manager. In fact, users do not need to concern themselves with which method is employed.

Central Management. Central management is used for remote dial-up and VPN access phone numbers. According to Microsoft, its ITG has found that one of the most common support problems traveling users face is determining and managing local access phone numbers. This problem translates into one of the principal reasons for support calls to Microsoft's user support centers. Using the Connection Manager Administration Kit (CMAK) wizard (which is part of Microsoft's remote access solution), Microsoft's ITG preloads each client PC with an electronic phone book that includes every dial-up remote access phone number for Microsoft's network. The Windows solution also allows phone books to be centrally integrated and managed from a single remote location, and clients to be updated automatically.

WINDOWS COMMUNICATION PLATFORM

In order to provide a flexible and comprehensive network solution, the open extensibility of the Windows 2000 allows Microsoft's ITG to preserve its current hardware network investments while partnering with UUnet Technologies, Inc. According to Microsoft, the Windows platform enabled its ITG to integrate the best-of-breed network services and applications to best meet its client and network administration needs.

High-Speed Internet Access on the Road

Microsoft employees can also connect to high-speed Internet access by plugging into public IPORT[4] jacks in hotels, airports, cafes, and remote locations. The Microsoft ITG integrates the IPORT[5] pay-per-use Internet access features into its custom remote access solution. According to Microsoft, this high-bandwidth, easily available connection helps Microsoft employees be more productive and have a better online experience while on the road.

Secure Internet Access and VPN

Microsoft's ITG, like its counterparts at every enterprise, must ensure that the edge of its network is secure while still providing all employees with the freedom needed to access information worldwide. Microsoft's ITG has also deployed Microsoft Proxy Server to securely separate the LAN from the Internet to meet this need.

To ensure that no intruders compromise the edge of network, the Microsoft Proxy Server firewall capabilities protect Microsoft's network from unauthorized access from the Internet by providing network address translation and dynamic IP-level filtering. Microsoft's ITG uses the powerful caching services in Microsoft Proxy Server to expedite the delivery of information at the same time.

The Proxy Server is able to service subsequent user requests of already-requested information without having to generate additional network traffic by reusing relevant cached information. In addition, in order to operate at peak efficiency with the utmost security, ITG uses Microsoft Proxy Server to enable the Microsoft intranet and remote employees.

RAS Reporting and Internal Usage Chargeback (Billing)

Microsoft pays a substantial amount for remote access fees due to the need to maintain private leased lines and dedicated 800 numbers like many large enterprises with a multitude of branch offices and remote employees. In addition, according to Microsoft, the sheer number of LAN entry points and autonomy afforded its international divisions made centralized accounting and retail reporting for remote access use and roaming users important.

Microsoft's ITG is deploying a VPN solution — bolstered with centralized accounting and reporting of enterprisewide remote access and VPN use — by using Windows 2000, integrated user domain directory, and RADIUS services. As part of this solution, Microsoft is also deploying TRU RADIUS Accountant™ for Windows 2000 from Telco Research.[6]

Furthermore, Microsoft's ITG is also able to generate detailed reporting of remote access and VPN network use for internal cost-accounting purposes while using familiar Windows 2000 management tools by using Telco Research's product. In addition, Microsoft's ITG is able to quickly and easily deploy a turnkey reporting solution built on the intrinsic communication services of Windows 2000 in this manner. According to Microsoft, while maintaining the flexibility to accommodate future change, they receive better security, reduced implementation costs, and enhanced reporting to improve remote access management and chargeback service.

VIP Services: Economical Internet Access And VPN

By working with UUnet Technologies, Inc. (the largest Internet service provider in the world), the Microsoft ITG supplemented its private data network infrastructure and RAS with VPN services. Microsoft's VPN solution is integrated with the UUnet Radius Proxy servers through the Windows 2000 native support for RADIUS under this relationship.

Through the Windows 2000 Remote Access Service integrated RADIUS support, Microsoft's ITG made reliable and secure local access to UUnet Technologies' IP network available to all Microsoft mobile employees. This resulted in the delivery of high-quality VPN services over the UUnet Technologies, Inc. infrastructure at a reduced cost. The ITG conservatively estimates that this use of VPN service as an alternative to traditional remote access will save Microsoft more than $7 million per year in remote access fees alone. Additional savings are expected from the elimination of call requests for RAS phone numbers, and greatly reduced remote access configuration support.

The ITG utilized the integrated support for RADIUS-based authentication available from the Windows Directory in Windows 2000. This allowed them to retain all existing authentication rights for both Internet and LAN access, avoiding change or redundant replication of directory, and provided for enhanced network security.

According to Microsoft, their ITG was able to instantly extend network access to its more than 50,000 employees in more than 100 countries through its relationship with UUnet Technologies. So that Microsoft employees can access information locally anywhere with reliability guarantees and the support of UUnet, UUnet Technologies' transcontinental backbone provides access throughout North America, Europe, and the Asia–Pacific region.

PLANNING FOR THE FUTURE

Finally, Microsoft's ITG wanted to ensure that its current investment in the remote access infrastructure would not only be able to meet today's needs, but also enable it to make the most of opportunities provided by the digital convergence of network-aware applications in the near future. Evidence of an increased need for higher degrees of client/server network application integration is found in the momentum of Windows 2000 as a platform for IP telephony, media-streaming technologies, and the migration to PBX systems based on Windows 2000.

The flexibility needed to address current and future needs of Microsoft's ITG economically is provided through the use of Windows 2000 as the backbone of the remote access solution. Through partnerships with multiple service providers such as UUnet Technologies, the selection of a Windows-based solution allows ITG the freedom to both centrally manage and incrementally extend the Microsoft direct-dial and VPN infrastructure at a controlled pace and in an open manner.

In order to connect Microsoft subsidiaries, branch offices, and extranet partners securely to the enterprise network over private and public networks, Windows 2000 Routing, RAS, and VPN services — along with tight integration with Microsoft Proxy Server — are already enabling Microsoft's ITG to seamlessly extend its RAS–VPN infrastructure. Furthermore, to meet Microsoft's enterprise needs into the future, the broad application support enjoyed by the Windows communication platform ensures that ITG will continue to have access to a host of rich application services made available by developers and service providers such as ATCOM Inc., Telco-Research, and UUnet Technologies, Inc.

CONCLUSION AND SUMMARY

Windows 2000's native VPN services allow users or enterprises to reliably and securely connect to remote servers, branch offices, or other enterprises over public and private networks. Despite the fact that this communication occurs over a public internetwork in all of these cases, the secure connection appears to the user as a private network communication. Windows VPN technology is designed to address issues surrounding the current enterprise trend toward increased telecommuting and widely distributed global operations, where workers must be able to connect to central resources, and where enterprises must be able to communicate efficiently with each other.

This chapter provided an in-depth discussion of virtual private networking, and described the basic requirements of useful VPN technologies — user authentication, address management, data encryption, key management, and multiprotocol support. It discussed how Layer 2 protocols, specifically PPTP and L2TP, meet these requirements, and how IPSec (a Layer 3 protocol) will meet these requirements in the future.

Every VPN solution needs to address the technological issues cited in the preceding text and provide the flexibility to address enterprise issues like network interoperability, rich application integration, and infrastructure transparency. Enterprise infrastructure decisions need to be made in a manner that empowers client access to local connections and client utilization of the network in a transparent manner to bolster economy and productivity.

Furthermore, escalating remote access and telecommuting needs and an increase in the use of distributed enterprise models like extranets require pragmatic remote access solutions that are easy to use, economical, and flexible enough to meet the changing needs of every enterprise. To support its 50,000+ employees worldwide with best-of-breed remote access and virtual private networking (VPN) services, Microsoft capitalizes on the built-in communication services included in Windows®, integrated VPN firewall and caching support from Microsoft® Proxy Server, and complementary services from partners such as UUnet Technologies, Inc., Telco Research, and ATCOM, Inc.

The remote access infrastructure that Microsoft's Redmond, WA, headquarters uses for its 15,000 HQ employees consists of four dedicated VPN server computers running the Windows 2000 network operating system. Each machine runs three new 400-MHz Pentium III processors with 204MB of RAM, 3 × 3 GB of local storage, and three 200-Mbps network interface cards.

The UUnet Technologies, Inc. network that supports Microsoft's wholesale remote access and VPN services provides access to one of the largest IP networks in the world. UUnet's backbone infrastructure features a fully meshed network that extends across both the Atlantic and Pacific, and includes direct fiberoptic connections between Europe, North America, and Asia. UUnet also provides satellite access services for remote areas that lack Internet connections.

Telco Research's TRU RADIUS Accountant™ for Windows 2000 provides Microsoft's ITG with a single source for reporting internal usage and chargeback (billing) information required to control remote access costs. TRU RADIUS' easy-to-use applications provide a turnkey analysis of remote access usage and the data needed to proactively manage Microsoft's remote employee costs across its enterprise.

Microsoft's use of UUnet infrastructure to provision its VPN services to its sales force and mobile users is a testament to the quality and reliability of UUnet's multinational IP network. Using Windows 2000 integrated communication services, both UUnet and Microsoft ITG can centrally update Microsoft remote users with the latest local points of presence (POPs) and RAS connection points as soon they become available around the world.

Chapter 24
ATM Access: The Genesis of a New Network

William R. Koss

DATA COMMUNICATIONS HAS COME A LONG WAY IN THE PAST SIX YEARS. In 1992, many users deploying wide area networks (WANs) still thought in terms of 56K frame relay and fractional T1 speeds as state of the art. Some six years later, the World Wide Web (WWW) has driven most of us to deploy V.90 standard modems in the home or explore the possibilities of cable modems, ISDN, and xDSL technologies. In the enterprise network, network backbones are now primarily cell-based backbones. Long gone are shared backbones such as Token Ring, Ethernet, and collapsed router backbones. FDDI is still around, but nobody is getting excited about DAS and SAS. The network core within the enterprise and within the service provider (LEC, CLEC, IAP, ISP, IXC, and CIXC) networks is using a cell-based technology called ATM (asynchronous transfer mode).

One may have heard of ATM — it was all the rage in 1994. Everyone went through a few years of ATM Year One, Year Two, Year Three, and then the marketing hype finally fizzled out. In its place came tag switching, voice over IP, IP over SONET, xDSL, and terabit routers. All this occurred while the service providers were deploying ATM within their network cores, and most enterprise IT managers were completing the deployment of ATM backbones to service the growing demand for bandwidth and network performance. The purpose of this article is to explore the emergence of ATM into the wide area network and outline the challenges and techniques that will be used to deliver increased bandwidth for voice, data, and video services.

ATM TODAY: THE NETWORK CORE

For discussion purposes, existing ATM networks can be divided into two primary categories: (1) carrier class networks maintained by service providers, and (2) enterprise networks maintained by corporations and organizations and primarily intended for private use. To understand the deployment of ATM into the WAN, one can start by examining how ATM is deployed within each of the defined categories.

Not too long ago, many people referred to the phone companies as the Telco or carrier markets. Today, this is an injustice. Such a large number of companies exist that offer some type of network access that the market they are part of is now referred to as "service providers." Within the service provider market, many types of companies and market verticals can be defined. This article concentrates on the following acronyms that define the major vertical elements of the service provider market:

- **LEC:** Local exchange carrier that provides local access to telecommunication services; formally defined as regional Bell operating companies (RBOCs).
- **CLEC:** Competitive local exchange carrier. Companies that provide local access to telecommunication services. Most CLECs have emerged in past few years to compete for local telephone service against LECs.
- **IAP:** Internet access provider. Companies that provide wholesale Internet backbone transport and access. Examples are UUNET, ANS, SAVVIS, and Concentric.
- **ISP:** Internet service provider. Companies that provide local Internet access and Internet-based services.
- **IXC:** Interchange carrier. Companies that provide inter-LATA services between LATAs within an interstate or intrastate basis. AT&T, MCI Worldcomm, and Sprint are the primary IXCs within the United States.
- **CIXC:** For lack of a better term, competitive interchange carrier. This vertical represents companies that are developing inter-LATA services based on their own developing network infrastructure. The prominent players are Level3, Qwest, and IXC Communications.

Each service provider has developed a different type of network to meet its business and customer requirements. For discussion purposes, some generalities that encompass many of the service provider networks can be defined.

By the end of 1998, most service providers within the United States had deployed ATM within their networks. Those that have are typically well-established companies or aggressive young companies with a business strategy that depends on leveraging the technical advantages of ATM to compete for customers. For the purposes of this article, international service providers or small market players that have yet to establish themselves are not discussed . The reason is that each international market is different, and the smaller market players have yet to develop enough mass to affect the market dynamics.

The existing IXCs and LECs are confronted with a series of challenges that can best described as legacy challenges. Most IXCs and LECs have created several networks to deliver customer services. They have networks for voice services, packet-based IP, frame relay, local access, and ATM. The emerging CLECs and CIXCs have bypassed legacy technologies and deployed ATM as their transport technology. The competitive nature of the service provider market requires that service providers be able to address the following challenges.

- provide the client an end-to-end network solution
- provide application services such as Internet access, global intranets, web-based co-location, web-based transaction and application support, and data warehousing
- unify their networks and provide a single managed service
- reduce duplicated costs and overall operational costs
- deliver integrated voice, data, and video services to the client
- provide unified billing and service-level agreements (SLAs) that guarantee quality of service

The IXCs and LECs that are the incumbent service providers are challenged to (1) reduce cost, (2) improve performance, (3) unify services, and (4) deliver higher quality services. Their plan is to achieve these objectives by leveraging the strengths of technologies such as ATM. The reason ATM is important is that it currently exists within many of the service provider core networks, and ATM has several important elements that are driving ATM deployment, including:

- It solves the TDM problem; existing networks based on time-division multiplexing cannot support future network growth.
- It supports all services; ATM supports three primary services that service providers are concerned with: voice, data, and video.
- It exists and is proven; ATM is already deployed in the network core. It is proven, understood, and technical resources have been trained to support the technology.
- It reduces deployment and operational costs; ATM access equipment is less expensive and less complicated. The long-term support and operational costs of ATM networks is projected to provide significant long-term savings.

As the IXCs and LECs view ATM as the unifying network technology that is able to provide a solution from the carrier core to customer premises, the CLECs and CIXCs view ATM as the network infrastructure of choice. Nearly all CLECs and CIXCs have developed extensive network infrastructures built upon ATM. ATM provides the universal network foundation to support all services and provide scalable bandwidth. As much as the IXCs and LECs view ATM as the future, CLECs and CIXCs view ATM as the technology that enables them to compete without the encumbrance of legacy networks.

Within the enterprise market, ATM has been widely deployed as the universal backbone of the network. Some users have invested in desktop ATM solutions, but many users have built successful network backbones using ATM solutions that encompass Ethernet or Token Ring switching as the final connection to the desktop. Most enterprise ATM backbones are deployed within buildings and campus networks. Using a switched cell network architecture, ATM backbones running at speeds of OC3 are capable of supporting an immense amount of traffic and adequately dealing with the challenge of performance-intensive applications. Similar to the service provider market, several important elements are driving the deployment of ATM into the WAN within the enterprise market:

1. It solves the performance problem. ATM delivers the bandwidth demanded by customers in scalable increments and meters from the customer premise to the network core.
2. It provides support for all services. ATM provides the foundation for voice, data, and video support within a defined standard.
3. It exists and is proven. ATM is already deployed in the network core. It is proven, understood, and technical resources have been trained to support the technology

ATM TOMORROW: THE ATM WAN

The deployment of ATM into the WAN has been a slow process. At present, the majority of ATM WAN installations have been for DS1 speeds. Several reasons can be attributed the slow deployment of ATM technology:

- ATM has primarily been used as a backbone technology for high-speed network trunks
- emergence of possible alternative technologies such as "Everything over IP SONET"
- service providers have been slow to develop infrastructure to support ATM-based services
- continued strong demand of frame relay-based services
- market readiness and demand has been lacking as many enterprise users complete network deployments based on core technologies such as IP, 100BT, and gigabit Ethernet and frame relay
- Y2K readiness and WWW integration have consumed IT resources, inhibiting planned network upgrades

Despite these challenges, 1998 was an important milestone in the ATM WAN market. Major service providers began announcing the availability of high-speed ATM services. Sprint announced its ATM intentions in June

1998, and UUNET announced its intention to deliver high-speed ATM-based services in 1999. To understand why ATM will grow in the WAN, one must first define the business challenges that will compel both the enterprise users and service providers to deploy ATM in the WAN.

Enterprise Network Performance

This is an important concern for enterprise IT managers. Applications are driving the expansion of networks and the need to deliver high-speed services to an increasing number of users. Enterprise managers are continually faced with the challenge of supporting an increased number of applications that require an increased amount of network resources. Collaborative group documents, business presentations, and reports are routinely sent via e-mail or the Internet as the work environment becomes increasingly mobile. The demand for increased access to information at the desktop has created a demand for better network performance. Applications that have been driven out from the network core to the remote offices require more bandwidth for better performance. The ATM core of the enterprise network is and will be pushed out to the edges of the network.

Service Provider Network Performance

Service providers are challenged with the need to deliver more services other than bandwidth and network access. They are challenged to provide services such as data warehousing, data archiving, co-location of web services, business-class Internet access (T1), corporate intranets or VPN services, and workgroup services. To satisfy the demands of their clients, the service providers are challenged to move ATM for their network core to the customer premise. Even today, several service providers have tariffed their ATM services at the same rate that they provide frame relay service.

Quality of Service

An important technical feature of ATM is that it provides multiple levels of quality of service, and it has internal provision to address the needs of flow control and dynamic bandwidth allocation. Clients are demanding a guaranteed level of service for network dollar, and ATM enables the service provider a mechanism to provide service-level agreements (SLAs). The classic example is of two customers, one who has contracted for 6 Mbps service and one who has contracted for 10 Mbps service. The flow control capabilities of ATM allow the service providers to guarantee the expected level of service for each customer, although the customers may be sharing the same network link. Clients receive a different billing rate for their service, and the service provider can maximize their network bandwidth.

Inclusion, Unification, and Expansion

The conversion of the network core to ATM within the enterprise and service provider networks provides a strong impetus to expand these networks to include other users or networks. Unifying and standardizing network technology are methods of reducing cost and improving reliability and serviceability. In order to simplify the network, ATM will move from the network core to the edge of the network.

Scalability

ATM is dynamically scalable and it provides metered bandwidth that conforms to the infrastructure of networks maintained by service providers as well as enterprise users. ATM enables the service providers to deliver bandwidth that meets the demands of users. Other competing technologies such as xDSL require bandwidth services that are asymmetrical to the network. Users will demand bandwidth in increments of 1.5 Mbps, 3 Mbps, 6 Mbps, 12 Mbps, DS3, OC-3c, and OC12.

Integrated Services

ATM provides the only standardized method for unifying network services under a single network medium. ATM supports both data and voice, as well as video services. The unification of voice, data, and video is an important long-term objective for both enterprise users and service provides. Enterprise users long for the day when they can contract for one service and one bill that includes voice, data, and video services. By deploying ATM to the customer premise, service providers can begin to unify network functions and deliver the long-sought-after "single pipe" that provides all functions.

Outsourcing Services

ATM enables service providers to deliver high-speed, transparent LAN services (TLS). TLS services provide a simple means of providing LAN-to-LAN extension without the need to deploy a complex solution. This enables service providers to provide "outsourcing" services to clients seeking to bundle network WAN services to achieve an overall lower operational network cost.

When examining how ATM will be deployed in the future in the enterprise market, one is primarily concerned with the WAN. Although the rapid emergence of switched technologies such as 100BT Ethernet and gigabit Ethernet at low cost points successfully stunted the deployment of ATM to the desktop — ATM found a home in the network core — it was in the core of the network that ATM provided the speeds necessary to reduce network congestion. It is only now that the extensive investments made in ATM backbone technology can be leveraged to deliver "in-demand" services across the network topology. Exhibit 24.1 illustrates the deployment of ATM access services within the enterprise market.

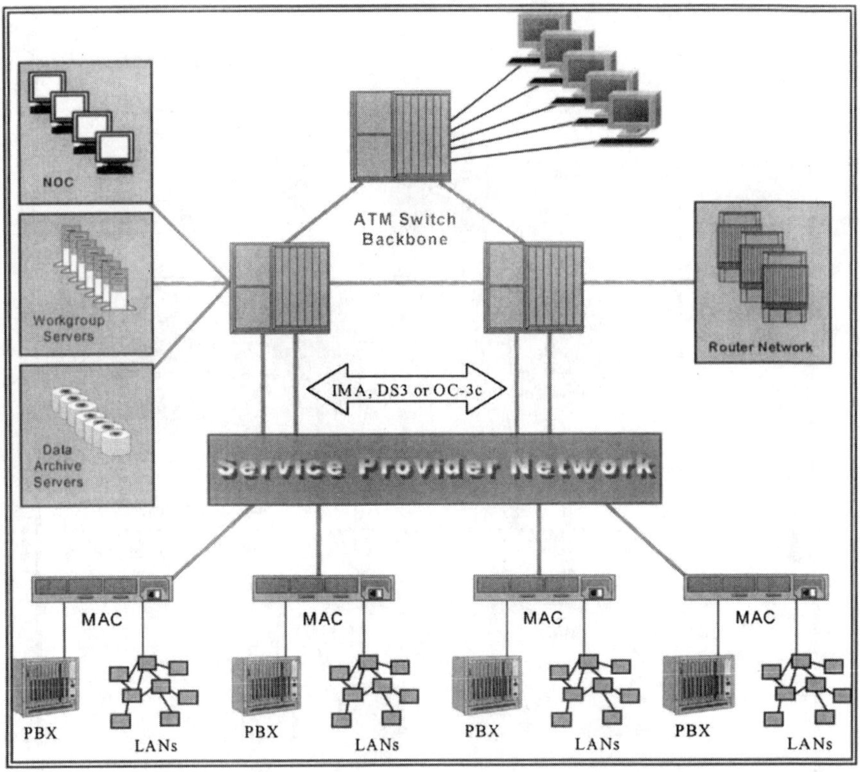

Exhibit 24.1 ATM Deployment in Enterprise Networks.

From the service provider perspective, the ATM network of tomorrow involves migration of the ATM network from the core of the network to the customer premises. By expanding the ATM as the network transport, service providers are presented with a viable strategy to modernize the components of their networks that generate profits — but at a considerable cost. The large incumbent interchange carriers such as Sprint, Williams, MCI Worldcomm, and AT&T have multiple networks under management. These networks include Frame Relay, voice, IP, and ATM. Expanding ATM throughout their networks allows the incumbents to (1) reduce costs, (2) deliver high-speed services, (3) unify networks, and (4) provide new CPE-based services.

Without exception, the new CLECs and CIXCs have built new networks almost entirely based on ATM as the core transport. Each of these new companies has seized on ATM as their preferred transport technology. Thus, none of these companies is encumbered with legacy networks such as Frame Relay or circuit-switched technologies. In short, when one thinks

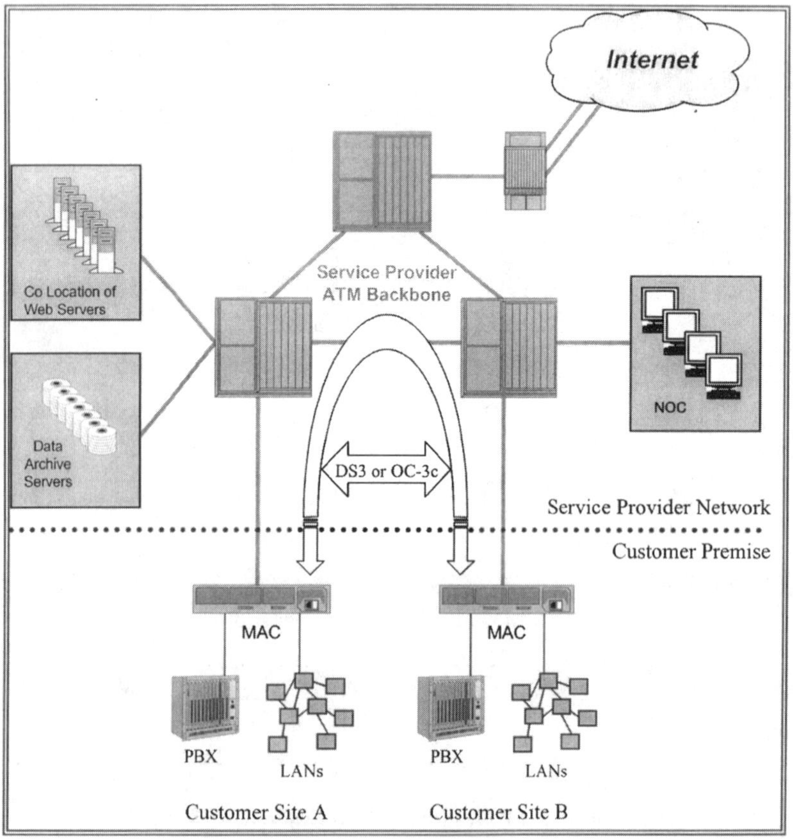

Exhibit 24.2 ATM Access from Customer Site to Service Provider.

about ATM access, one is thinking about a CPE device that is designed to be an access mechanism that transports CPE services into the ATM network for delivery across the service provider network. Exhibit 24.2 illustrates ATM access into the ATM service provider backbone network from two customer sites.

KEY TECHNOLOGIES FOR THE ATM WAN

The ATM WAN of tomorrow will be a combination of "core" ATM technologies and "transition" technologies required for supporting legacy networks. When examining the deployment strategy of ATM access equipment within the ATM WAN of tomorrow, one does so from two perspectives. The first perspective is the user or CPE side, and the second perspective is the network or service provider side. Using the CPE/service provider model, one can begin to define the "system" required for ATM access. On the service provider side, ATM services will be delivered at various speeds using a variety of interfaces. Typically, ATM access devices will support the following on the service provider side:

- DS1/E1 interface
- DS3/E3 interface
- OC-3c interface
- inverse muxing over ATM (IMA) interface
- add drop mux and add drop bypass

The service provider side of ATM access devices are simple compared to the variety of services that must be supported on the CPE side. The CPE side of ATM access devices must achieve two objectives: they must support the services that customers will use going forward, and they must also support a variety of legacy network technologies and services that clients have deployed over the years. It must be remembered that a motivating component of ATM is the ability to provide a consolidation of technologies and services. The following outlines the primary CPE side requirements:

- LAN interface: 10/100 Ethernet, 4/16 Token Ring, and FDDI
- IP routing support: static routing and default gateway forwarding support
- transparent LAN service support: bridging
- LAN client support: LAN client provides support for SVCs and bandwidth management and interoperability in enterprise networks with native ATM devices

Combining the service provider and CPE side features, one creates what can be called an integrated access device (IAD) or multiservice access concentrator (MAC). The objective of this device is to provide user-to-network connectivity. It is not a router and it is not a switch — it is an access device. In simple terms, it can be considered a fast segmentation and reassembly (SARing) engine that transports voice and data into an ATM stream. The primary objective of MACs will be to balance their cost and simplicity in order to leverage the benefits of ATM. Unified services or "single-pipe" strategy is a driving force behind the deployment of ATM; but if the devices used to provide ATM access are expensive or complicated, the compelling reasons to deploy ATM are diminished. Enterprise IT managers and service providers have a common objective that ATM and MACs must achieve — long-term operational cost reduction.

As enterprise networks and service providers begin the migration out from the network core, support for legacy technologies will be important in-demand features. Not only will the IXCs have legacy networks, but many enterprise users will have a mix of legacy technologies as well. The objective will be to provide an orderly transition to ATM, starting with the locations with critical needs. In order to achieve this process, MACs will need to provide support for legacy services such as Frame Relay. An essential element of this requirement will be for MACs to support Frame Relay connections from routers and FRADs, as well as FRF 5 and FRF 8 for support of Frame-to-ATM and Frame-to-ATM-to-Frame services.

As much as MACs must be data orientated to support the variety of services typically found at the remote site location, they must also be able to support a variety of voice services. A primary benefit of ATM is the integration of voice and data on a single pipe. The benefits of the technology are greatly reduced if all one can get is a faster pipe and still require a different connection for voice services. Thus, MACs need to possess strong voice services. These services include:

- circuit emulation services: structured and unstructured transmission of DS1/E1 streams for support of digital PBX interconnect
- FXO/FXS POTs: support for analog PBXs as well as dial tone, PLAR, and ring generator
- voice and silence suppression and compression
- AAL2 voice services that support dynamic bandwidth allocation through SVCs

As well as supporting a variety of services, MACs must also support a variety of network management services. MACs will be deployed by both enterprise users and service provider users. Thus, it is important for MACs to support features that enable each to manage and support the devices. The primary objective is for MACs to be "lights-out" devices that are installed, configured, and forgotten. Unfortunately, network devices rarely live up to such lofty goals and at times will be in need of management services. In this case, it will important for MACs to support a variety of "punch list" network management features. These features include:

- web-based management
- Telnet
- SNMP
- support for popular network management platforms with MIB extensions
- CRAFT interface
- out-of-band management via modem

An important tactical capability in the deployment of MACs will be support for inverse multiplexing over ATM (IMA). IMA leverages two market conditions for the service provider and enterprise client. At present, many service providers are forced to deploy full DS3 circuits for clients that desire to scale above the T1 range. Unfortunately, many clients do not require a full DS3. In other cases, clients are demanding T1 service in areas where

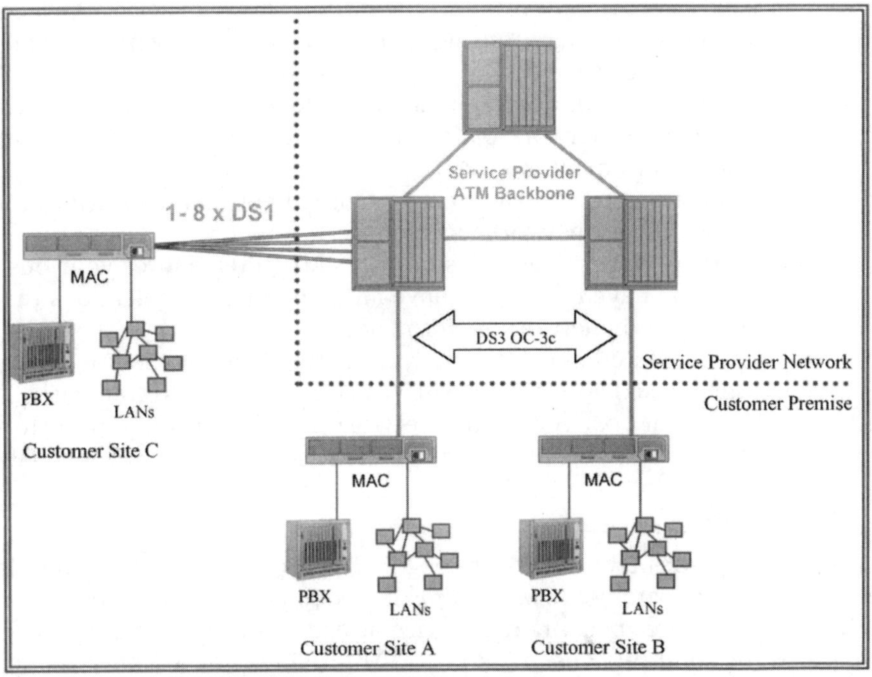

Exhibit 24.3 Deployment of Inverse Multiplexing over ATM (IMA) Services.

DS3 is not readily available. The solution to this challenge can be found in IMA. IMA allows T1 lines to be bonded together to deliver access services that are T1 but less than DS3. This elegant strategy provides a gradual scale in services and cost as network demands warrant. Exhibit 24.3 illustrates the deployment of IMA services.

ATM WAN: SHOULD YOU START NOW?

Why ATM will be important in the WAN is known; now one needs to ensure that one will be prepared to undertake the transition. From a preparation perspective, the following questions provide a starting point to determine whether or not one should initiate programs to begin transitioning the WAN transport to ATM-based services.

- Do you have remote locations that require improved performance or locations in your network that would benefit by extending your ATM backbone to include their location?
- Are you looking to reduce long-term costs and unify voice and data services?
- Do you have E-commerce, or web-based applications that require dedicated Internet access above T1 speeds?

- Can you benefit from leveraging the infrastructure of service providers to provide data warehousing, co-location of web servers, and high-speed intranet services?
- Do you have major application initiatives underway that will increase the demands for network bandwidth throughout your network over the next 12 to 24 months?
- Are you looking to contract for guaranteed network bandwidth for some or all of your network locations?
- Is network performance a mission-critical requirement for your business? Do you have or will you have mission-critical applications that are dependent on network performance?
- Are you interested in positioning your network infrastructure to support video-based applications such as conferencing and training?
- Are long-term cost reduction, network simplification, and performance long-term objectives for your IS organization?

CONCLUSION

The genesis of the new network has begun. The start has been slow, but the conclusion is inevitable. ATM is and will be deployed as the transport of choice for wide area networking. Service providers of all types are deploying ATM for solid business reasons. ATM provides key advantages not found in other WAN technologies: namely, quality of service, unification of services (voice, data, video), and superior scalability.

Enterprise networks are being deployed with public or private ATM backbones. This will be extended to remote offices and locations where ATM access solutions will be deployed for the following reasons:

- **Cost/performance balance:** increase network performance in a manner that allows for scalable migration of services based on a cost/need ratio
- **Simplified billing and service level agreements:** lower service cost, simplified billing statements, and guaranteed service that enable the enterprise users to purchase only what they require
- **Leverage investments in public and private backbones:** effective use of ATM core infrastructures that are already deployed will lead to improved services and lower long-term costs
- **Reduce acquisition and operational costs:** deploy less-expensive technological solutions that provide long-term, low cost, "lights out" operational costs

ATM is the only technology that is standards-based and can be used to deliver multiservice networking throughout the wide area network. In addition, ATM is the only technology that can be used by enterprise users and service providers to reduce their equipment acquisition costs and ongoing operational costs over the long term.

Chapter 25
Portable Network Access

Gilbert Held

THE STEADY INCREASE IN PORTABLE COMPUTER USE HAS HAD A POSITIVE EFFECT ON PRODUCTIVITY. It has also, however, challenged users to become educated on the vast array of networks and access methods that can facilitate the sharing of information. This chapter introduces users to several important modem terms, as well as a step-by-step guide to accessing today's popular packet-switching networks.

MODEM TERMS

Until the incorporation of microprocessors into modems, the selection and operation of a modem was relatively easy. At that time, the major concern was the type of modulation performed by the modem expressed in terms of Bell System or TSS standards compatibility.

Until the divestiture of AT&T and its operating companies, Bell System equipment was a *de facto* standard in the US, and third-party modem manufacturers built most of their products to ensure compatibility with Bell System modems. Popular Bell System-compatible modems used on the PSTN included the 103, which operates at 300 b/s, and the 212A, which operates at 1,200 b/s and can also operate as a 300-b/s Bell System 103-compatible modem.

The TSS, which is a standards recommending body, is part of the ITU, a United Nations organization headquartered in Geneva. Until the divestiture of AT&T, TSS modem standards were primarily followed in Europe, and their modulation methods were incompatible with Bell System modems. With the divestiture of AT&T, several TSS modem standards have been adopted worldwide, including the V.22bis and V.32 modem recommendations. V.22bis governs the operation of a 2,400-b/s, FDX modem for use on the PSTN; the V.32 recommendation governs the operation of a 9,600-b/s, FDX modem for use on the PSTN.

The incorporation of microprocessor technology into modems resulted in their use for a range of functions added to the basic data modulation function of modems. Among features added to modems are error detection and correction, data compression, and command set recognition.

Error Detection and Correction

Among the first modems to offer an error detection and correction feature were products manufactured by Microcom, Inc. This company created a revolution in modem technology by developing a modem protocol known as MNP.

The MNP Protocol. MNP is a communications protocol built into MNP-compatible modems that supports interactive and file transfer applications. In developing MNP, Microcom recognized that the first implementation of the protocol would not necessarily be the last, and structured it to accommodate changes in its implementation. To accomplish this, the major functions of the protocol are divided into classes. When an MNP modem communicates with another MNP modem, the two devices negotiate with each other to operate at the highest mutually supported class of MNP service.

Exhibit 25.1 summarizes the features associated with available classes. Until 1990, Microcom only licensed MNP through Class 5 to other modem manufacturers. In that year, they began to offer a full MNP license. Therefore, an MNP-compatible modem, though compatible with all other MNP modems, may have been compatible only with a subset of available MNP classes unless a third-party vendor obtained a full license and incorporated all MNP classes into its product.

Exhibit 25.1 includes such terms as V.29 and V.32. Both reference TSS modulation standards, with V.29 originally developed as a half-duplex 9,600-b/s modulation technique for use on leased lines. Microcom, as well as other vendors, modified that technology to work on the PSTN. In addition, using the intelligence of a microprocessor to monitor the direction of transmission, it became possible to quickly turn off the transmitter of one modem and turn on its receiver, enabling the half-duplex transmission. In comparison, a V.32 modem uses echo cancellation technology to enable transmission and reception of data to occur simultaneously on the PSTN, and also provides an inherent FDX transmission capability.

V.42 Recommendation. Although MNP error detection and correction is included in more than 1 million modems manufactured by Microcom and more than 100 third-party vendors, it is not the primary method of error detection and correction recommended for use in modems by the TSS. In 1990, the TSS promulgated its V.42. Unlike other TSS V-series recommendations that govern modern modulation techniques, the V.42 Recommendation defines a protocol

Exhibit 25.1 MNP Classes.

Class	Description of Functions Performed
Class 1	Asynchronous, byte-oriented, half-duplex transmission that provides an efficiency of approximately 70%. A 2,400-b/s modem using MNP class 1 obtains a throughput of 1,690 b/s.
Class 2	Asynchronous byte-oriented full-duplex data transmission that provides an efficiency of approximately 84%. A 2,400-b/s modem using MNP class 2 obtains a throughput of approximately 2,000 b/s.
Class 3	Asynchronous start and stop bits are stripped, enabling synchronous, bit-oriented, full-duplex transmission between modems. This provides an efficiency of approximately 108%, enabling a 2,400-b/s modem to obtain a throughput of approximately 2,600 b/s.
Class 4	This class adds an adaptive packet assembly to previous classes in which packet sizes are dynamically adjusted based on the number of retransmission requests. Data phase optimization, which provides a mechanism to reduce protocol overhead, is also included. The efficiency of Class 4 is approximately 120%, enabling a 2,400-b/s modem to obtain a throughput of 2,900 b/s.
Class 5	This class adds data compression to Class 4 service, which provides an average compression ratio of 1.6:1, meaning that every 16 characters are compressed into 10 characters for transmission. This increases the protocol efficiency of Class 5 to approximately 200%, enabling a 2,400-b/s modem to obtain a throughput of about 4,800 b/s.
Class 6	This class adds universal link negotiation and statistical duplexing to Class 5. Universal link negotiation enables MNP modems to begin operation at a common low-speed modulation method and negotiate the use of an alternative higher-speed modulation method. At the end of a successful link negotiation for Class 6 operation, the pair of modems operate at 9,600 b/s using V.29 technology. Statistical duplexing results in the monitoring of user traffic patterns to enable the dynamic allocation of V.29 half-duplex transmission to resemble full-duplex transmission. Under Class 6, 9,600-b/s operations MNP provide an average throughput approaching 19,200 b/s.
Class 7	This class adds an enhanced data compression capability to P, based on a Huffman statistical encoding technique. Under Class 7, a compression ratio of between 2.0 and 3.0 is achievable and increases throughput from 2 to 3 times the modems operating rate.
Class 8	This class is no longer marketed.
Class 9	This class adds support of V.32 modulation to Class 7, providing a throughput of three times the 9600-b/s full-duplex operating rate of a V.32 modem.

in which modems block data for transmission and generate and add a CRC to each block for error detection. Under the V.42 Recommendation, the flow of data blocks occurs according to a LAP, which differs from data flow under the MNP protocol. In recognition of the large installed base of MNP modems, however, MNP error detection and correction is supported as a secondary standard (i.e., a V.42-compatible modem will first attempt to communicate in its error-free mode using LAP). If the distant modem does not support the V.42 protocol, the V.42 modem will next attempt to communicate using MNP error control.

Data Compression

Although the TSS V.42 Recommendation supports MNP error control, the V.42bis Recommendation — which is a data compression scheme that requires the use of the V.42 protocol — does not support a secondary method of compression. This theoretically means that for a portable computer user's V.42bis modem to operate in a compressed mode, it must communicate with another V.42bis-compatible modem. Fortunately, most V.42bis modems also include MNP support through Class 5, enabling users to communicate in both a data compressed and error control mode with another MNP modem.

The key difference between V.42bis and MNP Class 5 or Class 7 is in their method of data compression. V.42bis uses a technique known as "Lempel-Ziv," which can operate on single characters as well as on strings. In comparison, MNP compression primarily operates on single characters and is slightly less efficient. When using a V.42bis modem that incorporates V.32 modulation, the 9,600-b/s operating rate of the modem may achieve a throughput of as much as 28,400-b/s when data is highly susceptible to compression.

Interface vs. Operating Rate. To use data compression effectively, users are required to set their modem's interface data rate to exceed its operating rate. To understand why this is required, Exhibit 25.2 illustrates the relationship between the interface speed and the operating rate of a modem.

Exhibit 25.2 Interface Rate Vs. Operating Rate.

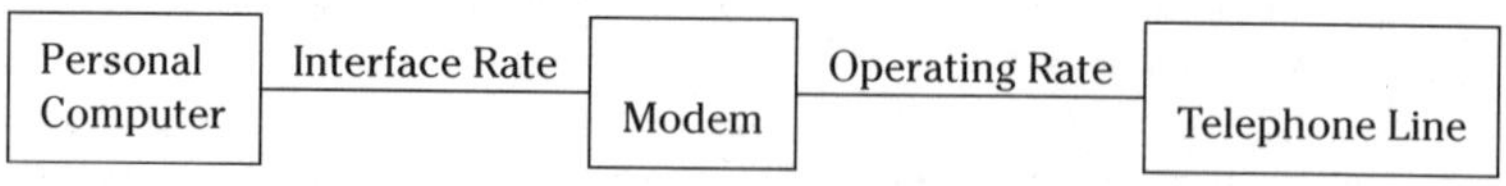

If the interface speed equals the operating rate, data compression will not increase the throughput of the modem. To increase the modem's throughput, users must set the interface rate to at least twice the modem's operating rate. For example, if the modem operates at 2,400 b/s, the interface rate should be set to at least 4,800 b/s. This allows more bits in the form of characters to enter the modem per unit of time than can be transmitted. Then, data compression can attempt to reduce the bits entering the modem so they can be transmitted at the modem's operating rate, increasing the throughput through the modem.

Flow Control. A second consideration users must take into account when using either error control by itself or with data compression is flow control. To see why flow control is necessary, Exhibit 25.2 shows the interface speed set to 4,800 b/s while the operating rate is 2,400 b/s. If the modem cannot effectively compress data, perhaps because of the composition of the data, the rate data enters the modem will exceed the rate it is placed onto the line. Although the modem has a buffer, eventually it will overflow and data will be lost. Therefore, a method is required to regulate the flow of data into the modem. This method is called flow control.

Most modems support at least two methods of flow control, referred to as XON/XOFF and CTS/RTS. The XON/XOFF method of flow control results in the modem transmitting an XOFF character to the portable computer to disable transmission, and an XON character to enable transmission to resume. Because the computer may require an ability to regulate data from the modem if users are printing received data or performing another mechanical-related operation, they can also use XON and XOFF from the portable computer to the modem to enable and disable transmission from the modem to the portable computer.

The CTS/RTS method of flow control refers to the use of clear to send (CTS) and request to send (RTS) control signs on the interface between the computer and the modem. When the modem regulates flow control, it will drop the CTS signal to inform the computer to stop transmission and raise the CTS signal lead to inform the computer to resume transmission. When the computer regulates the flow of data from the modem, it will toggle the RTS signal lead.

In setting flow control, users must ensure that their computer and modem are set to use the same method. Concerning which method to select, if users expect to transfer files that may contain control characters, including an XOFF, they should use the CTS/RTS method of flow control. Otherwise, the reception of an XOFF would inadvertently turn off the computer.

Command Set Recognition

To enable most of the features discussed in this section requires the use of the user's modem's command set. Although most modems claim Hayes Microcomputer Products command set compatibility, such compatibility only governs basic modem operations. To enable or disable MNP operations, V.42 error control, V.42bis data compression, or flow control requires the use of the modem's extended command codes. Because a modem's extended command set codes are not standardized, users should refer to their modem's reference manual to determine the codes to use to enable or disable the previously discussed features.

NETWORK ACCESS

The remainder of this chapter focuses on the step-by-step procedures portable computer users must follow to access several popular packet-switching networks, electronic mail systems, and an information utility. Included for each are the customer service telephone number users can call to obtain an appropriate local telephone access number; a WATS access number, if applicable; login procedures, including communications configuration parameter settings necessary to access each system; and a summary of commands when appropriate for accessing a special feature of the system.

British Telecom (BT) TYMNET

Customer service telephone numbers:

U.S. and Canada: (800) 336-0149

International: (703) 715-0045

The log-in procedures for British Telecom (BT) TMNET are:

1. Set the communications configuration.

Primary setting:	8 data bits, 1 stop bit, no parity
Secondary settings:	7 data bits, 1 stop bit, mark parity
	7 data bits, 1 stop bit, space parity

2. Dial the BT TYMNET access number. Dial customer service for telephone access or refer to the table at the end of this entry.
3. Type the terminal identifier. Use A to instruct the BT TYMNET network that no delay is necessary after transmitting a carriage return or line feed character.
4. Type the login string based on the following format: please login: [Ctrl]username:destination;password

where:	[Ctrl]	Optional control characters
	Ctrl-X	Transmit data flow control
	Ctrl-R	Receive data flow control
	Ctrl-I	Interactive data session request
	Ctrl-V	Volume bandwidth request
	Ctrl-H	Half-duplex request
	Ctrl-P	Even parity request
	Ctrl-W	Wipe login string
	username	User's network identification
	destination	Address of computer connected to BT TYMNET accessed
	password	security mechanism that verifies use of user name

Commercial Dialing from Hotels and Pay Phones

Many hotels and pay telephones subscribe to alternative operator services whose charges for long-distance calls can substantially exceed the cost of using a major communications carrier. To bypass alternative operator services, users can dial an access code to directly connect to a specific carrier. Common access codes include:

Communications Carrier	Access Code
AT & T	10-288-0
MCI	10-222-0
US Sprint	10-333-0

If the access code fails to work, users should dial 0 and ask the operator to connect them to the desired carrier.

CompuServe Information Service

Customer service telephone number:

US and Canada: (800) 848-8990

Log-in Procedures by CompuServe. To access CompuServe through CompuServe, the user must:

1. Set the communications configuration. 7 data bits, 1 stop bit, odd parity
2. CompuServe access number. Dial customer service for telephone access number or refer to table in this section.
3. Enter Ctrl-C.
4. Enter the user ID in response to the user ID prompt.
5. Enter the password in response to the password prompt.

Log-in Procedures by BT TYMNET. To access CompuServe through BT TYMNET, the user must:

1. Set the communications configuration.
   Primary setting: 8 data bits, 1 stop bit, no parity
   Secondary settings: 7 data bits, 1 stop bit, mark parity
   7 data bits, 1 stop bit, space parity
2. Dial a BT TYMNET access number.
3. Type A in response to terminal identifier request.
4. Enter CPS in response to Please login: message.
5. Enter CompuServe user ID in response to the user ID prompt.
6. Enter CompuServe password in response to the password prompt.

Log-in Procedures by SprintNet. To access CompuServe through SprintNet, the user must:

1. Set the communications configuration.
   Primary setting: 8 data bits, 1 stop bit, no parity
   Secondary setting: 7 data bits, 1 stop bit, even parity
2. Dial a SprintNet access number.
3. Enter two carriage returns at 300 and 1,200 b/s or one carriage return at 2,400 b/s.

4. Respond to Terminal = prompt with D1.
5. Respond to @ prompt with C 614227.
6. Enter the CompuServe user ID in response to the user ID prompt.
7. Enter the CompuServe password in response to the password prompt.

CompuServe Quick Access Reference

CompuServe telephone access numbers can be obtained by dialing the CompuServe customer service telephone number (800) 48-8990 and responding to the voice prompts. As an alternative, users can enter PHONES in response to the CompuServe Host Name: prompt when online.

Dow Jones News/Retrieval

Customer service telephone number:

(609) 452-1511

Log-in Procedures by BT TYMNET. To access the Dow Jones News/Retrieval Service through BT TYMNET, the user must:

1. Set the communications configuration.
   Primary setting: 8 data bits, 1 stop bit, no parity
   Secondary settings: 7 data bits, 1 stop bit, mark parity
   7 data bits, 1 stop bit, space parity
2. Dial TYMNET access number.
3. Type A in response to the request for a terminal identifier.
4. Enter DOW1 in response to the Please Log In: message.
5. Enter DJNS in response to What Service Please?
6. Enter the Dow Jones News/Retrieval password in response to the password prompt.

Log-in Procedures by SprintNet. To access the Dow Jones News/Retrieval service through SprintNet, the user must:

1. Set the communications configuration.
   Primary setting: 8 data bits, 1 stop bit, no parity
   Secondary settings: 7 data bits, 1 stop bit, even parity
2. Dial a SprintNet access number.
3. Enter two carriage returns at 300 and 1,200 b/s, or one carriage return at 2,400 b/s.
4. Respond to Terminal 1/2 prompt with D1.
5. Respond to the @ prompt with C DOW.
6. Enter DJNS and press Return in response to the What Service Please? prompt.
7. Enter the Dow Jones News/Retrieval password in response to the password prompt.

See Exhibit 25.3.

Exhibit 25.3 Dow Jones Information Retrieval Access Codes.

Description	Access Code
Online Help	
Systemwide Help	//DJHELP
Customer Information and Updates	//FYI
Menu of Databases	//MENU
Directory of News-Retrieval Symbols	//SYMBOL
Business and World Newswires	
Top Business, Financial, and Economic News	//BUSINESS
Japanese Business News	//KYODO
Top National and World News	//NEWs
Dow Jones Business Newswires	//WIRES
Dow Jones Text Library	
Current Business Articles	//DOWQUEST
The Wall Street Journal and Other Publications:	
Menu Version	//TEXTM
Command Version	//TEXT
Company-Industry Information	
Canadian Corporate News and Reports	//CANADA
Dun & Bradstreet	//DB
SEC Filing Extracts	//DSCLO
Earnings Forecasts	//EPS
Analysts' Reports on Companies and Industries	//INVEST
Statistical Comparison of Companies and Industries//MG	
Comprehensive Company Reports	//QUICK
Standard & Poor's Profiles and Earnings Estimate	//SP
Insider Trading, 5% Ownership & Tender Offer Filings//WATCH	
Worldwide Corporate Report	//WORLDSCOPE
Quotes, Statistics, and Commentary	
Current Quotes	//CQE
Historical Dow Jones Averages	//DJA
Mutual Fund Reports	//FUNDS
Futures and Index Quotes	//FUTURES
Historical Quotes	//HQ
Technical Analysis Reports	//INNOVEST
Forecasts and Analyses of World Markets	//MMS

Exhibit 25.3 (Continued) Dow Jones Information Retrieval Access Codes.

Description	Access Code
Quotes, Statistics, and Commentary (continued)	
Real-Time Quotes	//RTQ
Screening and 15-Year Historical Pricing on Securities & Indexes	//TRADELINE
Wall Street Week Transcripts	//WSW
Customized Information	
Clipping Service	//CLIP
Company and Industry Tracking Service	//TRACK
General Services	
Book Reviews	//Books
Encyclopedia	//ENCYC
Online Brokerage	//FIDELITY
Electronic Mail and Communications	//MCI
Movie Reviews	//MOVIES
Comprehensive Travel Services	//OAG
College Selection Service	//SCHOOL
Sports Report	//SPORTS
Electronic Shopping Service	//STORE
Worldwide Weather Reports	//WTHR

FTS2000 Offnet Calling

These procedures are for U.S. government employees dialing from hotels, motels, pay phones, and other non-FTS2000 telephones. Dialing procedures are based on the network servicing the employee's agency.

Network A (AT&T). To gain access to this network, employees must:

1. Use the AT&T Calling Card to dial the access number (800) 633-6384, and then wait for a dial tone.
2. Dial the 7-digit FTS or 10-digit commercial number to be accessed and wait for another dial tone.
3. Dial the 11-digit authorization code on the AT & T Calling Card.

Network B (U.S. Sprint). To gain access to this network, employees must

1. Use the U.S. Sprint FONCARD to dial the access number (800) 366-4000.
2. Dial the 11-digit FONCARD number shown on the card.
3. Dial the 7-digit FTS or 10-digit commercial number to be accessed.

If dialing from a rotary dial telephone using either card, users must wait for an operator after dialing the appropriate 800 number. For assistance, users can dial the THIS at FTS 295-8500 or (301) 295-8500.

MCI Mail

Customer service telephone number:

(800) 444-6245

Log-in Procedures. To log in to MCI Mail, users must:

1. Set the communications configuration.
   8 data bits, 1 stop bit, no parity
2. Dial the MCI Mail access number: (800) 234-6245
3. Respond to the user prompt by typing the user ID.
4. Respond to the password prompt by typing the password.

Mailboxes. Mailbox sections and their contents include the following:

Section	Contents
INBOX	Messages that have not been read.
OUTBOX	Messages that were previously sent.
DESK	Messages that were read.
PENDING	Messages being created.

Mailbox-related commands and their functions are:

ANSWER	For replying to the sender of a message that has been read.
ANSWER EACH	For replying to the sender of a message and everyone else who received the message.
CREATE	For writing an MCI Mail message. — To cancel a message, the user types a slash (/) at TO, CC, or subject. — To complete a message, the user types a slash (/) on a blank line after the body of the message.
DELETE	Enables messages to be deleted from any mailbox selection. The user enters the command followed by the scan numbers to be deleted. For example, DELETE 5, 7 causes the fifth and seventh message in the current mailbox section to be deleted.
FORWARD	Causes a copy of one or more messages in a mailbox to be sent to other people.
INCLUDE	Enables text from other messages to be inserted into a message. Type/INCLUDE/scan-number in the message being created. For example,/INCLUDE/2,4 & 5 inserts text from messages 2, 4, and 5 in the message being created.

NEXT	Causes the next message from the current section to be displayed after a READ command is issued.
PRINT	Enables a series of messages to be displayed without pausing between messages. Messages to PRINT can be identified by scan-number, mailbox section, or by one of the following: ALL ON [date] SUBJECT ["words"] AFTER [date] TO [lastname] BEFORE [date] FROM [lastname]
READ	Enables one or more messages to be displayed with page breaks between messages. Messages to READ can be identified by scan-number, mailbox section, or by one of the following: ALL AFTER [date] TO [lastname] BEFORE [date] FROM [lastname] After reading a message one of the following commands can be used: ANSWER NEXT FORWARD PRINT
SCAN	Enables users to obtain a summary of the messages in their mailboxes. One or more of the following options can be used with SCAN: INBOX AFTER [date] OUTBOX BEFORE [date] DESK SUBJECT ["text"] PENDING TO [name] ALL FROM [name] ON [date]
SEND	Enables a draft message to be transmitted for non-priority delivery.
SEND ONITE	Enables a draft message to be transmitted for priority delivery.
SEND 4HOUR	Enables a draft message to be transmitted by electronic mail and adds priority notification to the message.
/UPLOAD	Enables one or more files containing addresses into the envelope of a message to be transmitted.

Other MCI Mail commands include:

ACCOUNT	Enables a terminal's characteristics to be defined to MCI Mail.
DOW JONES	Enables users to access Dow Jones News and Retrieval Information.

EDIT	Allows users to make changes to pending messages.
EXIT	For terminating the MCI Mail session.
FIND	Permits users to search for an MCI Mail subscriber's name to obtain their MCI ID and other information.

The MCI Help facility. MCI Mail includes a built-in help facility that displays information about the use of commands and data about different topics. To use the help facility, the user types:

HELP { command-name topic

To obtain a list of topics, the user types HELP INDEX. Among the topics supported are:

Topic	Information Provided
ACCESS	Telephone access information.
ACCOUNT	Terminal settings.
ADDRESS	Available addressing options.
APPENDIX	Pricing, sales representatives, and directory searching.
BULLETIN BOARD	Owning and viewing bulletin boards.
COUNTRY	Countries that receive MCI Mail.
CUSTOMER SUPPORT	How to contact MCI Mail Customer Service.
DELIVERY	Mail delivery.
FAX MCI	Mail fax dispatch.
GRAPHICS	Letterheads and signatures.
HANDLING	Option commands that define how messages are handled and delivered.
LISTS	Creation of address lists.
MAILBOX	The MCI Mailbox.
PC	Offline file creation and uploading to MCI Mail.
PRICES MESSAGES	Rates for sending MCI Mail messages.
REQUEST	Ordering of reference manuals.
SPECIAL FEATURES	Additional features about MCI Mail.
TELEX	How to exchange messages with MCI Mail.
TEXT	Creating text in a message.
X400	Transmitting to an X.400 electronic mail address.

U.S. Sprint SprintNet

Customer service:

US and overseas WATS access: (800) 877-5045

International non-WATS access: (703) 689-6400

To log-in to this network, users must:

1. Set the communications configuration.
   Primary setting: 8 data bits, 1 stop bit, no parity
   Secondary setting: 7 data bits, 1 stop bit, even parity
2. Dial the SprintNet access number. The user dials customer service for the access number or refers to the completed table at the end of this entry.
3. Identify duplex setting. The user hits a carriage return for full duplex and hits a semicolon and a carriage return for half-duplex transmission. SprintNet can be informed of a computer' use of data bits, stop bit, parity, its operating rate, and duplex setting by the use of a 2- or 3-character sequence called Hunt/Confirm. The following items indicate communications parameter settings, data rate, and duplex mode identified by the Hunt/Confirm sequences supported by SprintNet.
4. Respond to the TERMINAL = prompt with the 2-character terminal identifier or press a carriage return. Use D1 for a computer.
5, Enter the area code (AAA) and the local exchange (111), followed by a carriage return — AAA, LLL if access is by in-WATS service.
6. Respond to the @ prompt with the connect command C followed by the destination host or terminal address and an optional ID/password using the following format:
   C DDDDDD,IIIIII [return]
   where: C is connect command
   DDDDDD is destination address
   IIIIII is the user identification
7. If required, the user then responds to the PASSWORD = prompt with the password followed by a [return].

US Sprint SprintNet quick access reference numbers: in-WATS service:

300.2,400 b/s: (800) 546-1000
9,600 b/s (V.32): (800) 546-2500
9,600 b/s (V.29): (800) 546-2000

SUMMARY

Knowledge of communications terms, features, and network access methods can provide portable computer users with the ability to efficiently and effectively communicate with other computers and information resources. In addition, preplanning trips to determine and record telephone access numbers required to use the resources of different value-added carriers and electronic mail providers can save users a considerable amount of time and effort to obtain that information when traveling.

Chapter 26
Cellular Digital Packet Data: An Emerging Mobile Network Service

Nathan J. Muller

IN THE WIRELESS DATA MARKET, MAJOR TELEPHONE COMPANIES ARE IMPLEMENTING CDPD SERVICE TO MEET THE NEEDS OF A MOBILE WORK FORCE. CDPD is an appealing method of transporting data over cellular voice networks because it is flexible, fast, available internationally, compatible with a vast installed base of computers, and has security features not found in other cellular systems.

INTRODUCTION

CDPD is a data-over-cellular standard for providing LAN-like service over cellular voice networks. CDPD employs digital modulation and signal processing techniques, but it is still an analog transmission. The CDPD infrastructure employs existing cellular systems to access a backbone router network that uses the IP to transport user data. Personal digital assistants, palmtops, and laptops running applications that use IP can connect to the CDPD service and gain access to other mobile computer users or to corporate computing resources that rely on wireline connections.

Because CDPD leverages the existing $20 billion investment in the cellular infrastructure, carriers can economically support data applications and avoid the cost of implementing a completely new network, as most competing technologies would require. CDPD also offers a transmission rate that is four times faster than most competing wide area wireless services, which are limited to 4.8 Kb/s or lower.

CDPD FUNDAMENTALS

Unlike circuit-switched schemes, which use dialup modems to access the cellular network, CDPD is a packet-switched technology that relies on wireless modems to send data at a raw speed of 19.2 Kb/s. Although CDPD piggybacks on top of the cellular voice infrastructure, it does not suffer from the 3-KHz limit on voice transmissions. Instead, it uses the entire 30-KHz RF channel during idle times between voice calls. Using the entire channel contributes to CDPD's faster and more reliable data transmission.

Underlying Technologies

CDPD is in fact a blend of digital data transmission, radio technology, packetization, channel hopping, and packet switching. This technology lets the cellular network carry the 1s and 0s of binary digital code more reliably than is usually possible over cellular voice networks.

Digital Transmission Technology. Digital transmission technology is reliable and more resistant to radio interference than analog transmission technology. The digital signals are broken down into a finite set of bits rather than transmitted in a continuous waveform. When signal corruption occurs, error-detection logic at the receiving end can reconstruct the corrupted digital signal using error correction algorithms. Digital technology also enables processing techniques that compensate for signal fades without requiring any increase in power.

Digital Cellular Radio Technology. DCRT is used for transmitting data between the user's mobile unit and the carrier's base station.

Packetization. Packetization divides the data into discrete packets of information before transmission. This approach is commonly used in wide area and local computer networks. In addition to addressing information, each packet includes information that allows the data to be reassembled in the proper order at the receiving end and corrected if necessary.

Channel Hopping. Channel hopping automatically searches out idle channel times between cellular voice calls. Packets of data select available cellular channels and go out in short bursts without interfering with voice communications. Alternatively, cellular carriers may also dedicate voice channels for CDPD traffic.

Packet Switching. Packet switching, using the IP, accepts data packets from multiple users at many different cell sites and routes them to the next appropriate router on the network.

APPLICATIONS FOR CDPD

The wireless industry consortium that funded the development of the CDPD specification includes Ameritech Cellular; Bell Atlantic Mobile;

Contel Cellular, Inc.; GTE Mobilnet, Inc.; McCaw Cellular Communications, Inc.; NYNEX Mobile Communications; AirTouch (formerly PacTel Cellular); and Southwestern Bell Mobile Systems. Three principles guided their efforts: that emerging CDPD recommendations could be deployed rapidly, economically, and in conjunction with technology already available in the marketplace.

More specifically, the consortium's stated objectives include:

- Ensuring compatibility with existing data networks
- Supporting multiple network protocols
- Exerting minimum impact on end systems; existing applications should operate with little or no modification
- Preserving vendor independence
- Ensuring interoperability among service providers without compromising their ability to differentiate offerings with service and feature enhancements
- Allowing subscribers to roam between serving areas
- Protecting subscribers from eavesdropping

EMERGING CLASS OF REMOTE USERS

CDPD allows traditional wireline networks to reach a new class of remote user: the roaming mobile client. With the establishment of a wireless link to the cellular carrier's CDPD network, remote users can operate their terminals as if they were located on the desktop in a branch office. Mobile workers, for example, can regain much of the productivity they lose while away being from their main offices by using CDPD to send and receive E-mail from computers or personal digital assistants.

Another application example is a debit card. Commuters could purchase a debit card to run through a card-reading device on a bus or another transit system, and the fare would be deducted automatically from the card's total. That fare information could be transmitted to a central processing center in less than a second for just a few cents. CDPD could also be used by service providers to monitor and control devices such as traffic lights, alarm systems, kiosks, vending machines, and automated teller machines.

SERVICE PRICING

As an overlay to the existing analog cellular infrastructure, CDPD networks are easy and economical for carriers to set up and operate. Carriers estimate that it costs only 5 percent over the initial cost of a cell site to upgrade to CDPD. Cell sites typically cost about$1 million to set up, including the cost of real estate.

Exhibit 26.1 CDPD Services.

Application	Bell Atlantic Mobile Systems (Per-User Pricing)	GTE Mobilnet (Per-User Pricing)
Data Base Inquiry	$23 to $27/month	$20 to $28/month, 25 sessions a day, 5 days a week
Electronic Mail	$40 to $60/month	$45 to $60/month, 14 messages a day, 5 days a week
Dispatch	$13 to $17/month	$10 to $20/month, 1 to 2 jobs per hour, 9 hours a day, 5 days a week
Alarm Monitoring	$13 to $17/month	$10 to $20/month, 1 transaction per hour, 24 hours a day, 7 days a week
Field Service	$23 to $27/month	$16 to $22/month, 20 transactions a day, 5 days a week

Note: Estimated per-user prices are based on sample applications and usage figures. All prices subject to change without prior notice.

Users are the beneficiaries of CDPD's resulting economies and efficiencies. For many applications, initial CDPD service pricing is competitive with that of the proprietary analog wireless services of ARDIS and RAM Mobile Data. Exhibit 26.1 indicates that CDPD is best suited for transaction-oriented applications. Although these services might prove too expensive for heavy database access, the use of intelligent agents can cut costs by minimizing connection time.

BENEFITS TO MOBILE USERS

Because CDPD uses the existing voice-oriented cellular network and off-the-shelf hardware for implementation, it is cost-effective. There are, however, additional benefits to users besides economy. These benefits include:

- **Efficiency.** CDPD transmits both voice conversations and data messages using the same cellular equipment. Using a single device, it is a versatile and efficient way to communicate. The digital data does not disrupt or degrade voice traffic, and vice versa.
- **Speed.** Having a maximum channel speed of 19.2 Kb/s — a four-fold increase over competing mobile radio technologies — CDPD is the fastest wireless technology available on the WAN.
- **Security.** With encryption and authentication procedures built into the specification, CDPD offers more robust security than any other native wireless data transmission method, preventing casual eavesdropping. As with wireline networks, users can also customize their own end-to-end security.

- **Openness.** Because CDPD is an open, nonproprietary standard, it promotes low equipment costs and broad availability of hardware and software.
- **Flexibility.** Because it uses existing cellular radio technology, CDPD units are capable of transmitting data over both packet- and circuit-switched networks, allowing applications to use the best method of communication.
- **Reliability.** Because CDPD uses existing equipment on the network (i.e., routers), as well as time-tested protocols based on TCP/IP, the highest quality of wireless data service is assured. CDPD also provides excellent penetration within buildings.
- **Worldwide Reach.** CDPD can be used in conjunction with existing cellular systems around the world. These systems already serve 85 percent of the world's cellular users.

Because CDPD allows the network to operate more efficiently by providing digital packet data over the voice network, carriers also realize maximum flexibility, simplified operations and maintenance, and cost savings. Carriers can offer enhanced messaging services such as multicast, cellular paging, and national short-text messaging. CDPD allows portable access to a variety of information services.

In effect, CDPD extends client/server-based applications from the LAN environment into the wireless arena. This extension provides nearly limitless possibilities for future wireless data services.

EQUIPMENT REQUIREMENTS

CDPD is not without its problems. Even though CDPD takes advantage of the existing circuit cellular voice infrastructure to send data at up to 19.2 Kb/s, existing cellular modems cannot be used on CDPD-based networks. Modems designed for CDPD networks are still larger and more expensive than those designed for circuit cellular.

CDPD-only modems cost about $500; modems that handle both CDPD and circuit cellular run about $1,000. When the cost of CDPD modems drops to the $200 range, expense will no longer be a barrier. Also, carriers are considering subsidizing the cost of Cellular Digital Packet Data modems, the way they currently do with cellular phones, when users sign up for service.

NETWORK ARCHITECTURE AND PROTOCOLS

The CDPD specification defines all the components and communications protocols necessary to support mobile communications. Exhibit 26.2 shows the main elements of a CDPD network.

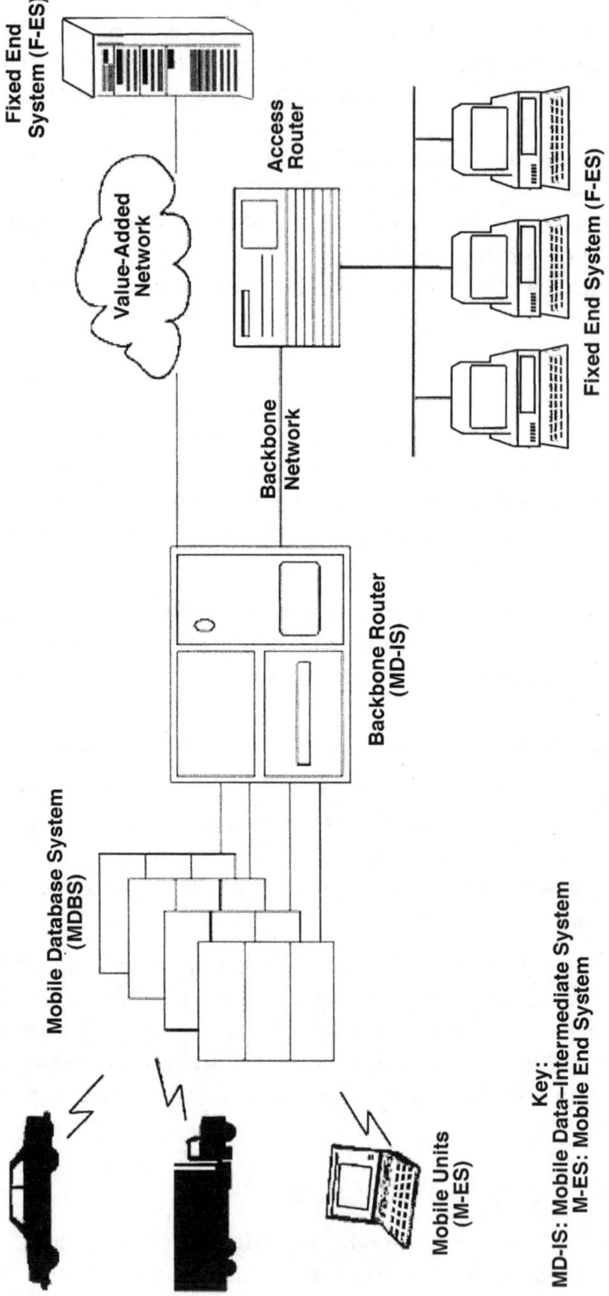

Exhibit 26.2 CDPD Network Architecture.

Mobile Data-Intermediate Systems

The backbone router, also known as the MD-IS, uses the location information derived from the mobile network location protocol to route data to the mobile units, which are referred to as M-ES. Information on the link between the backbone router and an MDBS is transmitted using a DLL protocol. Communications on the other side of the backbone router are handled using internationally recognized protocols. This ensures that standard, off-the-shelf systems can be used in the network infrastructure, and that computer systems currently in use can be accessed by CDPD networks without modification.

Mobile Database Systems

The MDBS provides the relay between the cellular radio system and the digital data component of the CDPD network. The MDBS communicates with the mobile units through radio signals. Up to 16 mobile units in a sector can use the same cellular channel and communicate as if they were on a LAN. This communications technique is known as DSMA. After MDBS turns the cellular radio signal into digital data, it transmits the data stream to its backbone router, typically using frame relay, X.25, or the PPP.

Mobile-End Systems

Although the physical location of a mobile-end system, or mobile unit, may change as the user's location changes, continuous network access is maintained. The CDPD specification stipulates that there will be no changes to protocols above the network layer of the seven-layer OSI model, so that applications software will operate in the CDPD environment. At the network sublayer and below, mobile units and backbone routers cooperate to allow the equipment of mobile subscribers to move transparently from cell to cell or roam from network to network. This mobility is accomplished transparently to the network layer and above.

OSI Protocols. The recommendations of the CDPD consortium were designed using the OSI reference model (see Exhibit 26.3). The model not only provides a structure to the standardization process, it offers recommendations regarding protocols available for use in the CDPD network.

Network Layer Protocols. The CDPD overlay network may use either the OSI CLNP or IP at the network layer. These protocols have virtually the same functionality: they both interpret device names to route packets to remote locations.

IP has been used for more than ten years and is one of the most popular protocols today. Its inclusion in the CDPD specification is intended to accommodate the vast number of networked devices already using it.

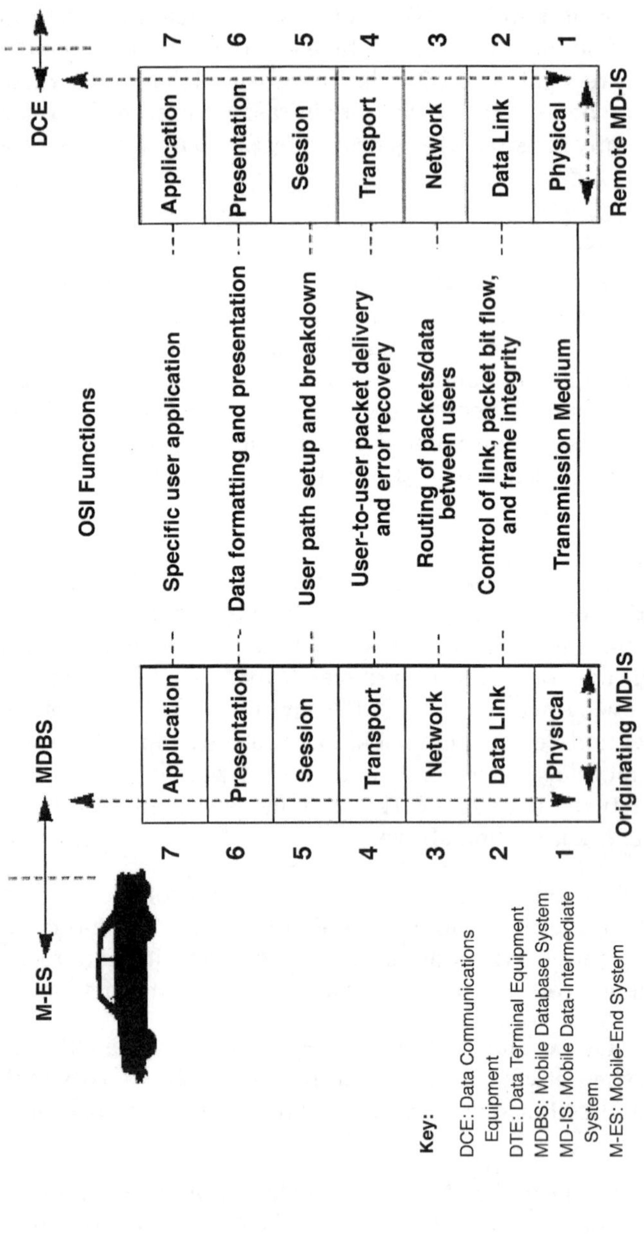

Exhibit 26.3 The OSI Reference Model.

Application Layer Protocols. Applications required to administer and control CDPD networks use OSI-defined protocols. OSI-defined application-layer protocols are widely accepted and have been tested to ensure robust, open communications among CDPD service providers. The use of these protocols provides a level playing field for manufacturers of the CDPD infrastructure equipment. Therefore, service providers can be confident that the various network elements will communicate together, and that no single manufacturer can exert undue influence on the market.

Examples of OSI protocols that operate at the application layer and can be implemented for CDPD network administration and control are explained as follows:

- The CMIP is the object-oriented management standard for OSI networks developed by the ISO.
- The X.400 message handling system is a global messaging standard recommended by the ITU-TSS, formerly known as the ITTCC, that defines an envelope, routing, and data format for sending E-mail between dissimilar systems.
- X.500 directory services is a standard for directory services recommended by the ITU/ISO that operates across multiple networks used to convey E-mail. It allows users to look up the E-mail addresses of other users they wish to communicate with.

MOBILITY MANAGEMENT

Traditionally, the network address of the end system has been used to determine the route used to reach that end system. CDPD is unique in allowing mobile units to roam freely, changing their subnetwork point of attachment at any time — even in midsession.

To find the best route for transmitting data to an end system, CDPD mobility management definitions describe the creation and maintenance of a location information database suitable for real-time discovery of mobile unit locations. Three network entities — the mobile units, the home backbone router, and the serving backbone router — participate in mobility management.

Mobile units are responsible for identifying their unique NEIs or network layer addresses to the CDPD network. As the mobile unit moves from cell to cell, it registers itself with the new serving backbone router. Each NEI is permanently associated with a home backbone router. The serving backbone router notifies the home backbone router of a mobile unit when it registers itself in the new serving area. Mobility management makes use of two protocols: the MNRP and the MNLP.

MOBILE NETWORK REGISTRATION PROTOCOL

MNRP is the method mobile units use to identify themselves to the network. This information is used to notify the network of the availability of one or more NEIs at a mobile unit. The registration procedure includes the information required by the network for authenticating the user's access rights.

The MNRP is used whenever a mobile unit is initially powered up and when the mobile unit roams from cell to cell. In either case, the mobile unit automatically identifies itself to the backbone router so its location can be known at all times.

MOBILE NETWORK LOCATION PROTOCOL

MNLP is the protocol communicated between the mobile serving function and mobile home function of the backbone routers for the support of network layer mobility. MNLP uses the information exchanged in MNRP to facilitate the exchange of location and redirection information between backbone routers, as well as the forwarding and routing of messages to roaming mobile units.

INFORMATION PROTECTION

To facilitate the widespread acceptance of CDPD by cellular service providers, the specifications define methods for ensuring the security of customer information while still providing an open environment for mobile users. Cellular service providers are legitimately concerned about protecting information about their subscriber base from each other, yet the nature of the service dictates that carriers exchange information with one another to provide subscribers with full mobility. For example, when a user who is usually served by Carrier X in Chicago roams to the Carrier Z service area in Boston, Carrier Z must be able to find out whether that user is authorized to use the network. To do that, Carrier Z queries the Carrier X database about the user's access rights using the network equipment identifier. Carrier X provides a simple yes or no response. The details concerning the identity of the user, types of service the user has signed up for, rates being charged, and amount of network usage are all protected.

CDPD NETWORK BACKBONE

The internal network connecting the backbone routers (i.e., MD-ISs) must be capable of supporting CLNP and IP. The backbone routers terminate all CDPD-specific communications with mobile units and MDBS, producing only generic IP and CLNP packets for transmission through the backbone network.

Mobile-End Systems Protocols

As noted, the requirement that mobile units support IP is meant to ensure that existing applications software can be used in CDPD networks with little or no modification. However, new protocols below the network layer have also been designed for CDPD. These protocols fall into two categories: those required to allow the mobile unit to connect locally to an MDBS, and those required to allow the mobile unit to connect to a serving backbone router and the network at large.

Digital sense multiple access is the protocol used by the mobile unit to connect to the local MDBS. DSMA is similar to the CSMA protocol used in Ethernet. DSMA is a technique for multiple mobile units to share a single cellular frequency, much as CSMA allows multiple computers to share a single cable. The key difference between the two, apart from the data rate, is that CSMA requires the stations on the cable to act as peers contending for access to the cable in order to transmit, whereas in DSMA the MDBS acts as a referee, telling a mobile unit when its transmissions have been garbled.

A pair of protocols permit communications between the mobile unit and the backbone router. The MDLP uses MAC framing and sequence control to provide basic error detection and recovery procedures; the SNDCP provides segmentation and header compression.

In addition to segmentation and header compression for transmission efficiency, other important features of S-NDCP include encryption and mobile unit authentication. While the cellular network provides a certain amount of protection against eavesdropping because of its channel-hopping techniques, the applications expected to be used on the CDPD network require definite security — competing businesses must have the confidence that their information cannot be seen by competitors. SNDCP encryption uses the exchange of secret keys between the mobile unit and the backbone router to ensure that there can be no violation of security when transmitting over the airwaves. The authentication procedure guards against unauthorized use of a network address.

TRANSPARENT OPERATION

Complete mobility is one of the key goals of CDPD networks. Because applications software must be able to operate over the network, the network itself must make any required operational changes transparently.

For example, the mobile units must automatically identify themselves to the network using the MNRP protocol, which recognizes the network addresses of mobile units whenever subscribers power on their computers or move to a new cell.

Data sent to a mobile unit is always sent through its home backbone router — another example of transparent operation. The home backbone router maintains an up-to-date table of the locations of the mobile units it is responsible for, thus making it possible to send connectionless data transmissions to a roaming mobile unit at any time. The home backbone router sends the data to the current serving backbone router. This scheme ensures that data reaches an end system regardless of its location, while keeping internal routing table updates to a minimum.

A connectionless service is one in which a physical connection need not be established in order to transmit data because the network is always available. In this scheme, each block of data is treated independently and contains the full destination host address. Each packet may traverse the network over a different path. A connection-oriented service, on the other hand, requires a destination address in the first packet only. Subsequent packets follow the path that has been established.

SENDING DATA FROM A MOBILE UNIT

Registration Procedure

Before a mobile unit can begin transmission, it enters into a dialogue, called the "registration procedure," with the backbone router serving the area in which it is currently located. This dialogue identifies the mobile unit's OSI network layer address to the CDPD network. The serving backbone router tells the home backbone router responsible for that mobile unit that it is requesting service. The home backbone router authenticates the mobile unit, checking such things as the user's access rights and billing status. The registration procedure must be performed whenever the mobile unit is first powered on or roams to a new serving backbone router.

Once the registration and authentication procedures are completed, the mobile unit begins sending data. The mobile unit is now on what appears to be a LAN connecting all such units operating within the cell of the telephone network. The LAN is really a single set of transmit and receive frequencies shared by the mobile units that access this cellular LAN using the digital sense multiple access technique.

The cells, or DSMA LANs, are interconnected by the backbone routers in much the same way that routers connect Ethernet or Token Ring LANs. The serving backbone router examines the data sent by the mobile units, looking for the destination address. By comparing the destination address with those in its tables, the backbone router can send the data to the appropriate destination by the best path available (see Exhibit 26.4). The user can now log on to the portable computer, access shared services such as CompuServe, or send information directly to other roaming mobile units. When sending data from a mobile unit to other computers, the CDPD network

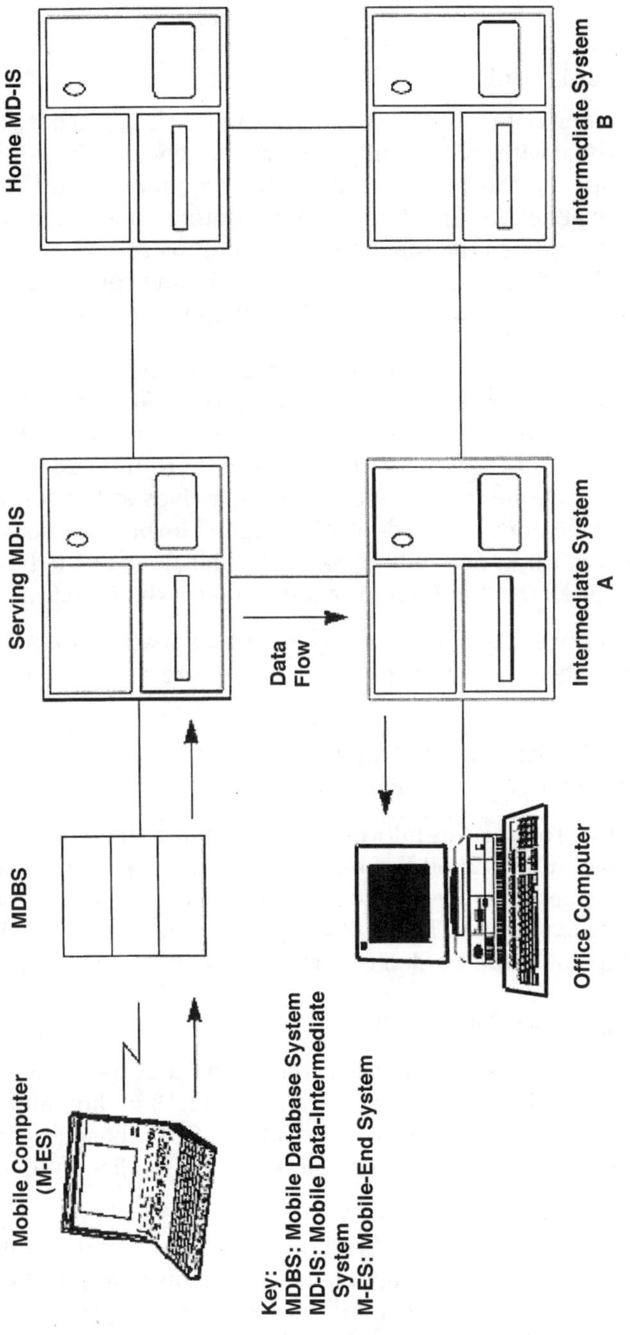

Exhibit 26.4 Potential Data Paths.

must only ensure that the user is allowed to transmit. Once the user is authenticated, data is sent in a manner similar to the way it is sent in current LAN internetworks.

Sending Data to a Mobile Unit

On the return path, when data is sent to the mobile unit, the CDPD network must be prepared to deal with mobile units that are actively mobile — moving in a car, for example. In this case, it is likely that the mobile unit would move from one serving backbone router to another during the session. The CDPD network accommodates the roaming mobile unit by always sending its data to its home backbone router. The home backbone router always advertises itself as the destination router for the mobile units it serves.

Redirect Procedure. The home backbone router knows the current location of the mobile unit because of the registration procedure. When sending information to a mobile unit, the home backbone router encapsulates it into frames using the CLNP protocol and sends them to the address of the current serving backbone router. Once the data arrives at the serving backbone router, it is de-encapsulated into its original form to be sent to the mobile unit. This method of handling data transmissions at the home backbone router is called the "redirect procedure" (see Exhibit 26.5).

The redirect procedure takes advantage of the identification done during the registration procedure. The registration procedure serves two purposes:

- To authenticate the user's access rights
- To identify the current location of the user

The redirect procedure uses this information to minimize network overhead. The alternative, in which all the backbone routers would update their global routing tables whenever a mobile unit moved, would saturate the network with overhead traffic. The CDPD network permits full mobility, but without imposing an undue burden on the network infrastructure.

IS THERE A MASS MARKET FOR CDPD?

Industry analysts estimate that the wireless data market could be worth $10 billion by the year 2000, providing service to about 13 million mobile data workers. Bell Atlantic Mobile, an early provider of CDPD-based services, predicts that as much as a fifth of its cellular revenues could come from data services by the end of the decade.

The eventual availability of low-cost CDPD modems does not guarantee a mass market for CDPD. For this to happen, commonly used applications must be adapted to the technology APIs. APIs are required to optimize new and existing applications for use over relatively low-bandwidth wireless

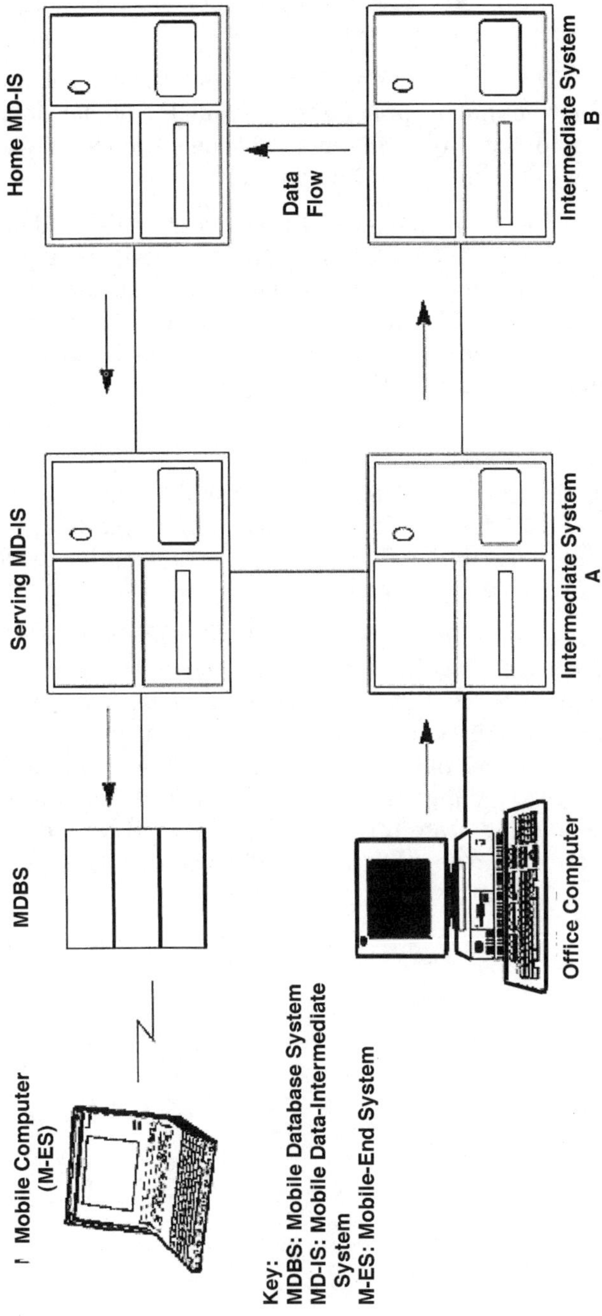

Exhibit 26.5. The Redirect Procedure.

links with their high overhead and delay. After overhead is taken into account, the wireless CDPD link will top out at 14.4 Kb/s. The average throughput falls between 9 Kb/s and 12 Kb/s, depending on the number of errors and retransmissions.

Although CDPD is ideal for vertical niche markets such as fleet dispatch and field service, the more popular applications include E-mail, facsimile, and RDB access. Several toolkits are available to give new and existing applications the capability to run over CDPD networks.

To improve application performance over low-bandwidth wireless links, middleware that uses intelligent agents is now available that allows laptop users to query a corporate database using a software agent at the corporate site. If the user does not want to wait for a response to a query, or the connection is lost, the agent collects the information and sends it over the wireless network when the user makes the next connection.

CONCLUSION

Mobile users who are already committed to wireless data services are among the early users of CDPD service. As the price of CDPD modems fall, coverage increases, and more applications become optimized for CDPD, the technology will have even wider appeal.

Cellular digital packet data networks are appealing because they offer seamless nationwide availability; work with the vast installed base of computers, applications, and data networks; and make use of existing private and public network infrastructures, encompassing all products and user equipment. The ultimate success of CDPD is, of course, closely tied to industry efforts to standardize its implementation. A universal standard for cellular packet data would facilitate terminal capability, allow users to roam between service areas, and simplify the introduction of wireless data services.

Section V
Wide Area Network

Chapter 27

An Overview of the Telecommunications Policy Process

Leo A. Wrobel

CORPORATE NETWORK MANAGERS GRAPPLE WITH MANY CHOICES IN NEW NETWORK SERVICES — AND GETTING RATES AS LOW AS POSSIBLE. A company cannot create a new communications network without understanding regulation and the new levels of competition ushered in by the Telecommunications Reform Act. This chapter discusses what communications managers need to know about the structure of regulation bodies, who sets service rates, and how to negotiate with commissioners and carriers.

INTRODUCTION

The current regulatory environment in the U.S. has built itself over the years piece by piece. By attempting to drive this process in the future, the public can foster innovation and bring the information superhighway to life. This chapter looks back at the past of the regulatory environment and ahead to what lies in store.

Most people know that Alexander Graham Bell invented the telephone. But Mr. Bell's invention may never have succeeded without the work of another man, Theodore Vail, who formalized the concept of the Bell System that controlled the telephone marketplace for three quarters of a century.

The turn of the century brought with it competition in the telephone business. The good news was that customers had a choice of telephone providers, and the bad news was that if customers did not subscribe to the same network, they could not talk to each other. A customer using the red network phone could not talk to someone on the blue network.

Under the Theodore Vail concept, one company would own all the telephone services and control the telephone call from one end to another. Vail had the concept of a vertical monopoly, in which the phone company would actually own the customer's phones, the telephone lines in their homes, and the long-distance lines.

THE BEGINNING OF COMPETITION

In the early 1950s, a rubber coupling was invented that fit over a telephone handset so that other people in the area could not hear what the person using the telephone was hearing. The telephone company sued the company that made the couplings. Because the phone company had tariffs in place that commanded the force of law, anyone using one of these "hush-a-phones" could legally have their phone service terminated. The manufacturer fought AT&T at the Federal Communications Commission (FCC) and won the right to have its equipment connected to the phone system in 1956. In some ways, regulatory activism was necessary even then.

Three years later the FCC took an even bolder step in deciding that telephone companies were "common carriers" and that they had an obligation to serve all segments of the population. Some companies required special lines or special services or wanted to have private lines dedicated to their specific use. (Because these lines were not available to the public, they were in fact called private lines.) The commission decided that it would be permissible for companies other than the monopoly phone companies to provide these private lines to corporations.

In 1968 another milestone was reached in the Carterphone decision. The Carterphone was an acoustically coupled device designed to connect a phone to a two-way radio. Even though the device was passive (i.e., did not electronically connect to the phone network), AT&T sued. The owner of the Carterphone company was an ambitious entrepreneur with an aggressive legal team. They convinced the commission that the Carterphone was not a threat to the market and that AT&T was trying to preclude competition. The commission ruled that the Carterphone could be connected to the network, as well as other devices that were "privately beneficial and not harmful to the public network."

Long-Distance Competition

Over the years, the FCC has allowed more inroads into what had been the exclusive telephone monopoly. In long distance, the commission allowed MCI Communications Corp. to tariff a service that allowed connection to the switched network on one end of a private line, allowing the caller to circumvent AT&T's long-distance network. To AT&T, this was an illegal intrusion into its market by "cream skimmers." To MCI, the actions AT&T took to preserve its monopoly on switched service was more evidence of an attempt to control the marketplace in violation of antitrust laws.

AT&T protested, the FCC agreed with AT&T, MCI appealed, and the Court of Appeals in 1977 ruled that MCI's actions were lawful. AT&T was compelled to connect MCI (and other competitors) to the switched marketplace.

In 1984, after a 10-year antitrust suit, Judge Harold Greene of the D.C. Federal District Court approved an agreement between the U.S. Department of Justice and AT&T that split the Bell System into AT&T and seven "Baby Bells" — NYNEX, Bell Atlantic, Bell South, Ameritech, Southwestern Bell, U.S. West, and Pacific Bell. This agreement molded today's calling environment, with competitive long-distance services and far greater consumer choice. Today the availability and cost of services is far superior to what was available ten years ago, before these profound changes.

Results of the Telecommunications Reform Act of 1996

Today, equally profound changes are taking place. The Telecommunications Reform Act of 1996 promises to be just as far-reaching as the AT&T divestiture was 10 years ago.

Telephone Companies in the Entertainment Business

AT&T, for example, is already carrying "Primestar" satellite direct television under its name. Many of the Bell operating companies (i.e., the Baby Bells) are delivering cable TV services in both their areas and in other companies' service areas.

Entertainment Companies in the Telephone Business

The near future may hold the "Internet channel" from cable TV providers. Motorola, for example, is rolling out hundreds of thousands of cable modems designed to allow users to surf the Internet at 10 Mbps, which is more than 150 times faster than an ISDN line, and more than 300 times faster than the fastest modem technology available today. Cable companies and competitive local exchange carriers are also making heavy investments in switching technology to go after the local telephone company monopoly.

Competitive Local Exchange Carriers (CLECS)

Users who want single providers can get one for local, long-distance, private line, and Internet services from competitive local exchange carriers. Billions are being spent building out local networks, usually with vastly superior technology. The availability of this leading-edge technology at truly competitive prices will be unsurpassed as a milestone in development of the nation's telecommunications infrastructure.

Foreign Entrants

More ventures, such as the recent announcement of British Telecom's acquisition of MCI, can be expected as position jockeying continues between the giant telecommunications companies worldwide. These events only scratch the surface at the highest levels when attempting to explain the major paradigm changes taking place right now. Smart technologists are honing their skills to take advantage of these remarkable opportunities before their competition does.

WHO CREATES TELECOMMUNICATIONS POLICY?

Some of players who create telecommunications policy include:

- AT&T
- The Baby Bells
- MCI, Sprint, and other IXCs
- Competitive access providers
- The FCC
- The courts
- State regulatory commissions
- Consumer groups (i.e., lobbyists)
- Congress
- Cable companies
- Cellular companies and wireless services
- Bellcore
- Large institutional users

THE RATE-SETTING PROCESS: WHO SETS THE PRICE?

For a company to take advantage of all the unique opportunities happening right now, it needs to understand the basics of how rates are constructed. Rate cases are divided into two parts:

- **Revenue.** Revenue determines how much money the telephone company should be allowed to earn.
- **Rate design.** Rate design determines who has to pay. In most rate cases, it is the utility vs. everyone else in the revenue portion of the case, because everyone believes that holding rates down in general will help their special interest when it comes to rate design.

In determining how much revenue the utility is entitled to, the commissions are restricted by court decisions that say that the return must be high enough to attract investors into the enterprise. The components of the revenue requirements part of the rate case can be broken down into the formula:

$$RR = (v - d)^{*}r + e$$

where:

RR = revenue requirement
v = the value of the rate base
d = the amount of depreciation of the rate base
r = the rate of return
e = expenses

This is a "cost plus" formula, in which the telephone company is granted the opportunity to recoup all of its expenses and get a return on its investment. In a major rate case, virtually every element and subelement of this formula is hotly contested. It can be expected that cost will be an overriding factor in any new custom network integrating advanced and emerging network services that a company designs and requests.

The Rate Base

The "v" in the "cost plus" formula is the value of the rate base. The rate base is the investment made by the utility in the plant that serves the customer. Thus, items such as telephone poles, wires, switches, and buildings are in the rate base. To be included in the rate base, the plant must be "used and useful." For example, if there is an old switch being kept in a switch building, but it is only used perhaps once every two years, it would not generally be considered used and useful. In most commissions, the evaluation of the rate base is made on an original cost basis — in other words, how much the company paid for the copper wire at the time of purchase.

Depreciation recognizes that as rate base equipment is used, it loses its value. The utility has to account for that loss in value and recover enough revenues so that the plant can be replaced at the appropriate time. In a time of rapidly changing equipment, depreciation has to recognize that the plant may still be able to function but have very little market value. For example, as the telephone companies move from copper to optical fiber, depreciation also has to recognize that it may be cheaper simply to write off the copper wire in the ground. Depreciation is a very complex area in a rate case and should be handled by a highly qualified expert.

RATEMAKING

Once the regulators determine how much revenue the regulated company should earn, they have to figure out who is going to pay that revenue. In telephone rate cases, there are literally hundreds of services that have to be priced, which can be unbelievably complicated. Moreover, companies that are not traditionally active in the regulatory process (unless they are telecommunications companies, of course) are almost always the ones who end up paying the most.

Ratemaking involves figuring up the costs of producing a service, including the profit for that service, and charging accordingly — this is the starting point for most regulators. However, the entire process is not that easy.

Many of the services that are offered by the telephone company are offered on an undivided basis. For example, a telephone company switch may be providing touch-tone service to one phone, business service to another phone, contacting the operator on yet another phone, and providing caller ID to another telephone. Yet each of these services needs to have a price determined out of the general cost of the switch. Thus, ways must be devised to split up the costs, which helps set the price.

Residual Ratemaking

To preserve universal service, regulators adopted a process known as residual ratemaking. For example, if a cost study demonstrates that the current rates for residential service only cover one-half of the cost, then a strict cost-based system would say that residential rates should therefore be doubled. In the real world, this is probably not a practical option. In addition to the political fallout of such a move, it would harm universal service. Regulators faced with this problem came up with the concepts of "value of service" and "residual ratemaking."

Value of Service

There are some products, such as three-way calling or call waiting, that do not cost a great deal once they are placed in a switch. Those services are optional. Therefore, those services can be priced based on the perceived value (i.e., value of service) to the subscriber, which is another way of saying that the provider can charge what the traffic will bear.

If the services are priced too high, they will not contribute as much toward maintaining universal service. The revenues obtained from those services allow the telephone company to remain whole, while the favored classes of service — residential and small business — get small increases in their basic line. If households have call waiting, they may see a small increase on their total bill, but most people will not take the time to drop their optional services.

Access Charges

Possibly the largest contributors to residual ratemaking are the long-distance carriers (i.e., the IXCs), who pay for residual ratemaking through their access charges. Access charges were devised at divestiture to make sure that the local companies were not harmed by the divestiture from AT&T. They were originally calculated on a revenue replacement basis, which included a hefty contribution to universal service.

Long-Run Incremental Cost (LRIC)

One of the ways of showing that a plan does not have a major impact on the concept of universal telephone service is by calculating the long-run incremental cost (LRIC) of service, and including a contribution to universal service on top of that cost. That is how commissions do it. However, LRIC costs are very different when the dynamics of fiberoptics are considered. In the world of fiberoptic transmission, there is no need to lay additional cable to provide additional capacity. The basic concept is to determine how much more one unit of output costs to produce.

For example, how much more does it cost to carry one more passenger on an airplane? There may be some cost associated with the additional fuel and the passenger's meal, but assuming that all of the other passengers' tickets have already paid for the flight, one additional passenger should have a very small cost. If a utility commission and potential network providers could be made to see such logic, consumers would all be better off.

DRAFTING A CUSTOM TARIFF OR AGREEMENT WITH A CARRIER

A tariff is the contract between the telephone company (i.e., the carrier) and the customer companies. Although it is debatable whether there will even be tariffs in the future (as of the time of this writing, the FCC is trying to get carriers to retract tariffs), they are still important. Companies need to have some kind of contractual agreement with their carriers, whether in the form of a contract, a tariff, or both.

It is important to remember that a tariff commands the weight of the law and preempts contractual law. That means the carrier can change a company's contract if a ruling at the state or federal level changes the rules for the tariff. An AT&T Tariff 12, for example, could change in 24 hours if the FCC rules against the terms in it as part of a broader issue, which is all the more reason to stay in the loop with regard to matters of policy.

Tariffs are similar to very detailed contracts. However, a tariff is actually a law. The carrier cannot lawfully deviate from its tariff, and the customer is bound to pay the tariff charge.

Although this seems reasonable on the surface, there can be some unexpected side effects. For example, the customer and the carrier agree on a specific price for telephone service to the customer's plants and the carrier promises to file that tariff with the appropriate regulatory bodies. The customer's bills change to reflect the agreement, and the customer operates for several years believing that it is paying the right amount. Suddenly, someone discovers that the revised tariff was never filed. The customer is now liable for all of the payments that should have been made under whatever tariff was the true lawful tariff. This is binding even if the provider company has gone out of business.

Customers should be especially careful in their understanding of the tariff because it is the rule book through which services will be purveyed.

INFLUENCING A REGULATORY BODY

Regulatory bodies include the FCC, state public utility commissions, Congress, and the courts. The most active are the state commissions and FCC. A wide variety of influences will apply to each commission, and they will each respond to those influences in a general manner. However, there are a number of factors that a commissioner may be taking into account that go beyond the narrow confines of any particular case.

The commissioner is influenced by many different parties that may include the general public, the governor, the legislature, the utilities, the interveners, the staff, the other commissioners, the press, or the courts. Commissioners are also limited — no commissioner can be an expert in everything their commission does all of the time. As one commissioner put it, it is like being the coach of the football team, the basketball team, the track team, the baseball team, the swimming team, the choir, and the debate team all at once. It can be done, but only with the help of many other people, and those people would have to be experts in their areas. Thus, it is helpful to know who helps the commissioners and what their interests and positions are.

In most commissions, the commissioners are assisted by aides who work directly for them. The aides help summarize the positions of the parties, privately assist the commissioners in understanding the issues, and may meet with people when the commissioner is unable to.

The amount of influence of the staff and the aides varies from state to state. In some states, the staff is extremely influential on the decisions of the commission; in others, the staff may be out of favor and other parties may have more influence.

Negotiating a Price with the Commissioner

A business trying to negotiate a price with the commissioner should show a return to the carrier over the price of the contract and a contribution to universal service. Commissioners will not approve a solution based only on reduced cost for fear that the business will pass the costs along to its customers. An organization should concentrate on networking solutions that not only grow the business, but also create growth, jobs, and national competitiveness. These are ideals that commissioners can endorse and approve. An organization should make its money through building business, not cost cutting.

Negotiating with the Carrier

Once a company has made initial contact with a carrier through a request for information (RFI) or request for proposal (RFP), the carrier goes back to headquarters and establishes what it can provide for the company and at what price. The business should do a little "homework" to determine what network elements are available and what prices are generally being imposed for them.

The interests of the business and the carrier are not necessarily the same; however, this is not necessarily a bad thing. The carriers want to maximize their long-term profits and market share, and the company seeks advanced services at rates as low as possible. If the carrier builds a custom network for a company, it may resell the same network to other companies, which creates new revenue opportunities. The company for whom the network was originally built has access to new, enabling technology first.

The company should make sure that at each stage of their cost studies, the carriers employ only long-run incremental cost principles, not just numbers from an arbitrary tariff. The company should start with the carrier's cost and negotiate from there. It is exceedingly difficult to get the carriers to release these costs, however. A basic nondisclosure agreement often sets the carrier's mind at ease.

The company should require the carrier to desegregate (i.e., unbundle) their services into the smallest possible network components. The company may not need all of the services that the carrier wants to bundle together. The carrier's progress should be monitored, and even after the services are approved by the commission, the company should ensure that someone is constantly checking the carrier's performance under the contract. The agreement between the company and the LEC may be in the form of a contract, a tariff, or both.

Converting "Feature Packages" Business Requirements to a Custom Tariff

All of the conditions are right for the new network when the company finds a willing trading partner, an accommodating utility commission, and an environment ripe for change. A detailed business analysis should be performed at this stage. All of the business processes should be condensed into a digestible white paper for executive management that can be used to justify the money required for the new network and measure the results of the innovations using traditional return-on-investment methodology.

For example, the team creating the white paper comes up with several business processes (e.g., imaging, multimedia, network management, and native LAN connectivity) that would make the operations departments a runaway success. Each of these processes should be described in the paper and numbered as Feature Package 1, Feature Package 2, Feature Package 3, and so on. This exercise will:

- Articulate the business need to management.

- Become the basis for the request for information that will be sent to the carriers.
- Articulate the business need to the carrier.
- Allow the carrier to craft a custom response based on business, not technology objectives.
- Show a utility commissioner why the business needs a certain technology and correlate it to something that equates to growth, jobs, and cost rather than just technology.

This features package can be crafted into the company's ideal network. It is possible to create a network that does not just spur revenue, but creates competition and new technology that everyone benefits from.

Chapter 28
The Future of Wireless Communications

Robert Rosenbaum

JUST A FEW YEARS AGO, WIRELESS LANS WERE SEEN AS A STRATEGIC MEANS OF MEETING THE DEMANDS OF TODAY'S HIGHLY DYNAMIC WORK ENVIRONMENTS AND MOBILE USERS. Today's 100 Mbps and 1 Gbps wired LANs have blunted this notion. Industry pioneer Robert Rosenbaum provides the reader his unique perspective on where wireless communication is going.

Today, the applications for wireless communications can be divided into three major categories:

1. wide area
2. metropolitan area
3. local area

Wide area wireless consists of applications that involve inter-city and longer distance communications. Examples of wide area wireless networks include cellular telephony, satellite-based networks such as Very Small Aperture Terminal (VSAT), paging systems, and such data-only systems as ARDIS from Motorola. Metropolitan wireless networks include competitive local exchange carrier (CLEC) bypass systems and building-to-building systems. Local wireless applications are those that are used within the geography of a building. Examples are wireless LANs, cordless telephones, and wireless peripherals like keyboards and mice.

WIDE AREA WIRELESS

Of the three broad categories of wireless systems, the wide area systems have met with the most success in the marketplace. It is in the wide area that wireless systems have the least competition from wire-based systems, and in many situations there is no wired alternative. For a person in a plane, train, automobile, boat, or any moving vehicle, wireless access is the only option. Even when the "road warrior" arrives in the hotel room, wireless access is generally much more convenient and not necessarily any more expensive. Dealing with the variety of hotel telephone systems and the various access codes needed to establish a network connection back to the home office can present a major obstacle that is solved using a nationwide wireless service. It is also almost impossible for a salesperson making a call at a client to establish a network connection to his home office. Once again, the wireless system easily solves the problem. Another major advantage that wide area wireless connections have is that the cost differential with a wired connection is not very significant. The entire wireless wide area situation is in marked contrast to that found in the wireless LAN market.

Typically, wireless data connections use the cellular telephone network in a similar fashion to how most wired modems use the voice dial network. A PC is connected to a jack in a cellular telephone, and the PC then dials the appropriate telephone number to establish a connection. Some companies, such as 3Com, even make cards that fit in the PC with a built-in cellular telephone.

Another example of a wide area wireless system is known as VSAT and uses satellite relays. These systems are used extensively by private companies that have a large number of widely dispersed facilities that they need to get information from or to on a regular basis. Many large retail chains make use of satellite systems to get inventory information from the various stores, and to send updated pricing.

The major challenge facing the wireless wide area market is global uniformity. A wireless link, either voice or data, may work in the United States or in Europe, or in Japan or in South America, but only in one of these and not the others. The problem is complex, involving not only different frequency bands, but also different modulation and encoding methods. It may be that a frequency band used in the United States for cellular telephone is used for an entirely different application in Japan. The allocation and usage of frequencies is discussed further in a later section.

One of the systems that is now trying to address the global uniformity issue is Iridium. Motorola was the creator of the system and the major investor. Iridium consists of 66 satellites in geocentric orbits 485 miles above the earth. The satellites act like airborne cellular towers. The control center for the satellite network is located in Virginia, with additional tracking stations in Hawaii and Canada. Gateways between the satellites and the public telephone system are located throughout the world to provide links between Iridium users and the rest of the telephone network. Exhibit 28.1 illustrates the Iridium wide area wireless network.

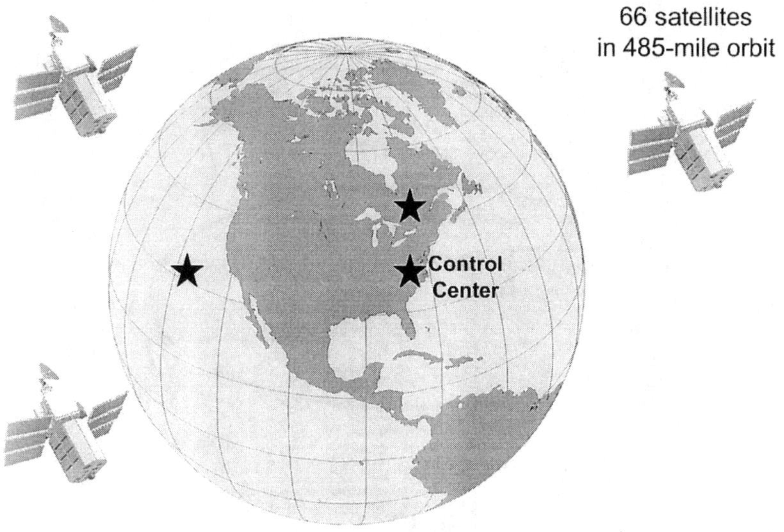

Exhibit 28.1 Iridium Wide Area Wireless Network.

METROPOLITAN WIRELESS

After wide area wireless systems, metropolitan wireless communications holds the most promise for large market size and growth. These wireless systems have the potential to break the stranglehold that the regional Bell operating companies (RBOCs), along with a few cable companies, have on local access (the local loop) for both voice and data. Virtually all communications from the home or office first have to traverse the RBOC-controlled local loop. All the competition to date has been among the long-distance connections and the Internet Service Providers (ISPs), both of which must rely on the local carriers to access the home or office.

Wireless technology is making good on the promise to provide low-cost voice, video, and data access directly to the end user. The key to the cost equation is the lack of expensive infrastructure. Typically, wireless local loop providers will install a high-speed fiber backbone loop around the metropolitan area where they intend to provide the service. In addition, the vendor needs to purchase the rights to use a block of frequencies in the locale as well. These are the only two major infrastructure costs associated with a wireless local loop provider. As the vendor signs new customers, an antenna would be installed on the roof of the customer's building. This antenna would be aimed at the closest wireless hubs attached to the fiber backbone. As the number of customers connected to a particular hub increases, the vendor has the option to add an additional hub and divide the

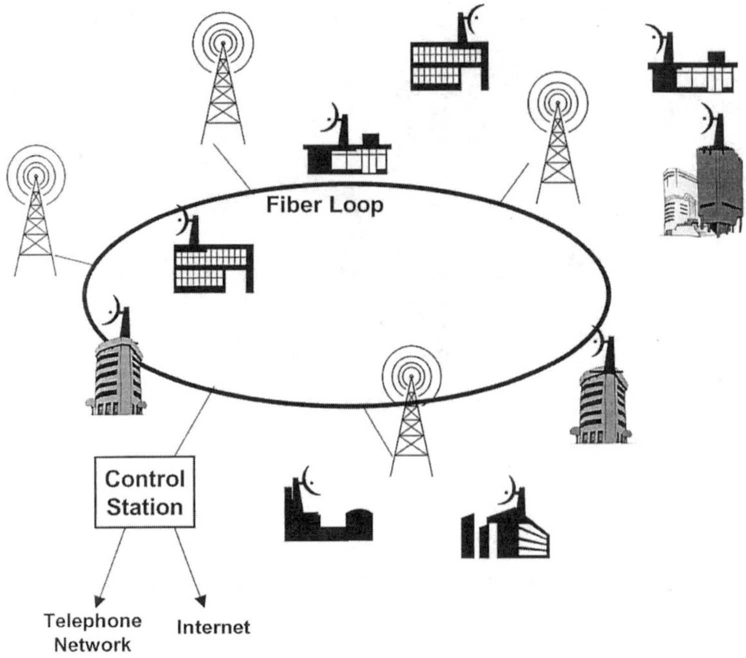

Exhibit 28.2 A Wireless Local Loop.

traffic to provide more bandwidth. Adding an antenna to a customer's building can be done in as little as several hours. Another advantage that wireless has is that, should the end customer discontinue service, the wireless vendor can redeploy the equipment to a new customer. Exhibit 28.2 illustrates a wireless local loop.

Most of these wireless local loop systems use radio-based technology at fairly high frequencies. In the United States, the bands being used for these purposes are 38 GHz, 28 GHz (known as Local Multipoint Distribution Service, LMDS), and 2.5 GHz (known as Multipoint Multichannel Distribution Service, MMDS). The architecture of these systems uses a point-to-multipoint topology, with the hub being the multipoint access node and the customer premise equipment being the end point. The coverage area these systems provide goes from about 0.5 to 20 miles, with the longer distances associated with systems using the lower frequency bands. All of these systems require line of sight from the CPE to the hub. A comparison between LMDS and MMDS is shown in Exhibit 28.3.

An interesting side note to the market potential for wireless local loop is that it is also a revenue generator for the U.S. government and other

Exhibit 28.3 Comparison Between LMDS and MMDS.

	MMDS	LMDS
Frequency band	2.5 GHz	28 GHz
Bandwidth available	200 MHz	1000 MHz
Radius of coverage	10 to 20 miles	1 mile
Affected by rain	No	Yes
Hub to CPE capacity	1 Gbps	1.5 Gbps
CPE to Hub capacity	1 Kbps	200 Mbps

countries. In February 1998, the U.S. government held an auction for the LMDS bands. This auction resulted in about $500 million in revenue for the government.

INBUILDING WIRELESS

Local or inbuilding wireless systems have seen moderate success in the market in the voice area, and very limited application in the data area. The most successful voice applications have been cordless telephones. The most promising wireless data application was wireless LANs. Unfortunately, the market never grew beyond niche applications, and it does not appear that it ever will. It may be interesting to note that the first local area network, the father of Ethernet, was AlohaNet, a wireless radio-based network developed by the University of Hawaii. However, the wireless LAN development effort suffered a long hiatus. The basic problem is that the wireless technology has too many limitations. The most significant restriction is speed when compared to wired LANs. With wired LANs now available at speeds up to 1 Gbps, wireless LANs are struggling to reach speeds of 10 Mbps. Today's battery-operated wireless LAN systems operate in the 1 to 3 Mbps range. Those requiring external power, and thus lacking mobility, are available featuring speeds in the 10 Mbps range.

Wireless LANs also do not fare well in a cost comparison analysis with wired LANs. Wireless LANs are approximately five times the price per connection. This is a very steep premium for mobility. In addition, trying to find a location within a building that does not have a wired LAN connection is rare. Wired LAN connections are ubiquitous because companies have been investing heavily in wiring their facilities. Since telephone wiring is needed to all offices, pulling extra pairs for data comes almost for free. One of the original justifications for using wireless LANs — that is, the ease of accommodating moves, adds, and changes — has disappeared. By using LAN switches with virtual network features, it is extremely easy to segment individual wired connections onto different subnetworks to respond to a

Exhibit 28.4 Tradeoffs associated with wireless LANs and infrared RFs.

	RF < 10 GHz	RF > 10 GHz	Infrared
Range w/ directional antennas (focused infrared)	Up to 10 miles	Up to 2 miles	Up to 3 miles
Range w/ nondirectional antennas (diffuse infrared)	Up to 1 mile in open space; up to 500 feet in areas with solid walls	Up to 1000 feet in open space; up to 100 feet in areas with solid walls	Up to 500 feet; open areas only
Speed using batteries	3Mbps	3Mbps	10Mbps
Speed using external power	10Mbps	10Mbps	50Mbps
Power requirements	Medium	High	Low
Solid object penetration	Yes	Some	No
Safety	?	?	Non-laser infrared believed to be safe
Interference sources	In the unlicensed band: microwave ovens, HAMs	None significant since mostly licensed bands	Sunlight
Cost	Medium	High	Low
Government regulated	Yes	Yes	No
Best applications	Buildings with walls; building to building	Areas with many interference sources; building to building	Conference rooms and lecture halls; building to building

network change within seconds. This switching capability also results in increased bandwidth and security. The ability to segment wireless LANs is difficult because overlapping cells can interfere with each other, and the speeds typically have to be reduced to accommodate direct overlapping.

Neither of the two technologies used in building wireless LANs and infrared RFs resolve any of these issues. Exhibit 28.4 outlines some of the tradeoffs associated with these two wireless techniques.

One of the "hot" issues surrounding the wireless LAN market has been that of standards. Some industry analysts have attributed the bucolic market growth to the lack of a standard. This is basically a red herring. Most purchases of wireless LAN equipment have been for specific applications: to provide network connections on a manufacturing floor, to connect two buildings located across the street from one another, or to provide patient monitoring in a hospital. All of these systems are supplied by a single vendor and, as a result, the need to provide intercommunications with other wireless devices does not exist.

Today, a wireless LAN standard finally does exist, although many would argue that it is totally irrelevant. Almost seven years in the making, the IEEE 802.11 group finally accepted a 2 Mbps standard for a particular radio-based technology called "frequency hopping spread spectrum." The effort began in 1990 with over 100 members representing several companies, ranging from as large as Motorola, AT&T, Apple Computer, NCR, National Semiconductor, and Sun Microsystems, to startups like Windata. Unfortunately, the agendas of each of the members were all over the map. Some members were primarily interested in voice applications, with data being a secondary concern. Other members had just the opposite view. Some members insisted portability was the primary issue; others insisted that speed needed to be the primary concern. A few members were only interested in wireless personal devices (e.g., Star Trek badges). Added to this confusion was the lack of any working products that had any market acceptance. Motorola had a product named Altair that met the speed goals but only operated in a licensed, U.S.-only frequency band that they virtually controlled. In addition, the band was not available in much of the rest of the world, and the equipment was bulky and expensive and had little hope of ever being portable. NCR had WaveLAN which held the promise of being battery-operated and hence portable, but operated at a speed of only 1 Mbps and the technology did not have much hope of increasing beyond 3 Mbps. Some will argue that the standard finally was able to be approved when most of the committee members lost interest either in the entire market (e.g., Apple) or decided to focus on vertical applications where the standard did not really matter. The next iteration of the IEEE 802.11 standard appears to be proceeding in a much smoother manner. The proposal is an 11 Mbps one that is being driven by Lucent and Harris Semiconductor. It is an example of how successful standards such as IEEE 802.3 Ethernet happen — namely, through cooperation. Nevertheless, even this higher speed standard may be too little too late to have any significant impact on the wireless LAN market.

Wireless LAN topologies are either peer-to-peer or point-to-multipoint. Most wired LANs, including Ethernet, tend to be peer-to-peer; that is, any station on the LAN segment can talk to any other station without having to go through a central station or access point. Exhibit 28.5 depicts a peer-to-peer wireless LAN environment. In this example, there are two workgroups: A and B. The transmissions of the two groups are kept distinct by using different frequencies or codes. However, it can be difficult to control membership in a group. In addition, as the picture shows, overlapping groups can interfere with one another.

Wireless LANs used in corporate environments tend to be point-to-multipoint, where all stations in a particular area or cell communicate through an access point. The access point generally does double duty in providing a gateway function for packets going between the wired and wireless LANs,

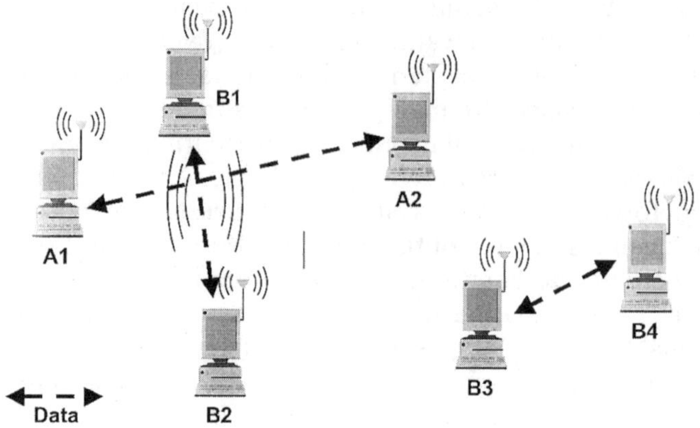

Exhibit 28.5 A Peer-to-Peer Wireless LAN Environment.

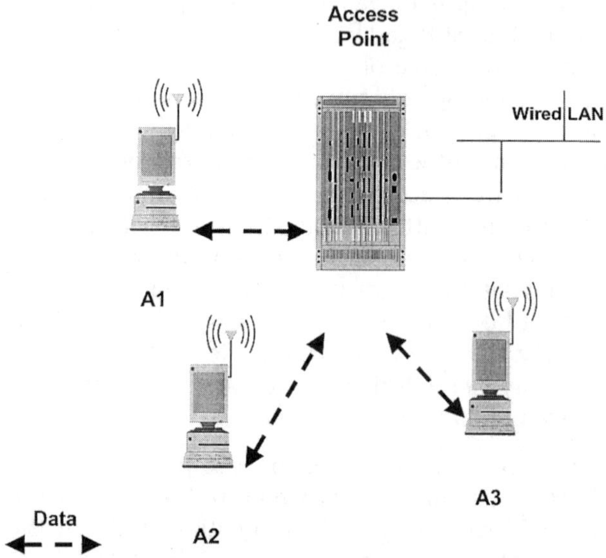

Exhibit 28.6 Point-to-Multipoint Wireless LAN Topology.

as well as acting as a relay and control point for all stations within its area. This works best in this environment because most packets from or to a wireless station go to or come from a resource on the wired infrastructure and there is little station-to-station traffic. Exhibit 28.6 depicts a workgroup utilizing a point-to-multipoint wireless LAN topology.

The access point provides some very important services to create order out of what could be a very chaotic wireless world. These services include:

- **Security:** The access point can easily maintain a list of station addresses with passwords that are allowed to access the overall network. Thus, if a wireless device that is not supposed to be a member station comes within range either accidentally or intentionally, the access point can refuse to forward any packets originating from it.
- **Multicell conflict resolution:** In situations where multiple companies are located in the same area (e.g., adjacent floors or office suites in a building) and each has a wireless LAN installation, the access points can detect this situation and use different frequencies or wireless encoding schemes to minimize interference. Combined with the security services, the access points could also ensure that Company A stations only communicate with the Company A access point and do not accidentally gain access to the Company B network.
- **Reception problems:** Unlike wired systems where every station can reliably hear every other station, wireless LANs cannot practically make such guarantees. The "near–far" problem, as it is called, is common in the world of wireless communications. Station A may be able to hear station B, and station B may be able to hear station C, but that does not mean that station A can hear station C. Using the access point, all stations within a cell need only be able to hear and be heard by the access point.
- **Store and forward services:** In the event that a wireless station moves out of range or is temporarily shut down for power conservation, the access point can store packets addressed to the station, thus eliminating the need to retransmit them over the entire network from the original source.
- **Expanded wireless range:** Access points can be strategically located throughout a building to provide expanded coverage in a similar manner to the way cellular telephone systems are built. Thus, as a station moves out of range of one access point and into range of another, an automatic handoff can occur, making an interruption-free session for the user of the wireless device.

Peer-to-peer wireless networks tend to be used more in ad hoc networks. Situations where a group of people with wireless devices get together for a meeting and then disperse lend themselves to peer-to-peer architectures. Contact information and documents tend to be broadcast around the group where more than one person may be interested in receiving the information. In addition, there is generally not a larger network available for use; hence, there is no need for the wired-to-wireless gateway function.

It was a goal of the IEEE 802.11 standard to support both types of topologies. A wireless device supporting 802.11 could operate in either environment. Thus, if the device found that there is an access point in the area, it would use it in a point-to-multipoint topology. If there is no access point, the wireless device would communicate with other wireless devices in a peer-to-peer mode.

ALLOCATION OF WIRELESS FREQUENCIES

Wireless technology can be divided into two major types: radio (or RF) and infrared (or IR). The similarities and differences between the two can best be understood by looking at where they fit into the electromagnetic spectrum. The electromagnetic spectrum is measured in units called Hertz, indicating number of cycles per second or frequency of a signal. The term Hertz was named after one of the early radio researchers, Heinrich Hertz. The electromagnetic spectrum is divided into a number of segments: very low frequencies from 0 Hz to 3 kHz, which includes most of the human audible range; that portion of the spectrum designated as radio goes from 3 kHz to 300 GHz (299,999,997,000 Hz of bandwidth); infrared ranges from 300 GHz to almost 10^{15} Hz; the visible light spectrum is a relatively small amount of spectrum and is squeezed in between infrared and ultraviolet; the ultraviolet, named because it is adjacent to the frequency associated with violet light, goes up to 10^{17} Hz; next comes X-rays, followed by gamma-rays, and finally cosmic-rays ending at 10^{25} Hz. Due to historical reasons, all the "radio" frequencies from 0 Hz to 300 GHz are controlled and regulated by each country. In the United States, the regulatory agency is the Federal Communications Commission. The regulatory body for the European countries is CCITT. Japan has its own agency to control the various radio frequencies. In contrast to the radio frequencies, the light bands, including visible, infrared, and ultraviolet, are not subject to any government controls.

The primary reason that governments regulate the use of the radio bands is to prevent interference. It would be of little use to the listener to have two FM radio stations located in the same city attempting to broadcast on the same frequency. In order to operate a radio station in the United States, one must apply to the FCC, which in turn grants a license to broadcast at a particular frequency and power. The FCC determines the frequency and power based on where other radio stations using the same frequency are located, and what their service area is intended to be. There are bands set aside in the radio spectrum for unlicensed use. Devices using these unlicensed frequencies are generally restricted to very low power, typically under 1 watt, and have other rules to which they must adhere. By unlicensed, the FCC means that the user of the equipment does not need a license to operate it, which is different from the rest of the radio spectrum. Instead, the manufacturer must submit the equipment to the FCC, which tests it to ensure that it meets all the rules for operation in the particular unlicensed band. Examples of equipment that use these unlicensed bands include garage door openers, cordless telephones, and most wireless LAN equipment. Although it might appear as though cellular telephones operate in these unlicensed bands because the end user of the telephone does not need to apply for a license, they in fact operate in licensed bands. In this case, the owner/operator of the network must apply for and be granted a license for a block of frequencies within a geographic area.

Because of this government control, radio-based systems have an ease-of-use disadvantage compared to those based on light. Anyone can build or operate a light-based wireless system. There is no need to be concerned about homologation issues. An infrared keyboard purchased in the United States can be used anywhere in the world. A wireless radio mouse built for Japan cannot be legally used in other countries. Another advantage that light technologies have over RF is the huge amount of bandwidth that infrared has compared to RF, almost 10,000 times. This translates into very high speed capabilities. Unfortunately, light technologies suffer from some technical drawbacks, as explained later.

To understand the operational aspects of wireless systems, it is useful to be aware of some additional characteristics of the frequency spectrum. One of these is that as the frequency of a signal increases for a fixed transmission power, the ability of the signal to penetrate a solid object decreases, and the effective distance at which the signal can be received decreases as well. For example, an AM station can be received outside, in a building, and in a tunnel. An FM radio station, which uses a higher frequency, can be received in wood structures but has more difficulty being received in a metal building, and cannot be picked up at all in a tunnel of any length. Light, an even higher frequency, cannot penetrate any solid barrier. These various characteristics are used by wireless system designers to achieve the desired range for the end product. For example, light can only be used in systems designed for use in open, small areas. On the other hand, it is easier to secure a light-based wireless system.

WIRELESS DESIGN ISSUES

One of the major issues beyond what frequency is used to transmit the information is how the information is packaged or encoded into whatever frequency is chosen. The primary goals are reliability, speed, cost, security, and safety. These issues all have major design consequences on the wireless product.

Reliability is of extreme importance when dealing with data, and less so if only voice is involved. The “static” one hears on a radio station may be annoying to the listener. This static, or interference, has much more serious consequences in regard to data transmission. Speed is important for both data and voice applications; higher speeds mean the ability to accommodate more voice channels. Cost has a major impact on market acceptance.

Security is generally a major concern of a buyer of wireless systems. In many ways, this is a perception issue. The general feeling is that wired communications are confined to the wire and the wire itself must be physically tapped to listen to the signal, whereas wireless signals are unrestricted. The fact that most wired systems are basically antennas themselves and the signals can quite easily be monitored from a great distance is generally ignored. To address the security issue, wireless designers have gone to great lengths to protect the information carried in the signal. Much of the technology in today's commercial systems has origins in the military. A rather interesting story relating to wireless security involves the 1930s movie star, Hedy Lamarr. Ms. Lamarr, along with a friend George Antheil, who composed music especially for player pianos, have one of the first patents related to a technique that is now known as frequency hopping spread spectrum. They devised the scheme to ensure that radio-guided torpedoes used by the U.S. Navy during World War II could not be jammed by the Germans. The patent for a "Secret Communications Systems" was granted by the U.S. Patent Office in 1942 (#2,292,387) and expired in 1959. The invention went unused until 1962. Ms. Lamarr received much more recognition for her nude appearance in the movie *Ecstasy* than for her invention, which is still being used today. It was finally in 1997 that the Electronic Frontier Foundation gave the two inventors public recognition for their scientific and patriotic contribution.

One of the big unknown issues surrounding wireless communications is: How safe is it? The safety issue raises its ugly head every once in awhile, such as what happened with the man in Florida who sued on behalf of his wife who had died of brain cancer. The man claimed that the reason his wife got brain cancer was because of her frequent use of a cellular telephone for business. The cellular telephone manufacturers, led by Motorola, presented a case supporting the safety of these devices and the man was never able to present solid evidence to prove otherwise. The question still remains as to what the effects of exposure to radio transmissions are to the human body. The medical school at Johns Hopkins has done some studies, as have some other researchers, but the money to fund these studies has been meager. One can almost come to the conclusion that people now depend on these wireless devices so much that no one wants to know if there is a problem — ignorance is bliss!

WIRELESS OUTLOOK

The prospects for wireless data networks are excellent, with a few reservations. There is little question that growth in the wide area and metropolitan area networks will continue at a very rapid rate. End-user demand in the wide area market is the primary driving force for growth because there is no other convenient way to connect. Low deployment costs will fuel the growth of the metropolitan markets, with wireless local loops presenting very attractive cost alternatives to the small-office/home-office (SOHO) users. Wireless LANs, on the other hand, would appear to have lost the race as a significant alternative to wired LANs. Wireless LANs will continue to be deployed in niche applications such as hospital and factory floor situations. Overall, wireless will only amount to an extremely small percentage of devices connected to LANs.

Chapter 29
Frame Relay Testing and Training

Steve Greer, Peter Luff, and Sean Yarborough

BECAUSE IT PROVIDES EFFICIENT, COST-EFFECTIVE TRANSFER OF BURSTY, BANDWIDTH-INTENSIVE APPLICATIONS, FRAME RELAY IS NOW THE MAIN ACCESS PROTOCOL FOR WIDE AREA NETWORK COMMUNICATIONS. This chapter discusses the advantages and disadvantages of frame relay, as well as the reasons why training and testing are important.

INTRODUCTION

Frame relay — a network access protocol for high-speed, bursty data applications — has been a data communications buzzword for the past five years, but only recently has it begun to serve its purpose. According to Vertical Systems Group, a consulting group in Dedham, MA, the number of U.S. public frame relay network subscribers, which was only 590 in the early 1990s, is projected to increase to 8,210 by the late 1990s. The main catalyst for this explosion is the need for high-speed LAN interconnection.

Frame relay networks provide companies with cost-effective and efficient transfer of such bursty bandwidth-intensive applications as file transfer, E-mail, graphics, and imaging applications. Frame relay decreases the cost of LAN interconnection through the use of statistical multiplexing, which allows many subscribers to utilize the bandwidth on a single circuit. The cost savings is a direct result of the reduction in the use of dedicated leased lines, which tend to be underutilized because bursty transmission is sporadic.

LANs have become so ubiquitous that subscribers expect to be able to communicate across multiple interconnected LANs just as easily as when they communicate over a single LAN. Frame relay addresses LAN interconnection extremely well (i.e., up to T1 speeds), so subscribers can enjoy the same quality data transmission over the WAN that they have come to expect over the LAN.

It is important that service providers and subscribers understand frame relay technology and its advantages and disadvantages. For example, frame relay has no inherent error correction. It assumes that a network successfully transmits data and avoids errors. As a result, users need to test the network for errors and comprehensively train employees on frame relay technology to reduce errors. This chapter briefly reviews the features of frame relay technology and explains some of the important tests to perform on frame relay devices and the network.

ADVANTAGES OF PACKET-SWITCHING FRAME RELAY NETWORKS

The main advantage of frame relay is that it is a packet-switching technology. Packet-switching networks send data from source to destination based on each packet's unique destination address. Once the data is packetized, it can be statistically multiplexed.

Statistical multiplexing allows many subscribers to share the same bandwidth by assuming that not all subscribers will be using the bandwidth at the same time. This avoids high-cost point-to-point connections such as dedicated leased lines that employ circuit switching. Leased lines are expensive because they are rented and dedicated for exclusive use 24-hours a day, seven days a week. With leased lines, subscribers pay for bandwidth whether it is being used or not. Statistical multiplexing provides multiple data connections through the network simultaneously, and no single customer pays for exclusive privileges. This results in a significant cost advantage over circuit-switched networks.

A second advantage of frame relay is that its variable-length frames and its low overhead provide excellent network throughput and low delay of data. The variable-length frames allow frame relay to encapsulate protocols well. Frame relay is protocol independent, so its payload can carry a variety of higher-layer LAN protocols, such as the IP. Because the network does not concern itself with error correction and flow control, overhead in the Frame relay network is low. Therefore, the network uses most of its resources switching user data.

Frame relay realizes another advantage through the CIR. The CIR represents the data traffic level that the network plans to support under typical network conditions; it is agreed upon by the service provider and the subscriber. The advantage of a CIR for the customer is that once it is agreed upon, the service provider should be capable of transmitting at or below the CIR. For example, if an average of 56 Kbps throughput is required between two sites, then the CIR should be equal to or greater than 56 Kbps. Typically, data traffic sent below the CIR passes through the network at a high priority, and data sent in excess of the CIR has a lower priority. This low-priority traffic is the first to be dropped when subscribers create network congestion. The CIR serves as the basis of a well-planned network as well as a billing mechanism.

Finally, frame relay has worldwide industry support from manufacturers, standards organizations, and service providers. This is important because it ensures that there is a high level of interoperability between devices of different manufacturers in different countries.

DISADVANTAGES OF FRAME RELAY

Although frame relay carries data more efficiently because it has no inherent error correction utilities, this lack of frame management can be detrimental. Frames can be errored because of transmission impairments in the network (i.e., a bit error corrupts a frame, which is then dropped by the network).

In frame relay, the error correction is left up to the user's intelligent devices (i.e., routers), which can discover discrepancies and request a retransmission. In addition, the reliable deployment of technologies such as T1, DDS, and fiberoptics decreases the need for error correction within the network.

Network problems can occur when subscribers exceed the CIR. As previously mentioned, the CIR is the amount of data traffic, agreed upon by the service provider and the subscriber, the network is planned to support under normal network conditions. When too many subscribers exceed the CIR, then a situation may develop in which the network becomes congested and begins to drop frames to alleviate the congestion. For example, 20 banks are all connected to a frame relay network. During the day, they all transmit under their CIR for routine electronic communication. At 2:00 p.m., however, all of these banks send their daily transactions to another location for processing, which causes network congestion. If only one bank transmits over the CIR, network congestion probably will not occur. It is the aggregate of many banks transmitting over their CIRs that causes congestion and dropped frames.

Frame relay frames are variable in length; therefore, they cause variable delay in the network. For example, a short frame can be switched quickly by the frame relay network. However, longer frames take longer to process and switch. This creates a variable transit time between long and short frames. Integrated data, voice, and video applications cannot be delayed; so these applications are not best suited to frame relay. They are better suited to integrated switched digital networks (ISDN) or cell relay technology such as ATM.

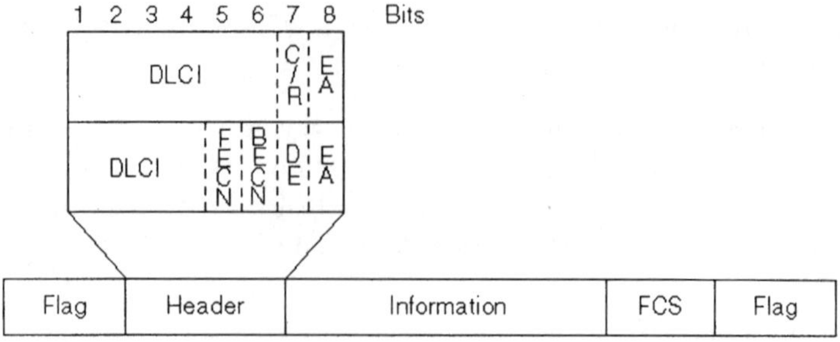

Key:
BECN backward explicit congestion notification
C/R command/response
DE discard eligibility
DLCI data link connection identifier
EA extended address
FCS frame check sequence
FECN forward explicit congestion notification

Exhibit 29.1 The Frame Relay Frame.

ANATOMY OF THE FRAME RELAY FRAME

Frame relay's packet-switched technology is based on its older relative, X.25. In a packet-switched network, data is subdivided into individual packets, each with a unique identification and destination address. These packets, or frames, as they are called in frame relay, are variable in length. Each frame contains a header, information field, frame check sequence (FCS), and two flags. (See Exhibit 29.1.)

The Frame Header

The header of a frame contains information on connection identification. Connection identification is managed by the data link connection identifier (DLCI). The DLCI is an identifier (i.e., address) associated with each permanent virtual circuit (PVC). A PVC is the path that is set up by the service provider routing data from point A to point B. Once a PVC is defined, it requires no setup operation before data is sent, and no disconnect operation after data is sent.

The DLCI

The DLCI is usually 10 bits long, but it can be extended. Each DLCI indicates a different PVC or end point for data. In other words, the DLCI must be unique for specific destinations. For example, for San Francisco, the DLCI is 45; for New York it is 49. These values ensure the proper routing of information within a network (see Exhibit 29.2). The DLCI can be extended to increase addressing capabilities. There are two, one-bit extended address

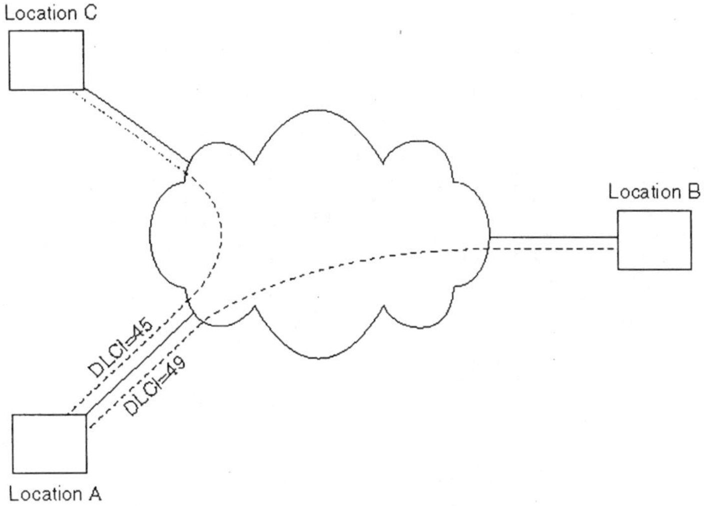

Exhibit 29.2 Sample DLCI Values.

fields in the header that indicate if another octet has been added for extended DLCI purposes.

Congestion Control

Also within the header are congestion control bits, which identify whether congestion is present. These congestion bits are important because they inform the user of potential errors in the network. If congestion overloads a particular switch, frames will be discarded by the network. There are three main types of congestion control bits: forward explicit congestion notification (FECN), backward explicit congestion notification (BECN), and discard eligibility (DE). (See Exhibit 29.3.)

The header also includes the Command/Response field. This one-bit field is used by many HDLC-based protocols to indicate whether a frame carries a command or a response. It is passed transparently through the frame relay network.

The Information Field

The information field contains variable numbers of octets (up to 4,096 octets for some implementations) and encapsulates many protocols, including TCP/IP, internetwork packet exchange (IPX), sequenced packet exchange (SPX), SNA, and X.25. The information field is the largest part of the frame and contains the encapsulated protocol header as well as user data.

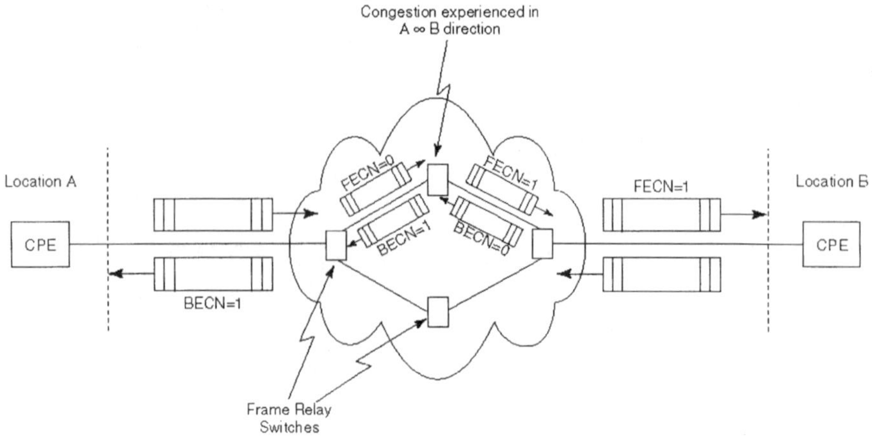

Exhibit 29.3 Congestion Control.

The FCS Field and Flags

The FCS field is used for error detection. The transmitting end devices (i.e., routers) apply a complex algorithm to the data in the frame and then place the result in this field as the frame is sent to the receiving equipment (i.e., the router at the far end). Each node along the way recalculates the FCS based on the received data and compares the result to the FCS within the frame. If the results are not identical, then the frame is discarded by the switching node or the end device.

Flags serve as eight-bit idle codes that delineate frames on the circuit.

Local Management Interface

The local management interface (LMI), a signaling protocol, was instituted by Cisco Systems, Inc. (Menlo Park, CA), Digital Equipment Corp. (Maynard, MA), Northern Telecom, Inc. (McLean, VA), and StrataCom, Inc. (San Jose, CA). It allows the subscriber and the network devices to share information on the status of DLCI on the link.

In addition, the LMI provides a mechanism for the network to recognize that end devices are operating correctly. The LMI proposal was accepted with minor additions by the ANSI and the ITU. These standards organizations slightly modified and standardized the LMI, which is now represented by ANSI T1.617, Annex D, and ITU Q.933, Annex A.

THE IMPORTANCE OF TESTING

Frame relay networks sacrifice function for speed by relying on intelligent devices and an efficient network to prevent errors and retransmissions.

With such a reliance on the network, it is imperative that users test the physical medium, the PVC, and the internetwork, both out of service and in service, when implementing frame relay. Specifically:

- Out-of-service testing is performed when a circuit is taken offline, or when no subscribers are active on the circuit being tested.
- In-service testing can be done while a circuit is active.

The testing of the transmission facilities, as well as testing the internetwork service, ensures that the service itself, the internetwork devices, and the physical media provide seamless network performance for customers.

Testing the WAN Facilities

Users need to carefully test the transmission facilities before implementing a frame relay network. There are three parts to a comprehensive test strategy: testing the transmission facilities between the subscriber and the POP switch, testing the PVC through the network, and conducting a lost frames analysis.

Bit Error Rate Testing. To test the transmission facilities between the subscriber and the POP switch, an out-of-service BER test should be performed. This test examines the transmission facilities by evaluating the percentage of bits received in error compared to the total number of bits received.

Typically, the BER test set generates one or more complex sets of pseudorandom bit patterns, and then transmits these patterns along network segments at rates dependent on the facilities. The BER test set provides the stimulus as well as the receiver to monitor the response of the transmission segment being tested. This test can pick up bit errors, bipolar violations (BPV), cycle redundancy check (CRC) errors, and other errors that may cause the network to drop frames. When these errors occur, an end device requests retransmission, which costs money and consumes valuable bandwidth.

Because the pseudorandom patterns used in the BER test do not pass through the switches, this test is useful only between the customer and the POP switch. However, because of frame relay's lack of error correction utilities, BER testing is vital. A clean transmission line must be ensured before any additional investigation can be done.

PVC Testing. Once the transmission facilities are verified, the PVC should be tested. This test verifies end-to-end connectivity between two or more subscribers on a specific PVC. This test is performed by generating and receiving a variety of frame relay traffic loads with the proper DLCI and LMI support over each valid PVC. The test should be able to transmit frames with control of various header bits, like DLCI, FECN, BECN, and DES.

Test results indicate that the transmission facilities are operating with minimal errors and the correct link management parameters, and that the network is properly routing frames across the PVC being tested.

When testing transmission facilities, PVC, and LMI, monitoring of Layer 2 and Layer 1 transmission facility and PVC phenomena is also important. For example, users can determine that a bipolar violation (BPV) at the physical layer (Layer 1) coincides with a FCS error on a Layer 2 analysis, which indicates a transmission facilities problem and not a frame relay problem.

A Lost Frames Analysis. Another important feature of Frame relay testing is determining if any transmitted frames are lost. This test is performed by placing a sequential counter in the data field of the test frames. The test set receiver tracks gaps in the test frames' sequence count. There are usually several reasons why the network may be dropping frames: FCS errors are occurring, the attempted throughput exceeds the CIR, too many frames per second are being transmitted, or the network may be experiencing congestion.

This test also discovers dribbling errors that cause network bottlenecks, and the determination of critical thresholds of frame relay devices. The lost frame analysis gives customers a final assurance when verifying network performance because it can discover errors that previous tests did not catch.

Frame Relay Internetwork Service Testing

The most important reason to test the internetwork service is the increasing number of premium services being offered. These premium services, such as the end-to-end management of network services, rely on LAN/WAN analysis at the higher-level functions that only internetwork testing can verify.

Internetwork service testing begins with the frame relay devices themselves. These devices generate and receive the frame relay frames. Simultaneous monitoring of both sides of these devices can be done to verify that LAN traffic to the router is correctly forwarded and encapsulated in frame relay. This verification of connectivity across the router also ensures correct configuration. The user should test all routers before installation on the network, and before calling the provider's wide area support group when switch problems occur.

The Echo Test. Another internetwork device test that can be performed to verify connectivity across the network is the echo test. The echo test's key applications are for implementation and installation of routers when establishing connectivity, for in-service measurements on in-service and out-of-service equipment, and for performing round-trip delay measurements through the network.

The echo test eliminates the need to take down the service when testing network connectivity. In an echo test, the test equipment bounces (pings) a packet across the network to the far-end router (or host) and awaits a response. It checks the connections through the WAN by monitoring both the outgoing ping packet and the returning response from the remote host.

Simultaneous Monitoring of the WAN and LAN. Simultaneous monitoring of the WAN and LAN should also be performed. By looking at encapsulated LAN protocols, users can verify that customer traffic is correctly addressed, encapsulated, and transmitted over the WAN. In the opposite direction, users should track packets to ensure that Frame relay information is removed, and that they are correctly routed to their destination on the LAN.

Multiport analysis provides the ability to test an internetwork at various access points simultaneously. For example, users could access LAN segments. Once the link is established, users can stress the circuit using a traffic generator.

Commissioning the Frame Relay Service

Once the testing of the frame relay devices and network is complete, users can concentrate on commissioning the frame relay service. This application focuses on emulating the DCE and DTE by generating frame relay packets to see how the network reacts to those packets.

First, the test should provide for Annex D ANSI signaling protocol LMI emulation to imitate network conditions. This action ensures that the network can process the frames across the LAN and the WAN. Once this is accomplished, the user can monitor Layer 2 and LMI signaling statistics. The test equipment should then stress test the line by generating frames to ensure that the network stands up to agreed communication rates. The commissioning process verifies and sectionalizes inside and outside the frame relay cloud.

End-to-End Analysis

Finally, users want to take advantage of the verification of end-to-end LAN/WAN connectivity. End-to-end analysis ensures the integrity of the entire internetwork. With this application, users can generate traffic to remote LANs to verify connectivity or capacity handling. In addition, users can monitor both local and remote trends in network operation to plan for network changes and expansion.

TRAINING ISSUES UNIQUE TO FRAME RELAY

One aspect that is often overlooked in technical articles, books, and magazines is the training of personnel in technology, network applications, and troubleshooting methodology. Yet training is extremely important when implementing a new technology such as frame relay.

THE PROVIDER'S PERSPECTIVE

From the service provider's perspective, training is critical in overcoming a technician's lack of protocol or packet-switching knowledge. These technicians, who must install the frame relay equipment and services, may know how to install DDS or T1 links, but they may have limited protocol experience, and most likely have never been exposed to frame relay. Frame relay training provides the technology background these technicians need. It also provides the installation and troubleshooting background they need to maintain the service.

Training can point out to the technician the prospective problems that occur when testing frame relay and how to solve them. For example, the physical layer may work but the service may not. This is often caused by switch configuration problems such as the switch not being configured, being configured incorrectly, or the provider and subscriber configuring their switches incompatibly.

THE SUBSCRIBER'S PERSPECTIVE

From the subscriber's point of view, training issues on Frame relay technology are less critical, but training on its test applications may be more important. These technicians are comfortable working with circuits and services focused on protocols. For example, an IS department professional may have extensive experience with the technology and the testing of X.25. However, frame relay, unlike X.25, has no error correction, which introduces a series of testing issues. Although the technology may be similar, the test applications are different.

For subscribers, frame relay training provides technology specifics, such as how frame relay applies to their particular network, taking issues like CIR, DLCI connections, and link management into account.

The traditional classroom setting is still popular in many companies. If the company does not have a training department, the data communications manager should consider sending technicians on staff to an outside organization whose specific purpose is to increase workers' comprehension or knowledge level of technology or product use. These sessions can take the form of seminars, classroom sessions, field training, video, distance learning (i.e., teleconferencing), or auditorium-style lectures. Computer-based training might also be available, which provides a self-paced environment that can cover multiple knowledge levels at the same time. As microprocessor technology and memory storage become more advanced, users are able to run complex instruction disks on their personal computers. They can take in the training at their own pace at home, in the office, or on the road.

The Benefits of Training

Training offers a critical advantage to an organization through increased production, reduced cost, and maximized efficiency with respect to frame relay network performance parameters. Training helps employees avoid pitfalls encountered during a new installation. It does this by increasing employee knowledge levels about frame relay technology, and then having them apply this knowledge to everyday work and even future upgrades.

In addition, training enables users to test and manage a frame relay network from the first day of implementation. This kind of knowledge about the network decreases downtime and outages, as well as return trips to the subscriber's premises.

The key to reducing cost and downtime of a frame relay network is to plan and implement training before circuit implementation, because the training offers insight on potential problems before they actually occur. Subsequently, service providers and subscribers save money as downtime is reduced because the job was completed efficiently the first time. Training is an expense, but it provides a company with a return on investment through knowledgeable employees who can efficiently run a frame relay network.

CONCLUSION

Frame relay is now the main access protocol for WAN communications and will likely stay that way for the rest of the decade. But that is not the end for frame relay. The future of frame relay lies in its capability as an access network for ATM-based services, which many consider to be the communication services of the future. Hybrid frame relay/ATM networks will provide the transition between the two networking technologies. Eventually users will run integrated multimedia applications and services over their ATM networks, making full use of technologies for which frame relay helped pave the way.

Chapter 30
Transporting Legacy SNA Traffic Over Frame Relay

Andres Llana

FOR YEARS, IBM MAINFRAME INSTALLATIONS THE WORLD OVER HAVE FORMED THE BASIS FOR HOST-CENTRIC NETWORKS. These networks support many of the world's largest business enterprises (e.g., airlines, financial firms, government agencies, transportation companies, and utilities). All of those networks share communications with remote locations.

Early networks, for the most part, were composed of slow, asynchronous, analog point-to-point or multidrop lines. To guarantee the quality of data transmission, IBM developed a telecommunications protocol known as Systems Network Architecture (SNA).

Although SNA is a successful and reliable transmission solution, users must still manage with a multitude of low-speed lines, each of which represents a single point of failure. However, because of its reliability, network managers still use SNA to support mission-critical applications while they consider other alternatives.

SNA TRANSPORT STRATEGIES

Some mainframe users have attempted to upgrade their front-end processors and remote controllers to support frame relay access, but have found the process to be costly. Furthermore, this upgrade may not represent a long-term solution as IBM migrates its mainframe product line toward distributed client/server environments. In keeping with this trend toward mixed networks—which must support multiple protocols including Token Ring, Ethernet, and Novell (IPX)—routers have become the solution of choice because of their ability to encapsulate SNA traffic in the IP.

Devices built around IP-based architectures use one of two basic methods:

- SDLC to LLC conversion combined with TCP/IP encapsulation
- Direct TCP/IP encapsulation of the SDLC frame to carry native SDLC traffic from a 3270 SDLC-type device through the network

The name for these processes may vary by vendor and the protocol overhead may also vary with each device. Overhead approaching 70 bytes per packet is the mean, while data link switching (DLSw) can add as much as 50 bytes per packet.

Once these SNA packets have been encapsulated, they can be encapsulated into frame relay packets. This is not true of frame relay switches, which eliminate the IP overhead by placing SNA frames directly into frame relay frames.

IP routing has not proven to be satisfactory for some users because running SNA through routers has resulted in unpredictable performance in some networks. Another problem is that the cost levels for routers have been higher than some frame relay switches, which also integrate voice traffic.

WHY FRAME RELAY?

Frame relay, which can trace its origins to ISDN, has matured into a practical and affordable replacement for low-speed, dedicated, private-line networks. For example, typical frame relay service is provisioned at 56 Kbps as compared with 4.8 or 9.6 Kbps for an SDLC multidrop line. In some cases frame relay service can be provisioned at T1/E1 levels ranging in channel speeds from 128 Kbps to 1.53 Mbps (fractional to full T1).

Frame relay for the user is a straightforward service in which there are logically defined data link connection identifiers (DLCIs). In addition, permanent virtual circuits (PVCs) or permanent logical links (PLLs) provide for a virtual permanent connection. Frame relay transparently passes most protocols if the endpoint frame relay access device (FRAD) supports encapsulation (DLSw or RFC 1490).

Guaranteed Bandwidth

Frame relay networks provide a guaranteed bandwidth to each PVC, however, any unused bandwidth can be shared among several active users. The guaranteed bandwidth of a PVC is specified as the committed information rate (CIR). A user's traffic may exceed the CIR rate and is referred to as the "burst rate" of the PVC. Traffic that exceeds the CIR rate may be eligible for discard.

Frame Relay Switching for More Efficient Network Operation

A FRAD located at the user's node is required to transform data into frame relay packets. A frame consists of a flag, which indicates the beginning and end of a data packet. A frame header contains the destination of the data in the packet. The user data is the data to be transmitted. An FCS validates the integrity of the data. Frames may be variable in length.

The frame relay network is composed of a group of interconnected switches that relay data across the network. The DCLI information contained in the header is used to forward the data across the network to its destination. The frame relay switch uses a two-step process to forward a frame of information. First, the network management tool checks the integrity of the frame using the frame check sequence (FCS). If the network manager finds an error, the frame is discarded. Second, the network manager validates the DLCI destination address. If the address is invalid, the packet is discarded. Any frame that is not discarded is forwarded to its destination. Because the network manager makes no attempt to correct a frame or request a retransmission, the network can operate more efficiently, leaving error recovery to the end stations.

More Performance for Less Money

FRADs are a more cost-effective approach for SNA networks because they simply convert SNA traffic into a frame relay packet, leaving the SNA data intact. Because the FRADs do not have to support multiprotocol routing (as do routers), they are more efficient and cost less. Furthermore, packet-based transport mechanisms offer a predictable performance for SNA networks. For example, in an IP network, packets containing time-sensitive information could get delayed behind a router queue while waiting to get transferred across the network. Frame relay transport mechanisms, on the other hand, offer a predictable performance by providing a connection-oriented session between devices in a remote site and the mainframe host. Furthermore, prioritizing mechanisms ensure that priority packets get through.

HOW SNA TRAFFIC IS CARRIED ACROSS FRAME RELAY

A frame relay network frames native SNA packets and sends them on their way. This is accomplished by a well-defined encapsulation process. Two possible encapsulation technologies for transporting information across a frame relay network are DSLw and RFC 1490.

DSLw

DSLw only supports the encapsulation of SNA, APPN, and NetBIOS traffic across a frame relay network. DSLw is an effective solution in an IP-based backbone supporting SNA and APPN multiprotocol environments. This is particularly true when a network has some native TCP/IP traffic.

The strength in the DSLw technology is its ability to dynamically locate SNA, APPN, or NetBIOS destinations using the destination's Layer 2 MAC address. DLSw conducts the destination location process using a search protocol. This protocol is an optimized version of source-route bridging's (SRB) broadcast search mechanism across a wide area network (WAN) using TCP/IP.

The DSLw technology involves the administration and management of 32-bit IP addresses, which requires defining the IP and the address at all remote local area network (LAN) bridge/routers, as well as at each source DSLw bridge/router. In addition to defining DSLw-specific IP addresses, it is necessary to specify all SRB-related parameters such as LAN segment numbers, bridge numbers, and MAC addresses at all destinations. This is definitely not a "plug-and-play" process.

DSLw does not support SNA or advanced peer-to-peer networking (APPN) routing or any kind of session layer switching between mainframes or APPN nodes. DSLw does support virtual point-to-point connections between pairs of MAC addresses across a TCP/IP WAN.

RFC 1490

RFC 1490 is the industry standard and an enhancement to the original RFC 1294 standard for frame relay encapsulation. RFC 1490 provides a low-overhead, highly optimized means for transporting SNA and APPN traffic. It also supports other multiprotocol traffic across a WAN. RFC 1490 is best suited for packet-switched networks, particularly when a network is totally frame relay-oriented.

RFC 1490 does not require the same amount of predefinition as DSLw. For example, in contrast to DSLw, RFC 1490 administration only involves Layer 2 PVC setup and definition to begin operation over a frame relay network. In contrast to DSLw, an RFC 1490 solution provides a greater level of service (e.g., bandwidth allocation and traffic prioritization). Some RFC routers can complement the inherent transparent alternate routing within a frame relay network by serving as frame relay switches. These switches also support dialback, which provides a vital disaster recovery resource.

RFC 1490 encapsulates SNA/APPN traffic directly within a frame relay with just a 10-byte-type control and logical link control Type 2 (LLC2) header prefixing the SNA message units (with no other protocols or headers involved). RFC 1490 is a Layer 2 minimum overhead, with an ultralight encapsulation technique including frame relay flags, two-byte address fields, and frame relay sequence. The total overhead for any SNA/APPN message unit is 16 bytes. Therefore, RFC 1490's overhead is minimal compared with that of DSLw, as DSLw relies on TCP/IP and requires two TCP connections between every remote bridge and router and the central site bridges and routers. Thus, a network with 250 remote sites would need 500 TCP connections at the central site. For this reason, if a network manager was to use frame relay to support the WAN fabric of a network with SNA or APPN traffic, RCF 1490 would be more efficient for integrating it with other multiprotocol traffic.

FRAME RELAY EFFICIENCIES FOR SNA

The technology of frame relay devices has gone through several developmental evolutions that have made it possible for the end user to realize an expanded application of frame relay. In comparing traditional SNA WAN technology, it is easy to see the advantages of transporting SNA traffic over frame relay. For example, because frame relay uses the same framing and CRC bits as SDLC, all FEP SDLC line interface couplers (LICs), data service unit/channel service units (DSU/CSUs), and modems can be used with frame relay.

Improved Throughput

With typical network access ranging from 56/64 Kbps to T1/E1 (1.55 Mbps), network throughput is greatly improved over the more traditional 4.8 to 9.6 Kbps line speeds associated with SNA networks. Because SDLC and frame relay are software implementations on an FEP, no hardware changes are required to support frame relay transport. In addition, migrating SDLC host applications to frame relay requires no software changes. Unlike SNA, frame relay can transport multiple protocols providing support for mixed or hybrid client/server networks. Frame relay also supports point-to-point or many-to-many connectivity.

Increased Bandwidth

Guaranteed bandwidth is ensured through the PVC's established committed information rate with the ability to support a burst mode above the CIR rate. Traffic such as voice or video can be prioritized over regular data traffic to ensure that it reaches its destination. Some carriers provide high-priority PVCs (priority PVC services) which can be dedicated to mission-critical SNA, voice, or video applications. In this scenario, a high-priority PVC might be sampled four times as often as a low-priority PVC.

Interoperability

Newer releases of IBM's Network Control Program (NCP) software provide an SNA boundary network node with connectivity between a PU4 to a PU2 device across a frame relay network. It is possible to establish connectivity across a frame relay network between an SNA node (PU2) and a front-end processor (FEP). This software also allows connectivity between a 3174 and an FEP.

SNA Network Interconnect (SNI) provides for the interconnection of two SNA networks over frame relay via an SNA link. This capability allows users on one SNA network to access resources on another SNA network.

With IBM's incorporation of the RFC 1490 standard across its current product line, interoperability between IBM and other IBM frame relay products can be established. In this scenario, interoperability requires SDLC to be converted to LLC2.

Unlike a dedicated network, meshed frame relay networks provide the user with a number of alternate routes, guaranteeing that all traffic that is not discard-eligible can reach its destination.

Better Functional Support

Although not all FRADs are the same, more advanced frame relay switches provide a much higher level of functional support. For example, if a PVC is experiencing high levels of traffic, the originating terminal can invoke flow control or metering to ensure that all traffic stays under the CIR level for the PVC being accessed. In addition, the network can send a backward explicit notification (BECN) bit to inform the endpoint device of a potential problem. In this instance, the endpoint device can regulate the flow of data to ensure that high-priority data can get through.

Network Backup

Frame relay provides fault tolerance with the inherent ability to reroute in the event of a failed route. Frame relay networks also tend to be more robust and less prone to error as opposed to more traditional network arrangements. However, outages can occur on the local loop that supports a remote location's access to the frame relay network. To meet this contingency, dial backup circuits may be used to restore service if the dedicated 56 Kbps link goes out of service. Dial-up ISDN with one B channel can support 64 Kbps service, which can be used to access an open port on the network. Regardless of the type of restoration that is in place, dial backup, address security, flexibility, and ease of installation are all important issues that network managers must address.

PRACTICAL APPLICATIONS OF FRAME RELAY

Frame relay provides a less expensive alternative than leased-line services while providing bandwidth for such added services as LAN, voice, and fax traffic. For example, if the right equipment is used to interface the network, the network planner can integrate voice, data, and fax. This strategy makes it possible to aggregate all information traffic across the frame relay network.

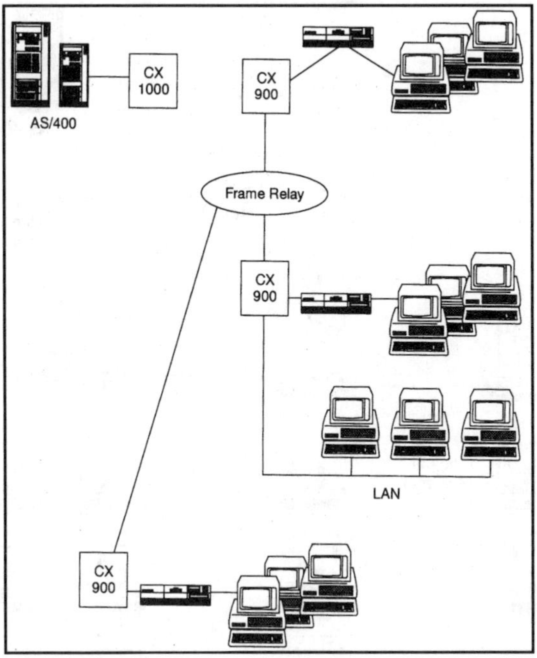

Exhibit 30.1 Star-Configured Network.

AS/400 Networks

Exhibit 30.1 shows a typical AS/400 network arrangement. The network links form a star topology, radiating from the hub to the outer periphery. In this example, CX 900 switches are at the periphery with a CX 1000 switch at the central site (mainframe). At each of the customer's remote locations there will be either an IBM 5394 or a third-party controller, as well as a LAN running NetWare for SAA. The CX 900s located at these sites convert the SDLC link protocol running from the cluster controller to frame relay protocol using the RFC 1490 encapsulation protocol. The central site CX 1000 converts the RFC 1490 protocol back to SDLC for presentation to the AS/400. The CX 1000 also "spoofs" the SDLC polling, appearing as a cluster controller.

In this solution, only the CX 1000 and CX 900 were added to establish a corporate network across a public frame relay network. No changes were required to the cluster controller parameters or other network equipment.

Exhibit 30.2 illustrates a multidrop or multipoint network architecture, possibly the most popular of all AS/400 network designs. Converting this type of network requires virtual multipoint mapping, which requires the CX 1000/900s to map the controllers from a point-to-point line to a multi-

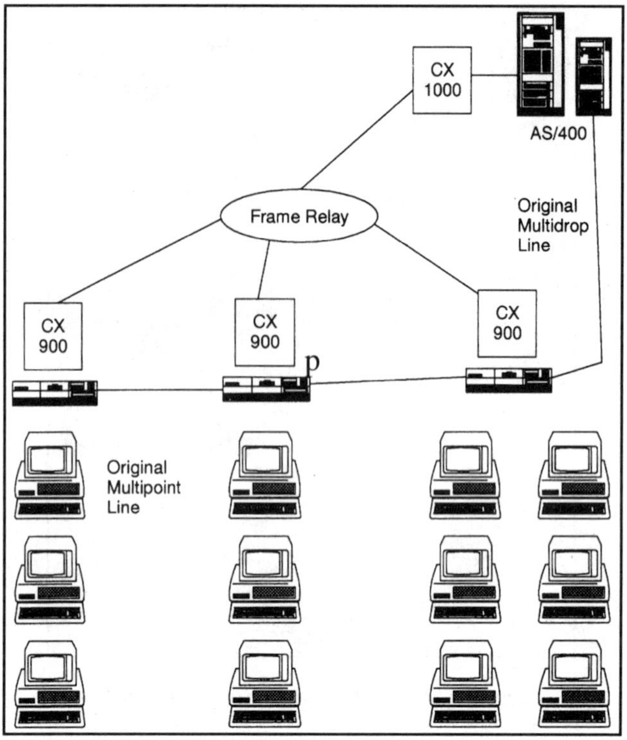

Exhibit 30.2 Multipoint, Multidrop Network.

point line. This remapping allows the AS/400 to think it is operating over a multipoint line when in fact it is running over public frame relay.

Support for Voice Traffic

Exhibit 30.3 shows another AS/400-based network in which a mix of IP/IPX/SNA and AppleTalk protocols is to be transported over frame relay. LANs as well as legacy nodes in several of the remote locations require integration into the network. In these locations, servers are running SNA emulators that allow them to appear as SDLC-attached controllers. TCP/IP and IPX traffic also comes from other servers in the same remote offices.

In this arrangement, the CX 900 was able to function as a router sending traffic over the frame relay network back to its ultimate destination. In this network, the CX 900 may also have to transport a mix of SNA, IP, and IPX over a single communications link (DCLI). It should also be noted that in Exhibit 30.2, voice traffic is integrated over the frame relay network. In this scenario, each of the remote locations has two voice channels configured on the CX 900. These voice channels are attached to the local telephone

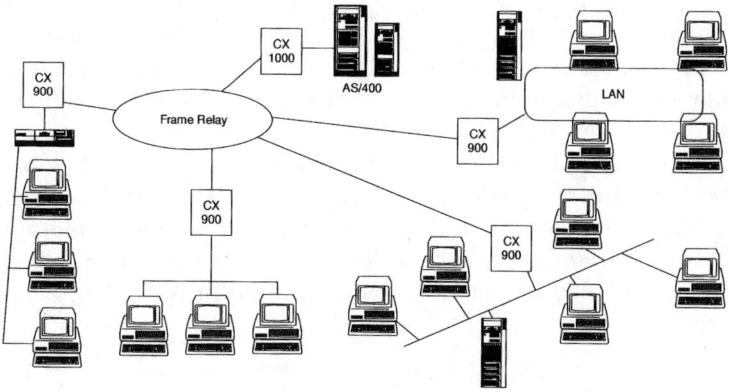

Exhibit 30.3 Hybrid Network with IP, IPX, and SNA.

system and appear as off-premise, extension-type connections. At the host, the CX 1000 provides support for each of the channels coming back from the remote locations. These appear as separate extensions on the PBX at the host site. This allows the remote offices to make calls to other locations or to the corporate headquarters. Companies using this strategy have saved a great deal of money on their voice traffic by piping it over the data network. Some users have reported tremendous savings in overseas calls by riding over their international frame relay network.

Special Application Servers

AS/400s are frequently used to support LANs as special application servers. In this scenario, the clients might access the AS/400 over an Ethernet. The CX 900 can be linked directly to the Ethernet via an LLC2 interface. The CX 900 can link to other remote CX 900s that interface with a remote cluster controller via an SDLC interface. A transparent Token Ring-to-Ethernet bridge might be used to link a Token Ring to an Ethernet backbone. Again, the CX 900 would be used to indirectly link traffic between the Token Ring and any of the other remote locations over the frame relay network.

CONCLUSION

In these times of belt tightening, network managers are under pressure to protect their investments in existing network infrastructures. As SNA network managers strive to improve their networks' performance, they must do so without any risk to the core business.

SNA over frame relay has obvious cost benefits. The application of frame relay to transport native SNA directly allows network administrators to continue the application of their SNA devices without reconfiguring their network devices. Eliminating the overhead associated with polling minimizes congestion and provides more available bandwidth for actual data transportation. This strategy allows the end user to realize greater network performance without buying more bandwidth. Furthermore, reducing local polling removes an overhead burden from the FEP, freeing processing cycles for other tasks. This reduces expensive FEP upgrades while providing much better network throughput. The key here is the selection of the right level of FRAD equipment to provide the maximum advantage of frame relay service. Not all frame relay access devices are created equal. When provisioning a frame relay network, terminals should provide the following capabilities:

- Termination of an SDLC protocol
- SDLC mapping to eliminate tasks from the AS/400
- Supporting the AS/400 SDLC host
- Supporting RFC 1490
- Supporting Annex G
- Supporting data compression for bursty LAN traffic
- Supporting IPX and AppleTalk routing
- Supporting LAN routing
- Supporting voice/fax
- Supporting international connections

Using these criteria as a guide, migrating to frame relay need not be done on a wholesale basis. Selecting a frame relay switch that provides a flexible platform is a key to upward expansion. This strategy allows network managers to plan for the future as their application mix begins to expand requiring such high-priority services as video and voice integration.

Chapter 31
Maximizing Mainframe-based WAN Resources

Gale Persil

IF YOU ARE READING THIS CHAPTER, THEN IT IS NOT NECESSARY TO TELL YOU THAT THE IP REVOLUTION IS TOTALLY TRANSFORMING THE HOST-CENTRIC SNA WORLD. As often quoted, IDC predicts that by the year 2000, 96 percent of MVS installations will be running some sort of IP stack on the mainframe. Like most organizations, one has likely already dabbled in IP by providing access to a remote office, or installing an MVS-based web server for a small intranet, or are already depending on IP network traffic to service critical business applications. The network likely contains some of the complex aspects shown in Exhibit 31.1.

But business is not standing still. According to IDC,[2] over 60 percent of mainframe organizations are looking to build or extend their customer support services to remote or Web access. This might be as simple as providing access to a problem database to remote offices or a Help Desk outsourcer, or as sophisticated as providing full customer access to all or part of the information database. Good examples of leaders in this area are the major package shipping vendors such as FedEx and UPS that offer package look-up services via Web access.

In close second, some 54 percent of IS organizations are looking to expand the access of enterprise server-based systems to new employees, both local and remote. To accomplish this, many corporations provide remote or local gateways, introducing TN3270 traffic to the network. While desktop-to-application implementations such as this are still very common, the introduction of the MVS-based web server has become an exciting prospect for many businesses. One U.S. insurance company introduced a local intranet in this fashion to communicate at its local headquarters. Word of its success spread so quickly through the corporation that the CIO

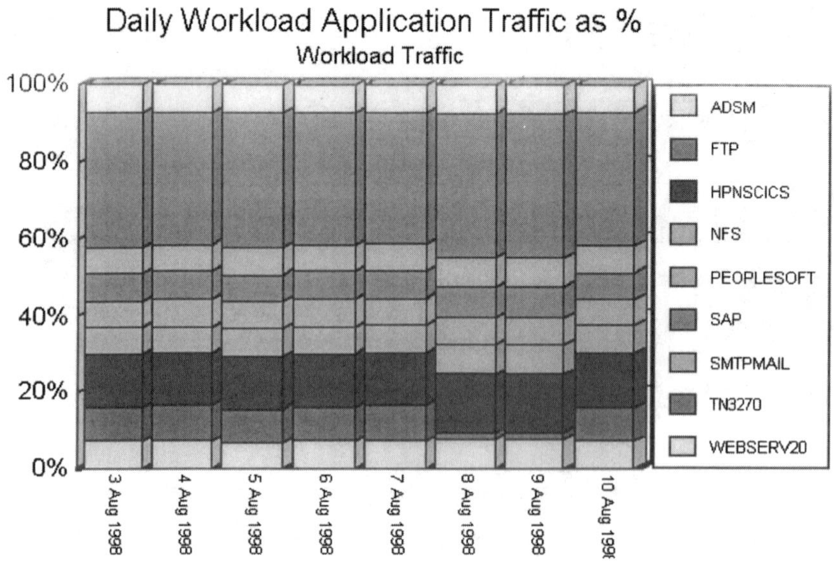

Exhibit 31.1 Daily Workload Application Traffic.

mandated that this "internal, local tool" be rolled out worldwide in the following 30 days!

Technology is moving at a quick pace. But probably more importantly, it is critical for businesses to be able to respond to competitive and market issues by being able to quickly adjust their technology to respond to the challenges. Building and tearing down extranets to work with a corporate partner, absorbing new technologies and systems due to acquisitions and mergers, and providing access to product information via the World Wide Web (WWW) are a few examples of the techniques that businesses are employing to maintain their competitive advantages.

And within the network management scope, the big question among network managers is: how am I going to ensure the same levels of service to this new generation of end users as I did with my SNA network?

SNA VS. TCP/IP NETWORKS

The task ahead is a bit like being an air traffic controller at a major airport — all sizes and priorities of planes are scheduled to land at the airport, and the job is to determine who gets to land and who circles around using up fuel until there is a free runway. However, air traffic controllers have several advantages over network administrators. First, as traffic increases, they are usually supplied with sophisticated equipment to perform their duties. Second, although they are about as unpopular as the person allocating network resources, every pilot and civilian knows that no one argues with the decisions of the controller. Managers of network infrastructure and their support staff rarely get the same vote of confidence about their decisions.

In a pure SNA environment, there is an entire operating system set up to monitor and adjust traffic traveling across the SNA bandwidth. It is called the network control program, or NCP. This traffic controller sets the rules for traffic coming into the mainframe, and all pilots are expected to follow the rules. However, because traffic usually originates from within their own country (another NCP), protocol violations are rare.

With IP, there is also a set of rules. However, this set of rules was developed by a committee and is executed by vendors that are not required to go through any training or certification in order to enter the network. To the analogy, they also might speak various languages. And when a protocol problem arises, the two parties are left to their own devices to work out an agreement. If they cannot, the plane is turned away and likely never seen again. A committee is then formed to evaluate the sad occurrence.

Luckily, most vendors who provide an IP solution to the market are hard-working, intelligent, and well-intentioned people who have not only a genuine interest in making their products work, but a genuine need to make them work. Simply put, the product does not work, no one will buy it.

Technology Decisions

On the positive side, this type of competitive atmosphere has long been the suitable environment where major advances can be made because even the smallest players can make significant technology contributions, pushing other vendors to follow. Technologically speaking, choosing the right solution for a particular application, regardless of the vendor, is now very possible and desirable.

For example, since TN3270 traffic was growing so dramatically at one CPU-constrained life insurance corporation running about 150 MIPs, the decision was made to install the Cisco Channel Interface Processor (CIP) found on the 7500 router (this same functionality on a smaller capacity version is found on the 7200 Series gateway). This allowed them to shift processing requirements for their 5000 sessions to within the routed network hardware, saving processing cycles within the MVS TCP/IP stack. This meant, however, that the organization now maintains both IBM and Cisco hardware, and needs to apply management tactics that support both vendors.

Of course, there are factors other than technology that might make sense in one's implementation. For example, while having the printing capabilities of Microsoft's SNA Server might make sense in one instance, and the web connectivity with low maintenance on the client PC provided by the IBM Communications Server in another, this results in having to negotiate separate maintenance contracts and learn two separate management interfaces. Still, the benefits of being able to choose a solution that closely fits one's organizational needs should not be ignored.

Different Camps

Technology is only one of the issues one faces when migrating to a wide-area IP network. The political issues can sometimes sabotage success more effectively than any router choice one might make. Merging of the legacy data center with an enterprise network brings together not only different protocols, but often different cultures, disciplines, biases, and management chains.

Think about putting an air traffic controller from a structured, standardized location such as London, England, together with a controller from Bogota, Columbia. The function of these two people on a high level is the same, and there has to be a basic agreement of the signs, protocol, and commands they will use because traffic often flies over the continental boundaries. Still, everything from the basic primary language to the complex internal queuing procedures will likely be quite different between the two towers.

In any application — system or network deployment — there is always the issue of dealing with the end users or customers and their requirements versus their expectations of the new system. The combination of SNA and IP adds another complexity — the integration of the mainframe and distributed network management camps.

Making Purchasing Decisions

Overcoming these issues requires a fair amount of patience, training, and political savvy. Keen network managers should:

- **Evaluate corporate management.** Evaluate the corporate management structure supporting the network infrastructure to make sure it is streamlined to make enterprisewide decisions. In this author's experience, there is no one successful structure, only people who successfully or unsuccessfully work within the various structures.
- **Multiple high-level managers.** Spend more time negotiating and getting buy-in from various factions with different needs. It is recommended that this time be spent up front. If this is not done, one will be left with having to choose lowest common denominator solutions.
- **Streamlined management structure.** Be cautious to make decisions based on sound business justifications, instead of just choosing the alternative that is attractive to your technical evaluators. Return on investment (ROI) evaluations, whether strictly implemented or satisfied

during an Q&A session with the staff who will be using the tool, can be very helpful in this process.

- **Upfront decisions.** Decide up front whether to choose a single vendor or best-of-breed orientation.
  - —**Single vendor.** Select management strategies that provide the *depth of functionality* required. This does not necessarily mean choosing the same vendor for management software as one did for IP hardware. In fact, the better software management solutions are often found from a non-hardware vendor because hardware vendors tend to be focused on producing efficient devices. Invest in training and cross-training staff in support of this environment. Motivate the Help Desk staff to learn these tools by positioning them as an opportunity to advance their skills and value to the organization.
  - —**Best of breed (or multi-vendor).** Select management tools that use standard practices such as SNMP and CNM to monitor and report on networkwide status. Make sure these tools interface with strategies within the enterprise such as enterprise management platform (EMP) implementations or Help Desk standardization.
  - —**Do not invest in strategies that will be costly for years to come.** It may seem obvious, but given some of the trends happening in IS organizations in the last few years — with senior management looking for the "pot of gold" at the end of the IT rainbow — this is worth stressing. Even if one has highly trained staff capable of developing new ideas in-house — and enjoy it — the next few years especially will require significant resource investment in projects that have nothing to do with day-to-day maintenance of the network, primarily Y2K. Staff turnover and crisis response are going to be common in turn-of-the-century data centers, so plan ahead. Any tools or strategies one invests in should be easy to implement, customize, use, support, and modify. Generally, that means using out-of-the-box solutions instead of choosing to implement new ideas in-house.
- **Complement your strategies.** To complement one's strategies, choose management tools that can provide or integrate with providers of a full range of options. Specific to an installation, this might include multiple protocols such as TCP/IP for MVS and TCP access, IP device vendors such as Cisco, IBM, Bay, and others, and multiple disciplines such as problem reporting, diagnostics, and capacity planning.

WHAT DOES PERFORMANCE REALLY MEAN?

Building and expanding the network infrastructure is an ongoing exercise. Once the best infrastructure decisions for an organization have been made, one will want to move on to the day-to-day issues of providing good service to the end users. Although bandwidth and its utilization have always been a hot topic in distributed network management circles, it is a fairly new concern for evolving SNA to IP network managers. Why does there not exist the concept of a "bandwidth manager" in SNA networks? And why are SNA network programmers not obsessed with trying to figure out which applications are using their bandwidth? In SNA, bandwidth is essentially dedicated to a single path (or redundant paths) for a single corporation. This makes bandwidth analysis fairly simple: when response time slows down and utilization is shown to be a problem, more bandwidth or a bigger front-end processor (FEP) is required.

In a TCP/IP wide area network, however, this is not always the case. With the use of publicly owned transport networks and other shared line configurations, as well as the fact that TCP/IP inherently and dynamically chooses multiple routes depending on current conditions, understanding traffic flow for different applications throughout the network becomes an important issue.

The air traffic controller example can be used to show why this is. Say you are a common destination within a small country for which there are two alternate routes to reach you. The number of alternate routes/airports depends on the fact that small airplanes are limited in their flying distance and, of course, will want to fly the most direct route whenever possible. Given this configuration, if the number of planes landing at your site is increasing on a regular basis, it would be good to plot and understand the growth rate so that you could plan when to either expand the facilities at your current location, or build a new location nearby to accommodate the growth. It does not matter where the traffic originated. There are only two routes to you; and when they become congested, you need to take action.

Within a TCP/IP network, however, the number of potential routes is expanded significantly. Imagine that you have just evolved from a local airport to one with international access. Now, planning for growth becomes more difficult because you do not have all the plans, schedules, and understanding of international traffic coming into your terminal. You do not know what their popular routes are, and which tend to be late — impacting your runway availability — until you are in the middle of a crunch and trying to accommodate all incoming traffic. If you could track all the information from a single international point, you should be able to detect patterns in growth and delayed arrivals so that you could adjust your runway utilization more effectively.

Although this analogy is basic, the same model easily applies to a wide-area TCP/IP backbone network. Because there are so many potential alternate routes, it becomes tedious and bandwidth-intensive to try to track the route of individual transactions through the network. Installing software to monitor performance on each and every PC within a network is a significant undertaking in terms of initial investment, as well as ongoing maintenance. Installation testing has to be done by the user or the IS department on every PC. Companies like Microsoft, Intel, and Tivoli now offer remote distribution capabilities to assist with the large task of disseminating new desktop software. Upgrades can generally be performed in the same manner, or specific vendors might offer dissemination from a centralized server on a scheduled basis.

Desktop PC solutions, however, often introduce what Gartner[3] in their total cost of ownership (TCO) model calls the "futz" factor. Intrusive solutions that use PC processing power to involve sophisticated users or give the users the ability to modify, interact with, or change the options of their system settings, contribute a significant amount to lost productivity at the individual level. Far from reducing downtime, this type of solution can actually reduce the amount of productive time during network uptime.

The Network from the Business Perspective

The thing that drives network usage is access to particular applications or data. Thus, it is not enough to understand that e-mail constitutes 20 percent of one's bandwidth. If there is a set of users (or node) accessing one application, the bandwidth method works just fine. But if an e-mail system is accessed from multiple destinations or hosts, or multiple applications from different sites, one needs to understand exactly what the relationship is between a specific destination and all the users accessing those applications. This information will allow one to make intelligent growth decisions instead of holding on to the SNA mentality of "surgically adding more bandwidth."

The most practical approach to this problem is to understand the types of traffic that are flowing in the network from an end-to-end point of view. Since that "end-to-end" term is extremely overused in today's marketing literature, take a look at what it means.

In a wide area network such as depicted in Exhibit 31.1, a comprehensive understanding of the actual usage of the network includes knowledge of traffic flowing into the mainframe, and the applications being accessed within the mainframe boundary — which may or may not be IP based. One indicator of application usage is port activity — because ports are generally dedicated to an application. So, for example, traffic connected to ports 1414-1417 indicates MQSeries activity, while traffic connected to ports 1364 and 1365 would indicate CONNECT:Direct usage. This is how many network management traffic reporters gather specific traffic statistics. When the mainframe is actually the server, it is also important to understand what specific CONNECT:Direct region is being connected to. The multiaddress space, multisystem, and load balancing capabilities of the OS/390 operating system provide a great deal of flexibility in managing traffic that does not involve simply "surgically adding more bandwidth" to solve capacity problems.

Many alternate solutions can be implemented that would be both reasonable and sensible from a corporate point of view. Take the case of a leading department store that noticed increased response times during prime working hours. While most of the time adding bandwidth would make sense, a careful analysis of network traffic showed that a particular department — the sales area — was the source of continuous large data transfers. Investigation showed that a popular web news page was being brought up in users' Windows startup and left running all day. A corporate policy was instituted that educated users on the damage of using active web pages as screen savers, and the IS department immediately saw a 30 percent reduction in overall bandwidth utilization, and a 23 percent drop in daily data transfer. Of course, with more bandwidth to handle other transactions, end-user response time for real-time activity benefited.

In another case, a mail-order catalog house began to notice response time deteriorating for their online customer support representatives. As a result, sales were beginning to suffer because customers became frustrated by holding on the line while representatives waited for their system to determine whether a particular item was in stock and available to ship. Careful analysis of the number of connections from this destination to the mainframe host showed that finance and administration personnel from this site had been recently logging onto the system during the day to obtain customer account information in "real time." Customer account information was usually updated to a local database once a month via a scheduled CONNECT:Direct transfer so that F&A would have reasonable information to work with. Recently, the F&A department had received numerous complaints from customers who had received late notices after having paid their outstanding invoices. In attempting to be responsive to these complaints, they were running extensive queries against the real-time data to make sure they had the up-to-date information for a particular account.

The "bandwidth view" would have suggested upgrading the local gateway to respond to the additional connections; and in truth, this is an option that would resolve this situation. But because there was room in the batch FTP window, the alternative chosen by this account was to update the local F&A customer database on a weekly basis instead of a monthly basis. This reduced the remote access requirements, freeing bandwidth for the online representatives and increasing their response time.

In addition to ports, effective network utilization can be achieved by understanding the following information on network usage.

- SNA applications (i.e., LUs and tasks) being accessed
- The number and type of connections (by port); for example, TN3270, FTP, CICS sockets, etc.

- Statistics on individual users and how they impact the system; specifically, how many and for how long they are connected, and how much data is being transferred while they are connected
- Reports on peak time periods, over a period of time; for example, what is the busiest day of the week? busiest week of the year?
- Given the current traffic load, what is the load and efficiency on the critical IP devices such as routers and hubs?

Because of the difficulty in using the basic functionality provided with OS/390 stack implementations — that is, the NETSTAT command — many organizations provide TN3270 access to remote sites, but have absolutely no idea how many actual users are taking advantage of this at any one time. In terms of providing service, and understanding the impact of an outage, this is critical to understand.

In the mail-order catalog example used earlier, the remote office had insisted that there were only 20 customer service representatives Telneted into the OS/390 system during a given work day. Implementation of a management tool provided visibility to almost double the number of expected sessions due to the addition of finance representatives and non-frontline customer support representatives who were Telneted into the system checking on orders for customers who called in to query on the status. In this case, management became aware of the increased usage of the system from this remote location, and was able to use this to build a case to upgrade several of their servers in response — and enlist the remote site to pay for a portion of the cost!

BUILDING THE MODELS

It is time to start putting this theory to work to build a baseline understanding of one's network and its traffic patterns so that one can make adjustments. Just like the air traffic controller, one must always have more than one solution for a particular bandwidth problem or growth requirement. So where does one start?

A Simple Baseline Approach

Quantify the traffic based on the number of connections and the bytes transferred. Correlate this with bandwidth and IP device utilization. Then determine the activity (load) of each OS/390 application on the system. This assumes an understanding of both port and MVS task access. This information needs to be further delineated by the contribution of each destination, and if possible, down to the user level.

Exhibit 31.2 may provide intelligent information that one could use to load-balance the system. Prior to this analysis, a retail food chain believed that the primary access to its systems was to access the MVS-based web

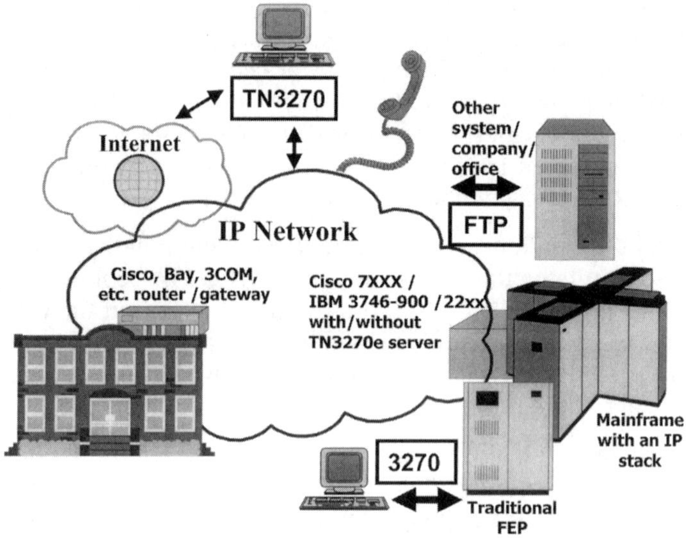

Exhibit 31.2 Bandwidth Utilization.

server it had recently implemented. Current information on specials and marketing programs was available from the intranet, and users had greatly praised being able to access this information in real time. Because of this, IS had budgeted for two servers to handle all the incoming traffic. But as Exhibit 31.2 shows, Web server access continues to be a small part of the actual activity in this system over a typical week. TN3270 traffic — which is currently being used to access inventory replacement and delivery — is a major driving factor for access to this system.

How does one address this? If there is no issue with response time and access (as with this organization), treat this information as a baseline of the system. If response time is being impacted by these real-time queries, one option would be to convert the inventory system to intranet-based access because the two servers were underutilized in this configuration. Another possibility is to move the server to support TN3270 access. Understanding the uses of a system provides the ability to choose from intelligent options such as this.

Another example is shown in Exhibit 31.2, where the greatest incoming traffic load can be determined by reviewing bandwidth utilization, shown in number of bytes. Understanding the breakdown of system access and utilization allows network managers to consider a “chargeback” or “charge for access” scheme where remote users jointly fund growth in the organization.

CONCLUSION

Of course, there are many more issues and metrics in the area of network utilization that can be explored to help one manage a system — from FTP usage to memory utilization on particular routers.

As stated earlier, the person responsible for maintenance of the network infrastructure has the responsibility to make sure that technology can adjust to meet the business needs of today's very competitive market environment. This article outlined key policies and metrics one will need to understand in order to be successful at that task. Although the tools one might have available to help will vary in capabilities, a person will be making the choices. By using real-world examples, this article has attempted to show that there is often more than one solution to capacity or growth issues. If one has access to the information needed to make intelligent choices, then one will be able to avoid crash landings, and enable end users to quickly and reliably reach their destinations.

Notes

1. IDC Computer Networking Architectures: Enterprise TCP/IP Requirements Intensify.
2. IDC Computer Networking Architectures: Intranet Migration Plans.
3. *The GartnerGroup Monthly Research Review,* April 1998. Web to Host: Lowering the Cost of Terminal Emulation.

Chapter 32
SONET for the Future: Satisfying Today's Users with Tomorrow's Technology

Leo A. Wrobel

WHAT ARE THE TWO MOST COMMON HOOKS THE NETWORK MANAGER HAS IN MEETING GOALS LIKE THOSE ABOVE? They basically can be boiled down to two statements:

1. "I can manage this better than you."
2. "I can provide you more capacity than you can get on your own."

The network manager can make good on these two pledges through the use of radically advanced network technologies coming available right now. Frame Relay, SMDS, and ATM are always bantered about in technical discussions. The point of commonality, however, for all of these technologies is a new, improved transport network technology called synchronous optical network (SONET). In addition to changing the dynamics of everything, SONET will be a 30-year technology (and how often is a planning window that long) and it is destined to replace T1 and T3 as the backbone transmission media in North America and worldwide.

The wide area network is getting smart. SONET is almost unbelievably smart by today's standards. For example, much of the work done for disaster recovery planning and network diversity is becoming an integral component of the network itself. This is not future stuff; it is here today.

The future network management system will need to read, interpret, and act upon the data provided it by these new intelligent network capabilities.

Because everything in a SONET network is smart except the glass, every component will be reporting its condition and performance. Other capabilities, such as the ability for a SONET user in Minneapolis to drop a circuit off at a node in Dallas without ever leaving his armchair, bode well for shortened circuit provisioning times. The day is rapidly approaching when a user will be able to call a carrier and say, "I need a 10 Mbps circuit to Atlanta," to which the carrier replies almost immediately, "Do you see it now?" Circuit activations to users with high capacity access will become as easy as turning on a new cable TV channel.

But first, some background. To truly appreciate these changes, look first at today's mainline transport system in the United States, the T1.

First, T1 is not new. Pulse code modulation technology for T1 actually was conceptualized in the late 1940s and saw widespread deployment in the public network in the early 1960s. When users began demanding larger pipes for data transmission and for other uses, T1 was the logical choice because, at the time, it was the biggest thing that the telcos could easily offer.

T1 was not designed to handle data. In fact, T1 was designed to run over copper cable primarily to carry voice, but was adapted by users and telcos to carry data. Everyone adapted their equipment to run over T1 because that was all the telephone companies had. Imagine a data processing manager voluntarily running his 10 or 16 Mb LAN interconnections at 1.544 Mb by choice?

In its most elementary manifestation, a channel bank was installed on the customer premise, as well as a matching channel bank in the central office. Twenty-four voice or data channels were carried over the T1 circuit. There were a few problems, however, with this. First, the technology is dumb. If things start going haywire, the only thing that usually can be done is to take the whole thing down to troubleshoot, usually putting at least 24 users out of business. The likely outcome of this action is known. Second, because the original channel banks were designed to carry voice, they robbed portions of the T1 for signaling information. This meant not getting the full 1.544 Mb for data and, in fact, usually not even coming close. Suffice it to say 20 years have been spent being shoehorned into a technology never intended originally for data. T1 was designed for interoffice toll telephone circuits meant to talk over, plain and simple.

Users eventually got around some of these problems by installing proprietary multiplex equipment on either end of the T1 at each customer location. The telephone companies made the first attempts at providing nonintrusive maintenance, namely extended super frame (ESF) capabilities. Users continued to develop applications that outpaced the capabilities of the telephone company by every measure — size, speed, network management, and reliability.

Enter SONET, the smart network technology. One of the first areas of impact by SONET networks will be in the area of network management. By integrating the capabilities of a state-of-the-art network management system with a feature-rich SONET transport system, the face of internetworking will be changed forever. SONET features and capabilities can be adapted and groomed to address pressing business issues, chiefly in terms of:

- Reduced expense
- Improved reliability
- Streamlined telephone company coordination and troubleshooting
- Higher bandwidth and capacity

SONET will change the face of networking, for example, the distributed networks of today will quite possibly evolve into a centralized client server. LANs will still exist, to be sure, but they will be centered chiefly around knowledge-based workers like attorneys, engineers, graphic artists, and others who need the functionality of a PC or require an oddball application. For repetitious tasks that are automated easily, a more centralized approach will make the best economic sense and will be the easiest to support in terms of operations cost. The graphical display interfaces will reside at the work location, but the meat of the system will reside in a secure, centralized location using the horsepower of the SONET network to pump graphical and multimedia information to the end user. The advantages will be greater security, greater standardization, reduced equipment cost, and drastically reduced personnel cost at the customer location.

The requirement to do this effectively will be communications capacity, lots of it, with an unparalleled ability to maintain and control it. That is one area where SONET comes in. It provides the capacity to install applications like these combined with the command and control and network management features up to par with mission-critical production systems.

SONET Technology

As already mentioned, this technology, designed to run over fiberoptic services, will be the backbone transmission medium for at least the next 30 years. Following are some of the other functions SONET provides as an integral part of its design:

Fault-Tolerant Ring Topology. Companies spend a lot of time designing fault-tolerant networks, pay a hefty bounty for diversity to avoid single points of failure, and still cable cuts are an all too common occurrence. Much of this effort will become part of the network itself as the network gets smarter and fixes itself in cases of major failure, in much the same fashion as switched telephone service does today.

Allows 10, 16, and 100 Mbps Circuit Sizes. Once again, why is T1 used? Answer: "Because that's all the telephone company has!" This is changing dramatically. Now, instead of trying to shoehorn data into what is convenient for the telco, they will begin custom tailoring for the customer. Virtually any size circuit can be mapped into a SONET frame. Many cost prohibitive applications like remote channel-attached mainframes, online vaulting, and other bandwidth hungry applications are well within reach, cost effective, and maintainable.

Drop and Insert Capability. Consider a transcontinental T3 (672 channels) circuit from New York to Los Angeles. To drop off a single circuit in Denver one would need 57 separate boxes. That is because the T3 must be broken to T1, and the T1 to individual channels, to get access to the channel wanted. Then it must be multiplexed back up again into a T3 before continuing on to Los Angeles. SONET carriers have the capability to extract a single channel without breaking down the transport carrier. Needless to say, the cost benefit here is remarkable not only for carriers but for users as well. The Fortune level customer in the United States today is already planning nationwide Optical Carrier Level 3 or 155 Mbps (OC-3) rings with drop and insert nodes in major points of concentration.

Advanced Diagnostics. How many times has a network control center or help desk chased an intermittently impaired T1, when the problem only lasts a few seconds, then clears for a day, only to repeat the following day? SONET literally will be able to look down the line and ask, "OK, who was the wise guy who made the errors?" And an intelligent repeater somewhere in Utah will answer back, "I did it. Looks like you got me!" The control center will then remove the offending component and restore perfect service. Contrast this to the half-dozen technicians who become involved now every time a problem of this sort occurs.

Ability to Burst. Later, when users map ATM over SONET, things really start to get interesting. For example, it would be great to have a red button that says "go faster" that could be pressed to temporarily turbocharge a network when the system begins to die under a severe load.

When beginning to map ATM over SONET transport systems, such a button will exist. And the user may pay only when pressing it. As users become addicted to gigacells in the conduct of their business, a pay-as-you-go pricing scheme will probably become the norm, as in long distance.

HOW DO WE BEGIN?

So how will all this capacity be affordable? First, a company must get comfortable in negotiating with carriers based on their long-run incremental cost (LRIC) of providing the service, not purchasing through the standard tariffs. Next, these carriers must be induced to install network nodes on site with the user. The abused term for this arrangement is reverse-colocation. In colloquial terms for the user, it means "make me a carrier, too." The approach makes sense. The owner of a factory would push for an entrance ramp to a major highway. Similarly, if one's product is information, the company obviously needs the electronic equivalent.

This new network-centric focus necessitates a radical transformation in the way network services are acquired, and how they will be utilized in enterprise networks of the future. It also means exploiting competitive opportunities in the changing regulatory scene to garner maximum impact for the company.

USING ADVANCED NETWORKS TO MANAGE CLIENT/SERVER COMPUTING

Think about all the issues and questions raised or implied earlier in this article, namely:

- Information Sharing — Insatiable User Demands for Capacity: Users want to share larger and larger files, taking a greater and greater toll on network capacity.
- Help Desk Support: There are not enough schools to attend to learn everything that is out there!
- WorkStation Support/Device-to-Technician Ratio: "Are there enough technicians in the whole state to support the number of devices I am planning to roll out next year?"
- Shared Capacity and CPU Cycles: "If I could find a way to broker the idle CPU cycles in my distributed environment, I could retire."
- Controlling Software Installations: "Who cares? Let 'em buy what they want but don't call me when it crashes."
- System Backups and Archival: "No problem. You can trust users to do backups." (This is hard to write without laughing.)
- Safety from Theft and Tampering: "The janitor hasn't spilled Clorox on my server yet. Besides, where else am I going to put it besides the closet?"
- Network Management: "When it breaks I'll fix it — oh … it's a T3?"

The key points you need to derive from the above items are the issues which haunted and taunted client server deployments and, until now, have not been adequately addressed: capacity, help desk, support ratios, resource sharing, software controls, workstation backups, physical security, network management and troubleshooting, and business resumption planning.

These are not laughing matters and everyone experiences these concerns to varying degrees. It just so happens, however, that a solution may be at hand, even for vexing problems like these.

The solution is connectivity on an unprecedented scale. With telecommunications competition will come new services, some of which will alter permanently the way business is done.

NEW OPERATING MODES

Client/server is an inherently inefficient way of providing network support. However, it presently is the only way to put the ergonomics, flexibility, and functionality on the desktop of the user. What happens, however, when telecommunications connectivity becomes available to project the same functionality in a cost-effective fashion? The answer is a return to a type of centralized processing, and an increased level of network management capability.

It will not happen overnight, but it is changing silently now. Today, bellwether users are beginning to partner with service providers to design advanced customized SONET networks, surgically tailored to the company's specific core business.

Superior Network Management, Security, and Control

There are sound business reasons for implementing a Johnson Space Center level of network management and control. First, a superior network management system can be leveraged as a unifying factor to reclaim renegade LAN managers, and provide a generally higher level of efficiency in the company by allowing users to concentrate on core business issues rather than technology. This includes providing (or selling) high-level network management services to client/server customers essentially blackboxing themselves to death. Many information executives and end users alike are having second thoughts regarding management of technology, including sentiments such as:

- Can they really afford to chase technology?
- Do they really want to chase technology?
- Would it be better to focus on core business instead?

Therefore, anything that will take the focus off maintenance, repair, and troubleshooting and allow the end user to concentrate on core business issues (the things that bring in money for the company — like selling things) will be an exercise in enhancing efficiency and the bottom line. The message to the end user: "We'll manage the network, you concentrate on building Chryslers." It is a powerful pitch.

Choosing the Right Network Management System

The traditional evaluation criteria of an effective network management system include capabilities in the areas of:

- Fault management
- Operations management

- Performance management
- Configuration management
- Change management
- Security management
- Accounting management

Remember, however, this system should not only make lives easier, it should show a concrete payback to the company by virtue of its implementation. This can be measured in hard dollars such as reduced circuit costs by virtue of being able to tune the network more efficiently. It can also be measured in terms of soft dollars, such as in greater customer satisfaction by virtue of proactive maintenance capabilities or better performance reporting. For example, if the system provides performance reports by circuit or by circuit drop, these can be used by a sales staff or technical analyst to chart performance by user. This is a power tool if used correctly and proactively.

To be effective in meeting user demand for network management services as well as other core business objectives, a network management system should also have the following characteristics:

- The ability to interface with at least 20 different and distinct equipment outputs. Eventually the system will encompass not only the LAN, but also mainframe, data communications, and PBX systems. The more of these systems that can be represented on a single display screen in a network control center, the greater the efficiency. Contrast this to shifting from screen to screen — or oftentimes from room to room — in present operating environments.
- The ability to satisfy the network management requirements of the user who is black boxing himself to death. The department can win these users back if it can manage the network better than the user can for himself.
- The functionality to be proactive rather than reactive in nature. It should be able to read network data and take action before a problem becomes critical.
- The system should be intuitive and easy to use. Specific events should trigger plain English error messages. Rather than something cryptic, plain English scripts often can be written to banner: "The Boston T1 circuit is about to go down!" This can be particularly useful if it happens at midnight on a Saturday night when only one college intern is in the control center.
- The selected system should be inherently adaptable even to the most proprietary systems because one does not want to limit its capabilities and often has no way of knowing in advance what platform a potential customer will be running. The more adaptable it is across a broad cross-section of products and inputs, the better.

- The system should sort and prioritize specific events proactively for long-term analysis, network tuning, and optimization. It should be able to support specific network performance data by customer for use in closer customer support by sales and account teams.
- Network management is extremely important to the overall operation of a technical organization. However, most end users can be expected to ignore the actual implementation of such a service insofar as installing it on their own. They will perform basic fire fighting on network problems, but generally lack the skills and tools to perform the other aspects of integrated network management. Consequently, an ongoing opportunity exists to provide these higher-level network management services, win back end users, and increase organizational efficiency.

An organization should endeavor to provide a level of network management that is materially superior to what is available to the end user on his own. If successful in this regard, expenditures on a first-rate network management system will prove not only a worthwhile investment, but one that will play a major role in the company's future success.

Leveraging the Wide Area Network to Win Back Users

Other SONET characteristics are easily salable to even the most skeptical users when packaged and presented properly in the context of solving core business problems. A few include:

Fault-Tolerance. Self-healing networks will become the norm, negating many of the engineering tasks on the wide area networks required for duplication and backup of facilities. Systems that switch to backup facilities in hundredths of a second are already in place in case of a cable cut or other disruption. Many functions for disaster recovery will become built into the network itself.

New Circuit Sizes. Leading-edge companies are already using native LAN services operating at 10, 16, 100 Mbps or more. Already, large users with on-site SONET equipment are positioning to bury T1 services forever.

Availability of Advanced Connectivity will Change Future Network Dynamics

Central Site Characteristics.

- Security
- Change control
- Trained help staff
- Remotely spawned processes

Other Advantages.

- Use of floating software licenses
- Shared database access
- Shared access to common software components/patches
- Superior back-up and recovery

Remote Site Characteristics.

- Low maintenance user device
- High device to support personnel ratio
- More concentration on core business, less on technology

Many companies already have configurations as portrayed in the previous lists. To take full advantage of the options discussed, the company will need a 622 MB SONET ring, perhaps with on-site ATM hubs after ATM standards solidify somewhat more. The basic transport fiberoptic ring will be installed on a custom basis through a joint partnership with one or more service providers. Relax, it is SONET, and it is going to be here for 30 years. The ATM part gets trickier with the operative question, "Whose ATM?" The SONET part is a sure thing, though. It is kind of like a user in 1980 ask, "Is T1 a good idea?" It is hard to mess up with basic transport technologies.

Because this network will become a revenue impacting system, bulletproof protective systems must be built in as well. This will probably involve tapping into network diagnostic data directly from the carrier and interfacing a Johnson Space Center level of network control, as discussed earlier.

The deal will also probably involve regulated carriers so the Federal Communications Commission, State Public Utility Commissions, or both will also get involved. That means beefing up regulatory support teams and extolling the virtues of this unique network in terms of company growth, jobs, technical prowess, and anything else except a means to save a buck or two.

Lastly, the bosses must be sold on making a large capital investment by articulating the business vision to myriad departments, both internal and external; filing paperwork required at the state and federal level to register as a lobbyist; and other concerns.

Sound complicated? It is. But think of the potential payoff. And if it is difficult and complicated to initiate, it is all the more difficult for competitors to copy. That means the company enjoys the strategic advantage longer before others jump in. And what is in it for the carrier? It stands to benefit most of all. The application can be licensed, shrink-wrapped, and sold again to another customer.

That benefit puts the carrier in a different class from those peddling long distance service at yard sales and swap meets. The more lucrative, tantalizing, and broadly adaptable the application is, the more accommodating the networking partner will be. This will be very important when getting down to the point of up-front concessions, who will buy what equipment, and what it will ultimately cost.

STRATEGIES

This radical new process demands many new skills. Because whole books can be written on just a single one of these topics, eight specific strategies that can be used to bring the custom network to reality have been selected. Although they do not represent everything, they will get a company well on its way.

Strategy #1

Keep improvements and capital expenditures consistent with the company's mission through effective linkage to its core business. A methodology for graceful integration of technology is:

1. Conduct detailed business process interviews.
2. Develop specific feature packages of services based on business need.
3. Divide packages into existing and emerging technologies.
4. Negotiate price on existing technologies in traditional fashion.
5. Develop short lists of potential emerging technology partners.
6. Establish executive liaisons and JV with selected partner(s).
7. Negotiate price based on LRIC.
8. Implement new technologies, share access to labs.
9. Release accomplishments to press. Both partnerships and emerging technology garner high investor interest.

Strategy #2

Find easy cost reductions that capitalize on economies of scale:

- Expand any present custom tariff filings such as AT&T Tariff 12s and state-sponsored efforts like Texas PURA Subsection 23.27.
- Evaluate packages normally reserved for telephone carriers like Feature Group D service (950 numbers) as cost-effective alternatives to local long distance.
- Continue to pursue volume and term discounts such as company-wide long distance contracts and special pricing on mobile telephone service.
- Shift focus to long run incremental cost (LRIC) methodology in negotiations with carriers. Start negotiations based on their costs, not on

an arbitrary number filed in a tariff. Tariff numbers contain a lot of other things besides the carrier's cost.

Strategy #3

Enhance and augment the company's shift into client/server computing through deployment of advanced network technology.

- Develop the capability to provide higher capacity services that support Native LAN connectivity at speeds of 10, 16, and 100 Mbps.
- Provide higher capacity for advanced applications such as televaulting, channel connection, video, and multimedia.
- Acquire the capability to burst during periods of heavy network traffic without affecting network performance.
- Develop shrink-wrapped packages of these advanced services, which can be easily assimilated, packaged, and sold by the sales forces.

Strategy #4

Purchase future services from custom rather than public tariffs to garner technology and price advantages. Procure telecommunication services in the future by negotiating with service providers based on long run incremental costs (LRIC) pricing methodology. Start with the carrier's actual costs.

- Train staff members in negotiating based on LRIC pricing methodology.
- Train the company's legal and contracts departments on the subtleties of negotiating custom telecommunications contracts.
- Develop a telecommunications regulatory affairs capability within the company to file custom tariffs on the company's behalf. Monitor and positively influence new legislation and regulatory rule-makings. Give the company the advantage of knowing about enabling regulation first.

Strategy #5

Tap into the potential market in local telecommunication services on internal customers' behalf. Take advantage of paradigm shifts in the telecom marketplace, principally, local competition and telecom reform legislation.

- Present newly acquired cost data and newly acquired services to internal customers. Make sure they understand the new tools they have available.
- The concept of a Tariff 12 for the local telephone company (often representing 50 percent of network costs) is an exciting and marketable concept. It will be an effective hook in bringing back many more autonomous users.

- Consider outsourcing telecommunications, both internally and outside the company, capitalizing on the advantages that have been created.
- Develop the capability to meter the telecommunications network just as many companies meter mainframe usage.
- Get customers addicted to gigacells. Because advanced networks tie in directly to the company's profitability (just like long distance calls produce tangible results like increased sales), the approach virtually assures a happy customer!

Strategy #6

Update overall telecommunications infrastructure through targeted capital investments aimed at specific core business goals.

- Explore establishment of a nationwide OC-12 (622 Mbps) SONET network with drop-and-insert nodes at major points of concentration. (Must be pursued using LRIC methodology to avoid being cost-prohibitive.)
- Upgrade the company's present T3 access facilities to SONET-based OC-12, based on LRIC pricing. Now.
- Buy the best network management system available, including diagnostic hooks into the advanced network described previously.
- Overlay advanced disaster recovery capabilities as an integral component of the new network to avoid additional expense later.

Strategy #7

Sharpen internal consulting capabilities. Make fundamental structural changes to the telecommunications environment and its relationship to the core businesses groups such as sales, production, operations, and others.

- Establish an emerging technologies liaison to the core business units.
- Become proactive in outlook.
- Foster linkages from technical services to core businesses.
- Strengthen liaisons between technologists and the people who build and sell.
- Gracefully integrate emerging technology into the business environment.

Strategy #8

Start looking for Microsoft Windows for Telecommunications!

A longshot to be sure, but how often does a company even get that? The methodology described will give a company a better-than-average chance of realizing such a goal, while producing innumerable benefits in the meantime. Face it, the next revolution will be in advanced telecommunications.

Nobody is the Microsoft Windows for Telecommunications yet. It is still out there for the taking by a foresighted and enlightened company willing to stretch the envelope.

Section VI
Data Center

Chapter 33
Data Center Security: Useful Intranet Security Methods and Tools

John R. Vacca

INFORMATION TECHNOLOGY (IT) IS NOW USED UNIVERSALLY TO SUPPORT CRITICAL ENTERPRISE BUSINESS DECISIONS. It has evolved beyond basic applications such as billing and inventory to the point where it directly supports customers and the manufacturing process. This sophisticated enterprise business information technology is entirely dependent on the diverse (and often incompatible) operating systems and hardware data center environments needed for their execution. All these systems must be managed and maintained if they are to continue to provide support for the ever-increasing applications on which the enterprise's data center depends. Information technology has a significant impact on the effectiveness of the intranet as a whole. Consequently, competitive performance of the enterprise is now directly affected by the management and control of data center computer resources. The measure of IT management's success or failure to support the enterprise is based on its ability to establish and meet required levels of performance, reliability, and availability — while staying within budgetary constraints.

The management and control of complex intranets and data centers are, however, daunting challenges to be met by IT professionals in the next decade. Some key areas that must be achieved are to:

- establish and consistently meet service-level agreements with end users
- control costs to meet service levels at the lowest possible level of investment
- protect the wealth of enterprise information and key resources that often span multiple operating systems and hardware platforms

In addition, IT management must not only achieve and maintain all these goals, but it must do this while ensuring complete system integrity at all times. An increasingly large role in enterprise computing is being played by intranet and client/server configurations of midrange and desktop computing environments, and open systems such as UNIX-based data center environments. Downsizing and decentralizing of processing resources is a result of the evolution toward a more global view — one that is replacing the traditional mainframe view of IT management.

In other words, IT management recognizes more and more that it needs both diverse and complementary information processing technologies if it is to meet the needs of the enterprise. Consequently, there has been substantial growth in complex heterogeneous systems that span multiple computing platforms. This leaves systems and data center intranet security management to contemplate some new concerns: it must now be recognized that midrange and desktop computing environments collectively represent a significant investment in information processing power. It must bring to each computing platform the standard systems management functionality that has been required on large mainframe systems: the need for the same level of automation, resource management, intranet security, and data integrity. Data center intranet security management is essential today for distributed systems, and this provides IT management with new challenges.

INTRANET SECURITY SOFTWARE AND SYSTEMS MANAGEMENT

In many ways, the challenges of distributed systems and data center intranet security management are the same as those of distributed applications. End-user applications, such as manufacturing and general ledger systems, are built on known database structures and application objectives. Distributed systems and data center intranet security management solutions, however, must address a very diverse and often perplexing variety of environments.

There are vast differences in the various operating systems of each platform, and management for distributed systems and data center intranet security must adjust accordingly. In order to present a unified whole, the goal is to insulate the administrator from the specific vagaries of each system.

Having the software solutions needed to provide comprehensive systems and data center intranet security management on each platform in the intranet is a significant part of mastering distributed systems and intranet security management. These solutions alone are only effective, however, if they can be tied together into a single point of management. Systems and intranet security administrators must be able to manage all or any desired part of the enterprise from any location within the intranet. Single point of management allows administrators to implement and enforce overall enterprise policies while continuing to provide the local controls necessary in a constantly changing environment to ensure responsiveness.

IT management must be able to provide the consistent, high service levels that are required by enterprises. Systems management and intranet security software should deliver integrated, total data center and intranet automation capabilities. At the same time, it should make possible the controlling of costs and ensuring protection of valuable data and resources. For example, enterprises provide distributed systems and intranet security management across multiple platforms. They also provide the ability to endorse and extend industry standards. This allows data center management software to be compliant with, fully support, and extend the capabilities of the Open Software Foundation's (OSF) initiatives for DCE (distributed computing environment) and DME (distributed management environment).

Systems and data center intranet security management software should provide a single point of management for complex heterogeneous systems. For example, data center management software should provide a single point of management through three strategic elements.

1. It meets the specific needs of the platforms and exploits the special characteristics and benefits of each by providing robust solutions that are appropriate to each platform.
2. It has a flexible manager–agent architecture that enables each solution to work cooperatively with other solutions in the intranet managing the flow of work, or performing work on behalf of other solutions in the intranet.
3. It provides a flexible user interface that enables each and every solution within the data center environment to be managed from a single location, or multiple locations if desired.

This third element is known as the systems and data center intranet security management workstation, as shown in Exhibit 33.1. It provides systems and data center intranet security management for a homogeneous or heterogeneous intranet.

INTRANET SECURITY AND SYSTEMS MANAGEMENT WORKSTATION

When managing all of the solutions in the data center environment, a systems and intranet security management workstation should provide the systems and intranet security administrator with a GUI (graphical user interface)-based user interface. Such a workstation could be used to administer a single UNIX system, a remote IBM mainframe, an OS/2 or Novell-based LAN, or even a heterogeneous intranet containing all of these

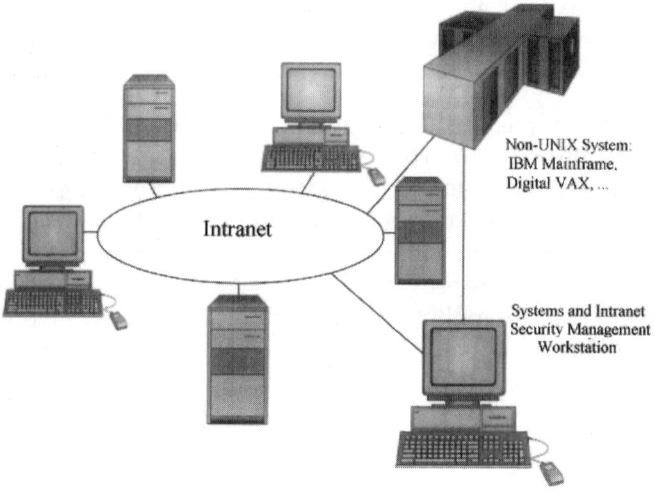

Exhibit 33.1 The Systems Management and Intranet Security Workstation.

components and others such as the IBM AS/400 and Digital VAX/VMS. The systems and intranet security management workstation should operate on a lower-cost X-terminal.

For example, in an enterprise's systems management workstation, the modern interface reduces the complexity of systems and intranet security management. It also provides an intuitive and logical view of otherwise complex issues. The GUI adjusts for the particular aspects of each environment where necessary. For example, the user interface is identical when managing a multinode job scheduler for defining jobs within job sets, or schedules and their relationships to one another. However, when opening up a job detail window for actually setting up individual jobs, the GUI would present a job from an IBM MVS system as a collection of JCL (Job Control Language) statements. A job on UNIX would display as a shell script, while a job on Digital VAX/VMS would contain Digital Control Language (DCL) in place of JCL or shell statements.

Control and flexibility of this type is essential for distributed systems and data center intranet security management. Being able to manage each platform from a common location with a common interface, and preserving the concepts and terminology across the intranet, are extremely useful — along with having the tools needed on each platform.

Investments in the training of systems and data center intranet security administration personnel can be leveraged into new platforms by maintaining the model for each solution across the system. For example, users familiar with ACF2 or TOP SECRET security systems on IBM MVS would find the workload management and intranet security provided through UNIX familiar and easily understood.

THE CONNECTIVITY FACTOR

An enterprise's systems and intranet security management software should cover a broad range of interrelated functions required to manage and control data center and intranet activity. In addition, it should utilize service layers to interact with each other, adding value to the intranet's information technology systems as a whole.

Furthermore, an enterprise's blueprint for a software architecture and its underlying guiding principles should try to provide a comprehensive strategy for software development for the IT community. This approach, where all components work together across multiple platforms, is essential in maintaining sufficient responsiveness to the continually evolving priorities of enterprise-driven information processing requirements.

Automating each area of systems and data center intranet security management does not in itself result in a complete and successful approach to systems management. Each component of the systems and intranet security management solution must support and communicate with every other component in order to achieve total data center and intranet automation. For example, intranetwide information security cannot be ensured if each systems management solution uses its own security tables. Problem and change management cannot keep pace with the dynamic activities of large and complex data center environments if software that addresses functions such as scheduling, report distribution, storage management, and security cannot automatically open and update problem incidents. Clearly, global workload management is unattainable unless each platform provides workload scheduling and resource balancing capabilities that interface with all other platforms. Therefore, complete and effective integration requires both a design for integration and an architecture that support development and enhancement of the completely integrated solution. As shown in Exhibit 33.2 for an IENF, an enterprise's services enable the integration of systems and data center intranet security management functions without an uncontrollable and unmanageable explosion of interfaces.

POWERFUL SOLUTIONS

The rule-based policies that govern data center procedures are particularly well suited to automation and can be handled by systems and intranet security management software. This automation capability is essential in addressing the complex activities of multivendor and multioperating system intranet environments where a variety of procedures are generally followed due to the variation in capabilities provided by the native platforms.

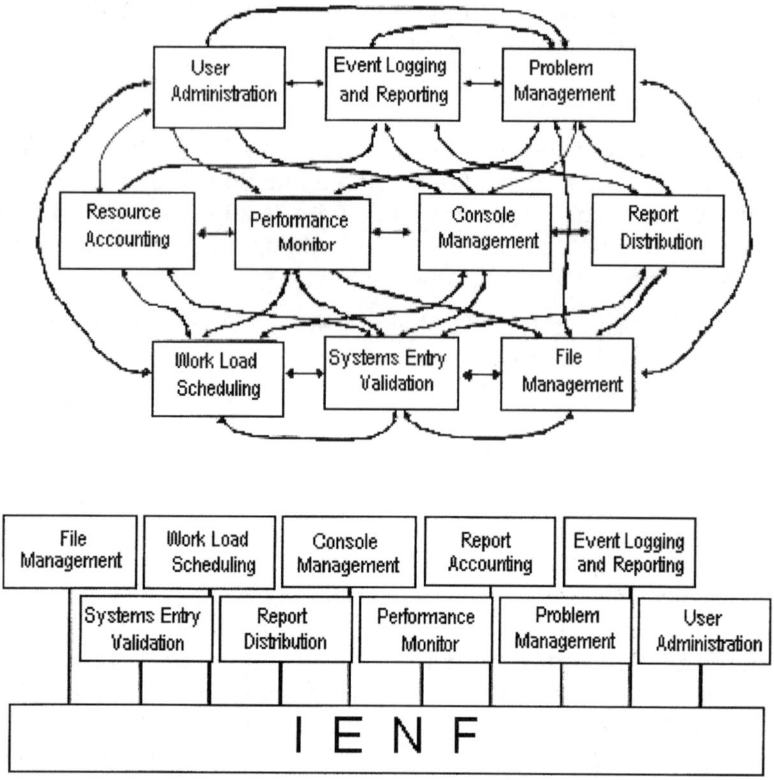

Exhibit 33.2 An Integration Event Notification Facility (IENF).

An enterprise's systems management and intranet security software should simplify the management of these complex data center environments by extending native platform capabilities wherever necessary to ensure operational consistency, regardless of platform.

The otherwise numerous and complex procedures are significantly reduced, enabling operations staff to become familiar with all aspects of a streamlined, unified system. This dramatically reduces the training required of operational staff, particularly in complicated subsystem procedures, and eliminates the need for multiple subsystem specialists to closely monitor data center and intranet activities. An enterprise's systems management and intranet security software should provide a robust, fully integrated, distributed solution that covers the essential disciplines of automated production control, automated storage management, performance management and accounting, data center administration, and security, control, and audit (see Exhibit 33.3).

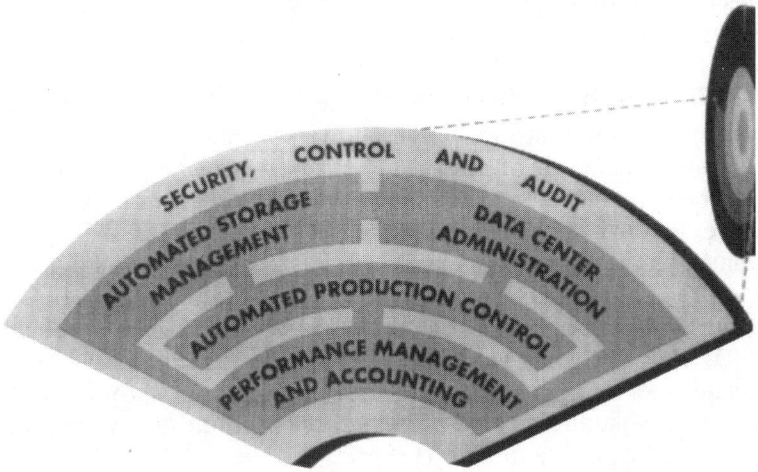

Exhibit 33.3 Systems and Intranet Security Management.

To completely automate information technology systems and derive maximum benefit, all of these components must be present for all platforms. Vendors that offer only a part of this functionality and limit the functionality to specific platforms cannot effectively assist IT management in meeting its service-level objectives.

AUTOMATED PRODUCTION CONTROL SOLUTIONS

A data center should provide an integrated set of solutions for automated production control. These solutions cover all areas of functionality, including workload management; rerun, restart, and recovery; console management; report distribution; control language validation; report balancing; and production documentation.

Automated Workload Management

Workload management is concerned with complete automated management of production workloads, including workload balancing, automatic submission and tracking of work based on user-defined scheduling criteria, priority, and system resource availability. This ensures that work is completed correctly and that critical deadlines are met.

Automated Rerun, Restart, and Recovery

Rerun, restart, and recovery automates the often complex, manual-intensive, and error-prone rerun and recovery process, thus enabling processing to restart at the optimum recovery point. In addition, it automatically handles the otherwise time-consuming manual procedures such as job setup, data set recovery, and backout.

Automated Console Management

Console management improves operating efficiency and reduces errors by automating the handling of console messages. It provides an advanced message/action capability that can alter, suppress, or reply to messages or initiate other actions, such as automatically issuing IPL/IMLs, alerts (through voice and pager notification capabilities), commands or invoking programs based on the content, and frequency and other characteristics of the message traffic. In addition, selective action based on specific console and terminal IDs allows the assignment of consoles to specialized applications such as intranet monitoring, system monitoring, or tape mount processing. A simulation capability is also provided to assist in the development and verification of these event/action criteria.

When used in conjunction with a programmable workstation, this software provides a single focal point for all console operation activities in a multi-CPU, multioperating system and intranet environment. In addition, remote access to perform console management is available through remote dialups (PC, remote TSO, CICS, session) or through the telephone using the latest voice and touchtone technology.

Automated Report Distribution Function

Report distribution provides extensive capabilities for the flexible and efficient production, tracking, and distribution of reports. This results in speeding the delivery time, increasing the accuracy, and improving the tracking of reports. Facilities are provided that automatically identify pages from existing reports, place them into bundles, and sort them by delivery location prior to actual printing.

These capabilities provide end users with the information they want, when and where they need it, while reducing or eliminating redundant information and the materials and efforts that are wasted in its distribution. Automated report distribution software also provides online viewing capabilities that can reduce the need for a hard copy of reports, as well as the option to select all or parts of reports for printing. Report archiving capabilities enable the storing of reports offline for auditing purposes, and for future viewing or reprinting.

A data center's report management software on intelligent workstations should extend report management capabilities by enabling end users to receive reports as files on their local computers. The ability to merge, annotate, or change reports using a familiar computer should be automated by the workstation-based software, and redistribution of these new reports should be provided through interaction with the software on the host system.

Advanced Control Language Validation

Virtually all systems in use today provide an interpretive control language for defining the execution of batch and online processing. Examples of these languages include IBM Job Control Language (JCL) and the UNIX Shell Script language. A data center's design for systems and intranet security management should include complete advanced control language validation capabilities that reduce or eliminate errors that can cause failures during production execution, and that aid the end user in diagnosing problems with control language programs. In addition, this software should enforce site-specific standards while providing the reports and cross-reference information that are needed for future maintenance.

Automated Report and File Balancing

Report balancing consists of extensive, automated report and file balancing capabilities that enable a quality level to be achieved that is unavailable through manual efforts. In addition to automatically ensuring the accuracy of reports after they are printed, this software can uniquely catch errors during the execution of production work cycles (enabling both fast and accurate resolution of problems), and prevent the completion of in-error production runs and the distribution of incorrect report output.

Production Control Documentation

Production documentation provides complete and consistent centralized online documentation system for the production control environment. Integration with other production control software enables documentation efforts to be automated and centralized, ensuring accessible and accurate information essential to data center operations, and particularly for contingency planning, disaster recovery, and future maintenance.

AUTOMATED STORAGE MANAGEMENT SOFTWARE

Automated storage management (ASM) software significantly extends the native operating system's capabilities of storage and resource management. This software optimizes performance and access to information. It ensures availability, integrity, and reliability — regardless of the various media device types and differing configurations of mainframes, midrange computers, PCs, and LANs that define the IT processing environment.

Backup Management

Backup management provides the ability to back up files based on creation date and version, as well as supporting backups of multiple versions of the same file. It keeps track of which volume each file has been backed up to.

It also enables users, system, and intranet security administrators to view the media (tape versus disk) and version of each backed-up file and easily initiate a restore if needed. Backup management eliminates the problem of keeping track of where backed-up files are kept, and how to find them when a restore is needed.

Archive Management

Archive management makes sure that files have been removed from the online disk system and stored on other media that are based on storage management policies and are available when needed. It ensures that enough storage is always available on the file system to keep users working.

Archive Transparent Restore Processing

Through an automated storage management (ASM) common file catalog, the archive management function can locate and initiate a restore of an archived file without user intervention. This function is known as "automatic transparent restore," or IXR. With IXR processing, the user request, process, or program attempting to access an archived file is automatically suspended, the file is restored by ASM, and the process is allowed to continue without failure. IXR helps to ensure a successful file management plan by removing the greatest fear that users have of any storage management system — not having their data when they need it.

Threshold-Based Archiving

A specialized capability of an archive management function is threshold-based archiving. Regardless of how careful systems and intranet security administrators are in defining archive policies, inevitably there will come a time when a process unexpectedly demands more storage from the file system than was anticipated. Sudden and unexpected file shortages can be disastrous to planned work and the users of the system — as the file system becomes exhausted and work halts. Up until now, the only answer was to run another backup as quickly as possible, and try to guess which files could be deleted without causing too much other disruption. This process was slow, disruptive, and error-prone at best. Threshold-based archiving utilizes a common file catalog to determine which of the next set of files eligible for archive are currently backed up. The catalog is then updated to indicate that the files have been archived, and deletes it from the disk file system. No additional backups are taken, no best guesses are made, and no costly disruption of work in progress results. IXR, of course, stands ready to bring back any file needed.

MULTIMEDIA STORAGE MANAGEMENT

Multimedia management uses a rule-based, policy-oriented design to provide comprehensive storage resource functions for a wide variety of media, both permanently mounted file storage devices and removable tape, WORM, and erasable optical technologies. These functions include space management, allocation control and management, I/O optimization, volume defragmentation, and mount management. Each of these capabilities is designed to provide the best possible utilization of the storage devices available, while maintaining the service levels defined by the enterprise's storage management policies.

Extended Data Storage Management

Extended data management enhances and fully automates all data management functions, regardless of platform. It includes reformatting, sorting, compression, and optimization of data seamlessly and independent of physical file organization and data format.

Performance and Error Management

Through data integrity and device failure recovery facilities, system throughput can be optimized, and disruptions caused by failures can be minimized. In addition, high-cost, high-performance options of disk devices can be exploited to their full potential.

Finally, take a look at yet another useful data center intranet security method and tool — xswatch.

WHAT IS XSWATCH?

Xswatch, like its predecessors, was built to watch log files for interesting information. Most log files that grow at a reasonably fast rate have a lot of data, but little useful information. There are several extant implementations that either scan an entire log file or use the equivalent of *tail -f* to monitor the file as new lines are added to it. They then cull log messages that are deemed important or interesting to the implementor. These implementations can be as simple as *tail -f | grep pattern* to ones that are as complex as a full-blown C program complete with its own macro language.

All of these programs have the same basic structure: match a line of text against a pattern/action pair. If the pattern matches, execute the associated action. The application is up to the end user; for example, watching /var/adm/messages for *file system full,* then executing a job to remove all core files older than one day from the file system. Another example is monitoring for authentication failures. A third might be to send a page to an operator in case of an intranet fault.

Xswatch extends this idea by creating a very general architecture that allows the end user to execute almost any arbitrary pattern/action pair. It takes advantage of the PERL programming language's ability to create code on-the-fly instead of imposing a specialized syntax. The result is a workbench for monitoring almost anything that uses a log file.

What Is the Motivation for Writing It?

The motivation for writing (yet another) log watching program was something that did not have a limited subset of the functionality of the language the engine was written in (PERL). Something was also needed that would monitor multiple files simultaneously. Also, an application was needed that had at least the ability to scale. Xswatch's architecture depends on forwarding syslog entries to a central server. On a large intranet, the number of syslog datagrams could consume a significant amount of intranet bandwidth. It seems better to have xswatch running on many machines, forward only the important data to a central server, and leave the uninteresting data on the local machine. Finally, writing xswatch is an experiment in code reuse and software integration. A primary design goal was to avoid reinventing the wheel wherever possible. For xswatch, there was at least partial success achieving these goals.

The Xswatch Components

Xswatch has four main components: an engine, snippets, support libraries, and log files, as shown in Exhibit 33.4. The engine is a small PERL program that essentially manages system resources for the user. The engine consists of four subparts: xtail, a signal handler, an event server, and a watcher function.

The Engine Itself

Xtail is a C program that tails off multiple log files at once. Xtail is the most operating system-dependent part of xswatch. It tracks each file for conditions such as rollovers, truncations, even appearance and disappearance of files.

The signal handler is mostly for managing xswatch in the background. Sending a SIGHUP signal will tell xswatch to kill the current xtail process, reset internal variables, and reload all of the snippets and resume operation. In this way, one can add or remove snippets without halting a running xswatch process. Sending a SIGQUIT signal will tell xswatch to write out configuration information to /tmp. This is useful for debugging snippets. Sending a SIGINT or SIGTERM will tell xswatch to shut down gracefully.

The event server is a PERL module that handles the general housekeeping chores of managing system resources such as I/O, child processes, signals, and timers. The watcher function is a PERL function call (like any other PERL function), except that the function itself is created on-the-fly, based on snippets. Watcher provides the structure that holds the snippets together. Watcher will be called once by the event server for every line of input from xtail.

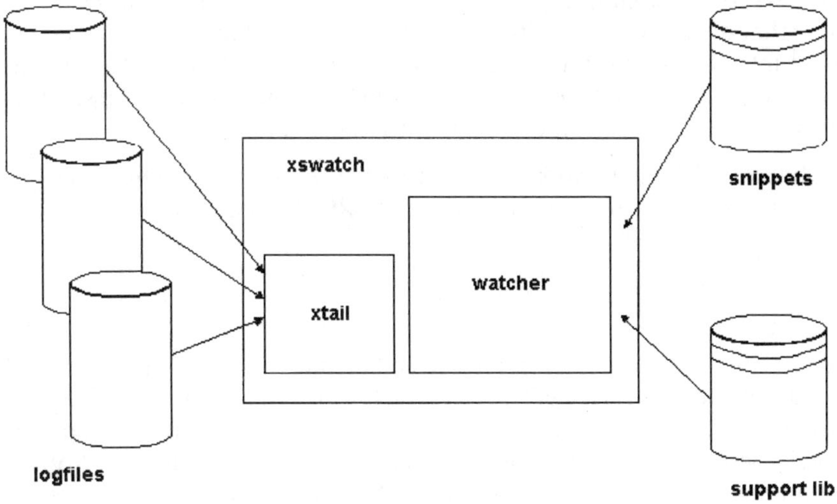

Exhibit 33.4 Xswatch Architecture.

Snippets

Just what are snippets anyway? The Webster dictionary definition of snippet is a small part, piece, or thing; specifically, a brief quotable passage. For xswatch, a snippet is a small piece of PERL code that handles pattern/action pairs. They have a very simple structure. There is an initialization section that uses the standard PERL *BEGIN* block and then some code, the simplest of which can be:

```
/... some regular expression ... / && do {
... some action ...;
}
```

Snippets are catenated together and compiled into the watcher function. Watcher will be called each time there is a new line of input data written to a log file. The watcher function provides just two pieces of data: an input buffer in *$_*, and the name of the log file where the data came from in *$logfile*. Anything else is up to the snippets.

The snippets are organized in a directory with file names that mimic the *System V rc.d* directories, (*nn.Description,* where *nn* is a two-digit string ranging from 00 to 99). The description can be anything (hopefully informative), with its length determined by the limits of the underlying operating system. This provides a simple way to order execution of the snippets without cutting and pasting. It is also an easy way to temporarily disable snippets. By changing the name from *10.Snippet to stopl0.Snippet*, or anything that does not begin with *nn.*, xswatch will not load the code and compile it into the watcher function. There are no restrictions as to what goes into a snippet. So, if a programmer wanted to put all his code in a single file, xswatch would work just as well as with many smaller files.

The *BEGIN* block is where each snippet should register which log files it wants to monitor. This is done with the PERL *push* call:

```
BEGIN {
   push@watchfiles,'/var/adm/messages';
}
```

Registering the same file more than once (that is, in more than one snippet), is allowed. Xswatch will reduce the log files to a unique list. This is because so many people can add or subtract snippets from a common directory and not worry about interfering with other snippets. Notice in the following code that there are a lot of *BEGIN* blocks, and they are actually inside the function call definition of *watcher.* Not to worry, though; when PERL compiles the function, the *BEGIN* blocks are executed immediately, once, and never again. When the watcher function is actually called by the event server, the *BEGIN* blocks are not touched.

Support Libraries

Since the snippets are simply PERL scripts, support libraries can be anything one needs or wants. For example, one can use the *Term::Cap* module for setting screen attributes, or one of the date and time parsers to convert the log file time stamp into seconds. Or one can include one's own modules to send a message to one's pager or e-mail.

Useful Features

Xswatch has a number of useful features. First, it is simple. Second, it uses no new grammars. Third, the use of snippets is widely accepted. Fourth, it can be extended using standard PERL libraries and contributed code. Finally, it is scalable.

A Simple Engine

The engine is simple. The main program is seven lines of code. It knows nothing about its inputs and does nothing with them except to pass them to the snippets in the form of the watcher function. This minimizes code maintenance:

```
$result = GetOptions(qw(help snippet-dir=s debug=i version));
usage("invalid parameters") unless $result;
usage("$0: v$version") if $opt_version;
parse_command_line;
```

```
init;
start_server;
exit(0);
```

No New Grammars

Instead of worrying about learning yet another macro language, parsing, and the accompanying errors, xswatch uses PERL itself. As previously mentioned, PERL is well-suited for creating code on-the-fly. It also cuts down on training time. If one knows how to code in PERL, one can write snippets for xswatch. One does not have to look around in the documentation for a syntax guide. Finally, PERL does a much better job at parsing and compiling code than one would ever want to write for a specialized tool like xswatch.

Shortening the Development Cycle

Prototyping and testing are simplified because snippets are written in PERL. One can use *perl -cw* to check the syntax just like any other PERL script. If what one is looking for in a log file has a time domain component, one can do several things: save a time stamp in a variable and set an alarm, or create a new event using the event server. The source code shows a more elaborate sample that checks that the number of authentication failures from login does not exceed a certain threshold. This is to handle cases where a user dials in on a noisy phone line. The login program will record an authentication failure, but this is not really something to worry about. On the other hand, if xswatch saw repeated login failures over a short period of time, one could guess that someone is trying to penetrate the intranet.

Finally, snippets are readable (well, as readable as PERL code gets). One can document snippets, put them under source code management, and essentially treat them like any other body of code:

```
BEGIN {
   push @watchfiles, '/var/log/authlog';
   require PagerTools;
}
if ($logfile eq '/var/log/authlog') {
TIMER_RESET:
   if (/INVALIDIREPEATEDIINCOMPLETE/) {
   if (!defined $auth-trigger) {
      $auth_trigger =
      register_timed_client([],600,sub ($hit-count=0});
   } else {
      cancel_registration($auth_trigger);
      if ($hit_count++ > 5) {
      Pager::call_pager oncall, 'Authentication alert';
```

```
            }
            undef $auth_trigger;
            goto TIMER_RESET;
            }
        }
    }
```

Scaling

As the number of systems to monitor increases, syslog packets flying back and forth might become a significant load on the intranet resources. By using xswatch to filter out information at the local machine level, one could create an intranet of xswatches that (using the standard PERL Sys::Syslog module) connects to a parent syslog daemon and forwards messages to a centralized server.

Distilling Data into Information

There are a number of applications for xswatch, and one of the most important is distilling data into information. One of a system and intranet security administrator's key duties is to reduce, or at least maintain, entropy. Systems and intranets have a natural tendency toward increasing entropy. System and intranet security administrators are constantly bombarded with log files, e-mail messages, pages, even voice mail. Anyone who has ever had to plow through */var/log/syslog* looking for some clue as to what went wrong knows that it would be far better if, somehow, the system knew what was important and notified a human being in realtime. Xswatch allows an administrator, or even a group of administrators, to consolidate empirical knowledge (about log files) into a single directory. Snippets contain bits of experience learned over time about what to look for in a log file, and what to do about it. Another approach is discard the expected; everything else must be a fault or of some interest. In reference to firewalls, one really does not care when the system resists an attack, or someone tries to access a blocked port; the firewall has done its job. What one really wants to log is when something goes wrong.

Downsides of Xswatch

Xswatch has its flaws. During the development process, a few issues arose that should not come as a surprise — but did anyway. PERL-centrism is seductive. During the early development of xswatch, and after several ineffective attempts, it was clear that although PERL may be portable, there are some things that really are better implemented in C or some other compiled 3GL. PERL is not cheap. A long-running PERL process will consume about 2 MB of virtual memory on a SPARCstation 5 running SunOS. The resident set size hovered around 1200 KB. Even so, the good news is that there need be only one of these processes running on any given machine. Also, if one considers that an xterm process consumes almost 500 KB, one xswatch process does not hurt so much.

The holy grail of code reuse also has its problems. The most recent version of the EventServer module was incompatible with PERL. Contributed modules tend to lag behind releases of the main body of code, so it is important to track each revision carefully. There are also many PERL modules that, although useful and publicly available, would not work without some effort. Hopefully, as the body of available code matures, these library modules will become more stable.

Xswatch shows promise as an incremental refinement in the publicly available collection of log-watching programs. Note that there are now commercial products in the field that offer more integrated and scalable services. They might even be more affordable! Xswatch makes an effort to encourage code reuse and integration. The maintenance costs for tracking admittedly diverse software packages seem to be less than the cost of maintaining an equivalent amount of homegrown code. Xswatch was also an experiment in toolsmithing — to the extent that the main engine is seven lines of code long. It is modest in scope. The entire xswatch program is six heavily commented pages of PERL code, including the 15 pages of online documentation. Using xswatch is as simple as creating a few lines of PERL script with only two assumptions and installing the file into a directory. Finally, because xswatch does not have its own macro language, it is only limited by the functionality of PERL.

CONCLUSION

Distributed systems and intranet security management solutions must address a very diverse and often perplexing variety of data center environments, each with its own needs and challenges. Distributed systems and intranet security management must also adjust for the vast differences in the various operating systems of each platform, insulating the administrator from the specific vagaries of each system in order to present a unified whole. Through the single point of IT management, overall enterprise policies can be implemented and enforced while continuing to provide the local controls necessary to ensure responsiveness to the constantly changing data center environment.

Finally, xswatch makes an effort to encourage code reuse and integration. Xswatch shows promise as an incremental refinement in the publicly available collection of log-watching programs.

Chapter 34
Why Parallel Sysplex is a Must for Today's IBM Data Centers

Anura Gurugé

PARALLEL SYSPLEX1 IS IBM'S COGENT AND HIGHLY STRATEGIC SERVER CLUSTERING TECHNOLOGY FOR CONTEMPORARY MAINFRAME SYSTEMS. Introduced in 1994, Parallel Sysplex is an integral and pivotal facet of IBM's highly successful initiative to dramatically reduce and redefine the cost of mainframe-centric computing. Parallel Sysplex permits up to 32 mainframes, with multiple copies of the operating system and application programs tightly coupled together in a fault-masking, workload-balancing mode, replete with concurrent read/write access to shared data from all of the clustered mainframes.

Parallel Sysplex delivers:

- Exceptional scalability capable of providing over 2,000 mainframe MIPS[2]
- Nonstop operation with 99.999% uptime per year, including planned outages for preventive maintenance, which translates to less than 5 minutes of potential, unexpected downtime over a 24 x 7 x 52 period
- Smooth, transparent, "not-even-a-noticeable-hiccup," total protection against processor, hardware, operating system, access method (e.g., ACF/VTAM), application subsystem, or application failures
- Seamless, policy-based load balancing between multiple instances of an application
- An intriguing capability for disaster recovery, known as *Geographically Dispersed Parallel Sysplex* (GEOPLEX), which permits Parallel Sysplex clusters to be separated by as much as 24 miles (i.e., 40 Km)

Parallel Sysplex, thus, persuasively and powerfully addresses the two fundamental issues crucial to the success of mainframe-based computing: unconstrained growth in capacity and ultra high availability. Therefore, it is not surprising that, within 5 years of its introduction, over 2,000 customers around the world, including huge financial institutes such as NationsBank, Fannie Mae, and BANC ONE Corporation, were already profitably exploiting the advantages of Parallel Sysplex, despite the fact that mainframe shops understandably are noted for their conservatism and deliberation when it comes to adopting new technology such as this. NationsBank, for example, was able, thanks to Parallel Sysplex, to increase its data center capacity from 720 S/390 MIPS in January 1995 to 2,030 MIPS in January 1998, using just its original staff of five systems programmers.

NOT YOUR FATHER'S MULTIPROCESSORS

Multiprocessor systems, where the combined processing power of multiple processors is harnessed by one copy of an operating system to execute its workload, are not a new concept when it comes to IBM mainframe computing. S/370 machines based on *Symmetric Multiprocessor* (SMP) technology have been available since the late 1970s. Today, SMP is a key and integral feature of IBM's latest mainframe family, i.e., the CMOS technology-based *S/390 Parallel Enterprise Server.* The word "parallel" denotes that parallelism is at the very core of these highly cost-effective mainframes. The fifth generation of these S/390 mainframes, referred to as the G5 models, started shipping in September 1998. In addition to uniprocessor models, there are G5 models for every multiprocessor permutation from 2 to 10, thus it is even possible to get G5 models with 5, 7, or 9 processors.

Parallelism is what enabled IBM to dramatically reduce the cost of mainframe MIPS in order to address the stern challenge it faced in the late 1980s with the client-server "Open Systems" movement inspired by the then highly cost-competitive RISC-based UNIX systems. The high cost of mainframes in the late 80s were not entirely due to high margin pricing on the part of IBM. The bipolar semiconductor technology then used to build its mainframe processors was in reality quite expensive compared to the CMOS[3] technology used to build nearly all other processors, including those for PCs, RISC systems, and minicomputers.

There were many factors that contributed to the high cost of bipolar technology. For a start, bipolar had limited economies of scale. Bipolar technology was in essence restricted to mainframe systems, whereas CMOS technology was nearly ubiquitous given its use in PCs and workstations. This meant that the innate cost of bipolar technology tended to stay on the high side whereas the cost of CMOS-based processors continued to decrease as volumes around the world multiplied. Moreover, bipolar, which is inherently faster than CMOS, was more expensive to fabricate. If this was not enough, bipolar systems generated considerably more heat than CMOS processors, thus necessitating expensive cooling systems such as the water and liquid cooling systems that were synonymous with S/370 mainframe systems in the 1980s.

Given the economic reality of bipolar semiconductor technology, the only realistic way that IBM could reduce the costs of mainframes to match those of RISC/UNIX systems was to abandon bipolar in favor of CMOS. That is exactly what IBM did around 1993 with the introduction of the S/390 Parallel Enterprise Server family of mainframes. A CMOS-based S/370 Central Processing Unit (CPU) on its own, however, could not generate as many MIPS as would have been possible if bipolar technology had been used. Thus, the only way that IBM could satisfy the ongoing, insatiable demand for mainframe MIPS was to resort to parallel processing, hence their name and the various models offering up to 10 processors working in SMP mode. (Hitachi Data Systems (HDS) is committed, at least in the short- to mid-term, to continue to develop bipolar-based mainframes for customers who want more MIPS than are possible with the S/390 machines.)

SMP alone, even in 10-processor mode, cannot address the number of MIPS required by the large mainframe shops. The increase in processing power with SMP is never linear. Thus, doubling the number of processors from, say, 4 to 8 will never produce twice as much power. Instead, the best one could hope for would be an increase in the 1.7 times range, with the "markup" decreasing as the number of processors increase. This propensity for diminishing returns as the number of processors increases is the ironic "Catch-22" of SMP. In addition, business processing is typically unsuited for *Massively Parallel Processing* (MPP), where hundreds of processors are harnessed together, since the processes cannot be broken into small components that can be handled by a separate processor. MPP is best suited for certain scientific applications such as 3-D model rendering, weather forecasting, and nuclear physics simulation. IBM's RISC technology-based RS/6000 Scalable Parallel (SP) systems that can have up to 512 processors cater to those customers who require MPP.

In addition to the disadvantages of the "law of diminishing returns," SMP systems, including S/390 systems, cannot provide for protection against the failure of the operating system or those of an application subsystem such as CICS or IMS. Consequently, while an SMP system will continue to work if one or more of its processors fail, the operating system, application subsystems, and the actual applications constitute single points of failure. Preventive maintenance and upgrades to the software, including the operating system, will also result in downtime, obviating the possibility of maintaining a 99.99% uptime record over a year or more.

GOING BEYOND BEING JUST LOOSELY COUPLED

Loosely coupled systems were the initial approach to overcoming the limitations of SMP. In a loosely coupled configuration, multiple mainframes, each running its own copy of the operating system, and application subsystems are grouped together via channel-to-channel connections. Each of the mainframes that are loosely coupled may in addition be SMP systems with multiple processors. Loosely coupled systems have been available in the mainframe world since way back in the late 1970s with IBM's JES2[4] and JES3 job and resource sharing schemes. With a loosely coupled system, jobs in a central job input queue would be distributed across the individual mainframes. In addition, files, printers, and operator consoles would be shared across the mainframes. Loosely coupled systems thus provided additional processing capacity over individual SMP systems, as well as protection against the failure of a single operating system or application subsystem. It was also possible to perform preventive maintenance or upgrade software on one machine without losing all of one's mainframe processing capability. Exhibit 34.1 illustrates the differences between uniprocessor, SMP, and loosely coupled configurations.

In September 1990 IBM introduced Sysplex for MVS-based mainframe systems. Sysplex, now referred to as "Base Sysplex" to emphatically differentiate it from Parallel Sysplex, set out to enhance the manageability and cross-system data sharing possibilities of loosely coupled systems. The key innovations of Sysplex, the *Cross-System Coupling Facility* (XCF) and the *Sysplex Timer*. XCF, which was in effect a feature within MVS/ESA SP4, allowed authorized applications to communicate with each other on a peer-to-peer basis across mainframe channels. Since the introduction of Sysplex coincided with the availability of the fiberoptic-based high-speed ESCON[5] channels, such XCF communications and data sharing could now be performed at speeds ranging from 10 MBytes/sec to 17 MBytes/sec (i.e., 80 Mbps to 136 Mbps). The Sysplex Timer enabled the system timers of all of the mainframes working together to be tightly synchronized.

Parallel Sysplex, introduced four years later, went further than Base Sysplex in two key areas. It enabled a larger number of mainframes to be clustered together, and permitted very high-speed, read/write data sharing to facilitate workload balancing within the context of a single system image. Exhibit 34.2 depicts the general architecture of a Parallel Sysplex configuration.

Today it is possible to have up to 32 mainframes working together in a Parallel Sysplex configuration. Though very much targeted at the CMOS S/390 machines, the mainframes making up a Parallel Sysplex cluster do not necessarily have to be just S/390 machines. One can use IBM's ES/9000 family bipolar mainframes in Parallel Sysplex configurations. Parallel Sysplex operation is also supported on certain non-IBM "plug compatible" mainframes, most notably by HDS and Amdahl. All the mainframes making up a Parallel Sysplex cluster do not have to be of the same type, hence it is indeed possible to have a cluster made up of S/390, ES/9000, and non-IBM mainframes.

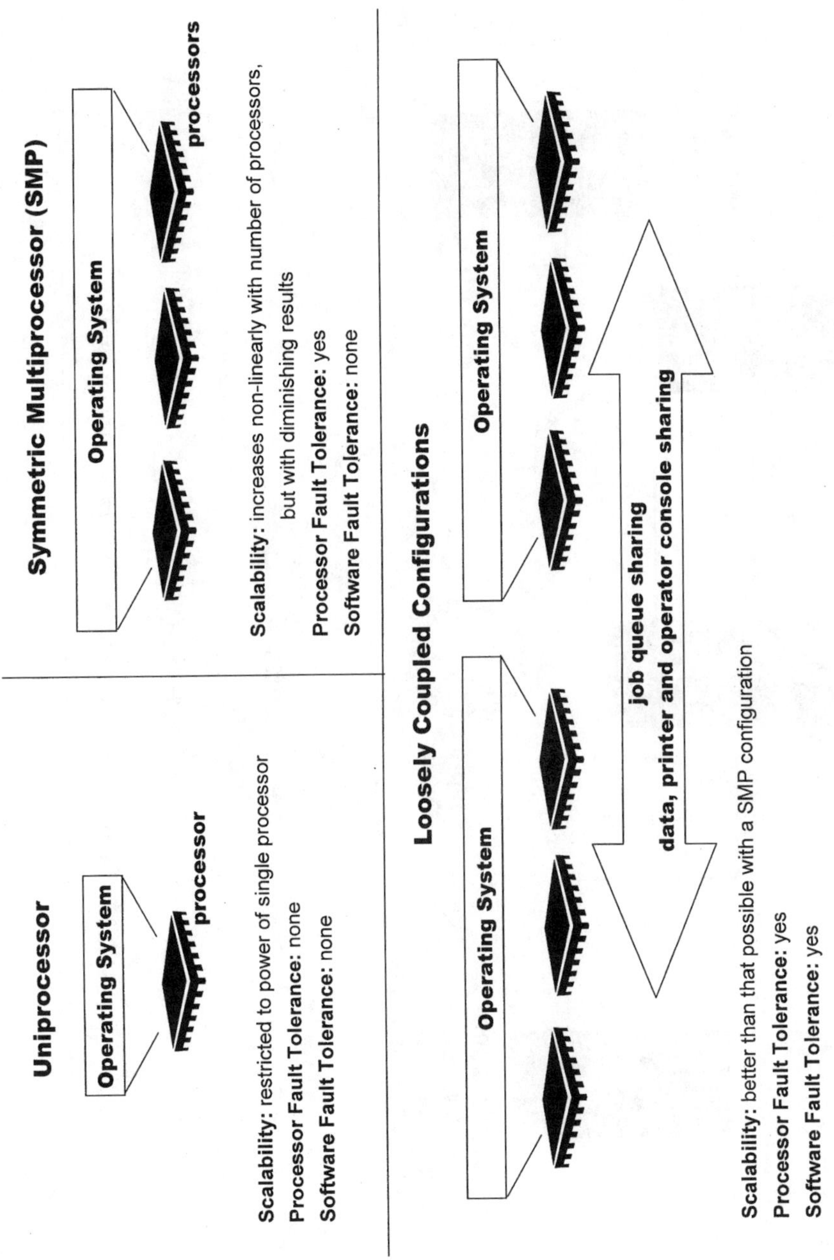

Exhibit 34.1 Differences between various Processor Grouping Schemes.

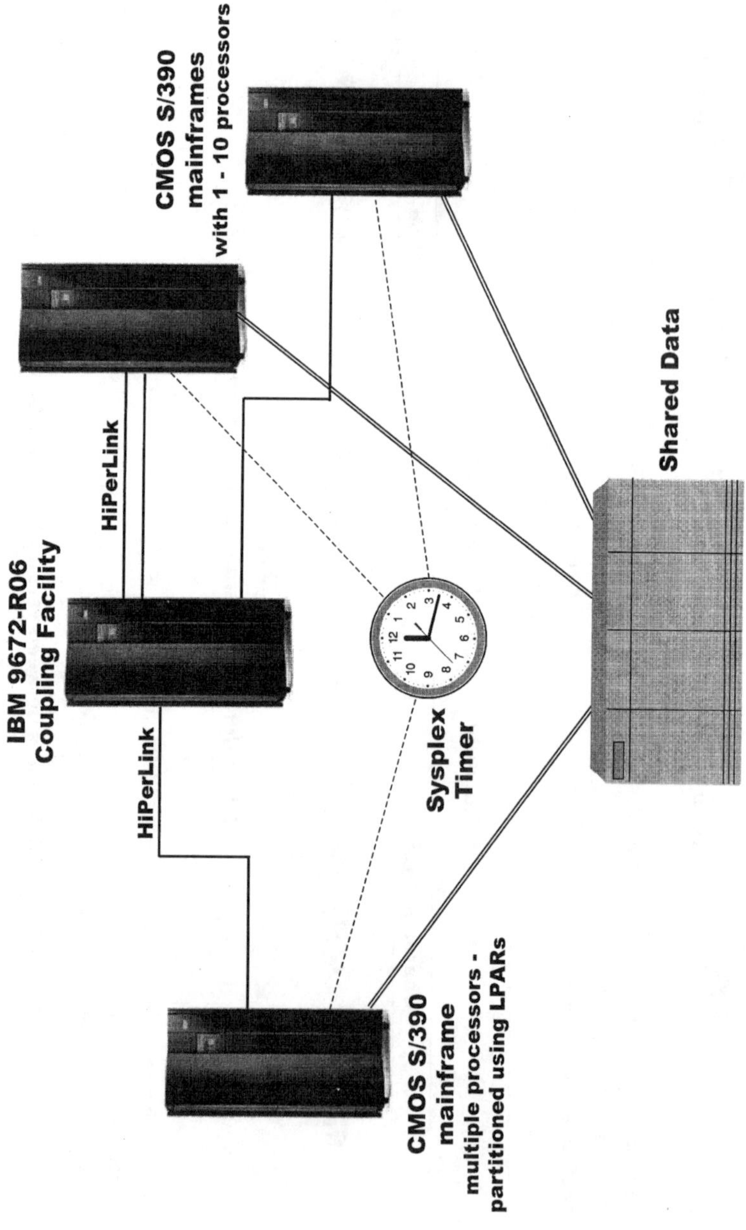

Exhibit 34.2 General Architecture of a Parallel Sysplex System.

In March 1999 the two most potent S/390 G5 models, the so-called 9672 Models RX6 and YX6[6], each contained 10 processors; i.e., they are ten-way SMP machines. Given that Parallel Sysplex supports up to 32 mainframes, it is indeed possible to have a "humongous" mainframe configuration with 320 separate processors by coupling together these top-end G5 models. Thus, it is no surprise that seamless scalability to effortlessly address the growing processor needs of its larger customers is invariably touted by IBM as one of the primary strengths of Parallel Sysplex. Given that transparent, nondisruptive fault masking is the other major attraction of Parallel Sysplex, the configuration and composition of these systems can obviously change dynamically, without any noticeable impact on the applications or end users. This also means that one can "Hot-Plug" additional mainframes into working Parallel Sysplex configuration "on-the-fly" and without any disruption to handle abnormal or peak workloads.

DIVIDE AND CONQUER

Though it is easier to think of Parallel Sysplex configurations as being made up of separate mainframes, this IBM clustering technology in reality is generalized enough to enable it to also be seamlessly used between processors within the same mainframe. IBM has had a hardware feature on its mainframes known as *Processor Resource/System Manager* (PR/SM) since the late 1980s. PR/SM, which is also available on the S/390 family, allows the resources of a single mainframe, including its complement of processors, to be subdivided into individual, autonomous "mini-mainframe" machines. This process of subdividing a mainframe into multiple units is referred to as *Logical Partitioning* (LPAR). (PR/SM is best thought of as a hardware-based equivalent to IBM's old Virtual Machine (VM) operating system that permitted multiple virtual mainframes to be cloned on top of a single physical mainframe.)

Separate LPARs can be included within a Parallel Sysplex configuration as if each LPAR was a *bona fide* standalone mainframe. This ability to transparently "slot-in" LPARs into a Parallel Sysplex configuration in "Hot-Plug-In" mode can be gainfully exploited as a means of adroitly dealing with peak workloads. For example, using LPAR, one processor of a five-way S/390 could be set aside for operating system testing while the other four participate in a Parallel Sysplex configuration, along with other mainframes, that is handling the production workload. At known peak load periods such as day-end or month-end processing, operating system testing could be suspended and the processor that was devoted to it "Hot-Plugged" into the Parallel Sysplex configuration to share the processing workload.

The ability to use LPARs with Parallel Sysplex also provides a very convenient and elegant way by which customers can evaluate and experiment with Parallel Sysplex just using a single multiprocessor mainframe, even a "small" two-way S/390 model. This capability of running Parallel Sysplex with just one physical mainframe can also be profitably leveraged to realize exceptional protection against operating system, application subsystem, and application failures. Rather than running their applications in SMP mode on an unpartitioned mainframe, customers can now create two (or more) LPARs, run separate copies of their operating system and application software on each LPAR, and then realize workload balancing across all the processors on the mainframe by using Parallel Sysplex. While there is an added cost for running such a Parallel Sysplex configuration, including the need to pay more for the software licenses (albeit with a Parallel Sysplex-specific "markdown"), this cost may be easily justifiable given the marked reduction in possible downtime.

The current 320 processor limit of Parallel Sysplex is also unlikely to be an unyielding barrier for long. IBM indubitably will increase, in the future, the maximum number of mainframes possible within a Parallel Sysplex cluster. Though IBM may not go too far beyond the ten-way SMP machines given the law of diminishing returns, IBM since 1994 has been significantly increasing the power of its CMOS S/390 processors on a yearly basis, hence the "Generation 5" (G5) notation. Therefore, it is safe to assume that it would indeed be a rare customer who would be able to claim that they are in imminent danger of running out of mainframe processing capacity.

NONSTOP WORKING

The two primary application- or software-related attributes of a Parallel Sysplex configuration are:

1. *Single System Image,* with the *Generic Resources* function, that ensures smooth workload balancing at an application subsystem (e.g., CICS) or application level across multiple mainframes (or processors if LPARs is being used).
2. *Multinode Persistent Sessions* (MNPS) that provide complete, nondisruptive, "not-even-a-single-keystroke-dropped" protection against processor, hardware, operating system, access method, application subsystem, or application failures. The term "persistent" in the name highlights the fact that the goal of this function is to ensure session persistence (i.e. continuity), while "multinode" describes that this persistence is realized by having multiple instances (i.e., nodes) of the components required to ensure session continuity in the event of a failure to a single node.

With Parallel Sysplex, every mainframe (or LPAR) in the cluster runs its own copy of the operating system access methods as well as all of the relevant software, in particular the applications or application subsystems that require load-balancing and fault-protection. Thus, there will typically be multiple copies of the same application subsystem or application running concurrently on a Parallel Sysplex system. Each application or application subsystem will have its own unique name, so if there are multiple instances of the CICS subsystem running on separate mainframes, each copy would have a name such as CICS1, CICS2, CICS3, etc.

The Generic Resources capability of Parallel Sysplex, however, hides these individual names from the end user. Instead, all that the user needs to know is the "generic" name associated with the application, for example, just "CICS." Thus all the CICS users associated with that Parallel Sysplex system will always logon to "CICS." The Workload Manager component of the Parallel Sysplex system, which in essence permeates across all the mainframes involved, will dynamically intercept each logon attempt to CICS. It will then determine which copy of CICS is best suited to handle that logon request based on current processing loads and implementation-specified policies, contingent on multiple criteria, dictating how the workload should be optimally balanced.

The Workload Manager will thereupon pass the logon to the appropriate copy of CICS, which will process the logon and, if valid, establish a session with the user. The user will be oblivious as to which copy of CICS it is talking to. In many cases, the end users may not even know, or care, about Parallel Sysplex and that there are multiple copies of CICS in play. For the duration of that session, however, the end user will communicate directly with that copy of CICS, unless, of course, there is a failure that impacts that copy of CICS.

In the event of a failure, MNPS comes to the rescue transparently and automatically. Parallel Sysplex relies on various "heartbeat" tracking mechanisms to continually monitor the availability of all of the components within the cluster. In its most basic form, a heartbeat tracking scheme could be one where each copy of a particular application sends an "I-am-alive-and-well" message to all of its peers on a preset, periodic basis. If such a message is not received from a copy of an application within the expected time limit, its peers and the Workload Manager assume that there has been some type of failure. With nonstop working being paramount, the Workload Manager and the other copies of the application immediately set out to recover the end user sessions that were being handled by the failed application. This recovery is nondisruptive and happens without any interventions from the end users, or for that matter the system operators. Invariably, the users will not be aware that there has been any type of failure at the data center. The only indication of the recovery process will be a slight delay in receiving the response to the last transaction entered by a user.

With Parallel Sysplex, the logon information related to every session that is established is "journaled" on disk storage, just in case the session has to be reestablished due to a failure within the system. This logon journaling ensures that end users' sessions can be automatically established with another copy of the application without the user having to re-logon, as is the case when MNPS is not being used. In addition to the logon information, all input coming into an application is mandatorily stored on a shared read/write disk volume before it is passed on to its intended application. This input caching guarantees that input submitted by an end user is not lost due to application failure. The Parallel Sysplex operating system software working in conjunction with the communications access method (e.g., ACF/VTAM) also monitors the transport layer acknowledgements being issued by the application for the transactions it receives to manage the contents of this input cache. With SNA and TCP/IP, any lost message units can be recovered at the communications subsystem level without end user intervention, provided, of course, that the recipient has not previously acknowledged the safe receipt of the message units in question.

During the recovery process, all the sessions that were being handled by the failed application will be switched over to other copies of the application. An installation could specify certain policies as to how this "cutover" should be handled so that certain sessions could get priority over others. Once the sessions have been redistributed across the working copies of the application, they are automatically re-synchronized to the state they were in prior to the application failure. The cached input and the acknowledgement status of the last transaction are used to realize this resynchronization. The bottom line here is that the recovery, whether from hardware, operating system, access method, application subsystem, or application failure, is quick, automatic, nondisruptive, transparent, and, above all, painless.

This powerful MNPS feature of Parallel Sysplex is in effect a further refined and generalized version of the *Extended Recovery Facility* (XRF) that IBM has offered since 1987 to enhance the availability of mainframe-centric, mission-critical computing. With Parallel Sysplex, mainframe customers in essence get "XRF+," in addition to load balancing and scalability. The Geographically Dispersed Parallel Sysplex feature extends the nonstop aspect of Parallel Sysplex to also embrace certain disaster recovery scenarios. With GEOPLEX, a Parallel Sysplex cluster can be spread across data centers that are 24 miles apart, connected via fiberoptics links running at speeds up to 800 bits/sec. Critical data, including the logon information, will be mirrored at each data center. Consequently, other remote sites can continue to maintain workload processing in the event of a total catastrophic failure to one of the data centers. This type of disaster recovery setup could be used to deal with data center failures precipitated by certain local disturbances such as earthquakes, hurricanes, terrorist attacks, riots, or accidents (e.g., plane crash).

COUPLING AT ITS BEST

The logical and physical connections between the mainframes and processors within a Parallel Sysplex cluster are realized via what IBM refers to as the *Coupling Facility* (CF). The CF in reality is the very heart and soul of the Parallel Sysplex technology and the basis for the read/write data-sharing that is imperative in order to achieve the single system image and MNPS characteristics. It is the CF that provides Parallel Sysplex systems with three key and prerequisite functions:

1. Very fine granularity *Data Locking* to permit data sharing and data updating between multiple copies of an application without contention and loss of integrity
2. *Data caching,* as in the case of I/O caching, on both a local and distributed basis, including high-performance data caching for the shared data access
3. *Queuing mechanisms* to support workload distribution, message interchange, and state information sharing

The CF per se consists of hardware and specialized IBM-licensed internal code (i.e., "microcode") known as "CF Control Code" (CFCC). IBM currently provides two distinct means of implementing a CF: an integrated S/390 solution, and one that requires a separate, standalone box. The integrated option is known as the *Internal Coupling facility* (ICF), while the standalone CF is implemented using either an IBM 9674 box or an IBM 9672 Model R06 machine.

With ICF one uses one (or more) of the processors in a multiprocessor S/390 to act as the CF and run the requisite CFCC software. The processor(s) used to run CFCC can be dedicated just for the CF function. IBM also provides the option whereby the CFCC code can run in shared-mode with processors being used to support a true application partition (i.e., an LPAR). The 9674 and 9672-R06 have their own processors to run the CFCC software; both boxes can have up to ten processors! The 9672-R06, though constrained so that it can only act as a CF box, is in reality just another S/390 G5 mainframe with one to ten standard S/390 processors.

Physical connectivity for data transfer between a CF and all the processors involved in a Parallel Sysplex cluster can be achieved using two very distinct mechanisms. If the CF and some of the processors are all resident within the same S/390 mainframe, high speed communications between those co-resident entities can be realized using an internal channel or bus. With G5 models the preferred option is to use the internal Self Timed Interconnect (STI) bus that runs at 333 Mbits/sec. External links between separate S/390 mainframes (9672-R06s or 9674s) are typically implemented using the so-called *HiPerLinks*. HiPerLinks are IBM-specific, fiberoptic-based connections that can sustain data transfers at speeds up to 800 Mbits/sec over distances reaching 24 miles. HiPerLinks are the basis for GEOPLEX configurations.

THE BOTTOM LINE

Parallel Sysplex is a pivotal feature of IBM's ongoing mainframe strategy. S/390 Parallel Sysplex brings the power, scalability, and cost economics of large-scale parallel processing, hitherto confined to the scientific community, into the commercial sector. With today's Parallel Sysplex it is possible to have a tightly coupled mainframe cluster that offers smooth workload balancing, and the nonstop computing strengths of Multinode Persistent Sessions that consist of 32 separate mainframes, each with up to 10 processors.

Parallel Sysplex synergistically synthesizes the best qualities of SMP and loosely coupled systems to provide scalability, 99.999% "up-time", nonstop operation, and workload balancing. With IBM's move to CMOS-based processors, Parallel Sysplex, as opposed to acquiring larger and larger SMP machines, is the cogent and strategic means for addressing the need for greater processing capacity. The fundamental difference between the original S/360 machines introduced in 1967 and the S/370s that came to be in 1971 was Virtual Storage. Today virtual storage is very much a given and not even thought of as a value-added feature. The key difference between the S/370 and S/390 is Parallel Sysplex. Just as with virtual storage, Parallel Sysplex in the near future will become an indelible, mandatory feature of mainframe computing. Of that there can be no doubt.

Notes

1. Sysplex = **Sys**tem Com**plex**
2. **M**illion **I**nstructions **P**er **S**econd
3. **C**omplementary **M**etal **O**xide **S**emiconductor
4. JES = **J**ob **E**ntry **S**ubsystem, job control system for the MVS operating system
5. **E**nterprise **S**ystem **Con**nectivity is IBM's current high-speed channel architecture which supports data transfers at up to 17 MBytes/sec.
6. RX6 has a 12 GBytes/sec memory bus while the YX6 is even quicker with a 15 GB/sec bus.

Chapter 35
Web-to-Host with VM/ESA: A Viable Platform for the Internet

Jim Foxworthy

THE RAPID-FIRE PACE OF TECHNOLOGY IS CHALLENGING US — AND CHANGING HOW BUSINESS OPERATIONS ARE MANAGED. Take the World Wide Web as an example. The days of providing web sites that only offer brochureware are gone. Acceptance of the Internet, intranets, and extranets has now been replaced by expectation. Today, expectations of a web interface include detailed exchange of business-critical information. Customers, employees, and business partners expect to seamlessly extract information from various platforms and use it on their desktops.

This problem is complicated by the latest revolution: E-commerce. Not only do E-commerce systems use new technologies, but these systems are also directly tied to the bottom line. Organizations are turning to the Web to allow them to streamline their businesses in order to maintain a competitive advantage. Before E-commerce can become mainstream, it will be necessary to satisfy another set of business requirements — providing continuous availability, peak performance, robust security, and high reliability. Thus, IT management is faced with a true conundrum: how to combine the flexibility and ease of use seen in desktop systems linked to the Internet with the system characteristics of RAS — reliability, availability, and scalability.

This dynamic puts pressure on IT. Nearly 30 years of application development resides on legacy systems that were not designed for the Internet. What must be done to these systems so they will meet user expectations and leverage the existing investment?

APPLICATION PROFILE						
Content	Text, images	Text, forms	Text, data, graphics, audio	Text, data, large graphic & image files, audio/video	Text, data, graphics	Large text & data files, graphics, images
Peak Hits per Minute	<10	250	300	500	1,000	1,200
MAINFRAME CONFIGURATION						
Total Cost ($000)	3,260.9	1,625.8	7,429.8	7,676.7	6,304.3	7,426.6
Peak Used Capacity (%)	0.3	1.4	0.6	1.1	0.8	1.4
Web Server Costs ($000)	9.8	22.8	44.6	84.4	66.4	104.0
ALTERNATIVE SERVERS						
Server Type	NT (1 CPU)	NT (1 CPU)	NT (2 CPUs)	UNIX (2 CPUs)	UNIX (2 CPUs)	UNIX (2 CPUs)
Web Server Cost ($000)	62.8	64.3	67.5	123.1	127.3	134.7

Exhibit 35.1 Platform Cost Comparison (Source: International Technology Group).

One misconception that quickly comes to mind is system migration. The supposition is that legacy systems should be moved to alternative platforms, because those platforms have characteristics designed for Internet applications, in order for desktop systems to connect to them. However, legacy systems can, in fact, connect directly into the Internet via the desktop. Many of the alternative platforms are also still evolving and, unlike the mainframe, deliver only minimal levels of RAS. Even those platforms that have achieved acceptable levels of RAS still lack robust systems management tools, creating high levels of total cost of ownership (TCO). See Exhibit 35.1.

According to the International Technology Group in Mountain View, CA, not only does the mainframe have the ability to support a high number of hits per minute, but it also has the lowest web server cost compared to the NT and UNIX alternative servers.

Every mainframe application represents a significant investment that should not be casually discarded. One alternative to migration is putting a new front end on the application. This allows for easy extension of its life, freeing one to focus resources on developing new applications for the organization.

And there is a light at the end of the tunnel. A solution is available today that allows one to provide access to legacy applications directly from a desktop system while those applications are stored on a platform that delivers high levels of RAS. This article will discuss the merits of bringing mainframe applications to the web browser using the IBM VM/ESA operating system as the platform for the web server.

VM/ESA: A PLATFORM FOR THE INTERNET

Over the past 25 years, VM/ESA has established itself as a premier mainframe platform for server applications on small and intermediate S/390 servers. Customers of VM/ESA have extraordinary loyalty to this platform, in large part because of its proven capabilities for application development and network computing. Many organizations continue to rely on VM/ESA to manage their vital applications — including order-entry system, finance, accounting, payroll, etc.

Although still reeling from the "politically incorrect" days of the early 1990s (before the merits of the Web were fully realized), use of VM/ESA as a web server has become more popular as web commerce applications become more prevalent. As organizations tie their day-to-day processes and applications to the Web, the mainframe is becoming a more logical platform due to its proven delivery of RAS. The mainframe is a natural for E-commerce because that is where the data typically resides — about 70 percent of corporate data according to the Meta Group.

The Yankee Group, an industry analyst firm, believes that two factors are driving the demand for mainframe-based web server applications among mainframe users. First, the Web is increasingly used as the business transaction medium. If businesses are to depend on the Web, then their web server must be able to meet all the demands that come from that dependency. VM/ESA has a proven history that has shown it can deliver reliable results.

Second, mainframes provide the most scalable and secure transaction processing environments. Web server applications placed on alternative platforms tend to quickly outgrow those platforms. The Yankee Group expects to see this expansion continue, which will create a significant demand within organizations for processing power, storage, and robust connectivity solutions.

VM/ESA offers extraordinary flexibility in these areas. Frequently implemented on mainframes as a hypervisor, or master operating system, VM/ESA has local access to multiple guest operating systems and their applications. As a web server, VM/ESA can act as a gateway to all the established legacy systems in an environment without creating new connections between systems, and without the addition of a new platform.

VM/ESA is well positioned to be a key component of an organization's World Wide Web strategy. With VM/ESA, organizations can exploit web technology to regain control and management of information, using existing infrastructure and current investments in hardware, software, and personnel skills.

Part of the reason VM/ESA appeals to organizations is due to its reliability, availability, and scalability that are still considerably ahead of alternative platforms such as UNIX and NT. In one sense, web serving on VM/ESA is simply the latest example of successfully capitalizing on these traditional strengths.

Today, keeping pace with the growing number of clients running graphical web browsers and the increasing volume of transactions occurring over the Internet requires more powerful web servers. In addition to taking advantage of highly scalable S/390 platforms, native VM/ESA web servers also enable one to serve existing mainframe data to a new generation of interactive graphical applications.

A summary of the key advantages of VM/ESA as a web server platform follows.

1. Easy to add a web server to VM/ESA
   a. VM/ESA supports TCP/IP
   b. VM/ESA stores all data types
   c. A web server is simply a service virtual machine
2. VM/ESA excels at handling volatile workloads
3. VM offers unmatched quality and RAS
4. VM/ESA is a mainframe system
   a. Powerful and scalable
   b. Storage capacity, bandwidth, redundancy
5. VM/ESA is extremely cost-effective
   a. Ever-decreasing costs per user and transaction
   b. Lowest administration costs of any platform

GOOD NEWS/BAD NEWS

Although legacy applications continue to offer reliability and high performance, they also present two obstacles to users:

- **Restricted access.** To use mainframe applications, users must connect to them through 3270 terminals or PCs running 3270 emulators. Users want access to information not just from the office, but also from home and on the road.
- **Antiquated interface.** Users find the text-only 3270 interfaces of mainframe applications antiquated and difficult to use. Today, users rely on a desktop or laptop system and demand graphical user interfaces (GUIs), preferably provided through the same web browser used to navigate the organization's intranet.

These obstacles represent the greatest drawback to E-commerce applications over the Web. The application logic, and the data locked away in legacy systems, represent one of an organization's greatest assets. And the mainframe represents a viable option to host a web server. If a solution

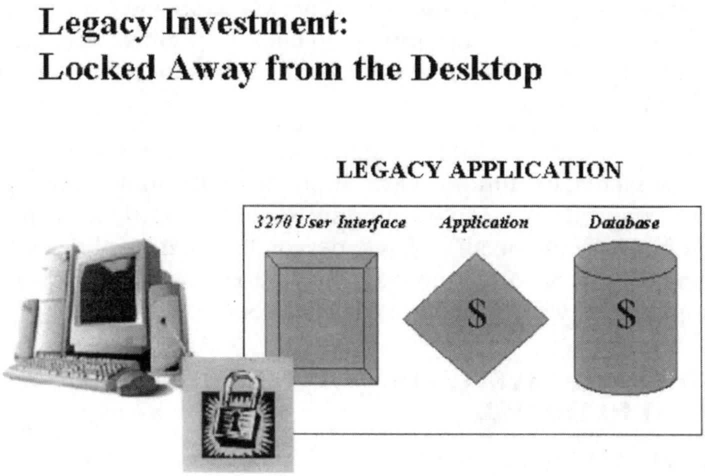

Exhibit 35.2 Legacy Investment.

---

could be crafted to improve the access from the Web to legacy assets, then the VM/ESA platform would represent a complete solution for E-commerce applications.

UNLOCK THE POWER OF YOUR MAINFRAME APPLICATIONS

So, how can organizations meet their users' needs for a graphical interface without the costs and risks of replacing proven mainframe applications? The answer lies in both web-enabling (providing web browser access to an application) and web-enhancing (exploiting browser capabilities to improve user interface) these applications. By putting a graphical web browser front end on mainframe applications, organizations can extend these applications and data to a wider audience in an easier-to-use format. This allows organizations to get additional returns on their VM/ESA investment (see Exhibit 35.2).

Web-enablement represents the fastest and least expensive solution. By use of a utility program, the 3270 data stream produced by legacy applications can be programmatically converted to HTML format. While the appearance of the screens is still antiquated, operations previously performed with PF-keys are now done through button clicks and scrolling bars. A legacy application with several hundred screens can be web-enabled by a single person in a matter of minutes.

Web-enhancement represents a solution that leverages the power in the desktop graphical user interface. Through the use of programming tools like CGI scripts, application programmers can quickly and simply create new screen displays that are more intuitive and more closely match the expectations of end users. Icons, point-and-click, hyperlinks, etc. can be added to the GUI; thus, it becomes irrelevant from the user's perspective where the data is housed.

Web-enhancement of a legacy application with several hundred screens can be approached in multiple ways. First, the entire application can be initially web-enabled; then, only those screens most critical or most often used can be web-enhanced. A single person in a matter of days can complete this process. Second, one may choose to web-enhance every screen in the application, which can take a few weeks.

CASE STUDY — LAFAYETTE LIFE: A MAINFRAME-BASED APPROACH TO THE WEB

The insurance industry, like other industries, faces an increasingly challenging environment. In life insurance, deregulation, mergers and acquisitions, and competition from banks are obliging companies to adopt new strategies. Customer demands for greater convenience are driving growth in call centers and online services. As other channels expand, the role of agents is changing.

One of the top ten U.S. life insurers, Lafayette Life, has diversified into a broad range of financial and insurance planning services, and has pursued an aggressive growth strategy resulting in average annual growth of more than 8 percent in premium income since the early 1980s. At the same time, the company has consistently maintained high ratings for asset quality, and has contained administrative overhead through a policy of rigorous cost control.

Management believes that the Internet will play an increasingly important role in the company's business. A range of potential benefits has been identified, including reduced administrative and telecommunications costs, faster product development and delivery, improved customer service, and increased support for agents.

Challenges

In 1997, the company decided to offer Internet access to its group life insurance customers as well as to its more than 1000 agents throughout the United States. The ideal solution, management believed, would be to adapt its New Online Administrative System (NOLAS).

NOLAS is a mainframe-based custom system developed in the early 1990s that handles all key business processes. Accounting for more than 90 percent of the company's transactional and decision support workloads, NOLAS integrates all functions from policy issue to claims processing, and employs a common relational database for all customer, policy, and claims information.

The system had played a major role in the company's success in the past, and its functionality was a good match with requirements for conducting business via the Internet. It has been easily adapted for internal and extranet users by deploying a VM-based Web server.

Solution

The company's IT organization reviewed several options. It became clear at an early stage that there was no need to develop new applications. NOLAS had proved highly effective, and it made more sense to leverage the system's competitive strengths via the Internet than to incur the costs and delays of building a new solution.

Use of a host access gateway running on a UNIX or NT server was also rejected. This was partly because of the poor quality of the interface, and partly because — as management put it — this approach would "add one more layer of complexity."

Because data and application logic were already mainframe based, the fastest and most cost-effective solution was to add a web server to the VM system. A pilot system was brought into operation in less than six weeks.

The pilot began in January 1998, and full NOLAS operations via the Web were realized by December 1998. In contrast, development of new applications would have taken twice as long, and would have required at least three to five times more developer time.

Benefits

Early experiences have been highly positive. By enabling group customers to input new policy data and changes electronically rather than by sending paper forms, administrative costs have been reduced, and problems caused by misspellings, illegible handwriting, and other errors have been removed.

Because delays caused by mailing and processing of forms have been eliminated, coverage now begins almost immediately, and customers no longer pay premiums for extra days or weeks after employees have left. For corporate customers with large numbers of insured employees, this adds up to substantial savings (see Exhibit 35.3).

Agents also experience faster, easier access to up-to-date information, and the company is able to reduce its own administrative costs by eliminating manual data entry. An additional benefit is that browsers provide more user-friendly access to NOLAS, which had been originally designed with a character-based interface. "Revamping our old 3270 applications eliminated cumbersome logon procedures, cryptic function keys and messages, and cluttered display screen," says Bob Nix, Technical Support for Lafayette Life.

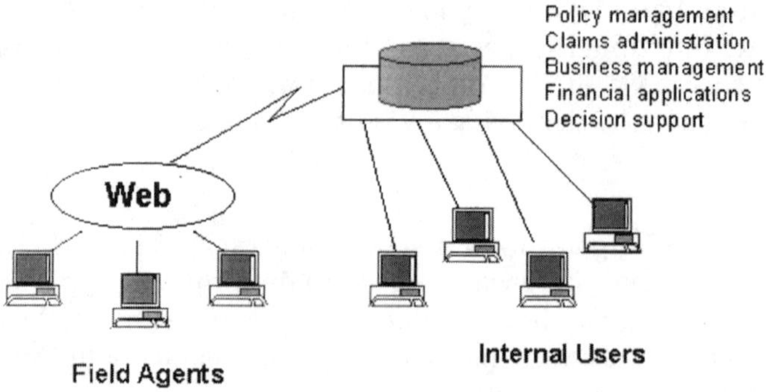

Exhibit 35.3 Lafayette Life Mainframe-Based System.

This approach has, again, proved faster, simpler, and less expensive than more complex techniques for delivering GUI capability. NOLAS is being extended to the Internet and intranet with essentially no changes to existing application code and data.

The mainframe-based approach also benefits from the high levels of availability provided by this system. The company is able to provide almost continuous 24-hour-a-day, 7-day-a-week (24 · 7) access to customers and agents who are spread across three time zones, and many of whom work outside normal office hours. The system is taken offline for only 30 minutes each month for scheduled maintenance tasks.

The Internet is an important new channel for insurance companies. But management believes that competitive success or failure will continue to be determined by quality of products, responsiveness of customer service, and financial discipline.

Most of the technological options reviewed by the company would have added complexity, cost, and delay for solutions that provided no functional advantage, and did not contribute in any way to the realization of the company's core business goals. Using VM made better business sense.

Here are some tips to consider when bringing mainframe applications to the Web. It should be noted that VM-based Web-to-host solutions typically meet the criteria in Exhibit 35.4.

CONCLUSION

With VM/ESA, organizations can cost effectively make mainframe-based business information more accessible and easier to use. VM/ESA cost of operating is less expensive per user than any other computing platform due to low storage and overall total costs. Studies by consulting organizations

Exhibit 35.4 Checklist for Selection of Web-to-Host Solutions.

1. Retain both data and applications on the mainframe and use web technology to improve accessibility and usability. This allows one to maintain extensive mainframe investment, while simultaneously addressing user needs for more friendly interfaces and universal access to information.
2. Maintain only one copy of information on a system that is accessible to any authorized user with a web browser. Changes to information and business rules are instantly available to everyone, so an entire organization is kept up-to-date with minimal effort.
3. Address applications on all mainframe platforms and handle full-screen applications — without rewriting the applications. The reason for beginning this effort in the first place was to avoid the risk of migrating to another platform and having a negative impact on the availability or performance of business-critical applications. Unlike VM, typical mainframe web servers cannot address full-screen applications unless one modifies the applications.
4. Provide both a web-enabling as well as a web-enhancing tool that offers complete flexibility to customize interfaces. It is imperative to avoid being locied into a solution that only "web-enables" applications and retains the old 3270 look and feel. Today, end users demand that their applications have intuitive, graphical interfaces.
5. Scale the web solution enterprisewide without requiring dozens of servers and a significant investment in hardware, software, networking, and administration. Windows NT and UNIX web-enablers can address mainframe applications, but they require an extra tier for the IT organization to manage.
6. Ensure security by leveraging mainframe authentication procedures and access control rules. The web solution must ensure that mainframe information is served only to authorized users on web browsers. By exploiting Secure Sockets Layer (SSL) technology, it is possible to encrypt all data transmitted from web browsers to the mainframe and from the mainframe to web browsers.

---

such as the International Technology Group show that per-user costs of PC/LAN environments are typically three to four times greater than per-user costs of mainframe systems such as VM/ESA. VM/ESA total cost of ownership is low because it has low administration costs and a robust set of systems management tools. As a consequence, a few people can administer a VM/ESA system that serves hundreds or thousands of E-commerce users.

Hosting a web server on VM/ESA, in combination with a web-enabling and web-enhancing solution, can provide the most rapid, reliable, and low-cost solution for E-commerce applications. Do not reinvent the wheel by recreating applications on alternative platforms simply to get them out on the Web. Extend and leverage investments on the mainframe by connecting them to the Web and giving them the fresh look and feel of a graphical user interface.

Chapter 36
Java on the Mainframe

Anura Gurugé

EVER SINCE JAVA WAS ANNOUNCED BY SUN IN 1995, IT HAS CHANGED THE LANDSCAPE FOR CORPORATE APPLICATION DEVELOPMENT. Java is a language that brings with it an operating environment (the Java Virtual Machine) that is inherently platform-neutral and object-oriented. These attributes, combined with Java's close association with the Web, have propelled Java into top mind-share in many IT organizations. And, in an era of anti-Microsoft sentiment, Java has garnered wide support from many major IT vendors, including IBM. In fact, IBM has made a very major and visible Java commitment to the extent that its corporate web site hosts jCentral, an entire site devoted to supporting Java developers.

IBM's support of Java has extended to the mainframe since 1996. The mainframe in recent years has proven its ability to act as a very scalable, reliable, and secure participant in client/server, web-based, and object-oriented environments. With the addition of Java technologies, the mainframe can play a pivotal role in the emerging web-based electronic commerce initiatives.

JAVA SUPPORT ON OS/390

Key to the support of Java on the mainframe is the availability of a Java Development Kit (JDK) for OS/390. A JDK has been generally available for OS/390 since September 1997. That initial version was JDK 1.1.1 The new version, based on licensed source code from Sun Microsystems and available since December 1998, is JDK 1.1.6. This JDK is supported on all releases of OS/390, from the early Version 1 releases to the latest and greatest releases, Version 2 Release 6 and Version 2 Release 7.

The current JDK 1.1.6 consists of:

- A standard, full-function *Java Virtual Machine* (JVM) equivalent to JVMs on other platforms; e.g., Sun *Solaris*
- A *Just-In-Time* (JIT) compiler

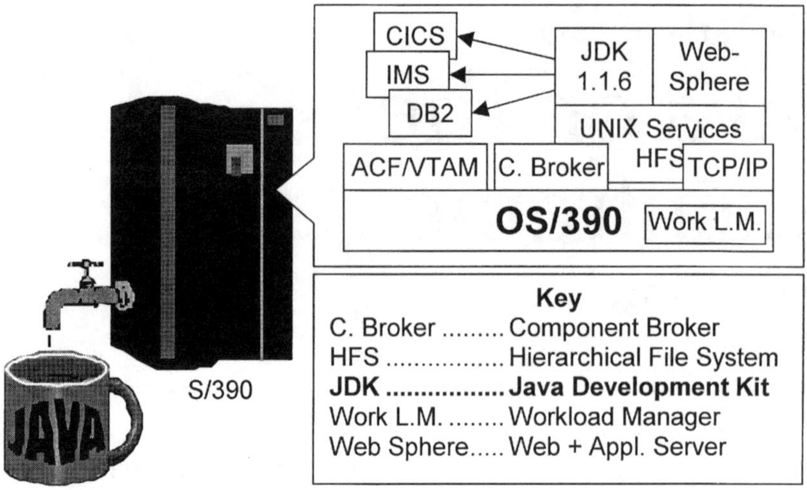

Exhibit 36.1 Components and Architecture of Java on OS/390.

- A standard set of Java Class Libraries
- A Java debugger

Exhibit 36.1 illustrates the components and architecture of Java support on OS/390.

There are two ways you can run Java code, either in *interpretative* mode where there is a software interpreter that interprets the Java instructions in the *bytecode* produced by a Java compiler the way BASIC is still executed on most systems, or by using a JIT compiler to translate the bytecode into a native machine language on the target machine.

With the JIT approach, the JIT compiler kicks into action when you try to execute the Java bytecode. The JIT compiler will read in the bytecode and translate it on-the-fly to a native machine language. In the case of the OS/390 JIT compiler, the native machine language ends up being S/370 "C" code, as opposed to assembler code or S/370 object code. It is this C code that is executed as the Java applet or application.

Having a JIT compiler on OS/390 in addition to the interpretative mode is a strong plus. A Java applet or application that is compiled in this fashion is likely to run much faster than if it was being interpreted. Exhibit 36.2 illustrates how this Java-bytecode-C process is actually structured on OS/390. The bottom line is that JIT compiled code inevitably gives you better performance than an interpretive process. The interpretive process, on the other hand, gives you complete platform portability.

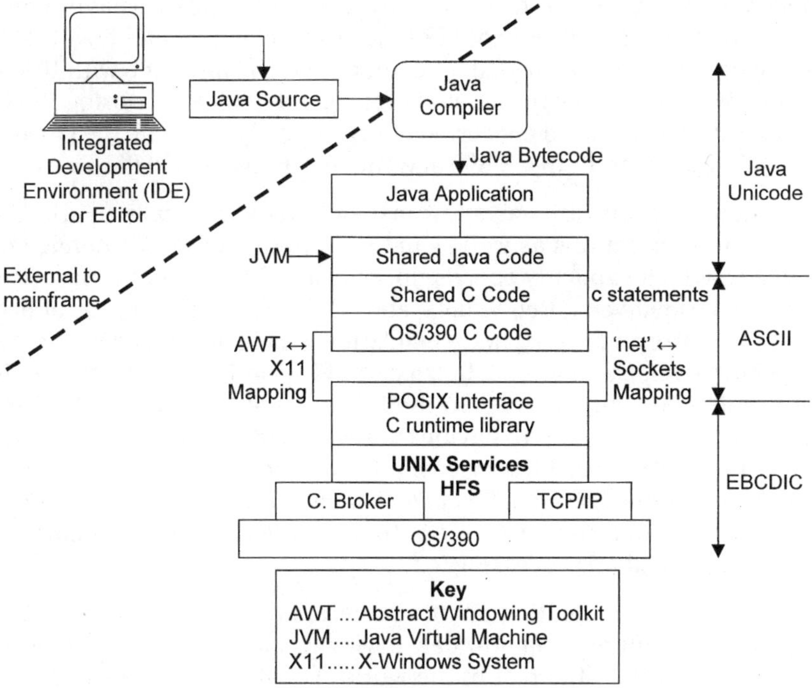

Exhibit 36.2 The JVM Implementation on OS/390.

In the early days of Java some vendors' JDK implementations lagged behind availability from Sun. That is not the case with the OS/390 implementation, however. As a testament to this, JDK 1.1.6 for OS/390 was available only one month after Sun initially released its JDK 1.1.6 for Solaris. There is a very good reason why the OS/390 JDK is in lockstep with the Sun Solaris JDK. From day one, OS/390 JDKs have been UNIX (or IBM's AIX) versions of SUN's JDK. This is also good news. OS/390 JDKs are not going to be out-of-date. IBM's goal, to which it has been true so far, is to port Sun UNIX JDKs to OS/390 as soon there appears to be a stable, general availability release on Solaris.

THE UNIX CONNECTION

Exhibit 36.1 highlights the fact that the OS/390 JDK runs on top of the OS/390 UNIX Services. UNIX Services is what, in the old days with MVS, was referred to as *OpenEdition*.

Having the JDK running on top of UNIX Services is not as incongruous as it may appear at first sight, and it should not be construed as being a problem or weakness. To start, a JVM by definition is its own environment, so it really doesn't make much difference exactly where and how that environment is realized, provided, of course, that the interface with the real machine hosting that environment is reasonably efficient. During the last few years IBM has made a concerted effort to tightly integrate the OpenEdition UNIX layer with both OS/390 and the new CMOS S/390 hardware.

The new G5 hardware has a few instructions specifically to speed up UNIX-related operations as well as native support for *IEEE Floating Point arithmetic,* which explicitly speeds up Java and C/C++ applications. (Prior to G5 and OS/390 Ver. 2 Rel. 6, Java- and UNIX-related floating point arithmetic was performed via software emulation which, as we know, is not that swift.) Moreover, OS/390 Ver. 2 Rel. 6 and Rel. 7 both contain UNIX performance enhancement features. So the bottom line here is that the JDK is well served when it comes to having true access to mainframe class performance. In addition, UNIX Services, and thus the JDK, can make use of OS/390's integrated *Workload Manager* which can provide dynamic access to additional mainframe processors (in a *Parallel Sysplex* configuration) and storage based on processing loads.

Representation of floating point numbers, interestingly enough, is not the only major difference in how data is viewed and represented in the IBM World as opposed to the rest of the world. We also have the whole ASCII (i.e., the rest) versus EBCDIC (i.e., IBM mainframe) issue. The good news is that you don't have to worry about that. The OS/390 JVM can automatically convert between ASCII _ EBCDIC if and when data is moved across the JVM boundaries, without the Java programmer or Java application having to be aware of this translation. That is yet more proof that IBM is dead serious about mainframe Java and is going to ensure that the mainframe-shunning Java crowd has no valid excuses why a mainframe cannot be used as a potent Java platform.

Before we move on, it is also worth noting that the new IBM OS/390 *WebSphere* web server cum application server also runs on top of UNIX Services as depicted in Exhibit 36.1. Just as with the JDK, this makes sense when you stop and think about it. Much of WebSphere's interaction is going to be done via TCP/IP (à la HTTP), Java, or Object-Oriented schemes. In reality, as we will see in Exhibit 36.3, much of the mainframe-related interactions performed by WebSphere will be realized using *Enterprise JavaBeans* (EJBs) and *JDBC* schemes.

THREADS AND GARBAGE COLLECTION

Here is another major plus of mainframe Java: it really and truly can be multithreaded. In multiprocessor environments, as is the norm now with the CMOS S/390s, threads can genuinely execute in parallel, thus giving you better performance.

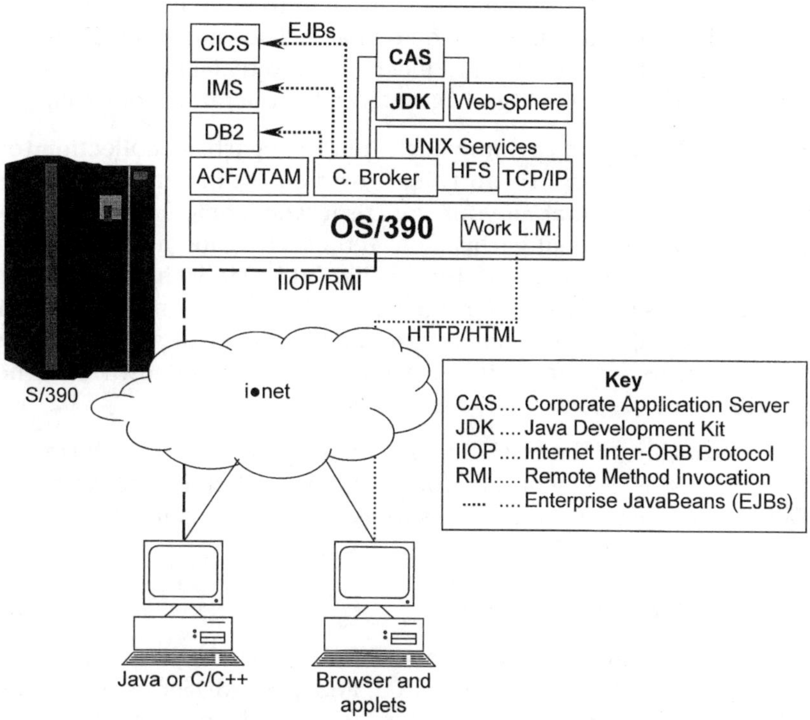

Exhibit 36.3 IBM's OS/390 Object-Oriented Paradigm Based around Component Broker and Enterprise JavaBeans.

Here is the deal. The OS/390 JDK uses a mainframe-specific threading scheme whereby Java threads are mapped onto native OS/390 threads and are run using OS/390's venerable *Task Control Block* (TCB) mechanism. I think you can get a feel of how established and proven TCBs are in mainframe environments if I just tell you that I was using them in 1975. Thanks to the use of TCBs, *all OS/390 Java threads can truly be multithreaded, preemptively multitasked, and definitely multiprocessor-capable*.

In this context, remember that multithreading, i.e., the ability of a Java program to spawn and run parts of itself concurrently, is based on many concepts that IBM had on mainframes way back in the early 70s, including *POST* and *WAIT* on TCBs, and the notion of *re-entrant* code. With S/390s offering multiprocessors, sometimes up to 12 processors per machine—and that is before you start clustering such n-way multiprocessors together using Parallel Sysplex—multithreaded Java applications can excel in OS/390 environments.

"Garbage collection" is the rather tacky term used in Java circles to describe the process of freeing up and reallocating memory (i.e., RAM). In Java, programmers are not allowed to allocate and free memory on-the-fly. Instead, the JVM allocates memory on demand. The catch, however, is that the JVM does not have a standard mechanism to reclaim unused memory. This is where this so-called "garbage collection" operation comes in.

Standard JVMs, on a periodic basis, perform garbage collection to reclaim unused memory. To do this, however, it suspends all running threads. This means that all your Java programs come to a grinding halt while the JVM goes about garbage collecting. The good news is that this, fortunately, is *not* the case with the OS/390 JVM. OS/390 has implemented what is known as *Concurrent Garbage Collection*, whereby memory reclaiming occurs in real-time *without* any threads being suspended. Obviously this improves performance since Java applications can now run without any periodic interruptions for garbage collection.

The bottom line of the OS/390 Java threading model and concurrent garbage collection is "win-win" all around: superior performance and better scalability on a platform that is unsurpassed when it comes to scalability to begin with.

JAVA ENTERPRISE CONNECTORS

The current JDK already provides connectors or gateways so that Java applications can freely interact with CICS, IMS, DB2, MQSeries, WebSphere, and SAP R/3 Access Builder. In addition, IBM is in the process of delivering connections to "Host On-Demand", ISPF/TSO, Tivoli TME 10 NetView, and IMS' TCP/IP Open Transaction Manager Access (OTMA).

Typically, there are many ways that an OS/390 Java application or WebSphere Servlet could interact with CICS or IMS. In the case of CICS, you could use either the OS/390 CICS Gateway for Java, manually code CICS EXCI calls, use MQSeries as an intermediary, or even use APPC LU 6.2 via an IBM-provided *Java Native Interface* (JNI). The *External CICS Interface* (EXCI) is similar to the old "client-server" programmatic (as opposed to 3270 "screen scraping") *External Call Interface* (ECI).

Interactions with IMS can be realized using either a direct sockets interface, using MQSeries, or via the JNI_APPC interface. Soon support will also be in place for IMS OTMA.

Relational database access via DB2 is realized using JDBC while most of the interactions with WebSphere entities are realized using the OS/390 Component Broker and other CORBA-compliant object request brokers. Think of the OS/390 Component Broker as the industrial strength version of Microsoft's Component Object Model (COM) and Distributed COM (DCOM). Component Broker, which is now an integral function within OS/390, is IBM's next generation model for all types of object-oriented programming, including Java. In time, the Component Broker will become the glue that binds the OS/390 JDK, WebSphere, and all the application subsystems (i.e., CICS, DB2, IMS, MQSeries) together. Enterprise JavaBeans are also going to play a prominent role. Exhibit 36.3 depicts the OS/390 Component Broker and how Enterprise JavaBeans fit into the picture.

Though not an enterprise connector per se, this is the appropriate juncture to bring in NetREXX. REXX, as many of you know, has been the preferred scripting language in the IBM world for many years, and was the successor to the very old Command List (CLIST) scripting scheme. NetREXX allows you to run REXX scripts on the OS/390 (or the IBM VM/ESA) JVM.

Yes, you had it right. It allows you to treat REXX scripts as Java applications! NetREXX in effect allows you to generate Java bytecode from a REXX script. Don't scoff. There are a lot of people out there who dream in REXX. Now they, too, can be Java programmers in their own rights. NetREXX obviously won't do IBM any harm; it further enhances the already powerful OS/390 Java brew. What NetREXX will do is to make the JDK more palatable and accessible to traditional mainframe systems programmer types. I am very much in favor of that. Systems programmers, a touchy clan at the best of times, can now feel that they are in the club rather than being threatened by Java. I can already see systems programmers of my vintage (yes, I was one myself for IBM at Hursley a very, very, very long time ago) staying late at work to write a NetREXX script and run it on the JDK so that they can feel that they have mastered this "Java thing".

LOGISTICS

IBM is so anxious for you to have the latest JDK 1.1.6 on OS/390 that it will even allow you to download the full JDK off the Web from *www.ibm.com/s390/java*. This applies to all OS/390 Releases, and you can even lay your hands on some of the older JDKs like 1.1.4. This being IBM, and mainframe software to boot, you can also get it on 9-track, IBM 3480 tape. The Program Number is 5655-A46. Provided you have an OS/390 license, there is no charge for this JDK.

A JDK 1.1.4 is available for VM/ESA, but as hard as I looked I could not find any mentions of JDKs for VSE or MVS. MVS is a conundrum of sorts. OS/390 is the strategic replacement for it, and IBM understandably is anxious to get people to migrate to it as soon as possible. Making things like Java only available on OS/390 is a good carrot given that mainframe types are loath to switch operating systems for the same reason that they have stuck with SNA for so long: mission-criticality.

According to the documentation, you require some OS/390 PTFs as prerequisites for the JDK. This could mean that it would not be too easy for an enterprising system programmer to try and port the OS/390 JDK over to

OpenEdition on MVS. I would not, however, rule it out. Mainframe types, contrary the outward façade, can be as creative as the Java crowd. I for one will not recoil in shock if I hear that some enterprising programmer has managed to get JDK 1.1.4 or 1.1.6 working on MVS.

Just to complete the record, so to speak, JDK 1.1.6 is also available on AIX and AS/400, and OS/2 right now has 1.1.7.

BOTTOM LINE

IBM has gone to considerable lengths to make mainframe Java potent, appealing, and accessible. The threading model, the JIT compiler, concurrent garbage collection, and hardware assist for Floating Point collectively ensure that the performance of Java applications and servlets on OS/390 are going to have an edge over most other platforms. And that is before we factor in the unparalleled scalability made possible by the multiprocessor and Parallel Sysplex capabilities of the new CMOS S/390 mainframes. Then, there is all that stuff of legends about reliability, manageability, resilience, and security.

There really are no two ways about it: OS/390 is a highly compelling, attractive, and strategic platform for large-scale Java applications such as Corporate Applications Servers.

Chapter 37
LAN-to-Mainframe Connectivity Choices

Lisa M. Lindgren

IN THE 1980S, CHOICES FOR ATTACHING THE MAINFRAME TO THE NETWORK WERE LIMITED AND IT WAS PRETTY CLEAR WHICH WAS APPROPRIATE IN A GIVEN SITUATION. Now, new types of connectivity devices are available and it is not always clear which solution is optimal in a given environment. In this chapter the choices are detailed and the strengths and weaknesses of each option are examined.

THE CHANGING MAINFRAME ENVIRONMENT

The mainframe has been the jewel in the crown of the enterprise network since its introduction in the 1960s. It is the repository for 70 percent of the mission-critical data and applications of enterprises. In the late 1980s and early 1990s most industry "visionaries" predicted the demise of the mainframe. In its place, enterprises would adopt client/server architecture implemented with low-cost servers based on industry standard technology like UNIX and Intel processors. In fact, one major industry publication actually ran a contest for readers to guess the date the last mainframe would be unplugged.

We now know, of course, that those visionaries were overly optimistic about client/server technologies. They seriously underestimated the costs and problems of downsizing mainframe applications to open system platforms. And they grossly overestimated the technical capabilities and scalability of client/server solutions. Finally, the visionaries overlooked one possibility in formulating their conclusions, that the mainframe would become dramatically less expensive and offer support for open technologies like UNIX and TCP/IP. Today's mainframe offers a lower cost, a smaller footprint, and support for even greater redundancy and fault tolerance than the mainframe of the past. Better yet, the mainframe leverages the vast investment already made in the processes for the comprehensive backup and disaster recovery capabilities that have as yet eluded the client/server environment.

Rather than selling their mainframes off for scrap metal, enterprises around the world have made the investment in retooling mainframe-based applications for Y2K. As IBM's rising stock price will attest, enterprises have continued to invest in new mainframe processors. Many large enterprises have installed TCP/IP on the mainframe to support new applications, the transfer of data to and from other servers, and host access by TCP/IP-based clients. Enterprises have deployed multiple mainframes and tied them together with Parallel Sysplex to support complete redundancy and fault tolerance.

In short, the mainframe is not dead. It is not even sick. It will continue to play a vital and viable role in large enterprises for many years to come, but it is playing a different role than it did in the early days. Rather than serving as the master of the enterprise IT infrastructure, it now participates on more equal terms with other servers. In addition to serving in its traditional role as an interactive transaction processor, the mainframe is now sending and receiving large amounts of data via file transfers and participating in client/server and object-oriented applications. Because of this changing role, there are new types of devices to provide connectivity between the mainframe and the enterprise network. The traditional channel-attached devices do a fine job of handling traditional traffic flows and volumes, but new types of devices are required to meet the new demands for high-volume data transfers and new software capabilities.

TRADITIONAL IBM (AND COMPATIBLE) DEVICES

IBM's Systems Network Architecture (SNA) is the predominant network architecture implemented in enterprises with IBM (and compatible) mainframes and midrange systems. IBM introduced SNA in 1974, and it gained widespread deployment through the late 1980s and early 1990s. SNA is a hierarchical architecture. ACF/VTAM, the SNA software that runs on the mainframe, is the master of the SNA network, "owning" all devices and serving as the primary end of all sessions. Traditional IBM devices that connect to the mainframe channel and communicate with ACF/VTAM are the cluster controller (e.g., 3174) and the communication controller, also referred to as a Front End Processor (e.g., 3745).

Since the late 1980s many enterprises have begun to integrate TCP/IP into the enterprise network including, in many cases, the installation of a TCP/IP stack and applications on the mainframe. IBM introduced the 3172 Interconnect Controller in 1990 to provide LAN-to-mainframe interconnection, initially only for TCP/IP traffic.

With the addition of the 3172, IBM had three different platforms that would provide connection from the network to the mainframe. The 3174 provided channel connection for a small workgroup of terminal devices or PCs emulating terminals. The Front End Processor formed the backbone of

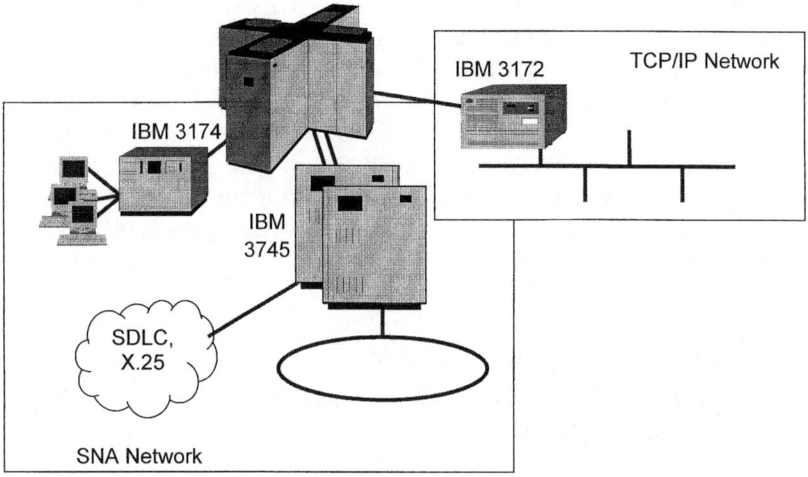

Exhibit 37.1 Typical Mainframe Data Center, Circa Early 1990s.

the SNA network and provided mainframe access to the majority of devices in the enterprise. The 3172 provided access to mainframe-based TCP/IP applications for TCP/IP-based workgroups. Exhibit 37.1 depicts a typical data center environment of the early 1990s.

Over time, however, each of the platforms evolved to the extent that the lines between the platforms have blurred. In fact, IBM spent a lot of time and effort in the 1990s trying to position the three platforms at user group meetings and in individual customer presentations. Although IBM is no longer actively marketing these traditional connectivity devices for new installations, it is still relevant to discuss their positioning, strengths, and weaknesses. There is a massive installed base of each device type, and organizations evaluating LAN-to-mainframe options often need to decide whether to upgrade, maintain, or replace these devices.

The 3174 Establishment Controller

The early cluster controller provided I/O, session, and presentation services to a limited number of coaxial-attached "dumb" terminals. The upstream connection was either a communication line (e.g., SDLC) or a channel connection. Since the early days of SNA, the controller (now called the 3174 Establishment Controller by IBM) has evolved to support LANs, PC clients, Advanced Peer-to-Peer Networking (APPN, SNA's successor), TCP/IP, Frame Relay, and other advanced features.

Despite the introduction of advanced features, both remote and channel-attached controllers have been and are being replaced by other equipment, primarily gateways and routers. This is due primarily to the diminishing installed base of traditional terminal devices and split-stack emulators that require a controller to complete the session with the mainframe. Many organizations find that upgrading an existing controller to support new LAN clients and new gateway functions is more expensive than implementing more modern solutions. The goal of enterprises should be to stabilize the controller installed base, keeping them only where they are required to support legacy devices or specific features. Over time, as the end systems are upgraded, the controller should be phased out.

The 3745 Communication Controller (and related products)

Prior to the proliferation of multiprotocol, bridge/router-based backbones, the 3745 Communication Controller (and its predecessors) formed the backbone of the majority of large enterprise SNA networks. The 3745 runs ACF/NCP software which is responsible for routing SNA traffic and managing the communication lines in the backbone network, among other tasks. These devices were deployed in the data center, channel-attached to mainframes, and in remote locations. The remote 3745s have largely been replaced by bridges, routers, gateways, and/or Frame Relay Access Devices (FRADs), but most large enterprises still have channel-attached 3745s, referred to as FEPs.

The FEP is an excellent platform for consolidating a large number of serial lines, and for handling terminal-oriented sessions and traffic. It also supports unique features for interconnection of two SNA networks (i.e., SNI), support for X.25 networks (i.e., NPSI), support for ASCII devices (i.e., NTO), and support for bisynchronous (BSC) devices and hosts.

However, as LANs and multiprotocol environments grew, IBM tried to adapt the FEP to the new environments with patchy success. Initially, IBM offered a Token Ring interface on the 3745. This has been quite successful, since most IBM-oriented data centers used Token Ring technology. The Token Ring interface on the FEP (called the TIC) was the dominant LAN-to-mainframe path for SNA traffic until the recent success of channel-attached routers.

IBM offered Token Ring support, and later Ethernet, directly on the 3745 frame itself. However, as the need for LAN support grew and the requirement for higher-speed channels (i.e., ESCON) and lines increased, the capacity of the 3745 could not really keep up with the demands. The 3746 Model 900 Expansion Unit (3746-900) was introduced to meet these demands. The 3746-900 is connected to the 3745 and provides support for additional X.25 and SDLC lines in addition to ESCON channels, Token Ring, Ethernet, ISDN Primary Rate Interface, and ATM networks. Originally the

3746-900 relied upon the ACF/NCP code in the 3745 for all software functionality, but it has recently been updated to support autonomous APPN and IP routing. The 3746-900 frame has sold very well, in large extent due to the ESCON channel and enhanced LAN and line support.

The ultimate goal in the IBM FEP strategy was to migrate customers from ACF/NCP-based 3745 FEPs to a standalone APPN and IP router. The 3746 Model 950 (3746-950) was this targeted platform. The 3746-950 is, quite simply, a 3746-900 that no longer has a connection to the ACF/NCP on the 3745, and that supports native APPN and TCP/IP routing. The migration story for large enterprises from the 3745 to the 3746-950 was outstanding but the implementation was terrible; consequently, the 3746-950 has had very limited success. In essence, the platform was too late, too slow, and did not support leading-edge interfaces and software functionality. While IBM still sells the product, it has turned its attention and marketing focus on its own channel-attached router, the 2216.

So where does that leave enterprises with an installed base of 3745 and 3746 devices? The goal should be to minimize the ongoing investment in them. The hardware maintenance fees on these devices can be quite high, and the ACF/NCP software on the 3745 incurs a monthly software fee that can be as high as $2,000 per CPU, depending upon the number of connections. Newer platforms (e.g., channel-attached routers) can accommodate most of the SNA and TCP/IP traffic and all of the connection types (plus more!) supported by the 3745/3746. However, certain functions can only be provided by a 3745 – e.g., SNI, NPSI, and NTO. Organizations should plan to move the majority of SNA and IP traffic to newer channel-attached platforms and keep the 3745/3746 devices for these special-purpose functions until the need for them no longer exists. It should be noted that the used market for 3745 and 3746 devices still exists (but values are rapidly falling), so organizations can sometimes partially fund new channel equipment through the sale of the used 3745 and 3746 devices.

The 3172 Interconnect Controller

IBM positions the 3172 Interconnect Controller as a LAN/WAN-to-mainframe gateway for TCP/IP and SNA devices. The 3172 is essentially a PC server based on the MicroChannel Architecture (MCA) bus that has been packaged for the rigors of the data center. The Model 3, which is the only remaining model of the product available to the market, was introduced in 1992. It offers a base 25 MHz 80486 processor and a 90 MHz Pentium processor option. This device is, for the most part, a straight passthrough device for TCP/IP and SNA traffic, performing little in the way of protocol processing. The exception is its TCP/IP Offload feature which offloads some cycles from the mainframe for TCP/IP traffic. The device supports parallel and ESCON channels; Token Ring, Ethernet, FDDI, and ATM LANs; and SDLC and Frame Relay WAN links. Its five slots can accommodate up to two mainframe channel connections and up to four LAN or WAN connections.

As is obvious from the product specification, the 3172 has not kept up with PC server technology. Its 90 MHz Pentium is grossly underpowered by today's standards. The MCA bus is anachronistic in today's PCI/Compact PCI world. The device cannot provide the raw power and throughput required for file transfers given the increasing sizes and number of file transfer operations to and from the mainframe. Organizations with relatively new mainframes, an up-to-date LAN switching architecture in the data center, and high throughput requirements should consider replacing existing 3172 devices with channel-attached routers (which emulate the 3172 but offer much higher throughput). On the other hand, organizations with older, smaller mainframes and light throughput requirements should hold onto existing 3172s until they become performance bottlenecks or until the end of their useful lives.

CHANNEL-ATTACHED GATEWAYS

As traditional fixed-function SNA terminal devices were replaced with PCs on end users' desktops, a new host connectivity market was formed, the emulation software market. Initially, the PC was outfitted with a coaxial card and emulation software, and then attached directly to the cluster controller via coaxial cable. The software was fairly limited and only mimicked a traditional terminal device. Over time the emulators became much more feature-rich, incorporating:

- APIs to allow the creation of new PC programs to access mainframe data
- User customization capabilities like keyboard and color remapping
- Automation features like scripting and record/playback
- Support for multiple hosts in a single package
- Support for various host connectivity choices and protocols

As departmental LAN servers (e.g., Novell's NetWare) proliferated, vendors began to offer new LAN-to-mainframe gateway software. This software is installed on a LAN server, and LAN-based PCs running emulators use the gateway to access the host rather than a traditional cluster controller. Novell's NetWare for SAA is a very successful example of this server-based gateway software. It is estimated that Novell alone had an installed base of more than 75,000 gateway servers in the early 1990s.

Several vendors in the early 1990s began to offer PC servers that were packaged in the factory with gateway software and mainframe channel cards. These channel-attached gateways appealed to organizations with a lot of departmental and remote gateways because they offered better cost-per-user figures, more complete manageability, a turnkey and pre-tested platform, and, often, redundancy features. The channel-attached gateways often offer greater scalability, supporting more sessions than remote servers because they are built using high-end components. There are a number of different solutions available on the market today, most supporting either Novell's NetWare for SAA (which is now marketed by IBM), IBM's Communications Server for NT, or Microsoft's SNA Server. Some vendors (e.g., Bus-Tech) offer a choice of gateway software.

These new channel-attached gateways, then, are very good options for organizations that have a large number of remote gateways. Organizations can consolidate on fewer, high-end gateways that are more manageable. These gateways do not, however, serve competently as general-purpose LAN-to-mainframe connectivity devices. They do not offer the same configuration flexibility, routing capabilities, networking protocol support, and high performance offered by channel-attached routers and switches. Another area of caution is scalability. While some gateway vendors claim to support tens of thousands of concurrent sessions on a single PC server, real-life experience suggests that these claims are boastful at best. Organizations planning to deploy new channel-attached gateways should carefully test the platforms with a traffic profile that closely mimics that organization's production environment in order to estimate equipment requirements.

A final niche set of channel-attached gateways must be mentioned for completeness. Early in the emulation market, some vendors focused on providing PC-to-mainframe access over a TCP/IP network. The IETF standard for this type of emulation is called TN3270 (or now, TN3270E). Some vendors opted to offer TN3270 client software (the PC emulator software) and a channel-attached TN3270 server (the gateway) in lieu of the full-function SNA clients and gateways. In the United States, OpenConnect Systems, Apertus, and CNT were the dominant TN3270 vendors. These products offer a rich set of features but are being surpassed in price/performance and connectivity options by IBM's and Cisco's channel-attached routers. In other instances, organizations are also choosing to forego external TN3270 gateways in favor of running that function on the mainframe itself. Organizations should maintain these products while they remain viable, but look to new platforms or the mainframe-based option for expansion of TN3270 services to new TCP/IP-based users.

CHANNEL-ATTACHED ROUTERS AND SWITCHES

The multiprotocol router became the dominant backbone-networking device by the early 1990s. Initially, routers were deployed to bridge and route TCP/IP and other LAN protocols. By the early 1990s, led by Cisco Systems, routers had advanced to support traditional SNA and other legacy protocols in a variety of different ways. As those integration technologies matured and became accepted in the enterprise, a wide variety of devices (remote controllers, gateways, and even 3745 Communication Controllers) were de-installed in favor of routers. Small, inexpensive routers were deployed in branch and remote locations; medium-capacity routers were deployed as distribution points in regional locations; and high-end routers with lots of slots and very fast buses were deployed in central locations like data centers.

Once the high-end routers were located in data centers and supported SNA and TCP/IP traffic, it was a natural extension to attach them to the mainframe channel. Cisco Systems led the way, announcing the Channel Interface Processor (CIP) for the Cisco 7000 router series in 1993. The CIP initially emulated the IBM 3172 Interconnect Controller, supporting TCP/IP traffic (with some optional mainframe offload for IP) and SNA. Later, a TN3270 server (i.e., gateway) function and additional mainframe offload were added. Cisco has extended the product line with the Channel Port Adapter (CPA) for the 7500 series. IBM was late to the channel-attached router game, but now offers channel attachment for its 2216 router. The 2216 offers support for SNA, TCP/IP, and TN3270 traffic.

The channel-attached router has become the conceptual replacement for Front End Processors. Like FEPs, these routers provide attachment to multiple mainframes and routing of traffic between multiple mainframes, mainframe partitions, and data centers. Like FEPs, routers support advanced software features and complex networking capabilities. Like FEPs, they are becoming the de facto standard solution for mainframe channel attachment.

There are instances, however, in which the advanced software features of routers are not required for mainframe-bound data. For instance, a TCP/IP-based file transfer from a UNIX system in the data center to a single mainframe requires simple connectivity over a high-speed LAN and no protocol processing. In those cases, network-based switches or integrated mainframe adapters are good solutions for channel attachment.

Channel-Attached Switches

Bay Networks announced the industry's first channel attachment for a switch with its 5745 Enterprise Switching Module (ESM) for the 5000 Switching Hub. Although the first to market, this product has had very limited success due to Bay's slipping enterprise router market share and technical problems with the product. IBM, on the other hand, has released a channel attachment for its capable 8265 ATM switch.

Because switches operate at a lower layer in the protocol stack than routers, these platforms cannot provide the protocol routing, encapsulation, and other software features that routers can and do provide. Therefore, organizations should look at channel-attached switches as possible solutions when simple and direct connectivity to the mainframe from the high-speed LAN and WAN switching environment is required.

INTEGRATED MAINFRAME ADAPTERS

All of the solutions discussed so far (even the traditional IBM devices) are external boxes that bring the network to the mainframe. IBM offers an integrated device that ships with all new G4 and G5 mainframes called the Open Systems Adapter. Conceptually, this device is a 3172 that is "under the covers", inside the mainframe. However, IBM has enhanced the OSA beyond its original 3172-like capabilities. The OSA supports Token Ring and Ethernet (including full duplex for switched environments), FDDI, and ATM (155 Mbps) for SNA and TCP/IP traffic.

The OSA is and will continue to be successful; it is a "no-brainer" for organizations buying new mainframes to use this integrated networking technology. As a "free" component for new mainframes, it offers immediate LAN-to-mainframe attachment and a direct pipe into the mainframe. Organizations should embrace the OSA, particularly in environments where simple and direct connectivity to the mainframe is desired. They should look to channel-attached routers, on the other hand, when offload of IP protocols from the mainframe is important, and when external routing of traffic is required.

MAKING YOUR SELECTION

Enterprise organizations now have a number of different choices when they are trying to integrate the modern high-speed LAN environment with the mainframe. They can continue to maintain traditional IBM (and compatible) connectivity devices where necessary. Or, they can augment or replace those systems with a new generation of platforms that offer higher throughput and advanced hardware and software capabilities. Exhibit 37.2 summarizes the strengths and weaknesses of the seven different types of platforms discussed in this chapter, and provides a recommendation on where or when to maintain or deploy each of them.

Exhibit 37.2 Seven Different Types of Platforms.

Platform	Strengths	Weaknesses	Recommendation
3174 Establishment Controller	Terminal device and split-stack emulator connectivity	Price/performance; lack of many internetworking features found on gateways and routers	Limit to traditional legacy device connectivity; attach downstream of router to gain advanced routing and encapsulation
3745 Communication Controller, 3746-900, 3746-950	Serial SNA line concentration; traditional SNA routing (PU_T4); unique functionality (SNI, NTO, NPSI)	Price/performance; support for modern LAN interfaces and complete TCP/IP function; low throughput	Isolate in data center to perform only those functions not performed by other platforms (e.g., SNI); move standard traffic to more cost-effective platform; minimize ongoing fees and expenses
3172 Interconnect Controller	Inexpensive platform offering basic LAN-to-mainframe connectivity	Low throughput compared to newer platforms; lack newest LAN/WAN connections; lack newer software features	Maintain until performance needs exceed throughput capabilities; look to newer platforms for advanced features
Channel-Attached Gateways	Less costly than multiple remote gateways; rich set of gateway features and protocol options	Not suitable as a generalized LAN-to-mainframe connectivity device; scalability can be an issue	Install to eliminate multiple remote gateways, especially when traditional gateway software features are required; if only TN3270 needed, consider using TN3270 in mainframe or channel-attached router
Channel-Attached Routers	Rich and up-to-date LAN/WAN connectivity choices; rich routing and security software features; cost-effective; high-performance	May be more functionality than required in certain instances	Look to as the conceptual replacement for FEPs; implement when routing between multiple mainframes/partitions and mainframe offload is required
Channel-Attached Switches	Simple connectivity from very high-speed, switched LANs and WANs	Unproven in market; lack of protocol offload features; lack of routing	Implement for specific needs (e.g., 155 Mbps ATM direct to mainframe)
Open System Adapter (OSA)	Bundled with new mainframes; integrated within mainframe; simple connectivity	Lack of protocol offload features; lack of routing	Implement where simple, direct mainframe connectivity is required

Chapter 38
Storage Area Networks

Jim Morin

THE ENORMOUS AMOUNT OF GROWTH IN DISK AND TAPE STORAGE IS GREATLY COMPLICATING THE LIFE OF INFORMATION TECHNOLOGY (IT) PROFESSIONALS assigned to manage their corporations' information assets. A new concept, called Storage Area Networking (SAN), has emerged to provide an architectural framework for storage-related applications. The SAN concept has encouraged exciting new hardware and software technology to focus on this storage management and capacity problem.

As this new technology emerges, it is important to implement tactical projects that will lead to a strategic, new SAN infrastructure. Obviously, this roadmap will be different for every corporation, depending on their business, size of company, and storage management requirements. This article provides guidance on the applications enhanced by a SAN to add to the IT manager's knowledge and experience toolkit.

NEW FOCUS ON STORAGE

Individuals, and corporations consisting of many people, creatively develop information related to their business every day. Often this is based partly on data the corporation collects about its sales, market, positioning, customers, etc. Then information is exchanged with other people to seek confirmation or even newer ideas based on many other sources of information. The driving force behind this creative process is to make the corporation more successful in its business.

Recently, this concept has shifted attention to how to manage the flow of information. This management process starts at the source. If the information is in hard copy form, the flow is relatively slow. Remember "snail mail"? Today, more and more information is digital, greatly increasing the potential velocity of data movement through an organization. It follows that more timely access to information and better data management can provide a competitive advantage. The "information at your fingertips" slogan epitomizes the concept. (If it was only that simple!)

As corporations rely more and more on this fast-paced digital flow of information, they realize that continuous investments are required to add capacity and streamline the arteries of transport. As the investments in capacity have geometrically expanded — from megabytes to gigabytes to terabytes and beyond — the transport and management infrastructure struggles to keep pace.

WHAT IS A SAN?

Storage Area Networks (SANs) were envisioned to solve these "big" problems surrounding data storage and management. The original computing paradigm has the CPU shuttle data from main memory to longer-term permanent storage over an input/output (I/O) connectivity architecture and communication protocol. The CPU "owns" the storage device, and the I/O path is relatively short to ensure minimum latency of data transfers. This model works well and can be scaled to fit the size of many applications. However, it does not facilitate the exchange or protection of information and can be very limited in environments with large applications or a large number of users.

So the computing model evolved to place multiple "servers" on a network, with storage facilities that could be partitioned and shared by many users. Faster networks and network switching attempted to solve problems of scalability created by expanding the number of users. However, each server still "owned" its storage devices, limiting affordable scalability. In addition, I/O traffic is forced to contend with user application access, resulting in slower overall performance.

The next evolutionary step is the SAN. The concept is simple: enable servers and storage to independently provide their services on the network. Just like an appliance gets its power from the wall plug, a SAN provides a common plug-in connectivity infrastructure for adding processing power or storage capacity. The SAN becomes an independent, high-performance network dedicated to I/O. (See Exhibit 38.1.)

The potential benefits of a SAN are appealing:

- independence of server and storage implementation, enabling the selection of the best technology for the application
- the ability to add new storage or server capacity as needed without impacting the existing production
- increased availability of the information with improved disaster recovery and high availability storage solutions
- improved overall performance of the information flow by implementing new channel I/O technologies that have better bandwidth, aggregate capacity, and processing speeds

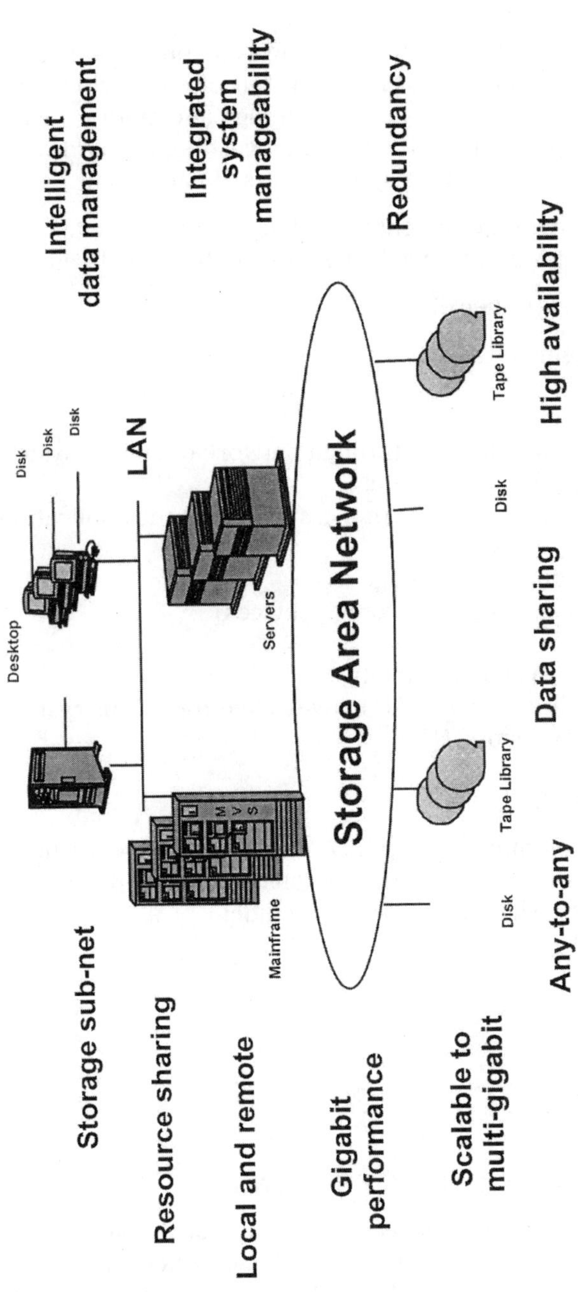

Exhibit 38.1 Storage Area Networks (SANs).

- a more streamlined approach to data management tasks that results in a more productive, efficient process with better security and lower cost of ownership

Since IT rarely has the luxury of building a computing and networking infrastructure from scratch, an important but often overlooked aspect of SANs is the need to integrate new technology and methods with existing systems. Integration protects the current investment in hardware and software and better leverages the capital dollars already spent.

Refer to the SAN requirements checklist below for several additional data management issues for SANs that need to be considered.

SAN requirements checklist:

- data management tools
- I/O protocols
- server file systems
- system-level optimization through network operating system, server interconnect, disk arrays, etc.
- record locking, transaction processing for shared applications
- virtualization of data storage
- shared data or shared storage
- device pathing (access to storage devices)
- security
- data protection and availability
- interoperability over the SAN between heterogeneous systems (S/390, multiple UNIX flavors, NT)

SAN ADVANTAGES

The advantages of implementing a SAN become apparent when applying the infrastructure to a series of applications that respond to various corporate requirements. These applications include:

- device sharing
- backup/restore
- disk mirroring
- data migration
- data sharing
- archive/retrieval

GETTING STARTED WITH DEVICE SHARING

Device sharing responds to the economic imperative that because money is scarce, an expensive resource is best utilized when shared by the optimum number of users. (See Exhibit 38.2.) This concept is as old as the idea

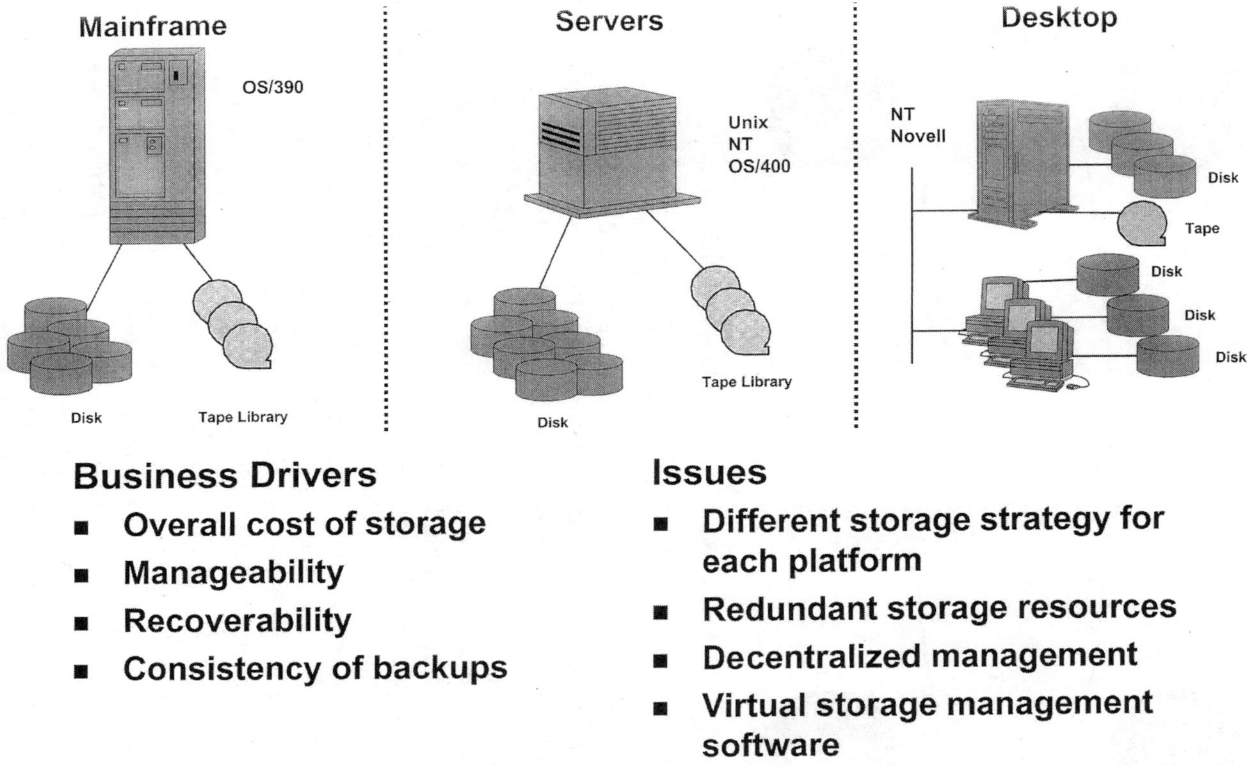

Exhibit 38.2 Shared Storage Requirements.

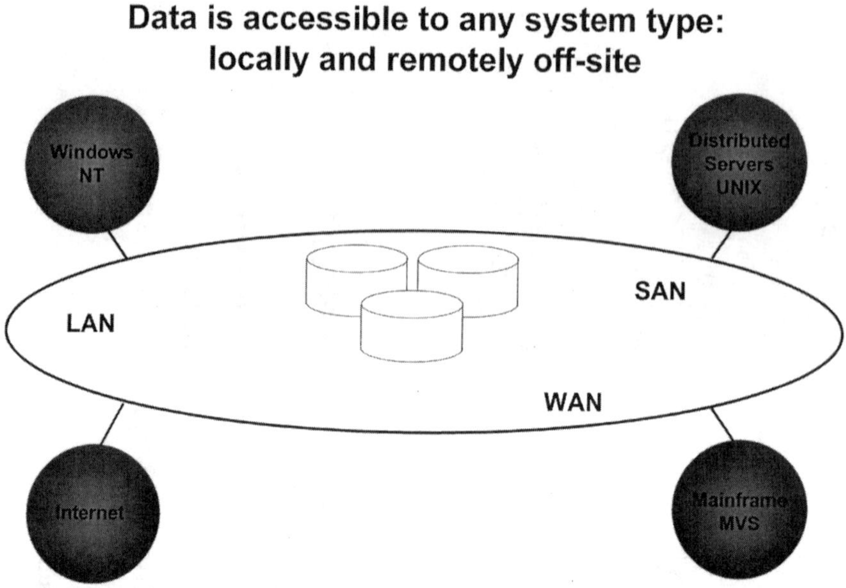

Exhibit 38.3 SAN: The Real Benefits.

behind the A:B switch that enabled a path to a destination to be manually switched from one source to another. However, the concept is complicated by one other important reality: diversity.

The fact is that corporate enterprise systems are diverse, comprised of mainframes, midrange systems, and popular open systems such as Windows NT and UNIX. Each kind of server has its own kind of storage systems, creating, in effect, islands of information. These islands are becoming more and more densely populated, growing rapidly (as much as 50 percent a year). By sheer weight, all have become critical, whether or not they were originally designed for mission-critical applications. The cost of growing these diverse storage resources and maintaining them with the same or shrinking IT staffs is a major problem.

SANs solve those problems. The SAN acts like a high-speed subnet that establishes a direct connection between storage resources and servers. (See Exhibit 38.3.) It presents a new infrastructure that liberates the storage devices, so they do not attach to particular server buses. Rather, they attach directly to the network SAN. This, in essence, externalizes storage resources and functionally distributes them to the organization. Storage resources such as tape and disk libraries can be centralized and servers can be clustered, which makes for easier access and less expensive administration.

Specialized networking devices interface servers and storage devices to the network. These are high-speed directors, multiplexers, and gateways that possess a high order of connectivity, performance, and manageability. Just as a LAN might use technologies such as Ethernet, Token Ring, or FDDI, a SAN can employ a wide range of local area and wide area technologies, including SCSI, ESCON, SSA, Fibre Channel, T3/E3, and SONET. All of this makes SANs suitable for corporate enterprise systems, which in addition to being diverse are spread all over the place.

Device sharing using SANs can be implemented with a slightly more automatic method than the A:B switch mentioned previously. However, the potential SAN value is greatly expanded through new software that automatically handles reserve/release protocol commands from multiple servers to the same devices.

BACKUP/RESTORE APPLICATION REQUIRES INCREASED COMPLEXITY

Once access to a shared device like a tape library is available, the first application to ensure data security is to put the data on tape via a backup application. In the event of a human-caused problem such as deleting a file, a data integrity problem like the fallout from data corruption, or a physical device problem disrupting access to the data, a copy can be restored from tape. Tape is also useful to restore files from a particular point in time, such as end-of-quarter sales figures.

A common problem in many corporate enterprise systems is varying degrees of discipline exercised in controlling and maintaining systems. Many LANs evolved from non-mission-critical applications, such as word processing, presentation development, and other personal productivity activities commonly resident on PC-based systems. Early on, these systems were indeed "personal systems" — outside the control of corporate IT staff. Individual users determined the level of discipline involved in maintenance and security. Activities like backup and restore, which were basic and integral to the maintenance strategies for mainframe-based systems and servers, were, for the most part, nonexistent on PC-based systems.

As PCs became common office tools, they evolved into LANs. Their importance and the importance of the information they generated and maintained increased. Now, many of the same backup disciplines of mainframe-based systems are needed for these "rogue" systems as well.

Faced with increasing responsibilities and shrinking resources, IT staffs can look to SANs to extend backup resources where needed in the corporate enterprise system. A shared backup resource in a SAN maximizes the use of tape and disk backup resources and provides consistency in the backup process.

MOVING ON TO DISK MIRRORING

Daily backups alone are not enough when companies are vulnerable to a substantial business impact from data lost between backups, or when non-stop data accessibility is required. In those cases, disk mirroring is often employed.

Disk mirroring is a method of duplicating data on separate disks in real-time to preserve the data's accuracy and make it available continuously, even in the face of disaster. Each time data is written to a primary disk, the data is duplicated — or mirrored — to a second disk. In the event that the primary disk goes down, the system can switch to the secondary disk in a matter of minutes.

Disk mirroring can be an important part of an overall disaster recovery strategy, along with backup/restore and archive/retrieval. It also can be part of a strategy to incorporate subsidiary functions, such as system development, testing, or maintenance, into a company's processing scheme without impacting mainline business operations. (See Exhibit 38.4.)

In any case, disk mirroring is employed by companies that simply cannot afford downtime for any reason, whether it is disaster-related or not. These companies must have continuous operation, seven days a week, 24 hours a day. The application requires duplicate storage resources, mirroring software, interconnect facilities, and, for many solutions, additional processor resources. This added expense requires careful consideration before and during implementation.

SANs are a natural host for just such a storage application, making the storage resources accessible anywhere in the enterprise system and allowing efficient use of this expensive facility.

The power of the SAN can be leveraged with disk mirroring systems beyond duplicating data for disaster recovery. Increasingly, additional software provides capabilities for moving data between disk volumes as a method of data migration, file transfer, data sharing, or data warehousing. Because the SAN can be a local, campus, or remote topology, disk mirroring now becomes a very efficient method for exchanging information across the enterprise.

ANOTHER STEP — DATA MIGRATION

Another storage application that suits SANs is data migration. Data migration accompanies business change, and business change is about the only sure thing left today — besides death and taxes. Businesses alternately expand and contract due to mergers and downsizing. That, along with continual development of new technology, requires data to be migrated to new disk technology or to new locations due to data center moves and

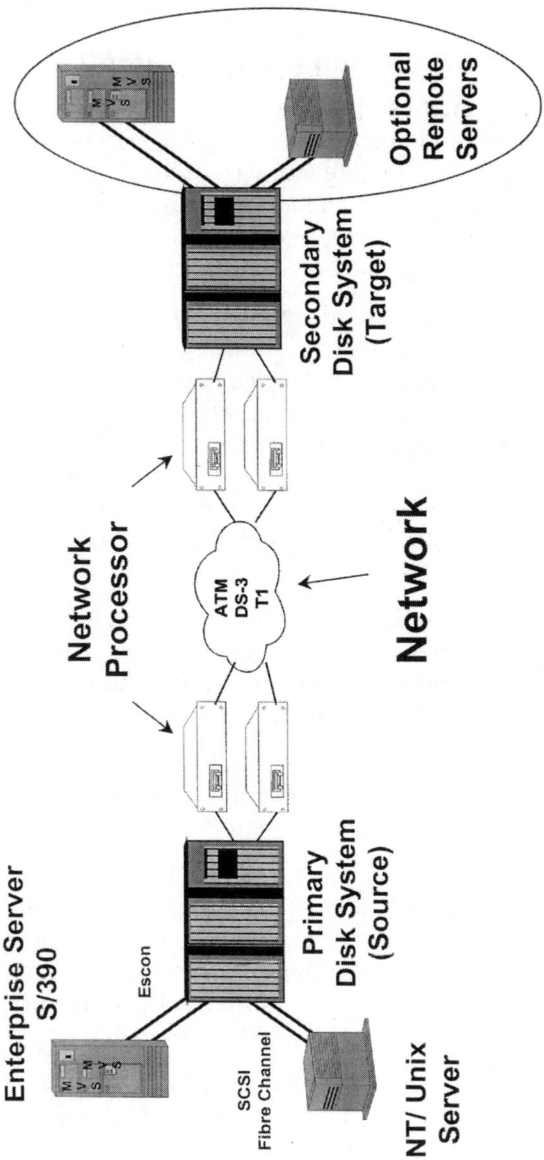

Exhibit 38.4 Typical Remote Disk Mirroring Network.

consolidations. Easy access to storage resources from remote and local sources, permanently or temporarily, is an important asset to IT managers; so is the ability to move large volumes of data quickly and safely. SANs allow all of that — quickly, easily, and safely — among diverse systems and through many types of interconnections.

ULTIMATE STORAGE APPLICATION — DATA SHARING

Once storage devices can be shared and data can be moved quickly and safely between servers and storage facilities, the way is paved for data sharing. Data sharing is the ultimate storage application. It enables enterprisewide data collection and dissemination on a real-time basis. Access to legacy applications can be extended to open systems. Intellectual property can be collected, organized, and accessed corporatewide.

Mainframes, which are good at collecting raw data from the sources or points of transactions, collect the data transaction-by-transaction in real ime. The data moves through interlinking facilities, becoming deposited in databases in open systems for access and analysis. This systematic collection, extraction, and transformation process is called "data warehousing." It is an emerging and widely acclaimed method of giving companies exactly the kind of relevant up-to-date information they need to run their businesses intelligently and competitively in constantly changing, market-driven environments. (See Exhibit 38.5.)

SANs have exceptional capabilities to store data efficiently, interface dissimilar resources over a variety of communication links, and move large volumes of data quickly and safely between servers and storage devices. They quite simply enable effective data sharing.

Data sharing can mean either multiple applications reading and writing to the same data fields, or it can mean sending a copy of the data from the source location to another location.

Data copy services, also known as file transfer, has existed with the first creation of a file. If one wanted to share the information with someone else, one either copied the file to a disk or tape and sent the physical media, or invoked a file transfer protocol and program to send an electronic copy. Today's programs are now specialized to utilize the I/O transport protocols for much faster transport.

True data sharing of the same data fields by multiple applications and heterogeneous environments is still mostly in the experimental development stage. The enterprise data management disciplines need to evolve significantly to enable true SAN interaction. Today's operating systems still expect direct ownership of the storage devices and do not have a referee to sort out allocation requests from operating systems competing for the same storage. The Storage Networking Industry Association (SNIA)

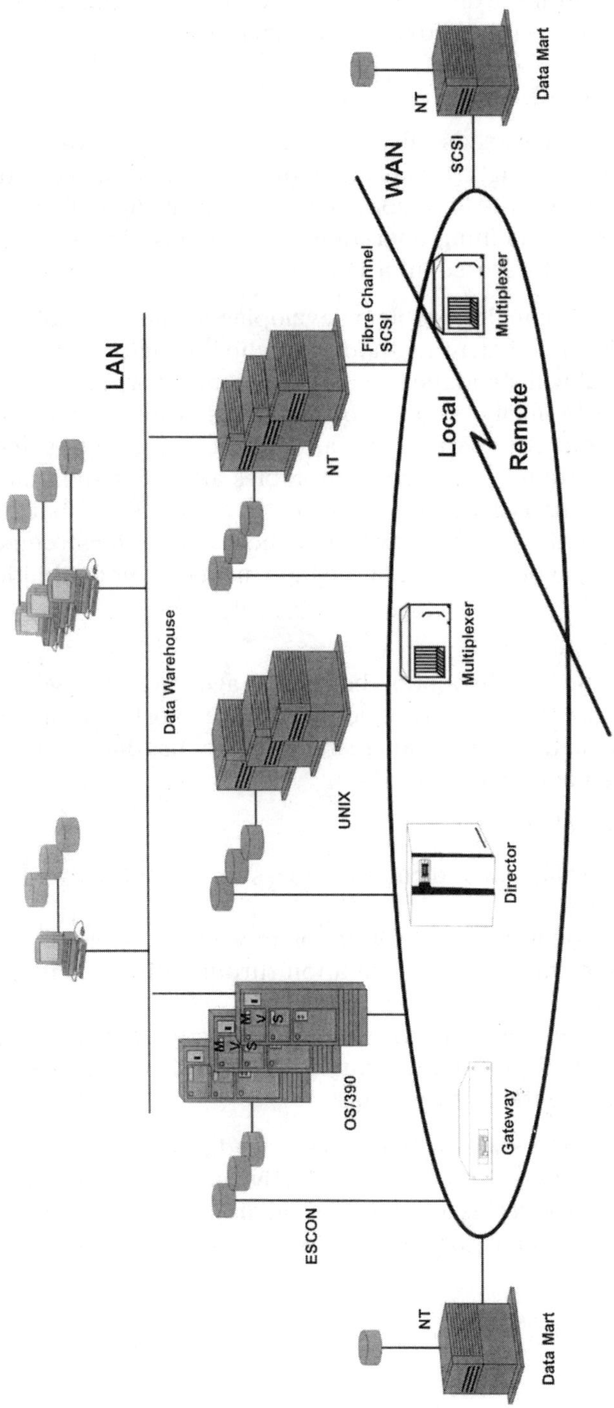

Exhibit 38.5 Building a Data Warehouse Network with a SAN.

was organized to promote development of SAN industry standards and has committees working on these data management issues.

ARCHIVE/RETRIEVAL

Finally, data sometimes needs a final resting place. Archived data is recorded on the least costly, highest capacity storage medium because the data is rarely accessed after the first month or so. Archiving data digitally offers the potential for streamlining operations, lowering costs, and improving the preservation of the stored information.

However, there is a major problem developing in many organizations as they grapple with an efficient method for archiving digital data. Selecting storage media that will stand the test of time is a major problem as tape formats and readers typically become obsolete within three to five years. For example, in the music industry, there are not too many eight-track audio tape players still around, and record turntables are becoming scarce. The fast-paced tape industry has the same problem as converter machines become difficult to find. The SAN may enable more data to be stored on low-cost disk systems, avoiding some of the tape media incompatibilities.

SUMMARY

If information is the key element of the current age, then Storage Area Networks, or SANs, are the key infrastructures. The explosive proliferation of information necessitates new and creative ways of handling that information and making it accessible.

SANs meet that challenge:

- SANs provide efficient and cost-effective storage for huge and growing volumes of data.
- SANs bridge the islands of information created by diverse systems.
- SANs smooth the flow of information through diverse communications facilities.
- SANs provide an effective, nonobtrusive structure for handling those storage applications so necessary for safe and secure operation of corporatewide enterprise systems.

Implementing a SAN starts with the basic need of sharing storage devices with multiple servers for more cost-effective storage. Each subsequent application can be viewed as another tactical project that leads to the strategic implementation of a Storage Area Network infrastructure.

References

1. ANSI site (http://www.t11.org)
2. Brocade (http://www.brocade.com)
3. Computer Network Technology (http://www.cnt.com)
4. EMC (http://www.emc.com)
5. Fibre Channel Association (http://www.fibrechannel.com)
6. Fibre Channel Loop Community (http://www.fcloop.org)
7. Legato (http://www.legato.com)
8. Storage Networking Industry Association (http://www.snia.org)

Chapter 39
Fibre Channel Architecture, Layers, and Services

Ed Frymoyer

FIBRE CHANNEL TECHNOLOGY COMBINES THE ATTRIBUTES OF A CHANNEL AND A NETWORK MEDIUM. This chapter explains how Fibre Channel works with existing interfaces, its use as a high-speed backbone in a LAN, and its messaging structure. Three topologies with guidelines for use are also explained.

INTRODUCTION

Fibre Channel is an OSI technology that provides flexibility, application layer friendliness, data integrity, and high availability. Fibre Channel technology provides a seamless application and systems interface without the need for the computer system to support the intricacies, buffering, and management of high-performance information interconnection features.

FIBRE CHANNEL LAYERS

Fibre Channel's OSI-style layering structure is based on practical separation of the functional layers (see Exhibit 39.1). Fibre Channel layers include:

- **FC-0.** This is the lowest level of the FC physical standard, covering the physical characteristics of the interface and media.
- **FC-1.** This is the middle level of the FC physical standard. It defines the 8- to 10-bit encoding/decoding and transmission protocol.
- **FC-2.** This is the highest level of FC physical standard, defining the rules for signaling protocol and describing transfer of frames, sequences, and exchanges.
- **FC-3.** This is the hierarchical level in the Fibre Channel standard that provides common services. Layer FC-3 is not currently used, but is available for future applications such as disk striping (i.e., multiplexing data

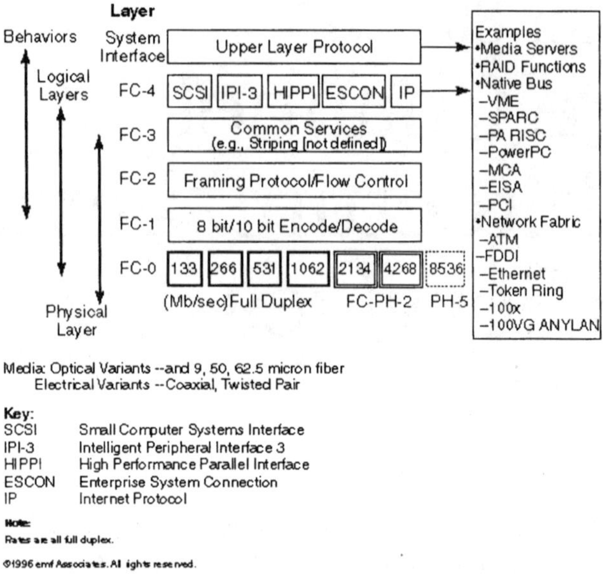

Exhibit 39.1 Fibre Channel Layers.

over multiple disk drives to improve performance) or data management.

- **FC-4.** This is the hierarchical level in the Fibre Channel standard that specifies the mapping of upper layer protocols to levels below.

The FC-0 (physical) layer and the encode/decode layers are physically and functionally separated from the link control and feature selection (FC-2) and application interface (FC-4) layers. The high-speed serial functions (i.e., optical and electrical) are separated from the parallel processed FC-2 and the software/firmware dependent FC-4.

In contrast to 802.x transport technologies that include a physical layer that is media connection only, and a MAC layer that is a combination of high-speed and low-speed functions, Fibre Channel offers a more practical approach by providing clean separations.

There are no sublayer physical definitions of the layers. A clean interface can be made to computer systems as well as to important peripherals such as mass storage, media storage, medical devices, process equipment, data gathering devices (e.g., real-time sensors), and other systems. The upper layer protocol is usually the systems bus or processor I/O function. This arrangement allows easy reuse of a system's software drivers with a minimal update to handle the speed of Fibre Channel, which can range up to 100 Mbps using optical fiber. Functionally, the information form and structure look the same. Specifically, current FC-4 layers address SCSI, IPI-3, HIPPI, block MUX, Enterprise ESCON, and traditional TCP/IP.

Future applications may use a streamlined form of Fibre Channel that provides tighter coupling between the Fibre Channel I/O and the native bus of the computer/processor. The technology is still developing. Clustering protocols are being developed with Direct Memory Access-to-Direct Memory Access connections. The protocols also offer the reliability of TCP/IP and less than 25 microsecond latency, which is much better than any other serialized system with the distance reach of Fibre Channel. This level of latency is useful in distributed systems, but not in very closely coupled systems such as backplane-connected processors.

Channel Networking

Fibre Channel provides links to other LAN technologies by means of straightforward bridging/routing methodologies. This combination of channel and network connectivity is called *channel networking* and is unique to Fibre Channel.

A Fibre Channel connection may support one or more of the defined sets of application layer (FC-4s). Fibre Channel implementations may be very simple — as in only serial SCSI — or very complex, as for multiple FC-4 support. Also, the link may support one or more transmission rates.

Chip Coding

A combination of hardware and firmware achieves the management, link control, buffering, and feature set implementation of layers FC-2, 3, and 4. Single and multiple chip implementations are available. The complexity of this adapter board hardware depends on how many FC-4s are supported of feature sets supported. Whether implementation is as a set of chips (i.e., microcontroller, memory, and the range firmware) or a single chip, the developer can use readily available CMOS processes.

The encode/decode function (FC-1) is the interface between the parallel process and the serial process. The 8-bit to 10-bit encoding provides a balanced set of 10 bits derived from a transmitted set of bits. The conversion provides a balanced set of bits in which there is only one more 1 than 0 for each set of 10 bits. Also, there are at most five 1s or 0s sequentially between each transition from 1 to 0 or vice versa. This coding has made it very easy for chip manufacturers and to build serializer–deserializer chips and hardware. In addition, all single-bit errors in a 10-bit segment are detectable.

Fibre Channel uses the 8-bit/10-bit methodology because it is easy to implement and provides an extremely balanced serial data stream for all possible idle inputs. The guaranteed transitions in the 8-bit/10-bit coding structure provide a multiplexed serial stream from which it is very easy to recover timing. The balanced serial data stream also reduces electromagnetic and radio frequency interference problems.

The output of the FC-1 layer is a parallel 10-bit signal. Some implementations have this layer on the physical interface chip, but this must combine functionality with the CMOS in FC-2 chips. The higher rates are in the extensions of the Fibre Channel standard; 2G bps and 4G bps are already approved for FC Physical Layer 2.

Physical, Behavioral, and Logical Subsets

The classification into physical, logical, and behavioral subsets is useful for the development of interoperability profiles. Because of the range and complexity of the Fibre Channel standard, profiles that define the physical, logical, and behavioral choices for a specific market or application are needed. Some of the first were defined and published by the Fibre Channel Systems Initiative (FCSI) and its participants (Hewlett-Packard, IBM, and Sun Microsystems) and provide the baseline for interoperable implementations by almost all of today's hardware for serial SCSI and TCP/IP. FCSI's goal was to create an open systems supply of interoperable pieceparts and systems.

MESSAGE STRUCTURE DEFINITION

In addition, Fibre Channel defines message structure to allow disassembly of and reassembly messages from the application layer of the transmitting side to the same level at the receiving side. These are called Fibre Channel exchanges, sequences, and frames.

Exchanges

An exchange refers to an application layer's functional communication between users. Examples are a SCSI connection or a TCP/IP session. Several exchanges can be active simultaneously.

Sequences

In Fibre Channel lexicon, a sequence is a related set of frames (one or more) within the context of a specific exchange. Sometimes a sequence is called the Fibre Channel information unit (though not specifically defined by the standard).

Frame

A frame is a set of bytes containing control (i.e., routing) and data elements. A frame is the atomic unit in Fibre Channel. Frame overhead is a fixed 36 bytes. Frames are variable from 36 bytes (control information

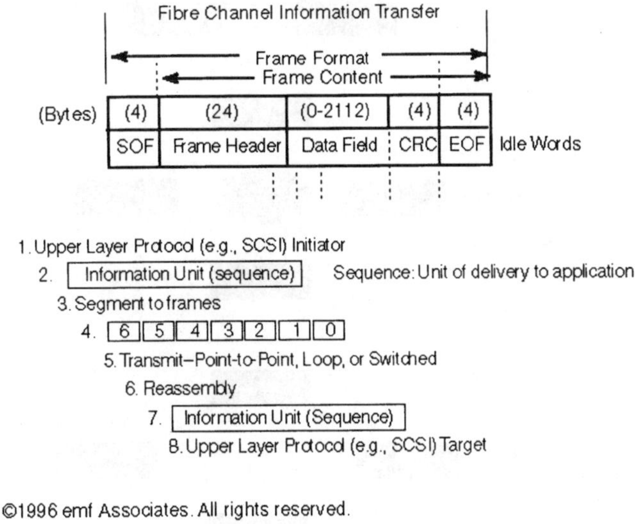

Exhibit 39.2 Fibre Channel Information Transfer.

only) to 2,112 bytes long, which allows it to be very efficient for normal-size exchanges between computers, as well as to and from mass storage connections. Exhibit 39.2 illustrates framing, message assembly, and message disassembly.

Fibre Channel further defines classes of messages that encompass both circuit connection (Class 1) and packet connection (Classes 2, 3, 4). Class 2 guarantees delivery and acknowledgment, a unique feature in packetized protocol technologies that improves transfer efficiency because:

- the receiver does not need to guess about delivery.
- packets cannot be dropped without notification to the sender; the receiver sends back an acknowledgment indicating the data has been received and whether or not it has been corrupted (i.e., a CRC Check).

Many small-scale implementations use Class 3 (i.e., packetized datagram-unreliable) because it is simpler to implement and relatively reliable.

SCALABILITY AND GUIDELINES FOR USE

Forms of Fibre Channel connection include point-to-point, loop, arbitrated, and switched fabric (see Exhibit 39.3). Users should select the connection that fits their application without device-cost impact. Topologies are designed to be interconnectable and upgradable.

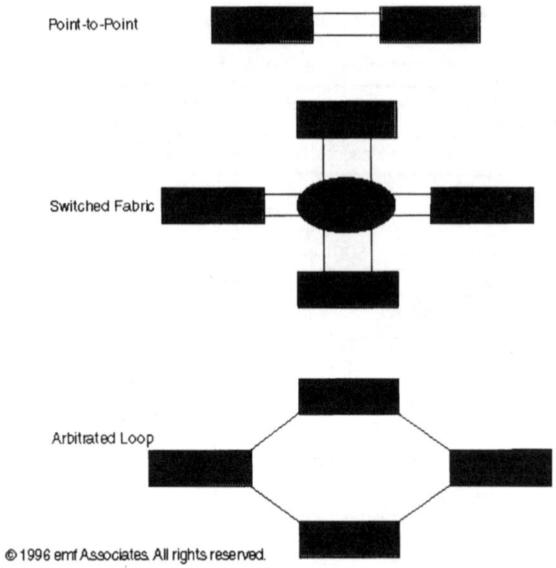

Exhibit 39.3 Fibre Channel Topologies.

Point-to-Point Connections

Point-to-point links provide a dedicated, high-bandwidth connection, such as the one from a disk array to a server. Some forms provide high availability, fault-tolerant access for two-server/two-disk-array applications.

Fibre Channel Fabrics

In the switched fabric topology, traffic between Fibre Channel ports passes through an intermediary switch called a "fabric" that combines cross-bar and packet-switching capabilities. Multiple switches can be linked together to form larger fabrics.

Like other switched media (e.g., Fast Ethernet switching hubs ATM switches), capacity in switched fabric topologies is scalable. New switch modules are added to connect additional devices, and the aggregate throughput increases to accommodate the increased load.

Fabrics can also interconnect devices with different speed and Fibre Channel media, so the speed and cost of the Fibre Channel link can be selected to match the anticipated traffic from the bridge or router.

Fibre Channel fabrics provide multicast services without requiring external frame replication servers, although the size of a multicast packet is limited to the maximum frame size of the fabric (generally 2 Kbytes).

Available fabrics support datagram, acknowledged connectionless, and connection-oriented service classes. In the future, some fabrics will support a virtual circuit-oriented class of service, in which multiple circuits with guaranteed bandwidth can be established by a single port.

Arbitrated Loop

In the arbitrated loop topology, up to 126 nodes can be connected in a shared-media topology. Instead of a circulating token or a collision sense mechanism, a simple arbitration mechanism ensures fairness. Star topologies with passive hubs provide robustness; some hub ports may be equipped with a loop bypass circuit to ensure continued loop operation if the port is offline.

Current implementations use TwinAxcess operating at a gigabit, which has ample capacity for interconnecting disk drives that can source, on the average, less than 20 Mbytes, or network connections with 10 Mbps and a certain number of 100 Mbps LANs (Ethernet).

Low cost is the primary advantage of the arbitrated loop topology. Only half the number of transceivers are required compared with fabric connections, and no switch is required. However, because bandwidth is shared among all nodes on an individual loop, the number of devices will be limited if many have high-traffic requirements.

Fibre Channel Application Interconnection

Fibre Channel can be viewed as an application-to-application connection using a common adapter port, such as a workstation, PC, storage array, disk drive, or any other appropriate peripheral (this port is known as the N_Port in Fibre Channel lexicon), and a common interconnection technology (i.e., point-to-point, loop, or switched). Exhibit 39.4 illustrates simultaneous multiplexed connections among applications using Fibre Channel.

Many applications can benefit from the economies of scale of other applications, with the incremental costs being small for both the hardware and software elements. For little additional complexity, a connection can be both serial SCSI and TCP/IP. Examples include multiple video connections (being investigated by digital video and digital movie studios), clustering and disk file access, and interactive medical applications that require streaming of data from an imaging device (e.g., MRI) and simultaneously sending the data via TCP/IP to a doctor's office in the same facility.

CONCLUSION

Because of its unique architecture, Fibre Channel can offer the high-performance, low-cost connection required by bandwidth-intensive applications. By combining the attributes of a channel and a network, Fibre

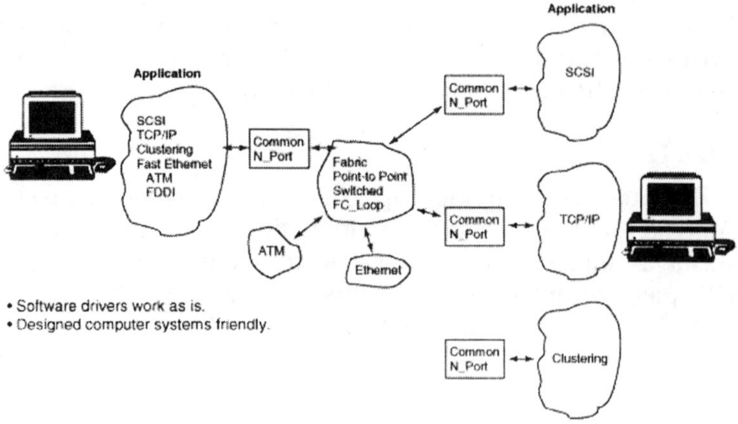

Exhibit 39.4 Simultaneous Multiplexed Connections.

Channel enables data transfer rates that are up to 250 times faster than many network protocols. Now that computers are faster and better able to handle large amounts of data, a network interconnection is needed that can handle higher speeds. Fibre Channel is emerging as the information connection technology that enables reliability and performance at record speeds.

Chapter 40
Mainframe Shops Falling Behind on e-Commerce Due to Y2K

Anura Gurugé

THE STATEMENT, "$1.5T BY 2003 AND GROWING", IS THE LATEST REVISED FORECAST by Cisco's highly respected CEO, Mr. John Chambers, on how e-Commerce—doing actual revenue-generating business over the Internet—will be affecting global commerce in just a few short years. The temptation to somewhat discount Mr. Chambers' comments as being self-serving has to be tempered by the fact that Cisco right now is doing *$15 million a day* in real business over the Internet, and that there are audited figures to show that nearly half of Cisco's burgeoning revenues are in the form of e-Commerce. Moreover, with 85% of Internet traffic going across Cisco equipment, Cisco, in general, has more than just the pulse on what is transpiring on the Internet, they, in effect, have a strapped-on, real-time electrocardiograph. Cisco is but another data point in this respect. The exploits of e-Commerce pioneers such as Amazon.com and eBay.com are already being etched into the folklore of the financial world.

There is an incongruity in the current e-Commerce landscape, however. E-Commerce is taking root outside the bounds of the traditional data center while the data center is bogged down in the throes of Y2K. In general, there is now an inverse correlation between a corporation's dependence on mainframes and the actual progress of its e-Commerce initiatives. Unfortunately, this is not going to change until around 2002, at which point nonmainframe-centric e-Commerce will be well on its way toward being accepted and proven.

There are no technical reasons whatsoever for mainframes to be out in the cold when it comes to e-Commerce, especially with IBM ensuring that

OS/390's built-in WebSphere web and application server has all the necessary features, in an unparalleled scalable form, to facilitate e-Commerce (e.g., Payment Gateway option and inclusion of the Secure Electronic Transaction (SET) protocol developed by VISA and MasterCard). Nonetheless, there is a very real danger that mainframes, at best, will become relegated to being occasional data servers vis-à-vis e-Commerce unless strategies are put in place today, despite the pressures of Y2K, to ensure that mainframes, with their critical data, applications, procedures, and, above all, scalability are fully leveraged as key components of an e-Commerce framework. The technology is already here. What is missing right now is the focus.

Ironically, Y2K in 1998 has had more impact on networking in the IBM world than any of the exciting, new web-oriented technologies on tap such as applet-based SNA access, 3270-to-HTML, Java Database Connectivity (JDBC), automatic user interface rejuvenation (e.g., IBM's ResQ!Net and OpenConnect's AutoVista), and Java-focused application servers. This is not to say that mainframe stalwarts such as airlines, financial institutes, banks, and service companies do not have a presence on the Internet. To the contrary, nearly all mainframe shops now have an intranet as well as a presence on the Internet. In addition, large mainframe customers such as American Airlines, Charles Schwab, FedEx, and Fidelity Investments already permit the general public to conduct secure, revenue-generating financial transactions over the Web, but, in many instances, access to mainframe data is not direct. Though proven technology replete with the necessary security features to facilitate such direct access is readily available from at least 40 different vendors led by IBM, many corporations beset by Y2K issues are postponing the provision of direct mainframe access by replicating some of the data on NT or UNIX servers. Such replication, though expeditious in the context of Y2K, is in most cases regressive since it results in mismatched data that has to be regularly synchronized and separately backed up via batch-oriented processes.

The remainder of this chapter looks at the current major trends in IBM networking and describes the pertinent technology, all of which is already available, that will enable mainframes to be better web-enabled so they can play a pivotal role in e-Commerce in the future.

Some current, unassailable IBM networking trends are:

- Nearly all mainframe shops now have a production intranet, as well as presence on the Internet in the form of a home page.
- Over 60 percent of mainframes running MVS or OS/390 already have TCP/IP installed, and this number will grow to at least 90 percent by 2001. OS/390 has TCP/IP as well as a full-function web server in WebSphere built-in—and Ver. 2 Rel. 7 of Communications Server for

OS/390, due around April 1999, will provide Parallel Sysplex support relative to TCP/IP, including WebSphere applications being able to benefit from OS/390's Workload Manager.

- At least 70 percent of corporate data still resides in the data center rather than on the Web or NT or UNIX servers.
- SNA-only and multiprotocol networks are being rapidly transitioned to TCP/IP-centric networks, with DLSw being widely used to transport SNA end-to-end across the network, and nearly 10 million users rely on tn3270(E) to access SNA applications without using any SNA software at the client.
- The scope of APPN and its successor HPR will be restricted to the data center—and in the case of corporations with multiple data centers, they will be restricted to certain inter-data center interactions such as bulk data transfer using ACF/VTAM Ver. 4 Rel. 4's *High Performance Data Transfer* (HPDT) option. IBM's Enterprise Extender capability on the IBM 2216, 2212, and 2210 bridge/routers, which permits SNA/HPR traffic to be routed across an IP backbone, is an alternative to DLSw that has the edge over the now-entrenched DLSw in two specific scenarios. The first of these scenarios is if SNA-based routing dictated by an SNA application name is sought in a network which has multiple data centers. The other is if there is a real need for SNA LU 6.2-based Class Of Service (COS) Traffic Prioritization from the desktop to the application.
- Though the future of mainframe SNA applications is secure well beyond 2005, these mission-critical applications are being routinely augmented with ERP applications from SAP, PeopleSoft, IBM, and BAAN that are being deployed in a three-tier client-server architecture with the mainframe acting primarily as a data server as opposed to the application server. Directly channel-attaching the application server, which is invariably an NT platform, to the mainframe using PCI bus-based 17 MBytes/sec (136 Mbps) ESCON II adapters (e.g., IBM's Netfinity ESCON or Bus-Tech's EnterpriseExpress adapters) is becoming increasingly popular. Using such an adapter, which has a price of around $12,500, can in many cases reduce response times by an order of magnitude, say from 35 seconds to 4 seconds.
- Though the installed base has not shrunk perceptibly as yet, Token Ring—like SNA-centric networking—is now in decline with most corporations already having made the decision to transition to the much more economical and ubiquitous 100 Mbps Fast Ethernet, despite IBM's, Olicom's, and Madge's eleventh-hour attempt to come up with a 100 Mbps High-Speed Token Ring (HSTR) alternative. This shift in mindset to Fast Ethernet has meant that Token Ring switching is not as widely exploited as it should be, even though third generation switches offer total functionality, wire-speed performance, and, furthermore, have driven the per-port cost to below $200.

- Invigorated by finally having a compelling and highly cost-effective, channel-attachable bridge/router in the 2216 Multiaccess Connector Model 400, IBM is aggressively trying to keep Cisco's penetration into the data center at bay as much as possible by making the 2216 the most cost-effective channel-attached tn3270(E) Server on the market. For configurations supporting 10,000 more concurrent tn3270(E) sessions, IBM's solution can in many cases be, at list, half the price of a comparable Cisco offering.
- Though over a million "seats" of browser-based access to SNA solutions have been sold in 1998, in some form or another the prevalent transition at the desktop was not toward browser-based "thin client" solutions. Instead, the major moves were from 16-bit to 32-bit desktop emulators, and from SNA-LAN gateway-based access (e.g., Novell's NetWare for SAA) to tn3270(E)-based, TCP/IP-centric access.
- ATM, though not the ultimate networking panacea that IBM tried to make it be in the 1993–1996 timeframe, is often being used at speeds between 155 Mbps–622 Mbps at the LAN backbone and LAN Server level to increase LAN performance. In North America, the public carriers are now offering "low-speed" ATM WAN services, with speeds starting at T-1 rates, as direct, highly scalable, multimedia capable alternatives to highly popular Frame Relay services.
- Encouraged by the availability of full-function Virtual Private Networking (VPN) capability on low-cost routers such as IBM's 2212 and 2210, increasing numbers of corporations are actively exploring the possibility of using the Internet as a ultralow-cost means of WAN bandwidth.

INTEGRATING SNA APPS WITH THE WEB AS A PRECURSOR TO E-COMMERCE

Browser-based access to SNA technology has now been around in proven form and with support for printing for over a year. The three primary advantages of browser-based access, over those of conventional emulator approaches, are as follows:

1. Greatly minimizes the cost and effort of client software installation, maintenance, and updating by providing a server-based, thin-client solution. In the case of applet-based solutions (e.g., IBM's Host On-Demand or Eicon's Aviva for Java), the applet can be dynamically downloaded from a web server. With today's browsers (i.e., Version 4 of Netscape Navigator and Microsoft's Internet Explorer) the applet can, moreover, be cached on the hard disk of a PC/workstation to obviate the need for repeated downloads across the network. Automatic version checking each time the cached applet is invoked ensures that caching does not preclude users from immediately getting new versions of the applet that are installed at the server.

This dynamic download-with-caching capability could save around $200 per desktop per annum in client software maintenance costs.

2. Economical and easy options for dramatically rejuvenating and streamlining the very dated and anachronistic 3270 user interface of most SNA mission-critical applications. User interface rejuvenation is a must if SNA applications are to be opened up for public use over the Internet for e-Commerce-related applications such as home investing, personal travel reservation, home banking, and online mail order.
3. It web-enables SNA applications and enables them to be readily accessed using just a standard browser across an intranet, an extranet, or the Internet.

Today there are four very different means by which to realize browser-based access:

1. 3270-to-HTML Conversion à la Novell's HostPublisher or Eicon's Aviva Web-to-Host Server
2. Applet-based tn3270(E) emulation à la IBM's Host On-Demand or OpenConnect's OC://WebConnnect Pro
3. Application-specific gateways such as IBM's CICS Web Interface
4. Programmatic solutions such as IBM's Host On-Demand JavaBeans or Bluestone's Sapphire/Web application server

3270-to-HTML, the very first form of browser-based access to SNA, is ideally suited for use across the Internet in that it always provides some level of automatic user interface rejuvenation, and furthermore does not require an applet to be downloaded. Consequently, this technology, readily available at relatively low costs from multiple vendors, should be evaluated as an expeditious means of Internet-enabling existing applications. This approach revolves around having 3270 screens that are dynamically converted to HTML pages being sent out by a standard web server. Security per se is thus integrated in that 3270-to-HTML benefits from the standard security features such as Secure Sockets Layer (SSL), available on today's web servers. SSL is a highly proven and widely used security scheme that provides authentication and arbitrates the use of end-to-end encryption. In addition to SSL, most of today's leading 3270-to-HTML solutions include session ID and sequence number facilities to ensure end-to-end session integrity to preclude the possibility of 3270 "screens" inadvertently being shown to the wrong user.

In addition to session integrity safeguards, early 3270-to-HTML conversion offerings lacked certain functions such as support for function keys, cursor-select (i.e., light pen feature), emulation of input field attributes, and unsolicited host output. The good news is that products such as Novell's HostPublisher and Eicon's Aviva Web-to-Host Server have overcome these limitations. Consequently, 3270-to-HTML at the start of 1999 is a full-function, "no-excuses" technology ideally suited for easily extending mainframe SNA applications to the Internet.

Applet-based emulation at the start of 1998 overshadowed 3270-to-HTML conversion by offering a broader repertoire of functions, including end-to-end integrity via the use of persistent TCP connections. Today, the gap in functionality is minimal—if any. Therefore, there is now a demarcation emerging as to how these two schemes should best be exploited. In general, 3270-to-HTML is the obvious choice for Internet access. On the other hand, applet-based emulation, which is in effect tn3270(E) emulation via a thin client, is ideally suited for intranet and extranet applications where there is still a need for high-performance, "green-on-black" 3270 emulation for "heads down" applications such as data entry. The primary advantage of using an applet-based emulator in this context is the elimination of client software installation and maintenance costs as described above. Some kind of user interface rejuvenation is possible with all applet-based schemes, with the latest trend being the provision of "auto-GUI" features such as IBM's ResQ!Net that will dynamically face-lift 3270 screens based on a pre-defined set of rules.

Both 3270-to-HTML and applet-based emulation will gain an increasing following in 1999 as easy to implement, "out-of-the-box" solutions for web-enabling SNA applications. However, it is now very clear that the strategic means for web-enabling SNA applications and leveraging them for e-Commerce, after the Y2K crunch, are going to be programmatic solutions.

PROGRAMMATIC ACCESS—THE FINAL SOLUTION

Most of today's venerable SNA applications will, by the year 2005, be front-ended or in the process of being "end gamed" by a new breed of applications. Though most people have heard such a prognosis vis-à-vis SNA for the last 20 years, this time the claim is not actually being made by detractors of SNA, but instead by those who intend to leverage the investment in SNA as long as possible. The current trend toward user interface rejuvenation of SNA applications is but a precursor to these applications being pushed into the background to act as transparent data servers to applications that will be performing expanded business logic functions in addition to interacting with the users via a web-inspired user interface. The SNA applications and the data accessible through them will also become back ends to the ERP and Online Analytical Processing (OLAP) applications that are now in vogue to help enterprises extract as much knowledge as possible from the data they possess.

Today's user interface rejuvenation techniques that permit one to record and capture the keystrokes required to establish an SNA session and then navigate through various 3270 screens will be expanded so that this "access logic" can be made into reusable JavaBeans. Many of the new "front end" applications will use such JavaBeans to obtain the mainframe data they require via the SNA applications. IBM's Host On-Demand JavaBeans that are now available pave the way for this new and exciting methodology.

Whichever way one looks at it, programmatic access to SNA applications from browser-oriented clients will be the final chapter in the eventful, three-decade-long evolution of SNA access. User interface rejuvenation, even when it involves considerable screen consolidation and rationalization, still relies on the original SNA application to perform all of the business logic. Programmatic access takes one to the next logical step: the ability to synthesize the data accessible through, as well as the business logic inherent in, an existing mission-critical application, or even multiple applications into the body of a brand new application, e.g., an e-Commerce application.

Programmatic web-to-data center access implies some client software on the client PC/workstation, be it browser, applet, or application, that is interacting with a data center-resident application or database across the network. Some form of intermediary software, regardless of whether it is strictly in the class of software deemed as "middleware," is always going to be required to make this happen. Today there is a very wide and disparate range of options as to what this intermediary software can be. These options include:

- 3270 datastream-oriented schemes such as IBM's "Host On-Demand" Host Access Class Library (HACL), the Host On-Demand JavaBeans, Blue Lobster's Stingray 3270 Software Development Kit (SDK), Attachmate's Enterprise Access Class Libraries (EACL), the Open Host Interface Objects (OHIO) endorsed by both IBM and Attachmate, and the old faithful HLLAPI interface which is universally available with tn3270(E) and 3270 clients
- LU 6.2-based program-to-program schemes through the APPC or CPI-C interfaces available with some ip3270 emulators such as Attachmate's EXTRA!, Wall Data's RUMBA, and IBM's PComm.
- Traditional, pre-web, middleware solutions such as IBM's MQSeries and CNT/Apertus' Enterprise/Connect that can be used with browser-based or browser-invoked clients to realize programmatic web-to-data center access
- New wave, web-oriented middleware solutions such as Bluestone's Sapphire/Web Application Server that comes replete with high-end database management features such as two-phase commit to facilitate the secure update of distributed databases
- ODBC- or DRDA-based capability, such as that available within Microsoft's SNA Server, or Cisco/StarQuest StarSQL, for directly accessing data located with DB2-like relational databases
- IBM VSAM file (i.e., "flat file") resident data access schemes such as that available within Microsoft SNA Server Ver. 4.0

- Application-specific solutions such as IBM's CICS Gateway for Java, Blue Lobster's Mako Server, and Interlink's ActiveCICX

Programmatic solutions will differ significantly from web enablement à la 3270-to-HTML or applet-based emulation in that they will require considerable amounts of programming effort. However, programmatic solutions are invariably the only way to ensure totally seamless integration of SNA applications with new wave web applications. While "out-of-the-box" solutions with rejuvenation capabilities can produce dramatic results across the Web, in the end there are inevitably some hesitations or "stutters" that betray their jury-rigged nature. With so much at stake with e-Commerce, and with competition fiercer than ever, corporations will indubitably want to present the best face possible to their clientele. The slick interfaces sought to expedite e-Commerce transactions will only be possible via programmatic solutions.

WHAT TO EXPECT IN 1999

Y2K efforts will certainly impinge upon networking initiatives, even more so than they did in 1998—with many corporations likely to suspend any major changes to the networking infrastructure during the July to December period. Given this background, the major networking themes of 1999 are likely to be:

- Relentless consolidation toward TCP/IP-centric networking backbones with use and significance of intranets and extranets continuing to grow and mature
- Further adoption of browser-based technology during the first half of the year as means of evaluating and experimenting with the web enablement of SNA applications
- Increase in the deployment of TCP/IP on mainframes, replete with mainframe-based TCP/IP management, à la Interlink's e-Control, to pave the way for a strategic, TCP/IP-centric, end-to-end, mainframe-to-desktop networking fabric for the future
- Interest in evaluating the OS/390 WebSphere as a highly scalable and centralized web server and application server—in parallel with existing NT- or UNIX-based web servers
- Growing momentum in the exploitation of the "near zero cost" Internet bandwidth for private corporate use for applications such as global remote access, telecommuting, remote office connectivity, and e-mail through the use of VPN or browser-based access solutions
- Marked presence of Fast Ethernet within what were hitherto Token Ring environments as corporations opt for Fast Ethernet as they procure new PCs/workstations, move to new buildings, or upgrade their wiring closet equipment to handle the additional data volumes precipitated through the widespread use of graphic-intensive intranet applications

- Initial evaluation of web-oriented application servers as strategic tools for kick-starting e-Commerce after the Y2K crunch
- Continued battle between Java and ActiveX for domination, with Java offering platform independence (which is not that crucial as corporations standardize on Windows) and ActiveX touting tighter integration with Windows. The outcome of this battle may eventually rest with measures taken by the U.S. Department of Justice vis-à-vis its antitrust case against Microsoft rather than on market pressures or technical superiority

BOTTOM LINE

Current Y2K efforts are yet again prolonging the life of mission-critical SNA applications. They are, however, doing so at the risk of sidelining these applications from the e-Commerce arena since the focus on Y2K is inhibiting data center professionals from being as active as they wish, or should be, when it comes to e-Commerce plans and strategy. The good news is that all the necessary technologies to enable SNA applications to play an active role in e-Commerce already exist and are proven. In the short-term, browser-based access solutions will pave the way for SNA web enablement with 3270-to-HTML optimum for Internet applications, while applet-based emulation is the ideal choice for intranets. Post Y2K, however, it will be programmatic access solutions that will rule the roost and usher SNA applications written in the 80s into the world of e-Commerce. By then, SNA-only networks, Token Ring, and ATM-to-the desktop will truly be legacy technologies. Nonetheless, given the conservatism and inertia inherent in IBM networking, one will still be able to find Token Ring and possibly even the odd SNA-only network ten years from now—if one looks hard enough.

Section VII
Clients and Servers

Chapter 41
TN3270 and TN5250 Internet Standards

Ed Bailey

WHY IS ACCESS TO SYSTEM/390 AND AS/400 TERMINAL APPLICATIONS USING THE INTERNET SO IMPORTANT? There are three basic reasons.

1. These terminal applications represent a large percentage of the data and logic of a typical enterprise.
2. There is a huge installed base of users who are currently accessing these applications productively.
3. New devices and end systems are being enabled with technology to access these applications.

ENTERPRISE DATA AND LOGIC

For larger enterprises, 70 percent or more of their mission-critical data resides on mainframes. While System/390 and AS/400 terminal applications are typically referred to as legacy applications, they collectively represent a vast majority of the business infrastructure upon which our economy rests today. And with a solid foundation of business logic that represents a history of investment, users of these applications rely heavily on their ability to access them to perform daily business transactions.

USER PRODUCTIVITY AND CONFIDENCE

Most of these terminal applications emerged in the late 1970s and early 1980s during the evolution from batch to online processing. Brought about by the introduction of the computer terminal and System Network Architecture (SNA), online processing gave rise to more direct access to enter, view, and alter information by the end user (see Exhibit 41.1). This empowered the individual and allowed productivity to soar. The ongoing reliability and availability of these terminal-oriented applications fostered confidence in the Information Technology (IT) professional and the end user to commit to completing ever-increasing volumes of transactions interactively.

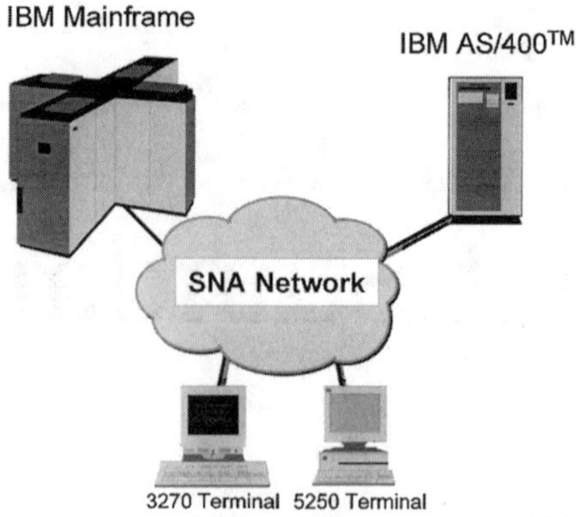

Exhibit 41.1 Terminal-Oriented Applications.

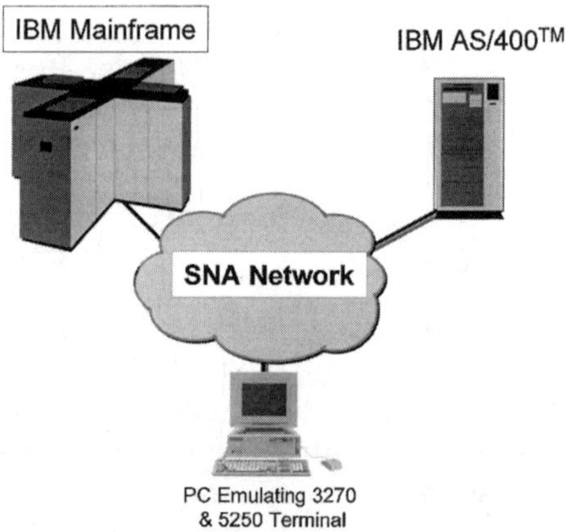

Exhibit 41.2 PC Emulation of Terminal Functions.

NEWER PLATFORMS AND DEVICES

Since their introduction, personal computers have been used to emulate the terminal functions without requiring change to the host application. This emulation, along with new personal computing paradigms for downsized solutions and client/server distributed platforms, has not displaced System/390 and AS/400 terminal applications. Rather, the technology that these approaches of the 1990s represented has propelled end user productivity to new heights. Greater end user mobility has also contributed to this technology, which enables the use of much smaller yet powerful devices such as laptops and PDAs (see Exhibit 41.2).

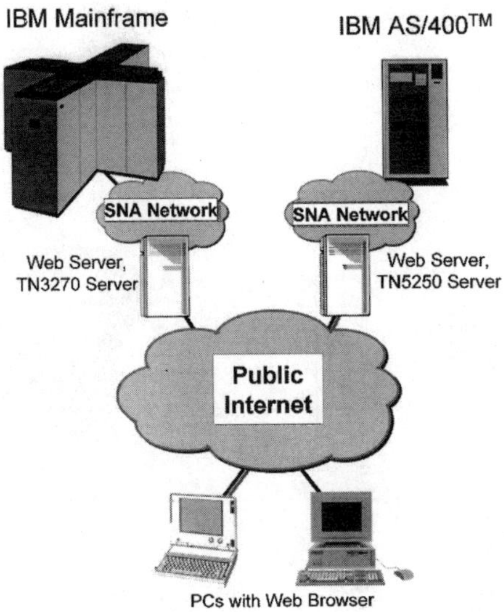

Exhibit 41.3 Internet Access to SNA Applications.

With more power at their fingertips, the demand by users for access to System/390 and AS/400 terminal applications remains high as even newer and more innovative approaches are developed to manipulate and present the information that these applications maintain.

The most recent contributor to this demand has been the increased number of users served by the advancement of the Internet and World Wide Web. Viewed as the open global information highway, the Internet is expected to provide worldwide access between the information provider and the consumer. Electronic commerce (E-commerce) has quickly become the model for conducting business involving the Internet in the next millennium. It then stands to reason that if one intends to participate in E-commerce, then System/390 and AS/400 terminal applications must be accessible to end users using the Internet. (See Exhibit 41.3.)

Basic to the Internet is use of Transmission Control Protocol/Internet Protocol (TCP/IP). Standards have been developed to enable access to SNA terminal applications using TCP/IP. These standards are TN3270 and TN5250. Client and server implementations of TN3270 and TN5250 provide access without requiring change to the host application. (See Exhibit 41.4.) These standards are actively evolving to adopt additional Internet technologies associated with web browsers and programming. The remainder of this article will address the TN3270 and TN5250 standards for using the Internet to access System/390 and AS/400 terminal applications.

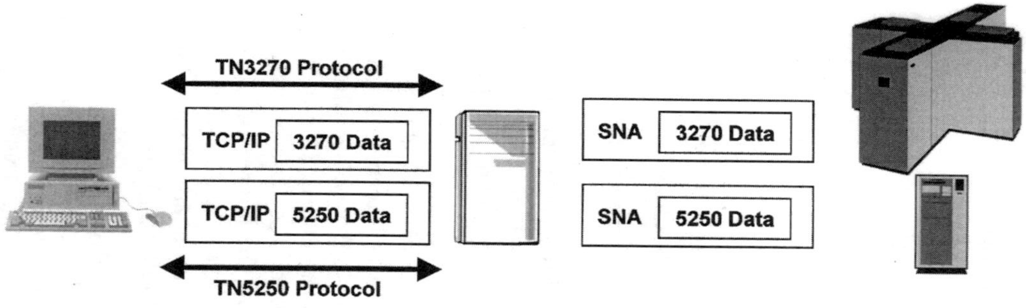

Exhibit 41.4 Internet Protocol to Access SNA Applications.

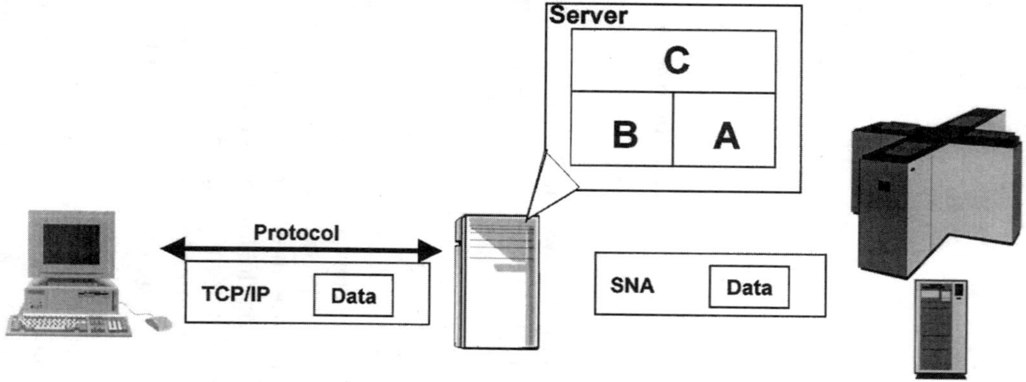

Exhibit 41.5 TN3270 and TN5250 Server.

HOW DOES STANDARDIZATION HELP?

The Internet consists of processors that are interconnected via transmission links. Some of the processors are designed to do a special task such as establish a path or route the information along to the next destination, while other processors provide services to end users such as security or information retrieval. Typically, the processor that the end user operates directly is called the client, and the processor that contains the application is called the server. To access a particular application, an end-user request may pass from the client through multiple processors before reaching the correct server. The manner in which these processors communicate is through the use of protocols.

Left to chance, it would be very unlikely that any of the independent implementations of hardware and software would operate well — if at all — with other implementations. This creates a multitude of business and technical challenges for IT professionals and their end users. Therefore, consumers expect developers of Internet products and services to follow specifications, known as standards produced by the Internet Engineering Task Force (IETF). The IETF is the organization that facilitates open participation in producing Internet standards and promotes interoperability among various Internet components from different sources. The benefit of greater interoperability among particular implementations is higher confidence in their use and the use of the Internet overall. More information on the IETF is available at the http://www.ietf.org web site.

PROTOCOLS

A closer examination of the various components of TN3270 and TN5250 allows one to better understand the benefits of standardization. One begins with the protocol. Simply stated, the protocol is the manner in which two components establish and maintain communications. As observed in Exhibit 41.4, there are two network protocols involved to allow end users to access SNA applications via the Internet: TCP/IP and SNA (Systems Network Architecture). Any enterprise with 3270 or 5250 terminal applications will have SNA protocol. Also, any enterprise with access to the Internet will have TCP/IP. Enabling users to be connected to the Internet using TCP/IP and have access to SNA applications requires two key components, referred to as servers and clients (Exhibit 41.4).

Servers

A server has the important role of initiating, managing, and terminating the SNA flows required by the OS/390 and AS/400 applications business logic. This is depicted as “A” in Exhibit 41.5. A server can include the entire protocol stack or use the programming interface provided by an existing stack. The SNA resources assigned to the server are used to support

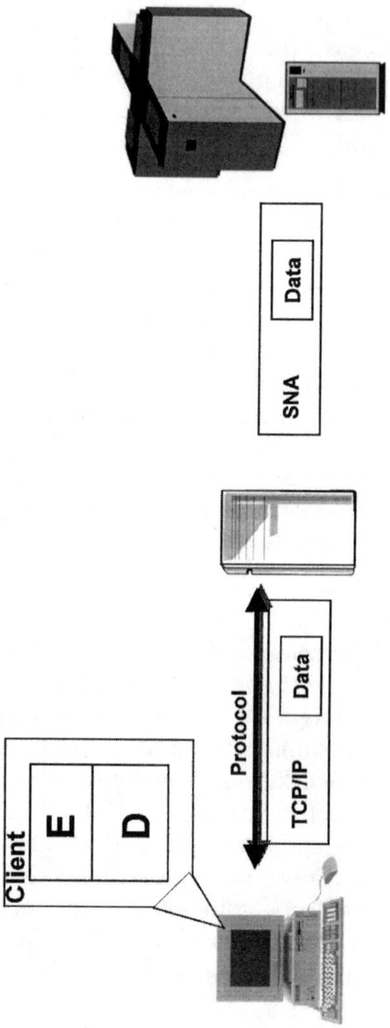

Exhibit 41.6 TN3270 and TN5250 Client.

requests from the end user. A server may reside on the same processor as the application, or on a different processor.

The server passes the SNA data stream received from the terminal application to the end user by supporting the TCP/IP flows to and from the client. Requests and responses flow to and from the user on the TCP/IP connection maintained by "B" in Exhibit 41.5. A server may provide the TCP/IP stack or rely on the use the programming interface of an existing stack. The role of "C" is to pass the data stream between "A" and "B." Early implementations of TN3270 and TN5250 servers conveyed only the data stream and very little information to the client about the SNA connection. The latest specifications include many more options for passing additional information to the client about the SNA connection and resources.

Clients

The client maintains the graphical end-user interface (GUI) and connection to the server (see Exhibit 41.6). Component "D" is responsible for initiating, maintaining, and terminating the connection to the server using TCP/IP. Just like the server, it can include the TCP/IP stack or use the programming interface of an existing TCP/IP stack. Component "E" sends and receives the data stream with the SNA terminal application and provides the GUI to the end user. The latest specifications include more options that enable the client to provide the user with many more choices for presentation of the information.

Because multiple client and server implementations exist in the market and the client implementation can be obtained from a different source than the source for the server implementation, one can readily see how interoperability could be an issue without the benefit of standards. Keep in mind that standards evolve and do not always address every unique circumstance. Therefore, they allow for options that are left to the discretion of the implementers. Although a TN3270 or TN5350 client or server makes a claim of support for the standard, it is the set of options that should be examined closely against the requirements to determine the best solution for a business. Options for supporting such technologies as security, management, and programming should receive particular attention.

INTEROPERABILITY

The IETF charters working groups to address particular Internet problems or requirements. The TN3270E Enhancements Work Group was such a work group chartered under the Applications Division. This work group has produced a number of specifications for enhancing TN3270 and TN5250. In support of the implementation of these specifications, the work group has conducted a number of interoperability tests.

Basic interoperability testing focused on:

- how well the server connected with the SNA application
- how well the server connected with the client
- how well the client connected with the server
- how well the client displayed the information

Results from the interoperability tests have been the consistent interpretation of the specifications, the increase in the number of new clients and servers, and the solidification of the protocol on which to base additional enhancements.

Enhancements

As end users access System/390 or AS/400 SNA terminal applications from the Internet, certain characteristics which existed in their prior access are expected. The work group has defined specifications to support these characteristics that include:

1. Security (encryption, authentication, authorization). This specification addresses the application level security based on the Transport Layer Security (TLS) standard and Secure Sockets Layer (SSL).
2. Management (configuration, response time monitoring). Two MIB specifications here address the configuration and response time monitoring for service-level management.
3. Performance (service location, session balancing, caching). This specification addresses the use of the Service Location Protocol (SLP) standard to identify services dynamically and learn their current workload factor.

Current efforts are under way by a number of vendors to implement these new specifications. These capabilities will enable IT professionals to deliver consistent levels of service to their end users when using the Internet.

Programming

With the growing popularity of E-commerce, independent software vendors (ISVs) want to deploy more Internet-ready applications. Most ISVs traditionally rely on the ability to build their services on top of clients and servers. Users want to integrate information from a variety of sources and formats with the data available from SNA terminal applications. Satisfying the need of the ISV and end user in this problem space requires new programming capabilities extensible to the Internet user. This leads to the use of object classes, object interfaces, and transform services. The support for Hypertext Markup Language (HTML), Java, and XML in browsers requires new programming interfaces which provide programmatic interaction with the SNA terminal application.

SUMMARY

The use of the Internet to access SNA terminal applications is considered to be rejuvenation as it opens up new markets and expands the reach of end users. TN3270 and TN5250 specifications are designed to support the needs of these applications. Interoperability and standardization have established confidence in their use. Migration to these technologies by the enterprise results in:

- no change to the terminal application
- reduced SNA networking complexities
- consistent user interface and tools
- E-commerce enabling

When selecting an implementation of a TN3270 or TN5250 client or server, always request interoperability information for the implementation from the provider. This information should state how well this implementation operates with similar implementations from other sources. An implementation that has not been well tested with other implementations may be lacking in capability and leave requirements unsatisfied after installation. If the provider is unwilling to produce such information, one should ask oneself the question, "Do I really want my mission-critical applications to depend on this implementation?"

INTERNET INTEGRATION

The TN3270 and TN5250 standards established by the IETF paved the way for the next generation of technologies for Internet integration of the System/390 and AS/400 SNA terminal applications. Comparing the role of the protocols to that of a typical package delivery service, the emphasis is on preservation of the package or data stream (sometimes referred to as the payload) from end to end. This gives the most consistency in behavior for presentation of information to the end user when compared to the original terminal or emulator. (See Exhibit 41.7.) When the application creates the data stream (1), it remains intact and unchanged as it travels to GUI (6). Although use of the Internet is made to carry the data stream to the end user, this does not support integration of the information with other Internet applications.

The web browser has introduced a new behavior that can enhance the presentation of the information to the end user. The use of a browser may require transformation of the data stream into another format — HTML and XML technologies use transformations. Java, on the other hand, does not require a transformation of the data stream. As an applet, it is designed to correctly process the original data stream. Typically, the transform occurs at Point 3 in the flow in Exhibit 41.7. However, some designs may actually not perform the transform until Point 6.

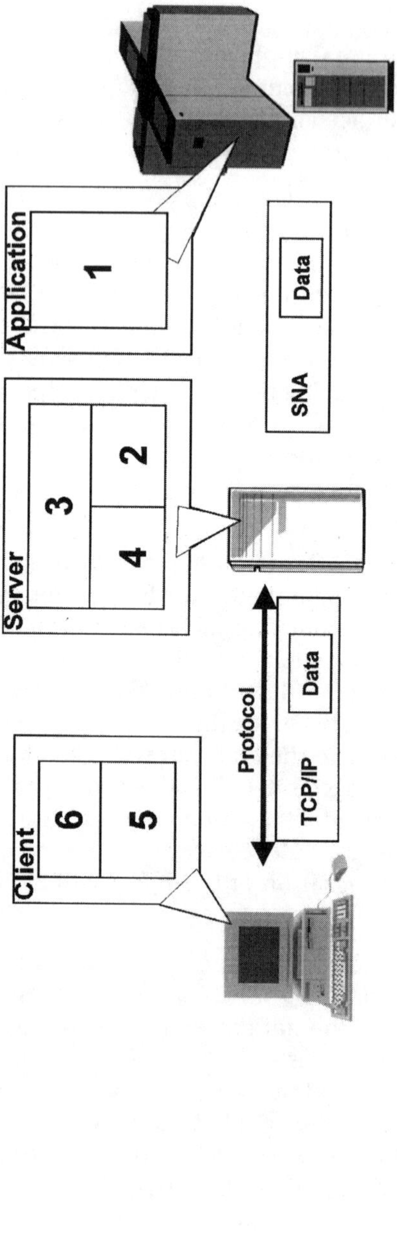

Exhibit 41.7 Data Stream Flow.

One can see that careful thought should be given to where the transform occurs. Performance and function can be significantly affected. The use of objects to assist in processing the data stream is advancing the use of transforms. This is the momentum behind such efforts as the Open Host Interface Objects (OHIO) specification to establish an industry standard. By placing object interfaces into Point 3 or 6 greatly enhances the capabilities of vendors and users to integrate the information from SNA terminal applications with other Web and Internet bases while maintaining function and performance. All the excitement of E-commerce has brought a number of new players to the market, providing SNA terminal application integration into the Internet. How consistent these implementers are in developing and using transforms and objects can impact E-commerce deployment within an enterprise. One should know where and how the transform is achieved before committing one's business to a particular implementation.

LOOKING FORWARD

Many of today's business processes (e.g., inventory, finance, claims, manufacturing, shipping) lack integration with the Internet due to a previous inability to offer services to end users that they needed and were accustomed to using. Slow progress is partly attributed to insufficient security and manageability. Although still maturing, recent enhancements in the market have produced better solutions that enable enterprises to speed up Internet integration of their host terminal applications and enter into E-commerce.

Note

System 390, CS390, and AS400 are trademarks of International Business Machines Corporation.

Chapter 42
Y2K on the Desktop: Where the Enterprise Meets Its Waterloo

Ron Angerame

YOU'VE SPENT MILLIONS OF DOLLARS FIXING THE YEAR 2000 PROBLEM IN THE GLASS HOUSE, and just as you're planning a well-deserved trip to a beach somewhere, you have a vision—are those thousands of PCs compliant? What about those desktops that will be accessing that clean and compliant mainframe data or client/server applications? Are your desktop hardware, operating systems, apps, and local and LAN-based data ready for the Y2K bug? Have you already seen failures across your enterprise-wide desktop systems? Are there failures coming that won't be obvious until some department uses an obscure application specific to that task?

Although desktop Y2K issues haven't received as much attention as those in the data center, that *doesn't* mean they're not as important. In fact, they're worse. At least you know which applications run on your mission-critical mainframe and midrange systems, who wrote them, and who has the source code. The PC is a different beast—in many cases these little rebel machines have user-specific software, nonstandard configurations, and who-knows-what-else that was downloaded from the Web. But they *can* be fixed, and because they don't require COBOL warriors to delve into mountains of code, it can be done affordably.

Think about all of your desktop and mobile PCs and servers. Think about one simple hardware question: how many makes and models do you have, and do *all* of them have a Y2K-compliant BIOS? Bear in mind that not all machines of the same model have the same BIOS, thanks to the manufacturers' practice of slipstreaming improvements during the manufacturing process. And even worse, due to the fact that modern PC workstations and servers use flash BIOS technology, even if you didn't update the BIOS, perhaps a savvy end user did.

What about the software? Your organization certainly has a recommended loadout, but in most companies that's only a recommendation, perhaps backed up with a "you load it, you support it" caveat. Because PCs are inherently harder to manage than centralized systems, you have no idea what's running on those systems. Some PCs have the recommended software at the recommended version level; some may be downlevel due to deliberate choice, oversight, or hardware limitations; or users might be using a newer version than you've officially deployed, beta software, or allegedly compatible alternatives.

There may be "special-case" software that's not on your approved software list but interacts with business-critical data. In some cases your technicians may not even know such applications are installed until a machine fails to boot or an app breaks or corrupts someone's local data or worse, until an app corrupts shared data files required by approved Y2K-compliant apps.

What can you do? The process of bringing your desktops into Y2K compliance may seem similar to that for the data center. But don't let that deceive you; the devil is in the details. To ensure that all the appropriate details are addressed, it helps to take a thorough, systematic approach. At Attachmate, we advocate a five-step approach conceived to help you ensure your desktop systems are ready for the Year 2000 and beyond: Discovery, Reporting, Resolution, Control, and Maintenance. These steps are practical and relevant to desktop Y2K before and after the turn of millennium.

STEP 1: DISCOVERY

Before you can develop a Y2K compliance plan for your desktops and estimate the time, manpower, and budget required, you'll need to know what you're dealing with, and that's harder than it sounds. The first step is to survey the environment, identifying trouble spots and setting priorities.

One of the first steps should be an inventory of standard company data files, templates, documents, macros, and so on. You can develop a list of critical data files which must either be replaced by new versions or migrated to a new format.

Next, you'll need to determine what's actually on the desktops—and be assured, every one of them can be different. Some companies may choose to conduct a physical inventory by sending an employee or temp from workstation to workstation to check the hardware as well as which applications are installed on each machine, and what versions.

A much smarter approach, however, would be to use software that can conduct such surveys remotely. Doing so would not only save time and human resources, but also yield the added benefit of automating the report process.

What else should you look for? For each user you may need to know where their company-critical files such as budgets or contact databases are stored. Are they all on a file server? On a local hard drive? On removable media? Take extra care to identify the non-Y2K-compliant files determined earlier. You might even need to interview users, especially those who might have varied their system configurations from recommended norms.

A particularly messy situation occurs when the user's equipment is located away from your main office. If you choose the physical inventory route, laptop-toting road warriors can be sent e-mail or login messages requiring that they either notify IT the next time they'll be in the office to schedule an inspection, or bring their laptops to a testing area upon their return. Workers in small field offices without on-site IT support are trickier. Unless their machines are already equipped for remote management, you may need to send a surveyor into the field, walk the local office manager through the inventory process, or hire a contractor. Similarly, the hardest situations are with telecommuters or contractors, many of whom may be using their personal hardware and software to perform critical company functions. Not only will this hardware and software be the most difficult to physically survey, it will likely have the greatest variance from company standards.

STEP 2: REPORTING

Whether you conduct a physical or electronic survey, the outcome should be a comprehensive database which lists hardware, operating systems, and application types and versions as well as the types, locations, and versions of nonY2K-compliant data files.

The advantage of conducting the survey electronically is that the inventory database can be generated for you automatically. From that database various reports can be automatically derived in whatever format necessary for analysis or integration with other enterprise reports. Reports generated can be compared, preferably automatically, to available information about the known (or unknown) Y2K compliance status of most major desktop business applications.

Based on this data you'll be able to see what's installed and its Y2K compliance status (compliant, non-compliant, or unknown). Determining the Y2K status of those "unknown" items should be parallel tasks, with research being conducted as soon as a new product is identified.

STEP 3: RESOLUTION

As soon as you've got all the information from the Discovery and Reporting steps laid out before you, it's time to determine what must be repaired in order to keep the business running smoothly. You need to think about when those repairs must be complete, what's required to make the repair, how long that'll take, and, of course, how much it'll cost.

Not all Y2K problems are created equal. If, for example, a given application tracks four-digit year information internally but only displays the last two digits on screen, that problem might go onto your "fix it later" pile. If you have a PC whose internal clock will not provide the right time in January 2000 but can pick up the correct time from a time server, that's not really a problem at all. But if you have an order entry macro that prompts for a two-digit year and automatically prepends a "19" in front of it, that needs to be high on the priority list.

Some departments may have applications running that fall outside of the corporate standards but fall into that business-critical category. If those applications aren't compliant and aren't going to be addressed by your department, then whose responsibility is it to contact the vendor for their compliancy statement, research fixes, etc.?

Here's an item that's often overlooked: file naming conventions. Many departments and individuals use dates as parts of file names. A status report generated on March 12, 1999 might be named "990312sr.doc." How should such files be named in 2000 and beyond? Should existing files be renamed to accommodate a new scheme? Must a new Y2K-compliant scheme fit within DOS's 8.3 filename constraints? For an individual's private files, that problem may not be hypercritical; but if CGI scripts on a web server expect files to use that convention, or if spreadsheets use macros to load other data files based on that format, it's pretty important.

Based on your analysis of the severity of those and other problems, develop a priority list sorted by the date that the work *must* be completed. From there, determine the time and staffing required to perform the repairs, including programming, field work, data conversion, IT training, and testing. Prepare a budget based not only on the staff and travel costs involved, but also for contractors and temporary labor. The biggest items may be hard costs such as buying new desktop or laptop PCs, or operating system or application upgrades. If you're upgrading user applications, don't forget to budget time and money for retraining your IT staff to install and support them.

A tricky issue for many companies will be where to budget the money. If you're replacing a battered 486-based laptop, is that a Y2K-related expense or just a normal upgrade? What does this do to depreciation schedules? Work closely with the CFO's office on such matters.

Once you've developed a preliminary timeline, it is critical to get buy-in from outside the IT department, from both line-of-business managers and corporate execs. Be up front in discussing which Y2K problems will be addressed first and which tasks will be delayed. Be sensitive to their needs and concerns, but also be realistic in quantifying the costs of accelerating the program.

Now that priorities have been set and funding has been approved, it's time to determine the exact fix for each problem. With desktop hardware you have four basic choices: ignore it, replace it, upgrade it (usually by flashing the BIOS or other firmware), or install software that patches or masks the problem.

If hardware is nearing the end of its useful service life, you may wish to take advantage of the Y2K situation to do an early replacement. Fortunately, with most PC hardware, users are barely affected by fixes other than by the inconvenience of having the equipment out of service for a short time.

Software is trickier. Far trickier. Once you've determined that a piece of software requires some form of Y2K fix, you run into issues of cost, data compatibility and migration, customization (i.e., identifying and replicating macros, templates, in-house applications), look-and-feel, policy-based management, the works. If possible, don't make software-related decisions in haste, particularly if there is already discussion going on about certain software standards. Moving to that latest version of a certain app for Y2K reasons may resolve the Year 2000 issue, but it may also make it more difficult to consider alternative products later.

Choosing a new piece of software, whether it's an upgraded version of an application already in use or an entirely new product, is relatively easy. The upgrade itself is hard. It will be necessary to test the product to see what else needs to be changed. All macros, scripts, and templates need to be tested. Many will need to be recreated, and that will need to be added to your deployment task list. If you're modifying data files you'll also need to touch every other app that uses that data, even if that other app is itself Y2K compliant. Test, test, test. Different research firms and analysts give different numbers, but there's no doubt that Y2K-related activity will create a number of *new* problems. The more you invest up front in finding those problems, the less trouble you'll have later.

What else will you need to change? Well, any network management scripts that push or monitor applications will need to be changed, as will user policies and permissions. How about network access? Don't forget the license or key servers. How will data be migrated and tested? How will you ensure that all users and their data are upgraded simultaneously or in such a way as to guarantee that some users aren't modifying old and obsolete data sets after the migration? (By the way, don't forget those branch office, mobile, and telecommuting workers!) How quickly can you train users on the new system? Plan for increased help desk activity surrounding new or upgraded apps or scripts.

Once these steps have been taken to determine the best way to deal with your desktop Y2K situation—and those ways have been fully tested—it's time to begin deployment. Every situation is different. You have a *lot* of deployment choices to make depending on your staffing level, physical distribution of IT resources and end users, and the urgency of the Y2K problems.

If upgrades will require on-site calls by IT staff or contractors, or major inconvenience (and loss of productivity) to enterprise end users, you may wish to consolidate the fixes and apply them all at once. On the other hand, if only a few major repairs must be completed to a small number of machines—such as firmware upgrades or upgrading one small group of holdouts who are using obsolete software—perhaps those could be scheduled separately. In general, however, it would be preferable to "touch" each machine only once.

What about sequencing? That's a matter of problem priority, resources, and, well, politics. Do you apply fixes department by department? Floor by floor? City by city? According to whoever uses a particular application or needs access to a particular data file? According to whoever has taken the new application training class? Again, every situation is different, so make sure that all contingencies are covered, such as determining when to upgrade people who work in multiple departments.

When it *is* time for that big rollout, how are you going to accomplish it? Sending IT staffers from office to office to load new operating systems and applications, flash firmware, patch executables, install new templates and scripts, and migrate data will be painfully slow, and should be avoided if possible. A better approach would be to install desktop management software once on each machine—a "push client," if you will—perhaps as part of the Discovery process. Then, whenever you deem it appropriate, fixes can be pushed out to the desktop or mobile clients from a central server. That's about as close to "simultaneous" as you're going to get.

STEP 4: CONTROL

Some analysts recommend "locking down" machines after a physical inventory. Others recommend that such a lockdown be imposed after all Y2K fixes have been made. Either way, it would be preferable not to have to go back and do either of those processes all over again. For example, you don't want to migrate a spreadsheet to a new format just to have someone save over it in the old format.

While you're deploying fixes, you'll need to do something to keep those machines clean, a step referred to here as "Control". Many experts have emphasized the importance of keeping systems in a Y2K-compliant state—and that includes file servers.

From an internal PR perspective, you might want to claim that you're isolating or quarantining "unclean" machines, but really the emphasis should be on setting up clean zones. The stricter your policies (especially if they're *electronic* policies, as in policy-based management), the unhappier many users will be. But the looser your policies, the sloppier your Y2K compliance program will be. It's up to you, the corporate culture, and "the powers that be."

Consult with users, managers, and executives while developing such policies. You're likely to find that grassroots approval can make all the difference.

STEP 5: MAINTENANCE

Achieving Year 2000 readiness is one thing, maintaining it is another. It's recommended that you regularly monitor your state of Year 2000 readiness by performing periodic inventories and spot checks. Also, late and unexpected patches are bound to become available from vendors between now and well into the new millennium.

As part of the Maintenance step, don't forget that things are going to go wrong. You'll need to provide lots of 7x24 coverage—the word "overtime" is one that budget hawks will get used to *very* quickly. Write lots of answers to FAQs for your intranet, and make sure that whoever answers the help desk line is empowered to do triage.

Consider setting up a Y2K SWAT (Special Weapons and Tactics) team with sufficiently broad authority (and budget) to throw whatever it takes at crises as they occur. And occur they will. Don't allow management to forget that although Y2K is a big issue, it's not the only situation your department will have to deal with. Printers will still break, laptops will still be dropped or stolen, web sites will still be hacked, and hard disks will still crash whether the clock says 1999 or 2000. Try to ensure that the Y2K issue doesn't suck up all of your troubleshooting bandwidth. Again, wherever possible, automated management might be the way to go.

THE DESKTOP AT CENTER STAGE

The Y2K problem outside the glass house is much different than the mainframe/midrange problem. The applications are myriad, the source code unavailable, and end user discipline often lacking (as the number of unapproved applications installed on your LANs will testify).

But while the desktop has been spared the media hype, your end user and server hardware and software still demand attention. Try a five-step process of Discovery, Reporting, Resolution, Control, and Maintenance. Exhibit 42.1 will help remind you of the steps. Get your users involved, and pick up the right sort of tools.

Exhibit 42.1 Step Checklist.

Step 1. Discovery

- Inventory of standard company data files, templates, documents, macros
- Inventory of each desktop (hardware and software)
- Determine location of all local / departmental versions of company data files

Step 2. Reporting

- Comprehensive database of hardware, software, operating systems, non-Y2K data files

Step 3. Resolution

- Develop a priority list sorted by the date the work must be completed
- Prepare a budget (include staff, travel, contractors, temporary help, hardware, software)
- Gain buy-in from IT management, line-of-business management, corporate executives
- Determine fix for each problem; remember to include migration and testing plan
- Deployment

Step 4. Control

- Set up clean zones
- Enforce compliance policies

Step 5. Maintenance

- Periodic inventories and spot checks
- Implement new fixes and patches from vendors
- 7x24 support—FAQs for intranet, empower help desk for triage, Y2K SWAT team

Chapter 43
Network Computers: Promises and Reality

Nathan J. Muller

A NETWORK COMPUTER IS A CATEGORY OF THIN CLIENT THAT ALSO INCLUDES WINDOWS TERMINALS NETPCS. Essentially, thin clients are scaled-down PCs that reduce the overall cost of ownership, yet provide flexible and secure access to essential business applications. All thin clients are network-oriented; that is, they rely on the network for running their applications. However, the degree of reliance on the network varies with the type of thin client, as does the degree to which they support traditional desktop applications.

The concept of the network computer originated with Oracle Corporation. Network computers are specifically designed for Internet and intranet connectivity. They rely on a Java-enabled web browser for the graphical user interface. Typically, a Java applet acts as the application front end. A network computer can optionally provide file transfer and terminal emulation capabilities. If file transfer is supported, it is accomplished using the file transfer protocol (ftp); if terminal emulation is supported, access to remote systems is accomplished using the Telnet protocol. Under Oracle's original concept, the network computer also could be equipped with Windows emulation software to interact with a Windows NT server.

Another type of thin client is the Windows terminal. The allure of Windows terminals is that they can access a variety of applications. They not only provide access to host applications via terminal emulation and Windows applications running on a Windows NT server, but they can run X-Windows software to access UNIX applications. Windows terminals are ideal replacements for traditional dumb terminals or legacy PCs, and they can be used to expand the reach of Windows applications into new environments, such as the factory floor. Windows terminals are also good for locations where desktop support is not readily available or is too expensive to provide.

Finally, there is the netPC, a concept developed by Intel and Microsoft. These devices are network-oriented and do not require much in the way of administration but, unlike network computers, they include a high-speed Pentium processor (100 MHz or better), a hard drive, and Windows operating system. This allows the netPC to support existing Windows software packages.

To one degree or another, Java plays a role in all of these thin-client architectures. Java was designed by Sun Microsystems to provide a cleaner, simpler language that could be processed faster and more efficiently than C or C++ on nearly any microprocessor. Whereas C or C++ source code is optimized for a particular model of processor, Java source code is compiled into a universal format. It writes for a virtual machine in the form of simple binary instructions. Compiled byte code is executed by a Java runtime interpreter, performing all the usual activities of a real processor, but within a safe, virtual environment instead of a particular computer platform.

Much of the Java applications development at major corporations hinges around the Web because the Internet is increasingly being considered as the foundation for network computing, providing an economical way to access corporate information from remote locations and mobile computers. Under this concept, network computers rely on the Internet or corporate intranet for running applications. A Java-enabled web browser is used as the interface. The use of network computers allows companies to save money by equipping their workforce with less-expensive hardware, minimizing the number of copies of application software, and streamlining software maintenance tasks.

With the advent of thin clients and the increasing reliance on network servers, the concept of the *thin server* has emerged. This is a dedicated, special-purpose server that is optimized for supporting thin clients. In addition to supporting a narrow range of network applications, the thin server supports localized services to reduce network traffic congestion and provide fast access to routine applets used for many database and spreadsheet programs.

Unlike traditional server-based systems that require an investment in separate hardware and software, the thin server comes complete with hardware and software at a fraction of the cost. These servers usually come ready to set up, install, and configure out of the box. Typically, all the user needs to do is plug in an Ethernet cable and set the IP address. A thin server usually comes with a browser graphical user interface (GUI), a Java-based management application, and an embedded HyperText Transfer Protocol (HTTP)-compliant operating system.

ADVANTAGES AND CHALLENGES

Businesses that have embraced thin clients are using them for a variety of applications. Most thin clients are used to access an office suite like Microsoft Office, but some are used to run mission-critical applications such as accounting, transaction-processing, and order-entry applications. To a lesser extent, thin clients are also running engineering, Enterprise Resource Planning (ERP), and medical applications.

Users of thin clients are usually task-oriented and prefer to do their work without being distracted by technology issues. These are front-line professionals such as doctors in HMOs, accountants, engineers, and salespeople of big ticket items such as industrial equipment and real estate. Thin clients are also used for back-office operations supported by clerical and administrative staff, low-level salespeople, and workers on the shop floor.

According to a 1998 survey sponsored by The Open Group, the primary benefits of network computers include:

- Cost of ownership: network computers lower the total cost of ownership for the IT infrastructure.
- Platform independence: the use of network computers allows applications to be written and deployed without regard for the desktop platform they will run on.
- Flexibility: the use of network computers eases the deployment of new applications, since they are installed and maintained at the server rather than every desktop.
- Security: in terms of administration and overall protection of the network, some security features such as virus filtering, are best implemented at the server rather than every desktop.

Even though some early adapters have been successful in implementing network computing, The Open Group survey also uncovered that security, standards, and compatibility are the key challenges facing users today. Among the challenges noted:

- Security: certain "must-have" security features such as authentication are still lacking.
- Incompatibility with existing applications/legacy issues: network computing must span all end user platforms and be compatible with popular proprietary infrastructures.
- Lack of standards: there are still too few standardized products available.
- Network availability: since network computers rely more heavily on the network for applications, bandwidth availability is a potential problem. Many companies would need to upgrade their existing networks.

OPERATION

The network computer is designed to operate without a hard drive. Its software and applications are stored on the server, which also provides booting service to the network computers when they are turned on. The server can be a suitably equipped PC, a RISC-based workstation, a midrange host like the IBM AS/400, or even a mainframe. The server connects to the LAN with an Ethernet or Token Ring adapter and supports TCP/IP for WAN connections to the public Internet or a private intranet.

Since all applications reside on the server, installation happens only once — not hundreds or thousands of times at individual desktops. Periodic updates to applications are conducted on the server. This ensures that every network computer uses the same version of the application every time it is accessed.

Network computers can access both Java and Windows applications on the server, as well as various terminal emulations for access to legacy data. Users accessing Java applications do so through a Java-enabled web browser, which also gives them access to applications on the Internet or intranet. For access to Windows applications, the server must run a multiuser implementation of Windows NT, such as WinCenter from Network Computing Devices, Inc. The server's operating system may also include terminal support for 3270, 5250, and X-Windows servers.

SECURITY ISSUES

Security becomes less of an issue because a company's data is not scattered across hundreds or thousands of desktops. Since the network computers have no floppy disk drive, there is no chance that sensitive data will find its way onto a disk and be taken off the premises. There is also less chance of contaminating the network with viruses, since all WAN connectivity goes through the server where virus scanning takes place. It is much easier to update a server with new virus protection software than to update each desktop.

Additional security can be provided through so-called *smart cards* that adhere to the OpenCard Framework developed by the OpenCard Consortium. Smart cards store such things as network log-on procedures, access privileges, and authentication information in a small, highly secure silicon chip. These cards have numerous applications, but when applied to network computers, they let users log onto the corporate network — either locally or remotely — by swiping it through an attached card reader. In addition, multiple users can have access to the same machine based on privileges that are encoded in the card.

In addition to authentication, the OpenCard Framework supports digital signatures. It functions with a variety of smart card drivers required by applications such as electronic commerce. The architecture also supports Java so that applications can be written only once, and will run across OpenCard-compliant network computers.

MANAGEMENT

Since thin clients are network-oriented, there needs to be a way to incorporate them into the overall management scheme. Many vendors have extended and enhanced the manageability of thin clients by making them compliant with the Desktop Management Interface (DMI), specifically DMI 2.0, issued by the Desktop Management Task Force (DMTF). DMI 2.0 provides a standard way to access management data about a system, which is a key element in reducing the total cost of PC ownership. Version 2.0 also conforms with the NetPC and Wired for Management (WfM) specifications.

The WfM Baseline Specification was developed by Intel to facilitate the central management of Intel processor-based systems over a network. WfM-capable systems can be remotely configured. This includes the remote installation of operating systems and application software, and after-hours remote maintenance.

Another management technology incorporated into netPCs is self-monitoring analysis and reporting technology (SMART). This tool monitors the performance of hard disk drives and identifies problems that could lead to failure and the possible loss of data.

Some vendors not only support management standards, but go beyond them, enabling system administrators to remotely boot up, configure, maintain, upgrade, and repair each system over the network. This usually entails equipping the system with remote management software such as Intel's LANDesk Client Manager.

Hewlett–Packard, for example, not only equips its Net Vecta PC with LANDesk Client Manager, but its own NightDirector software for turning the netPCs on using the Remote Power On feature available from within HP OpenView Workgroup Node Manager. HP OpenView Workgroup Node Manager is used to manage the Net Vectra PCs, providing security and alarm features, auto-discovery, automatic mapping, and monitoring capabilities.

MARKET ACCEPTANCE

To date, network computer sales have not met expectations. Sales in 1997 were below industry expectations by 50 percent. It was predicted that sales of network computers in their first year would hit 400,000 units, but less than half that number of network computers was actually sold. Other sales estimates were as high as 750,000 units.

The slower-than-expected adoption rate highlights the confusion that abounds in the industry at large. Vendors and application developers use different terminology to explain network computers. Standards among key players such as Compaq, Hewlett–Packard, IBM, Wyse Technology, and Sun Microsystems are lacking. It may take another three to five years for businesses to become acquainted with all the nuances of network-centric computing before they can make an informed decision about bringing any type of thin client into their organizations.

Despite its faltering start, some market research firms remain bullish on the technology. The Gartner Group, for example, predicts that 40 percent of businesses will deploy network computers by 1999, rising to 60 percent by 2001.

INDUSTRY STANDARDS

The success of network computing depends on a common, minimal but mandatory set of functionality for all network computers, as well as interoperability. One step that could potentially boost acceptance of the network computer is the establishment of a universal standard.

An industry standard was proposed in 1996 by a group of vendors that included Apple Computer, IBM, Netscape Communications, Oracle, and Sun Microsystems. The standard, called the Network Computer Reference Profile, is designed to be architecturally neutral; that is, not linked to any particular hardware, application software, or operating system. However, the standard specifies that the application programming model is Java.

In addition to being architecturally neutral, network computers that comply with the Reference Profile are intended to have the following attributes:

- They must offer a much lower total cost of ownership than personal computers.
- They must offer a lower entry cost than a typical personal computer.
- They must be significantly easier to use and administer.
- They must enable security.

The Reference Profile developed for the network computer specifies the following device characteristics:

- minimum screen resolution of 640 by 480 pixels
- compatibility with VGA or equivalent
- a pointing device
- capacity for text input
- audio output
- no hard drive
- operation over IP-based networks
- support for Web standards, including HTTP, HTML, the Java Application Environment
- support for electronic mail protocols such as SMTP
- support for multimedia formats such as JPEG

The Network Computer Reference Profile has since been extended to accommodate support for smart cards that comply with the OpenCard Consortium's OpenCard Framework.

The Open Group not only manages the Network Computer Reference Profile, it also runs a branding and certification program for network computer technologies. It tests the products of participating companies, as well as those who wish to have their products tested before registering them with The Open Group. Testing and registering of the products is required before companies can use the program's brand or logo. The goal of this program is to provide an open environment for the development of network computers.

Products that display The Open Group's Network Computer logo provide assurance to customers that the product is guaranteed to meet the industry-sponsored standard for conformance and interoperability with any other product that carries the Network Computer logo. These products do not have to be network computers — they can be any network device, from hand-held computers to kiosks. The goal is to provide a guaranteed level of security and interoperability to ensure the viability of the emerging net-centric infrastructure.

VENDOR APPROACHES

There are as many approaches to implementing network computers, Windows terminals, and netPCs as there are vendors.

Network Computers

Of the dozen or so vendors offering network computers, IBM offers the most comprehensive solution. Its Network Station Series consists of multiple products optimized for a different spectrum of use.

The Series 100 emphasizes access to server-based applications and data. The Series 300 provides multiple server access and enhanced Internet performance. The Series 1000 combines high-speed Internet and network application performance with Java capabilities. The management software, Network Station Manager, runs on the server and is used to administer the network computers and to set up their boot parameters.

The high-end Series 1000 runs business-critical applications and personal productivity tools written in Java, but can also run Windows applications via multiuser implementations of Windows NT on a PC server. The Network Stations can access multiple servers (IBM and others), including corporate intranet and Internet servers, via web browsers. Like series 100 and 300 machines, Series 1000 machines can also access 3270 and 5250 terminal applications and work with applications on AIX and UNIX servers using X-Windows server support. And, via a Domino server, the Series 1000 can run Lotus Notes.

The Series 1000 comes preloaded with Lotus eSuite WorkPlace, Java-based productivity software designed specifically for the emerging network computing environment. The eSuite WorkPlace provides a single point of access to everything most computer users need, including business productivity applets, integration with existing applications, terminal emulation, e-mail, calendaring and scheduling, and the Internet. IT managers and network administrators can easily customize the eSuite WorkPlace to bring the right tools to the right people. When a user needs additional capabilities, his or her WorkPlace is supplemented with the appropriate Java applet.

The Series 1000 sells for $999 and is powered by a 200-MHz PowerPC 603 chip, which is used in the mid- to high-end Power Macintosh. Base models of the Series 1000 ship with 32 MB of system memory (expandable to 64 MB) and 2 MB of video memory. They support display resolutions of up to 1600x1200, and provide LAN connectivity via an Ethernet 10/100 or Token-Ring 4/16 adapter. The units include one serial and one parallel port and support 16-bit audio. The Series 1000 also supports the OpenCard Framework-based smart card for security.

NetPCs

Unlike network computers, netPCs are designed to run programs locally, so testing the performance of these machines becomes important before committing to a bulk purchase. This type of thin client was introduced in mid-1997 and comes in many different configurations. Of the three types of thin clients, the netPC is the one that most closely resembles a traditional PC. Among the first vendors to offer netPCs were Compaq and Hewlett–Packard. Both companies offer netPCs that are compliant with the DMI 2.0 specification for desktop management.

Compaq offers a netPC called the Deskpro 4000N, which starts at $1,039. The DMI 2.0-compliant unit is powered by a 166-, 200-, or 233-MHz Pentium with MMX technology and comes with 32 MB of RAM, a 1.6-GB hard drive, and 10/100 Mbps Ethernet adapter. It also has a PCI slot, two USB 2 ports, two serial ports, and a parallel port.

Although it has a 1.6-GB hard drive, it lacks both a CD-ROM drive and floppy drive. A 15-inch color monitor and video card displays 800-x-600 resolution.

Hewlett–Packard offers a line of HP Net Vectra PCs based on the 166- and 200-MHz Pentium processor. Prices start at about $1,200. Key features include a 1-GB or 1.6-GB hard drive with SMART, 16 MB, or 32 MB of memory (upgradable to 192 MB), two USB connectors, two serial ports and one parallel port, one free PCI slot, and a 10 Mbps or 10/100 Mbps Ethernet adapter. Windows NT 4.0 is preloaded on the HP Net Vectra PC as is Intel LANDesk Client Manager.

Windows Terminals

Among the most prominent vendors of Windows terminals is Wyse Technologies. The company's Winterm 2000 Series provides local and remote access (e.g., Internet and intranet) to Windows, Java, web browser, and host-based applications. These machines are suited for retail, call center, financial transaction processing, manufacturing, and office sites. In the home office, they can access a remote server using standard modems.

Each model provides dual high-speed serial ports, a bidirectional parallel port, a Type II PC Card slot, and 10/100 Ethernet or Token Ring adapter. Depending on the model, the terminals have integrated 14- and 15-inch color CRT displays, and some have 16-bit stereo sound capability. All the models can be set up in five minutes and added quickly to the network. They boot locally, yet can be centrally administered and scripted to link up to 32 users — each with a unique configuration — to up to 32 servers.

Resident VT220, VT100, and VT52 ANSI terminal emulators make the Windows terminals ideal as character terminal replacements, ready for integration with Windows NT. Multiprotocol support gives users the ability to choose the one that best fits their needs. Users can also take advantage of Microsoft's Remote Desktop Protocol (RDP), as well as Citrix's Independent Computing Architecture (ICA) through the company's MetaFrame or WinFrame thin-client/server system software.

At this writing, Wyse is getting ready to launch the Winterm 3000 Series, which uses Windows CE 2.1, the Windows-based terminal version of the Windows CE operating system. These units will offer multisession and multiprotocol access to the Terminal Server Edition of Microsoft's Windows NT Server 4.0 operating system. The Winterm 3000 line will achieve this by supporting Microsoft's RDP, Citrix's ICA, and multiple terminal emulators.

The Winterm 3000 Series starts at $1,059 for the 14-inch integrated terminal, and $1,289 for the 15-inch integrated terminal. Current users of Wyse's Winterm 2000 Series terminals can convert their devices to Winterm 3000 Series models through a software upgrade.

COST OF OWNERSHIP

Thin-client installations tend to be small, falling between 50 and 500 seats. Most of the companies running thin clients tend to be small, having fewer than 1,000 desktop systems of any kind. Small companies migrate to thin clients because most big companies have already deployed vast and sophisticated PC systems and installations, whereas smaller companies are still in the early stages of information technology deployment. Also, small companies are most likely to be using UNIXWare or some other terminal application system, which makes them prime candidates for thin-client replacements.

As thin-client standards are worked out and servers for thin clients become stable and reliable, large companies looking to replace desktop systems in the near future will feel compelled to consider thin clients. The primary motivations for installing thin-client hardware are to reduce complexity and save on the total cost of ownership.

The cost for a thin client, such as a Windows terminal, is about $1,200, whereas the cost for a high-end Pentium-based PC starts at about $2,500. For a 100-desktop installation, the cost difference can be quite substantial: $120,000 for Windows terminals versus $250,000 for the fully equipped PC. Of course, the cost of a server to support 100 Windows terminals is higher but, even at $10,000, the overall cost savings amounts to more than 50 percent.

Moreover, supporting this many PCs is a full-time job for at least two people, but support for the Windows terminals can be a part-time job for one person. Each PC must be configured manually, whereas the thin clients can be configured remotely from a single server. If a problem occurs with a Windows terminal — or any other type of thin client — the whole unit can be swapped with a spare and be up and running in a few minutes.

Thin clients do not necessarily have the lower price tags that vendors originally promised. Prices range from about $900 to $1,200 (and more) per device, including monitor, but some low-end PCs are also coming down to this price range. To be a commodity product, prices for thin client solutions have to drop to $500 — which was first proposed by Oracle Corporation when it outlined its concept of network computing over two years ago.

IMPLEMENTATION CONSIDERATIONS

Thin clients are not intended to replace desktop PCs, but they do merit consideration as a cost reduction measure. Thin clients provide users with the flexibility of having PC and terminal functionality without the organization having to buy a high-end PC for low-end usage. Organizations can expect to spend more money on servers that are optimized to support thin clients, but they will only have to worry about upgrading a few servers as opposed to hundreds or thousands of desktops.

Network and IS managers who are already sold on the idea of a thin-client architecture must confront the myriad of choices. Although the market is still in its infancy, clear distinctions already are being drawn among the various thin-client hardware options, most of which are heavily tied to equally diverse application environments.

The netPC is tuned for easier network administration and will offer some characteristics of network computers, such as the lack of a floppy disk. The widespread familiarity with Windows and the immediate availability of thousands of applications makes the netPC a compelling choice.

Network computers, on the other hand, rely on Java applications. Yet until more Java applications arrive, however, many organizations consider Java-based network computers problematic. The IBM bundling of Lotus eSuite WorkPlace into its Series 1000 Network Stations provides a glimpse of what is possible with Java. Furthermore, Java has the potential of freeing developers from hardware and software lock-ins.

Windows terminals are a more clear-cut choice because they are aimed at a more specific user base — those who require terminal emulation to access legacy, but who also need access to Windows applications. Not only can Windows terminals provide access to applications on the LAN, but remotely over the Internet or a corporate intranet.

Yet even these fierce competitors now agree on several points, one in particular: in some situations, users are willing to trade flexibility and performance for networked desktop devices that cost less and are easier to administer.

CONCLUSION

Although the thin-client revolution is off to a slow start, this is to be expected given the initial hype from vendors who were eager to turn existing PC technologies into a new revenue stream with relatively little effort.

In 1997 — the first full year of product introductions — it was difficult for many organizations to determine exactly how the thin-client architecture fit into their existing IT infrastructures and how to quantify its benefits. In hindsight, it was probably wise for many companies to dismiss the first round of product introductions until the conceptual framework could undergo further refinement.

Now that things have settled down, it looks like traditional PCs and thin clients are really complementary architectures. In some cases, it will be difficult to distinguish between them. Overall, thin clients should provide the economies and efficiencies organizations are looking for, but these advantages will primarily come from among the user population that is task-oriented. Even if the ratio of single-task users to multitask users is 10 to 1, mid- to large-sized companies can expect a worthwhile return by implementing an appropriate thin-client solution.

Chapter 44
Handheld Network Devices to Access Intranets and Enterprise Networks

John R. Vacca

IN TODAY'S WORLD, COMMUNICATION IS THE KEY TO ANY SUCCESSFUL ENTERPRISE OPERATION. Fierce competition has driven the computer and telecommunications industry to new heights. People can no longer wait to access enterprise intranets or networks within the confines of their offices—they must be able to use their computers and phones on the go. This is where handheld devices in the form of palm computers, Personal Communications Systems (PCS), or a combination thereof have brought a lot of convenience to their users.

With the advent of PCS and palm computer technologies, users will enjoy more freedom with their handheld devices. They can not only use their phones to verbally communicate with other users, they can also connect their phones with their palm type computers to access and transfer data through the enterprise's intranet (either Local Area Network (LAN) and/or Wide Area Network (WAN)). Moreover, their communication mobility will not be limited to a predefined area set by the local operators, it will be continentwide or worldwide coverage. Roaming will be as easy as obtaining cash advances from ATMs around the world.

TYPES OF HANDHELD DEVICES

Modification and additional equipment are needed for existing telecommunications systems and infrastructures to become handheld (personal digital assistants [PDAs]) and PCS. In order to minimize the cost, new standards and equipment must have backward compatibility and a friendly interface with existing enterprise intranets, so custom application handheld device development is taking place in two distinct areas. The first focuses on developing the portion of the application that runs on the handheld device. Such tools are described next. The second area of development is on the portion of the application that accesses the enterprise's intranet and ports the data to and from the handheld device.

DEVELOPING APPLICATIONS WITH TOOLS THAT RUN ON HANDHELD DEVICES

The PalmPilot™ connected organizer from Palm Computing, Inc.[1] is the most successful handheld computer device or PDA on the market to date. More than 3.3 million units have been sold in the last three years according to an independent market communications industry analysis. Since bar code reading is the most effective form of data acquisition, it was only a matter of time before bar code scanning technologies were combined with the small form factor provided by the Palm Computing® platform from 3Com®. The results have been spinoffs of sophisticated handheld data access, acquisition and transfer devices that can be customized for a wide range of applications.

As you know, the primary purposes of most handheld devices are bar code scanning, data capture, and display. However, data has no value unless it can be transferred for use. Therefore, developing the portion of the application that runs on the device is of the utmost importance before data can be transferred to and from an enterprise's intranet. Such tools are described next.

HANDHELD DEVICES DEVELOPMENT TOOLS

The palm-size computing design has quickly become the favorite for personal productivity, and the PalmIII™ is the leading handheld personal productivity tool. With over 20,000 registered Palm™ solutions providers, what started as simple, basic phone/address information and daily organizer tasks has grown to include an impressive list of enterprise handheld device development tools. Let's look at a few of these:

Codewarrior™

CodeWarrior from 3Com® for Palm OS is a tightly integrated development environment complete with editor, project manager, C/C++ compiler, source- and assembly-level debugger, direct-to-device debugger, standalone assembler, graphical user interface (GUI) layout tools, and linker. It allows developers to take advantage of the enhanced Transmission Control Protocol/Internet Protocol (TCP/IP) support in the Palm OS 2.0 and design new applications to quickly and easily synchronize Palm OS data over enterprise intranets and in-house networks. Hosted on Microsoft Windows® 95 and 98, Microsoft Windows NT®, and Mac OS (Symbol does not support its Application Programming Interfaces (APIs) in the Mac OS version), CodeWarrior for Palm OS (www.palm.com) includes C++ support; increased support for standalone code resources and multisegment applications; debugger console, which allows the user to communicate with the target device; support for debugging directly from the integrated development environment; and, most importantly, a greatly improved version of Windows-hosted Constructor, a drag-and-drop design tool for Palm Computing applications.

Penright!

PenRight! from Penright! Corporation (www.penright.com)[2] for Microsoft Windows' cross-platform C code generator allows a single application design to be generated for Palm OS, Windows, DOS, and DOS protected-mode-based platforms. Designed as a tightly integrated environment, PenRight! for Windows utilizes easy-to-reference icons to activate nearly all aspects of application development, including an object-oriented project manager, graphical form designer, multifeatured code editor, property manager, and database mapper, allowing development time to be reduced.

Satellite Forms™

Satellite Forms Developer Edition 2.0 from Puma Technology™ (www.pumatech.com),[3] formerly SoftMagic, is a Rapid Application Development (RAD) front end tool similar to Visual Basic. It comes in two versions—a Developer version which does not include communications to the backend, and the Enterprise edition which includes interfaces that allow communication to back end tools such as NobleStar Scout, Advance Systems' ASL-Connect, etc. The App Designer visual development environment enables them to enhance applications with advanced functionality including control, actions, filters, scripts, and extensions. These extensions allow the developer to build additional hardware device drivers and create custom user interface controls. Satellite Forms applications interface with databases through ActiveX control and integration tables. ActiveX extends the 3Com HotSync® technology and gives developers complete control over data transfers.

EZTrak

EZTrak from Ideam (www.ideam.com)[4] is a 32-bit Windows 95, 98, or NT application generator that provides a completely user-definable application framework for mobile computing. Based on open system architecture, this cross-platform tool supports most major databases and relational database management systems (RDBMS), as well as most popular mobile devices. It is designed for developing tracking applications for inventory or assets. Administrative features help the developer define the tracking application, including database table creation and communication session definition. Form and table views let developers manage and query the data using standard functions for inserts, updates, deletes, etc. Data is processed through a wide range of fully supported mobile data collection handheld devices. EZTrak also supports Windows CE devices.

AvantGo Desktop

AvantGo Client from AvantGo (www.avantgo.com)[5] provides a web browser for Palm products and has been used mostly as an informational tool. Complementing the client with the AvantGo Server provides a good development environment for back end active server components. It allows users to easily define how sets of HTML pages, or *channels*, are managed on their handheld computers. It can collect data on a PalmPilot organizer and update a website, access the database, select specified information, and, on the fly, create a web page and present it to the user. Each channel begins with a top-level document identified by its URL, and includes all of the associated documents related by hypertext links. AvantGo automatically synchronizes selected content from the Web onto the handheld device using the 3Com HotSync conduit. It will also download information from any ASCII text file. The enterprise is now writing scanning widgets and adding Javascript™ capability to perform logic, such as compare and retrieve.

Pen Dragon™

Pen Dragon from Pendragon Software (http://www.webfayre.com)[6] is a forms development tool that has just recently added scanner support. It is directed toward less experienced developers, but it still has a good deal of flexibility in creating forms applications with a good interface to Microsoft Access databases.

CASL

CASL (Compact Application Solution Language) from Feras Information Technologies, www.caslsoft.com),[7] allows the developer to describe the visual objects the application will display with a few statements, and then provide instructions for the PalmPilot™ to perform when those objects are triggered. For developers who don't mind learning a new scripting language this may be an easy, flexible solution.

BACK END TOOLS THAT ACCESS AND PORT DATA TO AND FROM HANDHELD DEVICES

The second area of custom handheld device development focuses on the portion of the application that accesses and ports the data to and from the enterprise's intranet. Tools for this are described next. Among them are middleware servers that allow users to complete the data acquisition process with a bidirectional data transfer to the desktop or host computer.

Handheld Devices Back End Tools

Back end tool providers also support the Palm OS™ platform, facilitating synchronization, or transfer of data from a handheld device to a host database residing on the enterprise's intranet. Let's look at a few of these back end tools.

HotSync® Technology. The HotSync from 3Com® (www.palm.com) conduit product is available as part of a Palm Computing® software development kit (SDK) from 3Com®. It allows data to be exchanged between a Palm Computing platform device and a personal desktop computer. It depends on a PC or a Macintosh for synchronization and lacks other enterprise-level data distribution features. It is not easy to use, but for those who have the skills it provides basic back end capabilities and is supported by 3Com.

Conduit Development Kit - Java Edition. The Conduit Development Kit, Java Edition, from 3Com® is a collection of development tools that operate on a PC running Windows 95, 98, or NT. It allows developers to create conduits by writing them in Java™. It has similar limitations as the HotSync SDK, but for developers wishing to use Java this may be a good option.

Scout Server. Noblestar Scout MTS from Riverbed Technology (formerly Noblestar—http://scout.noblestar.com or www.riverbedtech.com)[8] is an NT server-based system for Palm Computing products that allows users to synchronize via a TCP/IP connection with other environments, including Lotus Notes and Microsoft Exchange. The Scout conduits implement the actual synchronization logic, giving enterprises customizable access to their data. As object linking and embedding (OLE) objects, Scout server conduits can be created with tool sets including Visual Basic or Visual C++, and can be configured to run in batch or real-time mode. To avoid synchronization problems associated with batch-only mode, the user or administrator can decide which data to retrieve in which mode. Noblestar just released a free public trial version of Scout that supports two simultaneous users. This process is also available to enable value added resellers (VARs) to provide better support for their customers.

ASL-Connect. ASL-Connect (from Advance Systems)[9] provides flexible, server-based connectivity for mobile handheld devices supporting Windows CE,[10] Lotus Notes, Microsoft Exchange, Open Database Connectivity (ODBC), Component Object Model (COM) OLE, and the Palm Protocol Independent Multicasts (PIMs) over any TCP/IP network. Easy to set up with the out-of-the-box Site Wizard, ASL-Connect allows centralized application installation, deletion and upgrades, backup and restore, handheld device reuse and ownership changes, and remote diagnostics. Based on an open plug-in architecture, this product is scalable with good connect session benchmarks. Its Visual Basic scripting allows for an easy configuration, extensibility, and integration with existing systems including NT connectivity and Business Logic as COM objects. Instead of the 3Com HotSync conduit product (which is not written to work in a multithreaded server environment), ASL-Connect (www.asl.com) provides a simple, intuitive, centrally-controlled user interface based on one program for all connection methods. The handheld device can also be run transparently from other programs. This gives the most inexperienced Palm users familiar applications and easy-to-use enterprise intranet and Internet communications tools.

Intellisync®. Intellisync 3.0 from Puma Technology (www.pumatech.com) is a back end product that includes a synchronization engine (based on 3Com HotSync) with logic to update and synchronize with most popular PC-based personal information management, contact management, and group scheduling applications all in one step. Optimized for Microsoft Outlook, Intellisync provides full conflict resolution between records on the handheld and PC, powerful filtering for large databases to optimize space, and complete customization with intelligent field mapping and date range support. Puma recently bought SoftMagic Corp. and will tightly integrate Satellite Forms with a new version of their synching engine. This will make it very easy to complete the data acquisition and transfer required by most vertical applications.

Oracle Consolidator/Oracle Lite. Oracle Consolidator from Oracle® (www.oracle.com)[11] enables data in native Palm format to be transferred to an Oracle Lite database residing on a backbone system. Consolidator protects the enterprise server from unnecessary wireless traffic by directing incoming information to specific handheld devices and independently replicating it throughout the enterprise. This product is useful for existing Palm applications (such as address books, etc.) and for new custom applications. The Oracle Lite database has a footprint of only 350-750 KB, yet provides a powerful, full-featured, object-relational DBMS, comprehensive Java support, a self-tuning design, and is web-enabled. Oracle plans to release a version of Oracle Lite which will reside on the Palm device itself.

SmartTran. EnterpriseLink's SmartTran from EnterpriseLink (www.enterpriselink.com)[12] software enables rapid customization of mainframe applications to be utilized for intranets, extranets, e-commerce, and supply chain automation applications. This authoring tool and application server works on three levels: first, out-of-the-box automatic translation for instant delivery of 3270 and 5250 applications via standard web browsers; second, sophisticated authoring capabilities for customizing mainframe applications without programming and without any changes to the host application; and third, integrating mainframe applications with additional data sources such as relational databases into a single, browser-based web application. SmartTran supports the PalmPilot (when used with a connected web browser), CE handheld devices, and even data-capable cellular phones.

IP Suite. This new product from Whole-Brain Media (www.whole-brain.com)[13] provides live interaction with back end data, allowing queries to ODBC databases. Custom enterprise applications can be developed in CodeWarrior and using Whole-Brain's C SDK, or a new graphical development environment. The Internet Protocol (IP) Suite solution includes encryption, key management, and user configuration based on NT domain, Lightweight Directory Access Protocol (LDAP), or internal user management.

WaveSync. WaveSync from WaveWare Communications (www.wavew.com) is a drop-in replacement for HotSync that enables synchronization of fleets of Palm devices with enterprise servers, and increases the speed of the data transfer over Internet or wireless links almost six times. It also provides an intranet management interface that allows IT managers to control enterprise computers.

Jini. Sun[14] recently announced its Jini technology with lots of fanfare about device federations. The term can be understood as a plug-and-play mechanism for handheld network devices that enables cooperation and direct communication between devices. This essentially means that you can plug a handheld device into an enterprise's intranet and it will be recognized and ready for instant use. Jini uses the term "discovery and lookup", a set of protocols by which devices are readied for this purpose. The distributed programming model of Jini is actually a coordination model which can be used by most handheld device software to *federate*. The model, or the concept itself, is orthogonal to the handheld device software implementation. That is to say, a well-designed coordination model can be incorporated into almost any handheld network-enabled device, allowing it to cooperate with other devices on the intranet. This idea essentially stems from the idea of coordination languages, first put forth in a programming system known as "Linda" at Yale University. The Tuple Spaces and concepts of Linda are reflected in Sun's Jini. Unfortunately, Jini is almost fully tied to Java, which makes Java or embedded Java almost essential on the devices.

JetSend. While Sun focused on the coordination aspect in device-to-device communication, HP[15] focused on the data content encoding, negotiation, and conversion aspects with its HP JetSend technology—a very important distinction. Jini does not define the data content types or the negotiations that can take place between handheld devices. Once two handheld devices engage, any dialogue between them is device-dependent. In contrast, a JetSend-enabled handheld device can communicate with any other JetSend-enabled device, negotiate and encode data in the best form of representation, and transfer this data over the network in a controlled manner—without having any other device (like a PC) coming in between. For example, you can take a photograph with a digital camera, walk up to a printer and *squirt* the image to paper. You can also scan a color image on a JetSend-enabled scanner and send it as a Fax to a destination across the world. What such a powerful protocol really misses is federation, which is exactly what Jini offers. Nevertheless, Jini and JetSend can work together to bridge that gap. HP is on the move with its embedded Java initiative.

Now let's take a brief look at some of the technology and standards issues for handheld network devices and their relation to personal communication system technologies. One can say that one technology cannot work efficiently without the other.

HANDHELD NETWORK DEVICE IMPLEMENTATION IN PCS

First of all, what is PCS? Most people think that PCS is just another type of cellular phone. In fact, a cellular system is just one of the elements in a PCS. A personal communication system should allow any individual to communicate in any form of information anywhere in the world. Ideally, the individual will be assigned a personal number which can be used at any location on the planet. The individual with the appropriate equipment can transmit and receive any form of information such as voice, paging, fax, or data. In general, the PCS can consist of both voice and data handheld network device interfaces, wired and wireless systems, cordless phones and mobile phones, Public Switched Telephone Network (PSTN) and Private Automatic Branch Exchange (PABX), and any kind of integrated services. Unfortunately, unless a global agreement is signed, the above idea will not be viable; after all, it could be just a dream.

In reality, most people would not care about the ideal capability of a PCS. How many people would actually have a job that requires traveling around the world 99 percent of the time? Maybe less than one percent! However, for those multinational enterprises that have point-of-presence in more than one country, a reduced-size PCS would interest them. Perhaps individuals may also be interested in PCS within a continent, a country, a state or province, or even a city or metropolitan area. In developing countries where they do not have a telecommunication infrastructure already in place, a PCS that consists of Wireless Local Loops (WLL) can be a solution. A global PCS may not be in place, yet, the demand for a smaller size PCS is here already.

PCS in this Country

In recent years the cellular industry within the United States has been growing at a very fast pace. Not only are more telecommunications enterprises trying to get into the competition, even the federal government sees it as an opportunity to make money for the national deficit. In mid 1994 the Federal Communication Commission (FCC) started the first round of auctions for the electromagnetic spectrum in the United States. As a result, some 300 MHz of frequency spectrum in the 3.8 GHz range has been reallocated to join the existing spectrum (in the 800 MHz range) for cellular telephone service and to be used by PCS. The FCC has defined three operational categories of PCS:

- Narrowband PCS
- Broadband PCS
- Unlicensed PCS

Narrowband PCS. As the name describes, the bandwidth of the Narrowband PCS is limited. The spectrum for it resides in the 1100 MHz range, next to the existing spectrum for cellular services. Since the bandwidth is limited, the ability of transmitting high-speed data or vehicular voice-oriented type service is diminished at this time. However, it could be used in homes and other buildings, and outdoors by pedestrians in residential neighborhood and urban areas. Wide area, low-speed data such as paging services or e-mail services can also take advantage of this spectrum.

Broadband PCS. The Broadband PCS spectrum is allotted at the 2.1 GHz range. With the increase of the bandwidth, it can be used to transmit higher handheld device (or PDA) voice quality and higher data rate in a duplex manner. Real-time video communications can also be done economically with further improvement on domain specific part (DSP) and electronics packaging. The Federal Communications Commission (FCC) issued 3185 licenses to cover the 120 MHz frequency block.

Unlicensed PCS. Private and public enterprises can take advantage of this type. Not only can they invest their capital and set up their own systems within the enterprise, they can also join the other public PCS providers or PSTNs to provide services to the general public. Even though they are only allowed to transmit signals in low power, they can save some money, or even earn money, by having their own systems. For example, a university can set up its own PCS and provide services to its students and faculty on campus. At the same time, it can team up with other PCS providers or local PSTNs to provide roaming services for people from outside of the university. The spectrum for the unlicensed PCS is allocated in the 1930 to 1950 MHz block.

Besides the preceding specified PCS bandwidths, the existing spectrum for handheld devices is shared by some cellular systems and standards. Those systems and standards can also be part of PCS. For example, the U.S. Digital Cellular Telephone standard (USDC), or IS-54, uses the frequency division multiple access/time division multiple access (FDMA/TDMA) technique with 869-894 MHz as the forward channel, and 824-849 MHz as the reverse channel. The Advanced Mobile Phone Service (AMPS) uses the same frequency allocation as USDC, but with a different multiple access technique. AMPS only uses FDMA with a Fiber Digital Device (FDD).

There is still a lot of room for new handheld device technologies. Now, let's focus on Digital Enhanced Cordless Telecommunications (DECT). DECT is a newly developed digital cordless telephone standard. As you will see, it is perfect for narrowband and unlicensed PCS.

Digital Enhanced Cordless Telecommunications (DECT)

DECT was first developed in Europe and was originally called Digital European Cordless Telecommunications. It began as an effort by European countries to unify and propose a digital radio standard for cordless telephones throughout Europe. Any enterprise can manufacture DECT base stations and handsets, and they should work with products of other enterprises. The standard was approved in late 1992 by the European Telecommunication Standard Institute (ETSI). Unlike its predecessors (Cordless Telecommunication 0 (CT0), Cordless Telecommunication 1 (CT1), and Cordless Telecommunication 1 (CT2)), the DECT standard provides better services, quality, and applicability, which attracted worldwide attention and acceptance. For instance, the CT2 standard has been adopted by China's Hong Kong and a couple of other Asian countries. However, CT2 can only provide one-way communication service. Most CT2 systems only allow handset-initiated calls within the limited coverage of a telepoint base station. On the other hand, DECT allows two-way calling, better quality, and enhanced mobility. It is a third-generation cellular system meant for providing handheld access to enterprise intranets of various types, from PSTNs to LANs. A comparison of CT1 and CT2 to DECT is shown in Exhibit 44.1.

How DECT Fits in PCS and PDAs

DECT is definitely implementable in PCS and PDAs. With its full Integrated Services Digital Network (ISDN), Global Standard for Mobile (GSM), data interworking capability, and its own capacity to handle high voice and data traffic density, DECT is the perfect fit in the concept of PCS and PDAs. The ongoing effort of improving the gap will also increase DECT's potential to be the all-around PCS and PDA standard element. Telephone manufacturers are all rushing to come up with dual-mode handsets for PCS and PDAs.

Exhibit 44.1 A Comparison: CT1, CT2, and DECT.

	CT1	CT2	DECT
Signal Type	analog	digital	digital
No. of Channels	50	50	140
Channel Allocation by	initiator	initiator	portable
Multiple Access (MA) Technique	FDMA	FDMA/TDD	FDMA/TDMA/TDD
Encryption	NO	NO	YES

THE FUTURE IS IN YOUR HANDHELD

Finally, PCS and PDAs will provide much cheaper services, and the number of users will continue to grow due to the ongoing effort of the electronics industry to make handheld devices cheaper and smaller, the introduction of competition by deregulating telecommunications industries, and the research and development by telecommunications engineers to improve existing systems and infrastructures. However, an improved two-way calling CT2 system and the Japanese Personal Handyphone System (PHS) that have similar characteristics as DECT may share the narrowband and unlicensed markets with DECT. In Hong Kong, a couple of tests have been done that show that DECT can coexist with PHS.

CONCLUSION AND SUMMARY

No longer is the mobile worker chained to time-consuming paper forms, inconvenient telephone calls, or bulky notepads while out in the field. No longer is it necessary for a driver to make daily rounds to all of the soft drink vending machines in a given city to see where refills are needed. A telephone enterprise need no longer fear the sudden loss of large amounts of customer data when disaster strikes its main computer facilities. A mechanic can now diagnose his client's ailing car from the comfort of his shop, even while the car is still on the road. The traditional, costly methods of data collection and management have gone the way of vinyl records.

This chapter basically focused on a new technology that delivers unprecedented freedom to professionals within a large variety of disparate fields, while dramatically reducing unnecessary expenditures of money and time. It is the handheld device or PDA embedded Web technology that is designed to make existing products accessible via the web, and access and transport data to and from the enterprise intranet—from vending machines to mobile phones to automobiles. Specifically designed for mobile, small memory footprint, electronic, handheld network devices, this new technology allows these devices to share and transfer data via the Web from one remote location to any other. Integrated directly into the hardware of a product or machine, the handheld device embedded web technology can easily be custom-tailored by any provider to a client's specific wishes or needs. Because it is pure software, this technology may be molded to fit the particular structure and internal circuitry of virtually any handheld device and make it accessible through the Web. When a customer incorporates this lightweight handheld device web server into any product or device (automobiles, phones, alarm systems, fax machines, televisions), he or she can easily access and control that product or device through the enterprise's intranet, or through the Internet from any standard web browser.

For example, with a handheld device server integrated into his handheld device, an electric plant worker can now log onto his enterprise's intranet and instantly send vital information about that particular station's weather conditions, static levels, or repair status. A trucking enterprise can assure its drivers are on time and check on individual trucking loads, weights, and status. A telephone enterprise with the handheld device embedded web server technology can redistribute its vast stores of customer data within an intranet, thus significantly lowering the chance of losing this data in the event of a physical disaster. A vending machine supplier can now dispatch drivers with soft drink refills only when a machine has run out of product to dispense—all he or she has to do is dial into "Vending Machine X" and download any information he or she needs via handheld device embedded web server technology.

Web servers today run on large Windows NT or UNIX boxes with several gigabytes of memory. The handheld device embedded web server sits in about 60K of runtime memory space and can easily run on handheld devices such as the Windows CE Operating System. The handheld device embedded web server uses standard HTTP web protocols and can share any HTML-based web page via wireline or wireless networks to enable handheld computers, in-vehicle systems, and other small handheld devices to access and transfer online content. Deployment of the handheld device embedded web server for an intranet requires little more than the selection of a TCP/IP-capable network. The handheld device embedded web server may be run with or without custom modules, allowing users to share both standard documents and custom information such as up-to-the-minute position or activity reports. For situations requiring fulltime access to mobile assets, the handheld device embedded web server can be used to maintain a virtual web presence for other mobile handheld devices roaming an intranet while causing an apparent uninterrupted connection to a mobile server, whether or not the server in question is actually online.

Finally, the chapter briefly introduced the concept of PCS and looked into the existing specifications for PCS in the U.S. It then introduced the

DECT standard and the technology. It also introduced a few DECT profiles. The chapter then showed the compatibility of DECT with existing technologies like PCS and PDAs, as well as how DECT can fit in PCS and PDAs.

Notes

1. Palm Computing, Inc., a 3Com company, 5400 Bayfront Plaza, Mail Stop #10112, PO Box 58007, Santa Clara, CA 95052-8007, 1999.
2. PenRight! Corporation, 47358 Fremont Blvd., Fremont, Ca 94538, 1999.
3. Puma Technology, Inc., 2550 North First Street, Suite #500, San Jose, CA 95131, 1999.
4. Ideam Corporation, 1560 1st Ave. S., Seattle, WA 98134-1402, 1999.
5. AvantGo, Inc., 1700 S. Amphlett Blvd. Suite 300, San Mateo, CA 94402, 1999.
6. Pendragon Software Corporation, 1580 S. Milwaukee Ave, Suite 515, Libertyville, IL 60048, 1999.
7. Feras Information Technologies, 315 Delaware Avenue, Lansdale, PA 19446, 1998.
8. Riverbed Technologies, 2070 Chain Bridge Road, Suite 475, Vienna, VA 22182, 1999.
9. Advance Systems Limited, 19 Portland Square, Bristol BS2 8SJ, United Kingdom, 1999.
10. The most widely used devices that primarily use Windows CE are the Handheld PCs (H/PC)and the Palmheld PCs (P/PC).
11. Oracle Corporation, World Headquarters, 500 Oracle Parkway, Redwood Shores, CA, 94065, 1999.
12. EnterpriseLink, 2542 S. Bascom Ave., Suite #203, Campbell, CA 95008, 1999.
13. Whole-Brain Software, 401 West 29th Street, Austin, TX 78705, 1999.
14. Sun Microsystems, Inc., 901 San Antonio Road, Palo Alto, CA 94303, 1999.
15. Hewlett-Packard, 3000 Hanover Street, Palo Alto, CA 94304-1185, 1999.

Chapter 45
Networking Features of the Java Programming Language

John P. Slone

JAVA IS THE FIRST LANGUAGE OF SIGNIFICANCE TO BE DEVELOPED IN WHAT COULD BE CALLED THE "POST-INTERNET REVOLUTION" ENVIRONMENT. Although the Internet had been under development for several decades, the advent of the World Wide Web in the early-to-mid 1990s sparked an explosion of growth of the Internet that could only be characterized as a revolution. Consequently, the Internet became a household word and a mainstay of business computing. Industry analysts project the number of Internet-connected computers to exceed 100 million in the very near future. Among those 100 million computers, one finds a variety of computing platforms, ranging from microprocessors costing a few hundred dollars to highly specialized mainframe or supercomputers costing millions. At the low end of the cost spectrum, hardware generation life cycles are measured in months.

For languages developed prior to Java, networking was a specialty area, and the technology of networking was typically handled outside the language itself, such as through system calls. For Java, networking is a fundamental assumption, handled within the language itself. In the next section, some basic concepts are examined related to networking in this environment. In the sections that follow, several networking mechanisms are reviewed that are built into Java for the purpose of exploiting this environment.

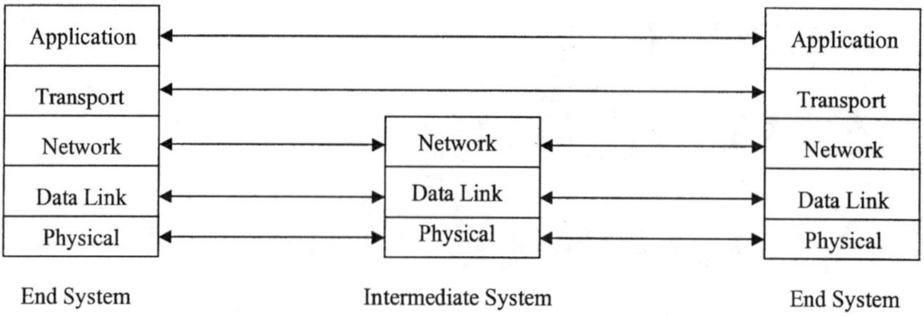

Exhibit 45.1 Layered Communications Model of the Internet.

NETWORKING BASICS

In today's era of enterprise LANs, the predominant networking architecture is the layered architecture developed for the Internet depicted in Exhibit 45.1. This architecture provides an abstract view of networking, arranging functions according to purpose, and providing a clean interface between applications software and the complexities of the network itself.

Briefly working from the bottom, the following basic functions are found provided by each layer:

Layer 1 — Physical

This layer is where the actual transmission of data takes place. Specifications for this layer address such things as the type of communication medium (such as copper wire, optical fiber, or radio waves), signal modulation, and bitwise encoding. This layer is almost always implemented in hardware, and is of no concern to the applications developer.

Layer 2 — Data Link

This layer ensures that transmitted data is actually delivered across the physical medium. Typical functions include error checking and retransmission requests as needed to ensure successful delivery. In a local area network (LAN) environment, this layer is often used to implement forwarding of data packets between LAN segments. This layer is also typically integrated in hardware, and is of no concern to the applications developer.

Layer 3 — Network

The primary purpose of this layer is routing. Given the size and complexity of networks such as the Internet and LAN-based enterprise networks, protocols at this layer have been developed to handle the routing of data packets as needed to deliver data from the originating end system, through an indeterminate number of intermediate systems, and ultimately to the destination end system or systems. The most commonly known protocol at this layer is the Internet Protocol, or simply the "IP" part of "TCP/IP." Protocols at this layer are implemented in a combination of hardware and software.

The only aspect of this layer that is of concern to the applications developer is the IP address. The IP address is a 32-bit number, more commonly expressed in a "dotted numeric" notation such as 204.5.208.18, where each of the four numbers represents the value contained in one of the four bytes of the IP address. The IP address is analogous to a telephone number; each system has a globally unique IP address which, when provided to the network layer, is sufficient to ensure that packets are correctly routed to the intended end system.

Layer 4 — Transport

This layer is responsible for the end-to-end delivery of data between end systems. This layer provides two fundamental services: connection-oriented and connectionless. The most common service is the connection-oriented service provided by the transmission control protocol (TCP). TCP provides for reliable delivery by maintaining a connection between the end points, and by ensuring that all packets are transmitted without error and in the proper sequence. When the systems are through exchanging data, the connection is torn down at the request of either end.

A less commonly known service is the connectionless service provided by the user datagram protocol (UDP). UDP provides an unreliable delivery mechanism by simply sending a packet toward the intended destination without taking any steps to ensure successful delivery. Although this concept may seem strange at first, it actually serves a very useful purpose. The classic example is that of time synchronization. If a packet is dropped, it is more important for the next packet to get delivered than it is for the systems to take the time needed to retransmit the missing one, thus throwing the time synchronization system out of synch.

Both of these services provide the primary interface between the application and the network. From the network perspective, the interface is known as a "port." A port is essentially a 16-bit extension to the IP address. It can be thought of as providing a capability similar to that of a telephone extension. That is, while a telephone number will successfully deliver a call to a business location, an extension may need to be specified to reach the intended party. In the same fashion, the IP address will deliver the packet to the end system, but the port must be specified to reach the intended application.

From the perspective of the application, this same point of interface is viewed as a socket. A socket is an abstraction that behaves similarly to other input–output abstractions such as files or printers. A socket is made

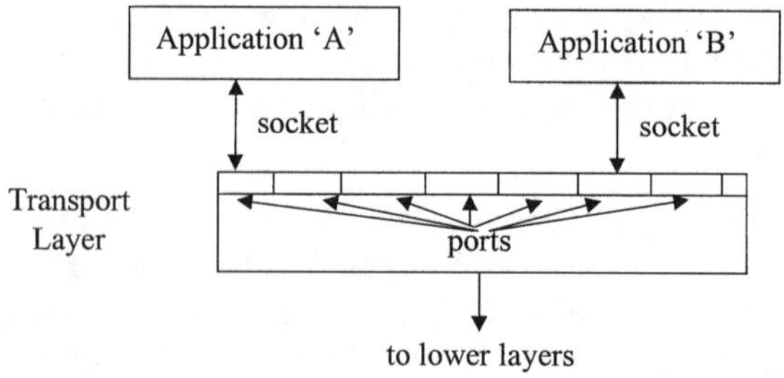

Exhibit 45.2 Socket-and-Port Transmission of Data.

available for use by the application by binding it to an available port. The application then transmits data across the network by writing to a socket, and it receives data from the network by reading from the socket. This notion of sockets and ports is depicted in Exhibit 45.2.

Layer 7 — Application

Programmers can effectively utilize network capabilities without being concerned with the application layer. However, a number of standardized network-enabled applications have been written, providing a level of network abstraction significantly higher than that of the socket. Perhaps the most familiar of such applications is the World Wide Web, which provides a high-level network abstraction known as the HyperText Transport Protocol, or HTTP. This protocol, along with other familiar protocols such as ftp, Gopher, and Telnet, is considered a Layer 7 protocol. Programmers taking advantage of these protocols can accomplish network programming tasks without being concerned with lower level details such as binding a socket to a port, opening a connection, reading from or writing to the socket, and ultimately closing the connection and unbinding the port. All this is accomplished by the Layer 7 protocol.

The Internet model is rooted historically in the OSI Basic Reference Model. In the OSI model, Layers 5 and 6 are used for "session" and "presentation," respectively, with Layer 7 defining the "application" layer. Although the Internet networking model does not generally use these two layers, the layer number designators have survived.

JAVA NETWORKING PRIMITIVES

With this application-oriented background on networking concepts, one can now discuss Java networking features within the context of these concepts. In this section, Java networking primitives will be discussed that enable the programmer to develop applications that interact directly with Layer 4. These primitives are found within the java.net package.

Connection-Oriented Networking

Connection-oriented networking employs a paradigm analogous to the telephone network. First, a connection is established between two parties. Once established, communications take place by having each party send and listen as needed. Once finished, the connection is broken. As discussed above, rather than using a telephone, each party "speaks" into and "listens" from a socket.

As with practically everything else in Java, the concept of a socket is provided as a class; thus, sockets are treated just like any other object in the language. For connection-oriented networking, there are actually two classes of socket objects in Java: "Socket," which implements client sockets, and "ServerSocket," which implements server sockets. The distinction is needed to provide server applications with the processing capability of listening for and accepting connection requests from clients. Both are subclasses of SocketImpl, which is where most of the low-level methods and variables are defined. The two subclasses hide many of these details from the programmer. For example, among the constructors for the client-side socket are the following:

- Socket (InetAddress, int) — creates a socket and connects it to the specified port on the host at the specified IP address. With this one, simple constructor, the program will actually create a socket, bind it to a local port, issue a connection request to the remote IP address/port combination, and wait for an acknowledgment.
- Socket (String, int) — creates a socket and connects it to the specified port on the host named in String. In addition to everything in the previous constructor, this one also resolves the name of the remote host specified in String to find its IP address prior to issuing the connection request.

Constructors on the server side include the following:

- ServerSocket (int) — creates a server socket and binds it to the specified port on the local host.
- ServerSocket (int, int) — same as above, but allows the programmer to specify the maximum allowable backlog of pending requests.
- ServerSocket (int, int, InetAddress) — same as the previous constructor, but allows programmer to specify which network interface on multihomed machines (e.g., machines sitting on firewalls).

Similarly powerful methods are available for these classes, including fairly high-level methods such as getInputStream and getOutputStream, as well as lower-level methods such as getInetAddress, getPort, and setTcpNoDelay. Thus, between the constructors and methods, the Socket and ServerSocket classes provide programmers with a fairly powerful assortment of tools for developing networked applications.

Connectionless Networking

In contrast to connection-oriented networking, connectionless networking is accomplished by sending independent packets between the parties involved. Rather than having the notion of a connection, which is established at the beginning of the exchange and torn down following the exchange, each individual packet, called a datagram, must be provided with the necessary addressing information. This paradigm is closer to that of the postal system, in which a letter, complete with destination and return addresses, is dropped in the nearest mailbox and routed individually to its destination.

Java provides a separate type of socket class for this purpose: the DatagramSocket. In this case, there is no distinction between a server socket and a client socket, since datagrams are sent and received outside the context of a connection. Fundamentally a simpler concept, the DatagramSocket has only three constructors:

- DatagramSocket ( ) — This constructor creates a socket and binds it to any available port on the local machine.
- DatagramSocket (int) — This constructor creates a socket and binds it to the specified port on the local machine.
- DatagramSocket (int, InetAddress) — This constructor creates a socket and binds it to the specified port/interface combination on the local machine.

Note that this class of socket is only associated with the local machine. Because there is no notion of a connection, there is no association with a remote machine.

Once a datagram socket object has been created, the programmer may invoke one of several methods. Of most importance are "send" and "receive," used to send or receive datagrams, respectively. Datagrams are implemented as a separate class, known as a DatagramPacket. This class has two constructors:

- DatagramPacket (byte[ ], int) — used to create a datagram packet for receiving datagrams of length int
- DatagramPacket (byte[ ], int, InetAddress, int) — used to create a datagram packet for sending packets of length int to the address and port specified

DatagramPacket methods are provided to get or set the datagram's address, port, data, or length as needed.

Multicast Connectionless Networking

Connectionless networking can also be accomplished between multiple parties, as opposed to the limited notion of point-to-point networking. Java provides a MulticastSocket class for this purpose as a subclass of DatagramSocket. Of particular interest are the methods joinGroup and leave-Group, which allow the system to join and leave a particular multicast group, and the methods getTTL and setTTL, which handle the datagram's time-to-live attribute.

High-Level Java Networking Abstractions

Up to this point, the discussion has been limited to fairly low-level networking concepts, focused at the transport layer service interface, the socket. Although extremely powerful when compared with the networking features of most other languages, these features are little more than primitives within the context of Java networking features. The remainder of this article focuses on several of the higher-level networking concepts provided in Java.

URL-Based Programming

The first such concept discussed is that of URL-based programming, a concept also supported within the java.net package. In general terms, URL-based programming allows the programmer to focus on the concepts associated with actually handling a remote object, rather than on all the lower-level mechanisms involved in creating a socket and binding it to a port, establishing a connection to the object's machine, locating the object, and retrieving information from or sending information to the object. How this is accomplished should become clearer in the following paragraphs.

The fundamental class provided for this purpose is the URL. This class, whose name stands for Uniform Resource Locator, uses a standard notation for representing a resource on the network. Popularized by the World Wide Web, the basic URL provides four primary components: a protocol identifier (such as http, ftp, or gopher), a machine name (such as www.yahoo.com), a port number (if not specified, a default port is assumed), and a file name (including the path to the file). The file name component can also optionally contain an internal reference to a specific named label within the file. The Java URL class provides four constructors to allow flexibility in the way a URL object is created:

- URL (String) — allows the creation of a URL object by specifying a complete, familiar URL specification such as http://www.yahoo.com/
- URL (String, String, int, String) — allows the creation of a URL object by separately specifying the protocol, host name, port, and file name
- URL (String, String, String) — same as the previous constructor, except that the default port is assumed
- URL (URL, String) — allows the creation of a URL by specifying its path relative to an existing URL object

Once the URL object has been created, Java provides a number of low-level methods, such as those that parse the URL and return specific elements, as well as several high-level methods, providing powerful capabilities to the programmer. Examples of high-level methods include getContent, which returns the entire content of the specified URL with a single line of code. Other high-level methods include openStream, which creates a connection to the URL and opens an input stream for subsequent reading of the contents, and openConnection, which creates a connection to the URL and opens a bidirectional stream for subsequent writing to or reading from the URL. This latter method is especially important for interacting with web-enabled applications, such as those implementing the Common Gateway Interface (CGI).

As implied by the previous paragraph, there is a concept of a URL connection, a concept that is also provided as a class known as a URLConnection. Once a URLConnection object has been created, the programmer has nearly 40 methods at his disposal for handling the connection. Among the capabilities provided by these methods are those of reading selected header fields, testing to see whether the URL accepts input, whether the URL is cached, when it was last modified, or when it expires. These methods are in addition to the methods for obvious concepts such as reading and writing.

Remote Method Invocation (RMI)

As discussed above, URL-based programming allows the Java programmer to interact at a high level with primarily non-Java resources on the network. RMI provides the Java programmer with the capability of developing truly distributed yet fully cooperative Java-only applications. These cooperating Java components can be peer applications, client and server applications, or client applets interacting with server applications. Compared with URL-based programming, RMI allows an order of magnitude increase in the degrees of complexity and sophistication of the resulting networked applications.

To achieve this level of sophistication, it is necessary to grasp concepts that go beyond the simple definition of classes and methods, although numerous classes are defined in the java.rmi family of packages. Instead, the exploitation of RMI requires a paradigm shift in the way one thinks about network-based programming. At a very high level, RMI requires the development of two components: a Java object that implements a method through a remote interface, and a Java object that remotely invokes that method. These two objects may be on the same machine or on different machines. Conceptually, all that is necessary to make this happen is for the calling object to obtain a valid reference to the called method in the form of a specially constructed URL. In most cases, the object reference is obtained either as a parameter or as the value returned by a method. The first such reference is typically obtained from an RMI remote object registry.

For clarity of discussion, the remainder of this section will assume that the object that implements the remote method is a server application, and that the object requesting the remote invocation is an applet.

Looking first at the server, the necessary ingredients are the definition of a remote interface, the definition of constructors for the remote object, and the definition of methods that can be invoked remotely. In addition, there are security requirements, but treatment of security is beyond the scope of this article. Finally, at least one of the remote objects must be registered in the RMI remote object registry on the server machine. At execution time, the constructor for the remote object creates an instance of the object and exports the object by having it listen to a socket for incoming calls.

Turning attention to the applet side, it can be seen that the calling object calls the remote method very much as it would any other method. However, instead of containing a reference to the actual remote object, it contains a reference to a locally implemented stub representing the remote object interface. The stubs are generated through the use of a special compiler tool called "rmic." Thus, the actual remote method invocation is abstracted in a way that isolates the programmer from the details.

The invocation and all the necessary semantics are actually handled by the remote reference layer and take place as follows. First, the applet calls a method making reference to a locally held stub. The stub then places the appropriate remote call, across the network if necessary, to its counterpart remote interface on the server side. That interface, in turn, invokes a method on the server side and passes the resulting return value back through the interface, across the network to the applet stub code, and on to the calling object. This interaction is depicted in Exhibit 45.3.

Thus, it is shown that RMI provides the Java programmer with an extremely powerful set of networking capabilities. As with other high-level networking features of the Java language, RMI allows the programmer to focus on the essentials of writing and invoking methods that accomplish a certain task, while ignoring the details of the underlying network.

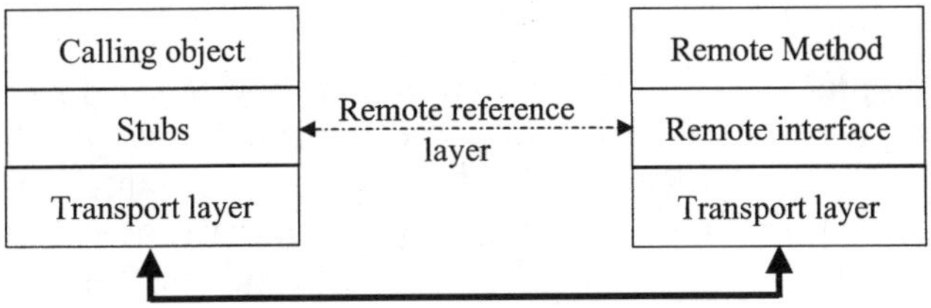

Exhibit 45.3 RMI Architecture.

Java Naming and Directory Interface (JNDI)

The two high-level networking features of the Java language discussed thus far, URL-based programming and RMI, have one thing in common other than the simple fact that they facilitate the development of networked applications. Specifically, both involve the notion of binding a name to a network-based resource. With URL-based programming, this resource is either a file- or a stream-based interface to an application. With RMI, the resource is a method in some remote object. Broadening the perspective to yet another dimension, it is found that network programming for any purpose will inevitably involve the binding of names to resources.

At the broadest possible level, network-accessible resources can be any type of object. For example, in addition to files, stream interfaces, and methods, other valid resources could include printers, calendars, electronic mailboxes, telephones, pagers, humans, conference rooms, control valves, remote sensors, or practically anything else, limited only by the imagination. To expand programming to this broad horizon, the concept of a generalized directory is needed. In the broadest sense, a directory can be thought of as a system that provides a mapping between the name of an object and one or more attributes that describe the object.

In practice, there are multiple directories in existence in today's networking environment. For example, the Internet Domain Name Service (DNS) is the directory that maps machine names (such as www.yahoo.com) to IP addresses (such as 204.71.200.74). Other directories, such as X.500, LDAP, NIS, and NDS, provide mapping between objects named in other name spaces and their attributes. Further complicating the situation, certain objects are actually identified by compound names — those names that exist in multiple, disjointed name spaces. URLs are a prime example, since part of the URL names the machine (named in the DNS name space), and part of it names the file (named within the name space managed by the particular machine).

To provide the simplest abstraction for programmers, what is needed is a mechanism that allows objects to be identified to programs by their compound names (such as a URL), while hiding the complexities of the underlying directory structures. This is precisely the objective of the Java Naming and Directory Interface (JNDI).

JNDI is implemented in three standard Java extension packages: javax.naming, javax.naming.directory, and javax.naming.spi. The first two packages comprise the JNDI Application Programming Interface (API), giving application programmers a suite of powerful classes and methods for handling object names and for interacting with the directory services. The third package makes up what is referred to as the JNDI Service Provider Interface (SPI). Conceptually, a Java application accesses the JNDI Implementation Manager through the JNDI API, while a variety of naming and

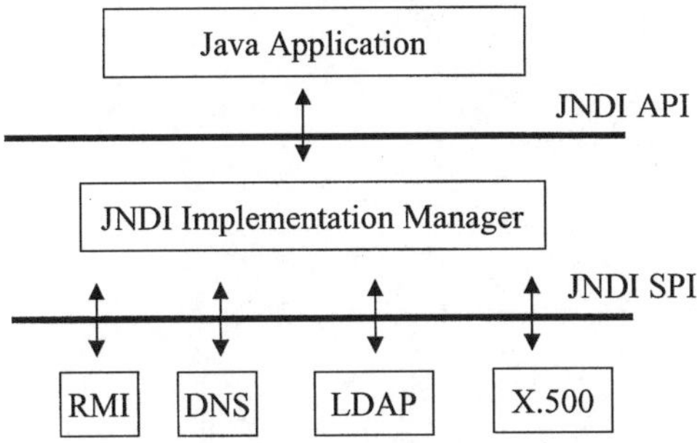

Exhibit 45.4 JNDI Architecture.

directory services sit transparently behind the JNDI Implementation Manager, plugged in through the JNDI SPI. This concept is depicted in Exhibit 45.4.

Of particular interest for this article is the naming and directory service shown in the lower left portion of Exhibit 45.4, RMI. As shown, the RMI object registry becomes a part of a substantially larger naming and directory service, and is implemented through the same classes and methods used for other network-based resources. In this manner, the entire world of network-accessible resources is placed literally at the fingertips of the Java programmer.

CONCLUSION

This article has barely begun to scratch the surface of network features built into the Java programming language. Many powerful features, those discussed in this chapter and many more, are built into the core Java packages; others are implemented as extensions. Through these features, Java provides programmers with the ability to write programs that implement networking concepts along a wide spectrum of abstraction levels, ranging from socket-level primitives to remotely accessing objects of any type by merely knowing the object's name. By building upon the object-oriented nature of Java, higher and higher levels of abstraction are readily accommodated.

In the opening, it was suggested that the ultimate success of Java as a language would depend to an extent on how purposeful the designers were with respect to designing the language in the context of its target environment. If one only considers the single aspect of networking as a fundamental component of Java's target environment, one can only conclude that Java is set to profoundly impact the computer science discipline in the years to come. As such, IT managers are well advised to assess the potential role of Java in their own environments, and to encourage and promote the development of Java expertise among their staffs.

Chapter 46
JavaBeans and Java Enterprise Server Platform

David Wadsworth

A MAJORITY OF THE WORLD'S DATA RESIDES ON MAINFRAME SERVERS. This legacy poses many challenges to the information systems (IS) community as it struggles with the demands of business units for new and innovative solutions to business problems. Organizations need to adopt a flexible, secure, and cost-effective architecture that will enable them to remain competitive and enable breakaway business strategies. Adoption of Java™ computing realizes these benefits by providing key technology enablers.

JAVA TECHNOLOGY REVIEW

The Java programming language was introduced to the public in May 1995. Key features of the language, such as platform independence and ease of programming, made it an instant success in the software development community. Other features such as safe network delivery and baked-in security have made the language the de facto standard for the development and deployment of web-based applications.

Applications written in the Java programming language are compiled to bytecode that can run wherever the Java platform is present. The Java platform is a software environment composed of the Java Virtual Machine and the Java Core Application Programming Interfaces (APIs). Portability of applications is achieved because there is only one virtual machine specification which provides a standard, uniform programming interface on any hardware architecture. Developers writing to this base set of functionality can be confident that their applications will run anywhere without the need for additional libraries. Core libraries include functional support for GUI development, I/O, database connectivity, networking, math, components (JavaBeans), multi-threading, and many others.

Sun's Java computing architecture is an implementation framework that uses standard, currently available network protocols and services to deliver the power of Java applications to the widest possible base of Java platform-enabled devices and users. With this architecture, transactions can be moved transparently to the most cost-effective, appropriate support channel within a network owing to the portable, Write Once, Run Anywhere™ nature of Java applications.

JAVA PLATFORM COMPONENT ARCHITECTURES

Designing and developing applications by means of components has been available for many years. The challenge has been to embrace and extend existing technology with new. Until recently such an approach has been proprietary and difficult to deploy. The Java computing environment with JavaBeans, a component technology, and server architecture solution, Java Enterprise Server, enables organizations to greatly simplify access to business systems. What follows is a description of the JavaBeans component model and an overview of the Java Enterprise Server platform.

JAVABEANS

A JavaBean is a reusable Java software component that can be visually manipulated and customized in a builder tool. These application building blocks are constructed so as to easily communicate with each other in a common environment. They also have the ability to store their state "on the shelf" to be revived at a later date. Since they are written in the Java programming language for deployment on any Java platform, JavaBeans are the platform-independent components for the network.

JavaBean components can range from simple GUI elements, such as buttons and sliders, to more sophisticated visual software components, such as database viewers. Some JavaBeans may have no GUI appearance of their own, but can still be manipulated in an application builder.

The JavaBean API has been designed to be accessible by builder tools as well as manually manipulated by human programmers. The key APIs such as property control, event handling, and persistence can be accessed by both handcrafted applications and builder tools. As well as event handling, property control, and persistence, introspection and customization are distinguishing features of all JavaBeans.

Property Control

Property control facilitates the customizing of the JavaBean at both design and runtime. Both the behavior and appearance of a JavaBean can be modified through the property features. For example, a GUI button might have a property named "ButtonLabel," which represents the text displayed in the button. This property can be accessed through its getter and setter methods. Once properties for a bean are configured, their state will be maintained through the persistence mechanism.

Persistence

The attributes and behavior of a bean are known as the state of the bean. The persistence mechanism within the JavaBean API supports storage of this state once the bean is customized. It is this state which is incorporated into the application and available at runtime. This externalization can be in a custom format or the default. A custom external format allows the bean to be stored as another object type such as an Excel document inside a Word document. The default is reserved for those instances where the bean's state needs to be saved without regard to the external format.

Event Handling

Event handling is a simple mechanism which allows components to be connected based on their production and/or interest in certain actions. A component or series of components can be sources of events that can be caught and processed by other components or scripting environments. Typical examples of events would include mouse movements, field updates, and keyboard actions. Notification of these events generated by a component would be delivered to any interested component.

The extensible event handling mechanism for JavaBeans allows for the easy implementation of the model in application builder tools. Event types and propagation models can be crafted to accommodate a variety of application types.

Customization

Changing the appearance and behavior of a JavaBean is accomplished through the customization features of the JavaBean's API. Each JavaBean contains a list of exported properties which an application builder can scan and use to create a GUI property editor sheet. The user can then customize the bean using this dynamically created sheet. This is the simplest form of customization.

Another layer of customization is possible by attaching to the bean a customizer class which acts as a properties wizard. This wizard will have a GUI which can be employed to tailor the properties for the related bean in a guided tour fashion. Such wizards are more likely to be found associated with complex beans, such as calculator beans or database connection beans. Once customization is complete, the properties will be stored using the persistence mechanism.

Introspection

The properties, methods, and events a JavaBean supports are determined at runtime and in builder environments by means of introspection. Introspection is a prescribed method of querying the bean to discover its inherent characteristics. Introspection is implemented using the Java programming language rather than a separate specification language. Therefore, all of the behavior of the bean is specifiable in the Java programming language.

One introspection model supported by the JavaBeans API provides a default view of the methods, events, and properties. This simple mechanism does not require the programmer to do extra work to support introspection. For more sophisticated components, interfaces are available for the developer of the bean to provide specific and detailed control over which methods, events, and properties are exposed.

Default, low-level reflection of the bean is used to discover the methods supported by the bean. Design patterns are then applied to these methods to determine the properties, events, and public methods supported by the component. For example, if a pair of methods such as setColor and getColor are discovered during the reflection process, the property "color" is identified by the application of the "get/set" design pattern for property discovery.

More complex component analysis can be built into the bean by the use of a BeanInfo class. This class would be used by a builder tool to programmatically discover the bean's behavior.

Security

JavaBeans are governed by the same security model as all other Java applets and applications. If a JavaBean is contained in an untrusted applet, it will then be subject to the same restrictions and will not be allowed to read or write files on the local file system or connect to arbitrary network hosts. As a component in a Java application or trusted applet, a JavaBean will be granted the same access to files and hosts as a normal Java application. Developers are encouraged to design their beans so they can be run as part of untrusted applets.

Runtime versus Design Time JavaBeans

Each JavaBean must be capable of running in a number of different environments. The two most important are the design and runtime environments. In the design environment, a JavaBean must be able to expose its properties and other design time information to allow for customization in a builder tool. In some cases, wizards contained in the bean may be employed to simplify this process.

Once the application is generated, the bean must be usable at runtime. There is really no need to have the customization or design information available in this environment.

The amount of code required to support the customization and design time information for a bean could be potentially quite large. For example, a wizard to assist in the modification of bean properties could be considerably larger than the runtime version of the bean. For this reason, it is possible to segregate the design time and runtime aspects of a bean so it can be deployed without the overhead of the design time features.

JavaBeans Summary

JavaBeans are the component object model for the Java platform. These device-independent components can be customized and assembled quickly and easily to create sophisticated applications.

JAVA ENTERPRISE SERVER PLATFORM

As organizations adopt Internet technologies to enable new business strategies, they are faced with the task of integrating all of their legacy applications, databases, and transaction services with web-based services. Traditional applications designed in the client/server model do not deploy well in an Internet/extranet environment. While not new, multi-tier architectures for application development and deployment are best suited for extending the reach of a company's infrastructure to partners, suppliers, customers, and remote employees. The Java Enterprise Server platform provides such an architecture in an open and standards-based environment that incorporates existing infrastructure while extending their reach to intranets, extranets, and even the Internet. An extensible architecture, the Java Enterprise Server platform contains the APIs, products, and tools necessary to construct new enterprisewide applications and integrate with existing systems.

Traditional mission-critical applications are written to the APIs of the underlying operating system, thereby tying the application to a single operating system. Porting of the application to a new operating system is both difficult and expensive. These same applications may rely on a service, such as a transaction monitor. Access to this service will be through the software vendor's proprietary APIs, creating another platform lock and presenting a barrier to moving to a different service provider.

The Java Enterprise Server platform is designed to address these platform-lock issues. It extends the notion of "Write Once, Run Anywhere" to include "and integrate with everything." Based on a layer and leverage model, the Java Enterprise Server platform can be built on top of existing legacy systems such as transaction monitors, database access, system

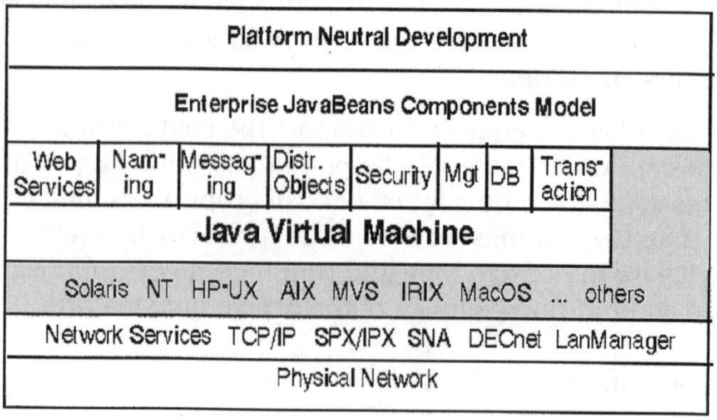

Exhibit 46.1 Java Enterprise Server Architecture. Copyright 1998 Sun Microsystems, Inc. Reprinted with permission.

management, naming and directory services, and CORBA (see Exhibit 46.1). Interfaces to these services, as well as a component model which provides for application encapsulation and reuse, are integral to the Java Enterprise Server platform. The component model includes JavaBeans, components for the client, and Enterprise JavaBeans (EJBs) components for the server.

All of the benefits of rapid application development, scalability, robustness, and security of the JavaBeans component architecture are extended to the Java Enterprise Server platform. EJBs also have the ability to provide transactional services. Coupled with these benefits is an open architecture capable of providing ease of development, deployment, and management.

Enterprise JavaBeans, an extension of the JavaBeans architecture, provide a distributed component architecture for developing and deploying component-based, multitier applications. Business logic is encapsulated in the Enterprise JavaBeans promoting a high degree of reuse. Access to low-level services, such as session management and multithreading, is simplified such that developers building applications do not need to deal directly with these functions.

Distributed applications developed with Enterprise JavaBeans can be deployed on any other platform without modifications. Support for transactions and messaging integrates with existing legacy systems and middleware.

The heart of the Enterprise JavaBean platform is the Enterprise JavaBean executive (see Exhibit 46.2). This runtime executive is used to execute the components that provide the services required by an application.

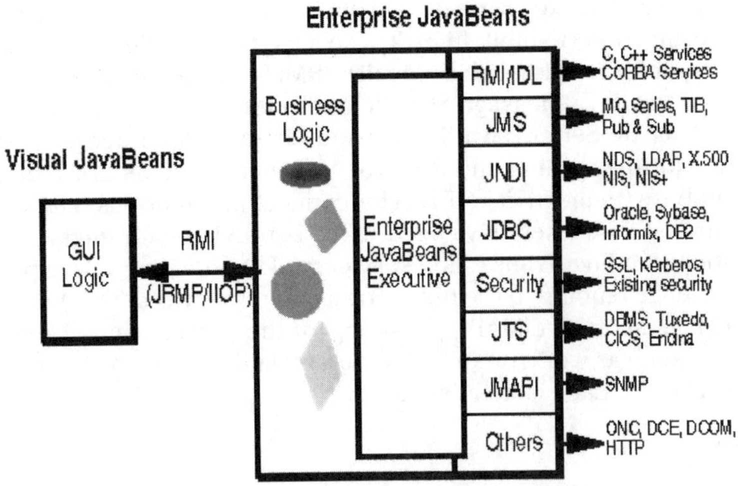

Exhibit 46.2 Enterprise JavaBeans Framework. Copyright 1998 Sun Microsystems, Inc. Reprinted with permission.

Through its components, the executive manages load balancing, and handles multithreading, transaction management, security, and connection management. This frees programmers so they can focus on developing the components that contain business logic.

Communication between the client and server in an application does not need to rely on any particular protocol. Both client and server sides of the application are coded using the Java programming language. At deployment time, the underlying communication stubs are generated automatically. The Java programming language introspection of the application class files is used to generate the communication stubs.

Unlike JavaBeans which use the Java event model, Enterprise JavaBeans use the distributed CORBA event model. The event model supported by the Java programming language is well suited for local, tightly integrated applications, but does not perform as well in a networked environment where high latency and insecure networks are common. Enterprise JavaBean events are propagated across the network over CORBA's Internet InterORB Protocol (IIOP) to other components.

Enterprise JavaBeans can be configured automatically as CORBA objects, and then accessed through IIOP by clients. These client applications do not have to be written in the Java programming language to access the components. EJBs can also function as COM/DCOM objects for Windows clients.

Access to several key services are offered as part of the Enterprise JavaBean specification (Exhibit 46.2). These services are offered through specific Java platform APIs, such as JavaIDL/RMI for accessing CORBA, DCE, or ONC services; Java Message Service (JMS) for access to messaging systems such as MQ Series; Java Naming and Directory Interface (JNDI) for accessing multiple naming and directory services such as LDAP and NDS; Java Database Connectivity (JDBC) for connecting to various relational and nonrelational databases; Java security APIs providing for encryption and authentication; Java Transaction services (JTS) providing a Java programming language binding to the object transaction services (OTS) of CORBA; Java Management API (JMAPI) providing for the management of networked resources such as workstations and routers; and web services through the Java Server API. Each is detailed below.

JavaIDL

The Java Interface Definition Language (IDL) provides standards-based interoperability and connectivity with CORBA. Through these interfaces, Java applications are able to access existing infrastructure written in other languages. This is one of the key interfaces for legacy system integration. JavaIDL is part of the Java Platform Core API set and is therefore available across multiple platforms.

Java Message Service

Java Message Service (JMS) provides an interface to messaging systems which provide publish/subscribe and message queue services. This platform-independent interface will also support the emerging push/pull technologies.

Java Naming and Directory Interface

Many different kinds of naming and directory services exist in today's enterprises. Directory services such as LDAP, NIS, and NDS provide network-wide sharing of information about the users, systems, applications, and resources that exist on an intranet or the Internet. User information can include login ids, passwords, security access, and electronic mail addresses. System information can include network addresses and machine configurations. The Java Naming and Directory Interface (JNDI) is independent of any specific naming and directory service implementation. Application developers can easily access multiple namespaces through JNDI. A single interface simplifies the access to composite namespaces and enables an application to be portable across different platforms.

Java Database Connectivity

One of the earliest and now core APIs is the Java Database Connectivity API (JDBC). This is an SQL-based, database-independent API which frees developers from writing database vendor-specific code in their applications. JDBC supports the common database functionality such as remote procedure calls, SQL statements, database connection, and result sets. Since JDBC is implemented via a driver manager which itself can be implemented in the Java programming language, applets can be delivered to the client with the database connectivity built in. Implementation drivers for all the major RDBMSs are already available for JDBC, and a JDBC-to-ODBC bridge is standard in the Java Developer's Kit Version 1.1. JDBC drivers for object-relational DBMSs, as well as the IBM IMS, are also currently available.

Java Security API

Security is an integral part of the Java platform and extends to the Java Enterprise Server architecture. There are four key areas that are supported by various security APIs: authentication, authorization, privacy, and integrity.

Authentication is the system's ability to verify or recognize a user. Typically performed at application access or system sign-on, authentication is the first line of defense present in a comprehensive security model. The JavaCard APIs allow smart cards to be employed as secure user authentication devices. These physical cards, combined with a secure personal identification number (PIN), enable users to be recognized by the target system. Digital signatures, another authentication method, are also supported through the Java Virtual Machine.

Authorization is the means of determining which data, systems and services a user can access. The Java Security APIs and access control lists (ACLs) are available for managing who can access what. ACLs can be built for each Enterprise JavaBean and consulted whenever the bean is accessed. Based on the user's role, some form of access can be given or denied. Transaction servers installed in the application enforce the ACL at runtime. Since ACLs are not static structures, they can be moved around the network with an EJB object. These embedded ACLs can then be accessed by the application developer.

Privacy concerns are raised in the context of transmission of sensitive data across public networks. To protect data, such as credit card numbers, encryption is typically employed. The Java language cryptography APIs provide application or session-level encryption. This interface can support any encryption implementation, including DES.

As data passes through a network, be it private or public, there is a chance for malicious or accidental modification. To prevent such actions, it is necessary to be able to guarantee the integrity of the transmission. The same mechanisms for ensuring privacy can be used for maintaining integrity of network communications, namely session and application encryption.

Java Transaction Services

Java Transaction Services (JTS) within the Enterprise JavaBean framework is a low-level API not meant as an application programmer interface. JTS programming is targeted to the resource managers and TP monitor programmers. Currently available implementations include BEA Systems Jolt product for Tuxedo access, or the IBM JavaCICS for access to mainframe CICS applications.

Java Management API

The Java Management API (JMAPI) is a set of interfaces for the development of distributed network, system, and application management applications. JMAPI is designed to be incorporated into a variety of devices, across diverse network protocols and numerous operating systems. With support for the Simple Network Management Protocol (SNMP), JMAPI can communicate directly with a variety of existing devices. In the future, device manufacturers will incorporate the JMAPI directly into their products. System administrators using applications developed on this foundation are able to easily manage their network, applications, or other systems from any Java platform located anywhere on the network.

Java Server API

The Java Server API is an extensible framework which can be employed to quickly develop network-centric servers. These servers are capable of providing network-based services such as web services, file and print services, proxy services, and mail services. To extend the functionality of a Java server, a developer can create servlets using the Java Servlet API. Java servlets are programs that can be local to the server, or downloaded across the network and then executed on the Java server. These servlets are perfect for processing form data from HTML pages, replacing the platform-dependent CGI-bin scripts in use by many organizations.

SUMMARY

The ability to integrate with legacy systems and extend enterprise services to the network with platform-independent technologies are key benefits of developing a Java Enterprise Server strategy. Enterprise JavaBeans, the component architecture for the Java Enterprise Server, provide a software- and hardware-independent method to access these systems and make them available to business components. These components can easily access services such as transaction monitors and message systems, DBMSs, and naming services with the assurance of the Java platform's "Write Once, Run Everywhere."

Chapter 47
A Quick Overview of Linux

Raj Rajagopal

LINUX IS A UNIX-TYPE OPERATING SYSTEM, originally created by Linus Torvalds in 1991, that has been enhanced by developers around the world. Linux is an independent POSIX implementation and is compliant with X/Open and POSIX standards. Linux is developed and distributed under the GNU General Public License. The GNU license specifies that the source code for Linux plus any Linux enhancements should be freely available to everyone. Vendors are free to charge for distributing Linux, and the availability of source code does not apply to applications developed on top of Linux. Linux features includes true multitasking, multiuser support, virtual memory, shared libraries, demand loading, proper memory management, TCP/IP networking, shell, file structure, utilities, and applications that are common in many UNIX implementations. A complete list of features is included in Exhibit 47.1. Linux is a candidate operating system to be evaluated by enterprise and data center managers who have any flavor of (or are considering acquiring) UNIX or Windows NT.

SUDDEN SURGE OF INTEREST IN LINUX

For an operating system invented in 1991, the interest in Linux has only recently surged. Like many other computer- and Internet-related phenomena, Linux has again proved that there is strength in numbers. Linux has reached the critical mass of users where vendors have started taking an interest in the operating system. Exhibit 47.2 shows the number of Linux users over the last five years. These numbers are estimates from Red Hat Software, one of the distributors of Linux.

The biggest customers for Linux include Internet service providers that use Linux along with the free APACHE web server, universities, and web developers. While Exhibit 47.2 shows cumulative numbers, analysts expect Red Hat to ship about 400,000 CDs this year. By contrast, Microsoft Corporation is estimated to have sold more than 100 million copies of Windows this year.

Exhibit 47.1 Linux Features.

Feature	Description
Virtual memory	Possible to add swapping areas during runtime. Up to 16 swapping areas, each of which can hold 128 MB and can be used for a total of 2 GB swap space.
Development languages	Supports most common languages including C, C++, Java, Ada95, Pascal, FORTRAN, etc.
UNIX commands, tools supported	Commands include ls, tr, sed, awk, etc. Tools include gcc, gdb, make, bison, flex, perl, rcs, cvs, and prof.
UNIX source/binary compatibility	Compatible with most POSIX, System V, and BSD at the source level. Through iBCS2-compliant emulation, compatible with many SCO, SVR3, and SVR4 at the binary level.
Graphical environments	X11R5 and X11R6. Motif is available separately.
Shells	All three common shells.
Editors	GNU Emacs, Xemacs, MicroEmacs; jove; ez; epoch; elvis; GNU vi; vim; vile; joe; pico; and jed.
Internationalization	Supports many localized and customized keyboards.
LAN support	Supports Appletalk server and NetWare client and server.
Internet communications	Supports TCP/IP networking including FTP, Telnet, etc.
File systems	Linux file system supports file systems of up to 4 TB and names up to 255 characters long. Also supports NFS and System V. Transparent access to MS-DOS FAT partitions via a separate file system. Partition looks like a normal UNIX file system.
CD-ROM	CD-ROM file system reads all standard CD-ROM formats.
Y2K compliancy	Compliant. Linux's 32-bit data representation should handle dates until the year 2038.

VENDOR SUPPORT FOR LINUX

The growth in users has resulted in renewed interest in Linux from many vendors. Intel Corporation and Netscape Communications Corporation have announced an investment in Red Hat Software. Two venture capital firms, Benchmark Capital of Menlo Park, CA, and Boston-based Greylock, have also made investments in Red Hat.

Besides financial support, vendors have also started providing product support. Major vendors, including Sybase, Oracle, Netscape, Informix, Computer Associates, Interbase, and Corel, have announced versions for Linux. Some of the popular applications available for Linux are summarized in Exhibit 47.3.

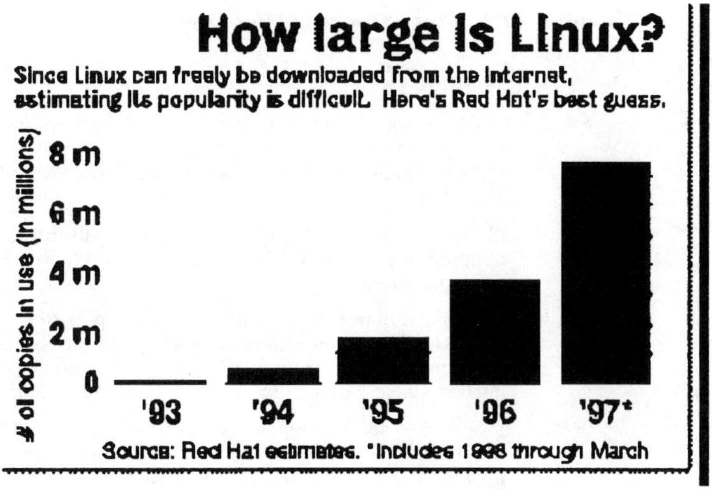

Exhibit 47.2 Linux User Base.

Exhibit 47.3 Applications Available for Linux.

Application Category	Popular Products
Office applications	Star Office, Applixware, WordPerfect, XessL:ite4 spreadsheet
PDF support	Adobe Acrobat Reader
Backup	Backup/9000, BRU 2000
Video player	MpegTV Player
Database support	Sybase Enterprise Adaptive Server, Oracle DBMS, InterBase, Informix-SE
Web software	Apache Web Server, Netscape Communicator, Red Hat Web Server

LINUX CUSTOMERS

Linux is used extensively at universities and by ISPs as web servers. Linux is also used in research and development environments and in government. The visual effects company Digital Domain used Linux in creating visual effects for the movie "Titanic." The U.S. Department of Defense uses Linux for DNS platforms and SMTP mail relays.

OPERATING REQUIREMENTS FOR LINUX

Most Linux versions run on most Intel-based machines, although other platforms are also supported (see Exhibit 47.4). Besides the web pages of

Exhibit 47.4 Linux Operating Requirements.

Component	Comments
Processor	Most Intel-based machines. Some implementations support multi-processor such as SMP. Other processors supported include Sun SPARC, Alpha, PowerPC, RS/6000, MIPS, and Macintosh.
Memory	4MB required to install most distributions; 5MB to 10MB for minimal setup; 8MB to 16MB required to run X. Requires swap partitions.
Hard disk	Hard drive required. 8MB to 16MB for swap partition. 24MB for basic installation with no X, development tools, or TeX. Most installations require 40MB to 80MB minimum, including free space. Complete systems need 60MB to 200MB.
Bus	Supports 386/486/Pentium machines with ISA, EISA, PCI, and VLB buses. VESA local bus and PCI are supported. SCSI support is available in some implementations.
Coexistence with other operating systems	Linux will coexist with DOS, MS-Windows, and OS/2 on the hard drive. Partitioning the hard disk is required for each operating system.

specific Linux distributors, the web pages http://www.linux.org and http://sunsite.unc.edu/mdw/HOWTO/INFO-SHEET.html provide more detailed lists of hardware, video cards, disk controllers, etc.

LINUX SOURCES

Linux is freely available on the Internet (see Linux-related FTP sites) along with the source. Linux is also available through distributors. There are many different Linux versions, or distributions. A distribution is the compiled Linux software and source code usually combined with extra features such as configuration, installation, Web, and other applications for the specific version. Most importantly, however, distributors provide support for their version. Some distributions are available for download at no charge, while others are available at prices typically in the range $50 to $100 on CD-ROM from Linux distributors/vendors/retailers.html worldwide. Keep in mind that some of the distributors offer multiple distributions, which differ in price and features. When picking distributions, enterprise and data center managers need to consider the features and other applications besides evaluating the operating system itself.

Other options include purchasing Linux preinstalled on a computer, or purchasing a book that includes a complimentary CD-ROM (see Exhibit 47.5).

Exhibit 47.5 Linux Distributors.

Organization	Product	Web Address	Features and Comments
Caldera	OpenLinux	www.caldera.com	Includes a new graphical desktop called K Desktop Environment (KDE), Sybase Enterprise Adaptive Server.
Work Group Solutions	Linux Pro	www.wgs.com	The Linux Pro Plus package includes a 7 CD-ROM set, a Linux Encyclopedia, a 1,600+ page reference manual, etc. Linux Pro is aimed at the professional developer.
S.u.S.E. LLC	S.u.S.E. Linux	www.suse.com	S.u.S.E. Linux package includes 4 CD-ROMs, a reference book, and about 800 preconfigured, compiled packages. It features a menu-driven installation, hypertext help, an X11R6.3-based graphical interface, and source code.
Red Hat Software	Red Hat Linux	www.redhat.com	The Red Hat Linux Operating System, which has won many awards, can be used as a web server, e-mail server, DNS server, or a news server for multiple sites, with virtual hosting. The package includes the Apache Web server, sendmail, publishing, calendars, Internet tools, X Window system, Netscape Navigator, and Netscape Communicator. The operating system includes disk partitioning with Disk Druid, autodetection of hardware, configuration for multiple window managers, graphical tools, and sound support.
Debian	Debian Linux	www.debian.org	Debian is not a company. Debian Linux is produced by volunteers and the primary focus is Intel-based machines. Although it is only supported through e-mail/third parties, support from the volunteers is normally very quick. Compared to other sources, the software is probably the cheapest (less than U.S. $5 for 2 CDs, or download for free).

Exhibit 47.6 Linux-Related Web and FTP Sites.

Site	Comments
sunsite.unc.edu	Free Linux implementation at /pub/linux
tsx-11.mit.edu	Free Linux implementation at /pub/linux
ftp.ncsa.uiuc.edu	Mosaic web browser and web server software for Linux
ftp2.netscape.com	Netscape web browser for Linux
ftp.blackdown.org	Sun's Java Development Kit for Linux
sunsite.unc.com/mdw/linux.html	Linux Documentation Project Web page

Linux-Related Web and FTP Sites

There are a number of web and FTP sites that provide the Linux code as well as applications for Linux. Some of the popular ones are shown in Exhibit 47.6.

LINUX DRAWBACKS

Popular products such as Microsoft Office and some common databases will not run on Linux. Unless the Linux distribution is from a reputable vendor, support for Linux can sometimes be hard to come by. The support situation may be changing, at least for some major players. At the recent Comdex show, Red Hat Software announced 24-hour enterprise support for its Linux users starting in 1999. A similar service is also due in 1999 from another Linux distributor, Pacific HiTech. Installations planning to use Linux should be prepared to hire technically adept Linux pros to ensure that problems can be handled in a timely manner. While there are a lot of enthusiastic programmers working on the operating system itself, the situation is different when it comes to peripheral device support. Peripherals typically need drivers, and details of the peripheral necessary to write the drivers are not commonly available. It is up to the peripheral vendors to come up with the drivers for the different platforms. Peripheral vendors do not always make a Linux version and, even when they do, it may require some effort to make the configuration work in an installation. Users accustomed to the Windows GUI may feel that the user interfaces available for Linux are not that user friendly. While Linux is UNIX-like and follows X/Open and POSIX standards, it has not been formally branded as UNIX 98 compliant. This is due in part to the fact that Linux is unlike a traditional UNIX offering from one manufacturer that is responsible for paying for the development and certification.

Exhibit 47.7 Linux Usage Guidance.

Situations for Which Linux Is a Good Candidate	Situations for Which Linux Is Not a Good Candidate
DP environment is very cost-conscious	DP environment requires UNIX branding
DP environment has skills to make things work	Mission-critical applications
Environment that has many UNIX users and programmers	Need a wide array of peripherals
Linux-proven applications and environments such as web servers, universities, and R&D establishments	Installations that need 24 · 7 support (this may change for some Linux vendors in 1999)

IMPLICATIONS OF LINUX

Windows NT has been able to make significant inroads into the low-end server market because it is typically priced lower than many UNIX offerings and is, in general, easier to use. Although Linux is UNIX-like in terms of ease of use, it does offer significant price advantages. This should appeal to cost-conscious customers, particularly those who do not mind the potential drawbacks listed above. While the ability of Linux to make significant inroads into the NT market is debatable, Linux is certainly bound to prove a very strong challenger to low-end UNIX solutions that are priced significantly higher than Linux.

On the high end, applications tend to be more mission-critical and customers are typically concerned about reliability, availability, and serviceability as much or more than the price. Hence, Linux may not have as much of an impact on the high end, unless one or more of the distributors establish themselves to the point that customers are comfortable about the support.

GUIDANCE FOR LINUX USAGE

Exhibit 47.7 summarizes Linux usage guidelines for enterprise and data center managers.

LINUX FUTURE OUTLOOK

The money from the venture capital and investments at Red Hat will most likely go toward creating an enterprise server group within Red Hat. Linux will follow the same path as other UNIX operating systems and Windows NT and transition to a 64-bit architecture. Intel has already indicated that the company intends to support a 64-bit version of Linux with its Merced

Exhibit 47.8 Linux-Related Newsgroups.

Newsgroup	Description
comp.os.linux.announce	Announcements of Linux developments
comp.os.linux.devlopment.apps	For Linux applications development programmers
comp.os.linux.devlopment.system	For Linux operating system development programmers
comp.os.linux.hardware	Linux hardware specifications
comp.os.linux.admin	System administration questions
comp.os.linux.misc	Miscellaneous questions and issues
comp.os.linux.setup	Installation/setup problems
comp.os.linux.answers	Answers to problems
comp.os.linux.help	Questions and answers for particular problems
comp.os.linux.networking	Linux network-related questions and issues

chip and is working on adding to Linux its Wired for Management features, features that are aimed at making Linux easier to install in a corporation. In addition, Intel has also already disclosed details of its universal driver initiative, which aims to make developing Linux applications much easier.

All these developments and the new interest by users and vendors bode well for Linux. The potential downfall is that Linux may face the same problems as UNIX where there are multiple versions and applications are not portable from one version to another.

LINUX RESOURCES

Help on Linux is available through a variety of sources such as newsgroups (see Exhibit 47.8), publications such as the *Linux Journal,* and books. Linux International is a nonprofit consortium of Linux users and vendors.

Books on Linux include:

1. *Linux Unleashed* by Sams. Over 1100 pages and includes a CD. August 1998.
2. *Running Linux* by Matt Welsh and Lar Kaufman. O'Reilly & Associates. 650 pages, 2nd ed., August 1996.
3. *Teach Yourself Linux in 24 Hours* by Bill Ball, Stephen Smoogen, and Ryan K. Stephens. Sams Publishing. 380 pages and includes a CD.

CONCLUSION

Linux is definitely worth a serious evaluation by enterprise and data center managers who are looking for cost-effective solutions. It is a rich operating system that is gaining momentum. It is free and is being used for some specific applications. The support situation is improving. It may not, however, be appropriate for enterprise and data center managers who are considering mission-critical applications, need a wide array of peripherals, or need UNIX branding.

Chapter 48
Evaluating the Performance of NT-Based Systems

Gilbert Held

COMPUTER PERFORMANCE represents a topic that in one area is very similar to a discussion of the weather — that is, many people can be expected to talk about both topics. However, unlike the weather, which is normally beyond one's control, one does have some latitude when it comes to tailoring the performance of a computer system. By changing hardware components, adding additional memory, swapping a fixed disk for a faster one, and tuning software components, one can usually enhance the performance of a computer. Unfortunately, until recently it was difficult to evaluate the benefits derived from changing hardware or modifying software settings. Fortunately for Microsoft Windows NT users, this operating system includes a built-in performance monitor that can be used to determine the existing level of performance of a computer, as well as for ascertaining the impact of hardware and software changes. Thus, by understanding how to use the Windows NT built-in Performance Monitor utility — either by itself to ascertain the performance of various computer components or in conjunction with hardware and software changes — one obtains the ability to evaluate the performance of an NT computer. In doing so, one might be able to determine that a simple and easy-to-perform operation may be all that is required to eliminate a performance bottleneck, alleviating the necessity to replace or upgrade an existing computer. Thus, in addition to providing a mechanism to examine the level of performance of existing hardware and software, the use of the Windows NT Performance Monitor may enable one to postpone or avoid an expensive equipment upgrade.

OVERVIEW

There are several key components of a Windows NT computer system that can affect its performance. Those components include the amount of memory installed in the computer, the access time and data transfer rate of hard disks, the type of processor or processors installed on the computer's motherboard, and the type of network adapter card used for a connection to a LAN. In addition to hardware, there are also several software component settings that can have a major effect on the performance of an NT-based computer system. Two of those software settings are the use of virtual memory and the operation of a screen saver. Concerning the latter, one of the most overlooked facts about computers is the fact that the use of a screen saver requires CPU cycles. This means that the simple act of removing a screen saver from a heavily utilized server will enhance its performance.

Unfortunately, the simplicity associated with enabling and disabling the use of a screen saver does not carry over to other hardware and software changes. That is, while the decision to activate or disable a screen saver is a binary decision, the decisions associated with altering hardware or changing system properties are more complex. Thus, in many instances, one will want to consider making an initial hardware or software change and then use the NT Performance Monitor to assess the results of the change on computer performance. Recognizing the old adage that "the proof of the pudding is in the eating," one can now focus on the use of virtual memory and its potential effect on the performance of a computer.

VIRTUAL MEMORY

Both Windows NT Workstation and Windows NT Server are similar products, with the key difference between the two being the fact that the server version is optimized to support more users accessing its resources. Thus, although this discussion of virtual memory is applicable to both versions of Windows NT, it is more applicable for Windows NT Server because the use of virtual memory has a greater impact on a computer running multiple applications accessed by many persons than a computer operating multiple applications operated by one or a few persons.

Virtual memory is a term used to reference the use of disk storage as a temporary storage area for program code and other information normally stored in random access memory (RAM). When the RAM storage requirements of applications programs and various NT operating system modules exceed physical RAM, NT will temporarily swap data to virtual memory. When that information is required again, Windows NT will retrieve it from disk and, if necessary, swap other information to virtual memory. Although virtual memory swapping is transparent to the user, it has a significant effect on computer performance.

PERFORMANCE EFFECT

When Windows NT uses its virtual memory capability, it transfers data to and from a special file on the hard disk referred to as a virtual-memory paging file. This file is also commonly referred to as a swap file. The transfer of information to and from disk occurs at electromechanical speed, with the movement of disk read/write heads over an appropriate disk sector contributing to a major portion of the delay in reading from or writing to a disk. Although modern disk drives are relatively fast devices, they still operate at 1/50th to 1/100th the speed of computer memory in terms of data transfer capability. While paging will always adversely affect the performance of a computer, the size of the paging file on the computer can have a more profound impact. If the size of the paging file is too small for the amount of activity on the server, one can experience a "thrashing" condition, with the operating system repetitively reading and writing small portions of RAM to and from disk.

As this occurs, the performance of the computer will be significantly impacted because additional input/output (I/O) operations are occurring that contribute to delays in processing application data. For example, if operating a database application on the server, the paging operations will delay access to the database. While an individual paging operation will hardly be noticeable, when thrashing occurs the delays can become significant, especially if the server supports hundreds of employees. Although one could alleviate this condition via the installation of additional RAM, one could also considerably enhance the performance of the server by making an adjustment to the size of the system's virtual memory paging file. How this is done is the focus of the next section.

CHANGING THE PAGING FILE

Since the introduction of Windows NT Version 3.5, Microsoft has considerably changed the interface and components of the operating system. However, one component that has retained both its location and general properties is the System Properties dialog box in the NT Control Panel. Through NT Version 5.0, one can use the Performance tab in the System Properties dialog box to review the computer's current virtual memory page file settings as well as to change those settings.

Exhibit 48.1 illustrates the screen display of the System Properties dialog box selected from the Control Panel under Windows NT Version 5.0. Note that the middle portion of the dialog box displays the current size of the paging file and provides the ability to change the size of that file.

Under Windows NT, a virtual memory paging file is automatically created on the computer when the operating system is installed. The size of the paging file initially created is based on the amount of RAM installed in the computer, and is typically set by the operating system to equal that amount plus 12 MB. However, the size of the file also depends on the amount of available free space on the hard drive when the paging file is created. Thus, if one installs Windows NT over a prior version of the operating

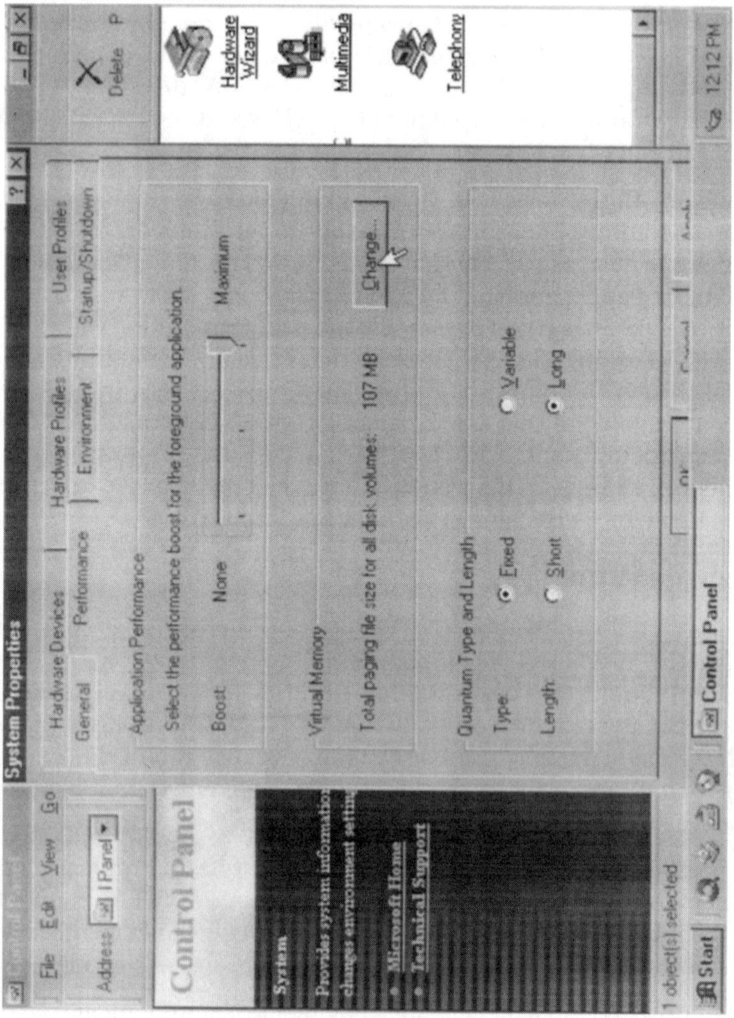

Exhibit 48.1 View the Size of the Windows NT Paging File from the Performance Tab in the System Properties Dialog Box.

system, or over a different operating system, and then changes the hard drive or removes obsolete files, the size of the paging file may vary from the Microsoft recommendation. In addition, the recommendation is merely a recommendation and may not be suitable for the traffic expected to be supported by the server. Thus, one can now examine how to change the size of the paging file and then observe changes in performance resulting from changing the size of the paging file.

Exhibit 48.2 illustrates the screen display resulting from selecting the button labeled "change" in the Performance tab in the System Properties dialog box. Note that the Virtual Memory dialog box displays the initial size of the paging file, which represents the recommended size of the file and its maximum size. The latter setting permits an area on disk to be reserved to enable the size of the paging file to grow. In this example, the paging file is shown set to 107 MB and can grow to 157 MB.

Although Microsoft's default settings for the computer's paging file will be sufficient for many users, it is a fact of life that very rarely are two computers used in a similar manner. This means that for some Windows NT users, an adjustment to the size of the paging file may be in order. However, how can one determine if sluggish performance is being caused by an improper allocation of paging file space, the inability of the computer's processor to support current operations, a lack of RAM, or another factor? The answer to this question can be obtained through the use of the Windows NT Performance Monitor which provides the key for evaluating the performance of an NT computer.

EVALUATING NT PERFORMANCE

Exhibit 48.3 illustrates the initial display of the NT Performance Monitor after the plus sign (+) button in the icon bar across the top of the display was clicked. That action results in the display of the dialog box labeled, "Add to chart," which enables one to display in realtime metrics associated with different counters, which are in turn associated with different NT objects. Under Windows NT, an object is a higher layer component, such as a processor or paging file. The counters represent performance elements whose values are tracked. If the computer contains two or more objects, such as a multiprocessor computer with two or more processors installed on the motherboard, each processor will have an instance associated with it. One would then be required to select the instance to identify the particular processor for which one wants a counter to be displayed.

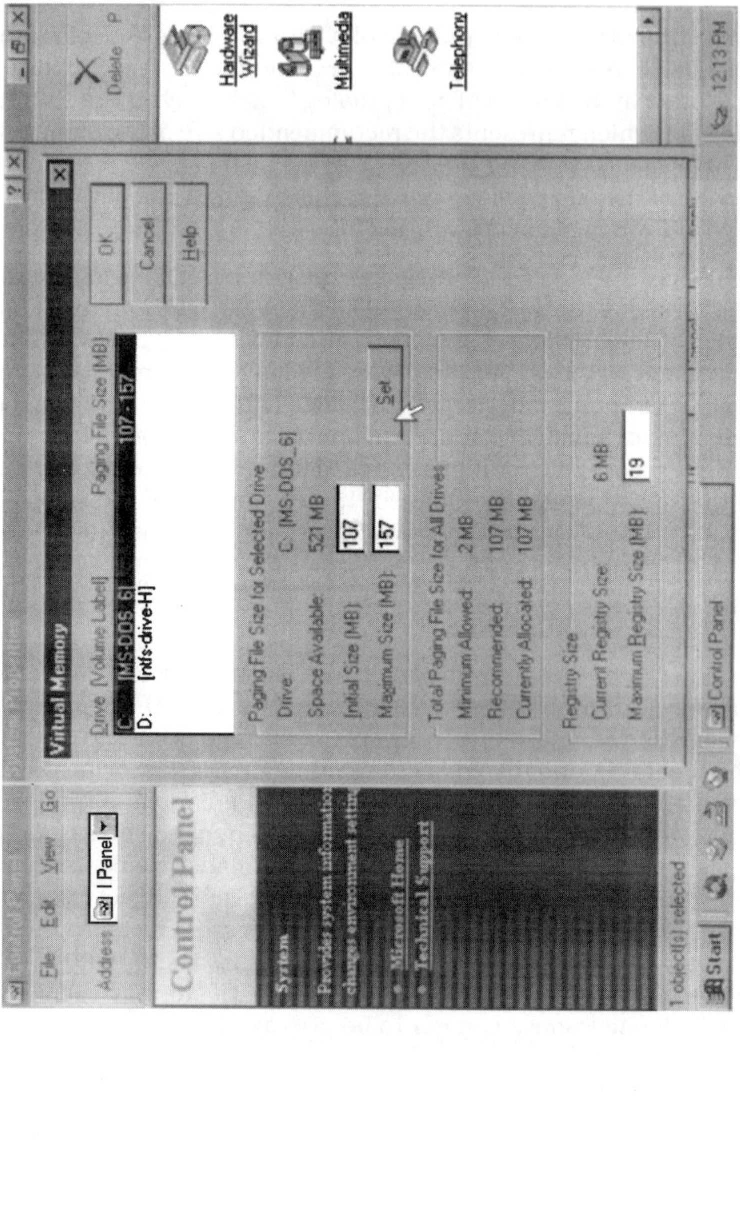

Exhibit 48.2 The Windows NT Virtual Memory Dialog Box Provides the Ability to Both View and Change the Size of a Computer's Paging File.

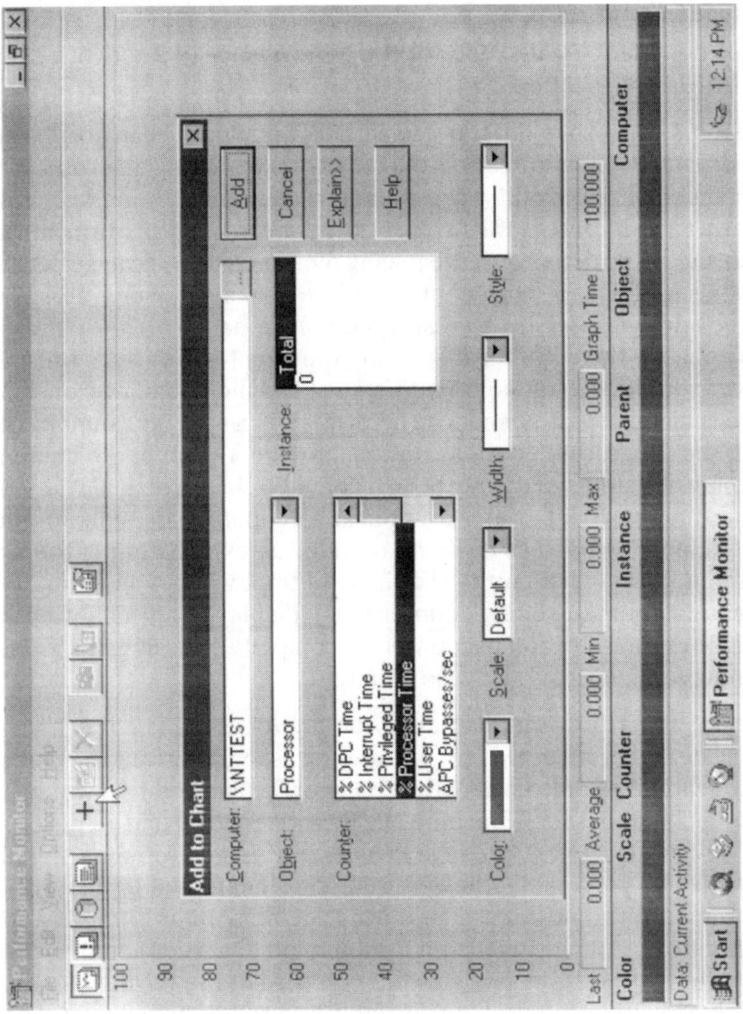

Exhibit 48.3 Selecting the % Processor Time Counter for Display in a Line Chart.

Returning to Exhibit 48.3, note that the counter "% Processor Time" was selected. This will result in the display of the percentage of processor time being used once the "Add" button is clicked. By observing this metric, one would be able to note whether or not the processor was a bottleneck. Also note that if that computer is connected to a network and one has an appropriate account on another NT computer, one can use Performance Monitor on the computer to monitor the performance of a distant computer. Thus, a supervisor or user with a series of appropriate accounts on different computers could periodically monitor the performance of a network of NT computers from one location.

The previous concern about the size of the NT paging file can now be addressed. Exhibit 48.4 illustrates the selection of the "Paging File" object in the Performance Monitor utility program. In examining Exhibit 48.4, note that Performance Monitor does not provide a direct indication of the amount of usage associated with the paging file. Instead, one can select the % Usage, % Usage Peak, or both counters to be displayed over a period of time. As one adds counters to be displayed, Performance Monitor will automatically change the color for displaying the counter; however, one can elect to override the default color with a specific color. Similarly, one can change the scale line width used for the display of a specific counter and the style of the line. Although those options provide the ability to customize the display of a series of counters on a common screen display, they do not alter the ability to display any counter, as all counters are built into the program and must be used as-is. If one requires an explanation of the use of a counter, one can click on the "Explain" button, which will result in a brief explanation of the use of a counter being displayed at the bottom of the screen. Once the selection of counters is complete, pressing the "Done" button results in the display of those counters.

Exhibit 48.5 illustrates the display of the % Processor Time, % Usage, and % Usage Peak, the latter two counters being associated with the paging file object. A careful examination of the colors of the lines plotted would indicate that the % Processor Time periodically spiked to 40 and 60 percent, while the percent usage and % Usage Peak associated with the paging file never exceeded 10 percent. If you were running a mixture of typical applications and monitored performance during the busy hour (which represents the hour of the day with the highest amount of server activity), this would indicate that one can safely reduce the size of the paging file. This could be valuable information that translates into the bottom line of an organization if it is low on available disk space and wishes to defer the downtime associated with the installation of a new drive.

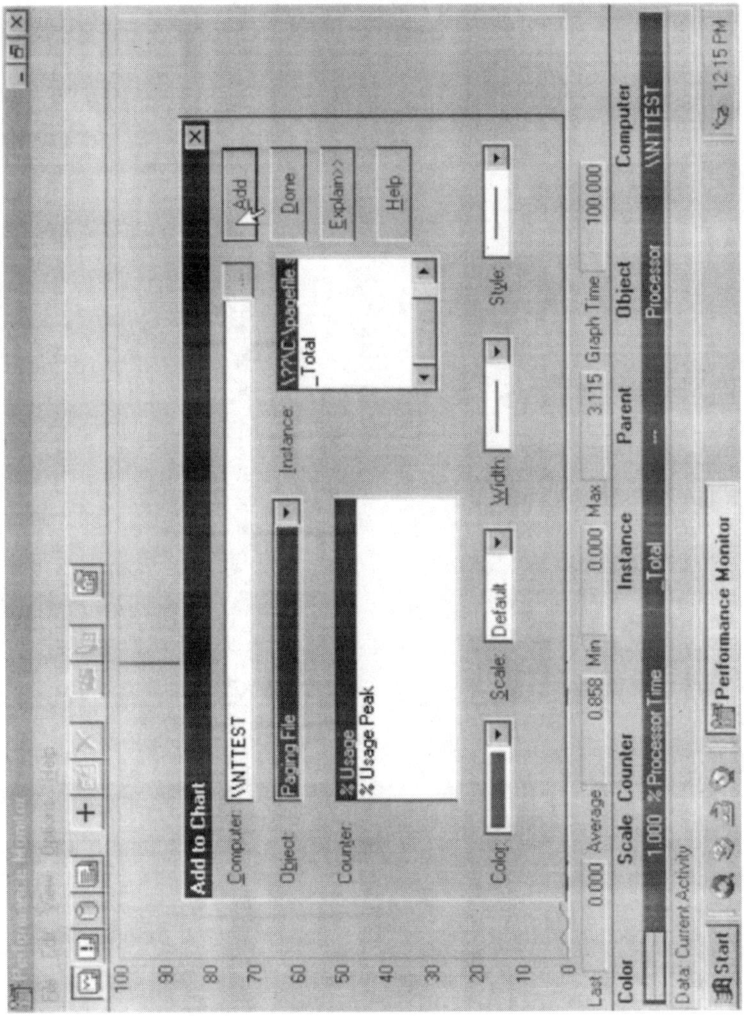

Exhibit 48.4 Selecting the Performance Monitor Paging File Object Permits the Display of Two Counters Associate with That Object.

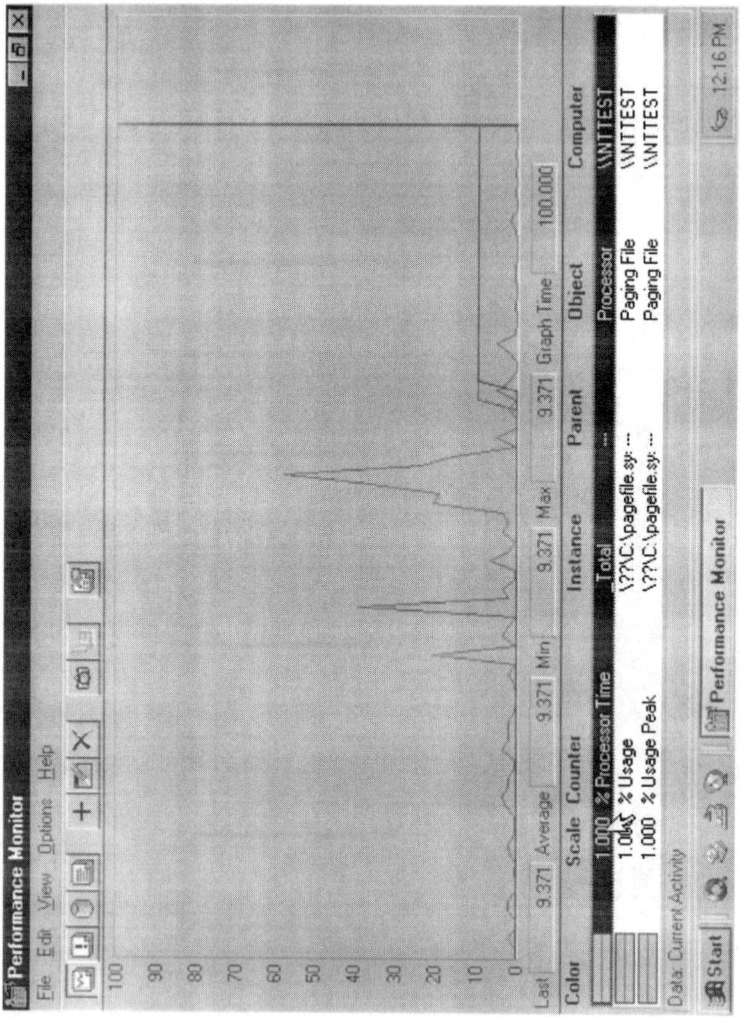

Exhibit 48.5 Using Performance Monitor to Display the Values of Three Counters.

As indicated in this chapter, Performance Monitor provides one with the ability to view the use of different hardware and software components. By using Performance Monitor, one can observe the existing performance of hardware and software as a mechanism for the adjustment of computer components. Then one can reuse Performance Monitor to evaluate the effect of those changes as well as to periodically monitor the performance of the computer. In doing so, one may be able to note potential problems before they occur, and initiate corrective action prior to the user community experiencing the effect of performance bottlenecks.

Section VIII
Security

Chapter 49
Secure Access to Mission-Critical Applications

David Johnson

EVERY DAY, MIS DEPARTMENTS IN ORGANIZATIONS AROUND THE WORLD WONDER HOW THEY CAN MAXIMIZE THE POTENTIAL OF THE INTERNET. One of the biggest obstacles still holding them back is their concern over security. Organizations face new challenges when attempting to interface legacy systems with the Web. Whether it is an internal intranet or external access for employees or business partners over the Internet, a series of security measures and policies can be put in place to ensure secure access to these mission-critical applications. This article will look at several levels of security that companies should address to help alleviate concerns over utilizing the Internet for business-critical activities. Areas to address include network integrity, authentication services, communication integrity, and protection of the information flowing across the network.

WHY USE THE INTERNET?

In today's information society, timely access to information can be critical. Many organizations adopted the IBM mainframe as their mission-critical computing platform for its reliability and transaction processing capability. Not only is processing the information key, but also getting it into the hands of those who can act on the information quickly. The Internet offers a low-cost network that can be used to share information with customers, business partners, or remote employees. That free flow of information can greatly increase a company's competitive edge. Great care must also be taken to keep this business-critical information from unwanted eyes. With the right security measures in place, the risk of someone accessing unauthorized information can be greatly reduced. This article focuses on how to provide secure access to mission-critical applications predominately running on enterprise IBM mainframes. While IBM mainframe applications are the focus, the security discussions relate to almost all mission-critical platforms.

THE CHALLENGES

Companies face several obstacles in relation to older security models that have sufficed up until now. In the past, most organizations utilized leased lines for communication to mission-critical applications. Local area networks (LANs) were used only for sharing devices such as printers; 90 percent of mission-critical applications were in use on the IBM mainframe. The mainframe security model was simplistic in nature, but very powerful. Dumb terminals were attached directly to a communication controller that was in turn directly attached via a leased line or hard-wire connection to the mainframe. A number of companies took this one step further and only authorized certain applications or user access from specific terminals. This is what is known as Logical Unit (LU) Nailing. A specific user's session or ID was "nailed" to a specific terminal. The only way to eavesdrop on an application conversation was to physically tap into the leased line or controller. Physically tapping into a phone line, while not impossible, was difficult. The eavesdropper would also have to record the session and play it back through a special decoder program designed to filter SNA traffic. Because all terminal traffic for a given controller was multiplexed across a single line, it could mean hundreds or thousands of packets to filter through to find any useable information.

This communication or session level of security was coupled with a series of host-based security products such as Resource Allocation Control Facility (RACF), ACF-2, or TopSecret, to name a few. These host-based measures were centrally controlled and administered. They determined what resources a user or a program being run by a user could access. One user may be able to view records in a DB/2 table based on his or her authorization. Another may have complete read/write privileges over that same information. This access control could also be location dependent. In some instances, access to top secret information can only be done from the terminals physically located in a secure room. These central control facilities allow MIS staff to easily monitor any unusual activity and greatly protect assets from both other applications running on the mainframe, and unauthorized users.

While these security methods are still available, the majority of companies have moved away from terminals to networked PCs, and from leased lines to switched networks based on frame relay, X.25, or asynchronous transfer mode (ATM). Most users no longer use SNA all the way to the desktop. An SNA gateway converts the traffic to IP, IPX, or some other LAN-based protocol. Also, LANs use shared media. There is one physical wire that all the devices share. To help troubleshoot this shared medium, numerous LAN-based protocol analyzers have been developed over the years. These analyzers make it quite easy to observe or record all the information flowing across the network.

Organizations have also developed remote access capabilities to support remote and mobile employees. Early solutions relied on user ID passwords and even dial-back methods to help protect the integrity of the network. With efforts to continue to reduce costs, many companies are looking to the Internet as a means of providing extremely low-cost dial access and communications with business partners. Mobile employees using the Internet for a remote access network make traditional dial-back solutions unfeasible.

While there are a number of new concerns for MIS departments as the world moves into a network-centric infrastructure, there are methods available to "plug" potential security holes and open access to authorized users. To provide access to one's mission-critical applications to business partners or other users via the Web, there are four main areas to focus on when providing access. These are:

- Network integrity
- Authentication/access control
- Application integrity
- Data protection

The remainder of this chapter takes a look at each of these areas in detail and provides an overview of the different options and methods to secure each area in the network.

Network Integrity

The first area to focus on is protecting the underlying network itself. Unauthorized users often try to gain access by locating and collecting information flowing over the network. While a number of companies have moved to dynamically assigning IP addresses to workstations, most servers and other fixed devices use a permanent IP address. This is often required so that users can always quickly find resources such as file servers, application servers, and printers. A fixed IP address also facilitates troubleshooting access to those systems. A skilled individual with access to a network can begin to learn the IP addresses of critical servers, sockets they communicate on, and other information.

Network integrity involves providing security at the network layer. For example, network integrity means preventing unauthorized users from using an Internet connection to gain access to one's internal networks. Firewalls are the most typical deployed method to prevent unauthorized access to the network. Firewalls are available in several varieties. While some vendors combine several functions into a single offering, the most common functions include:

- Filtering firewalls
- Application firewalls

- Proxy firewalls
- SOCKS firewalls

Some organizations have taken a shortsighted approach by effectively putting a Berlin Wall around their network and allowing no traffic in or out. While this might provide a sense of security, it is extremely limiting. The competitors of these shortsighted organizations are leveraging existing security solutions to provide access to information to business partners and customers. Those companies that do not provide access may feel safe, but they will suffer from competitors that offer better services to their customers.

Filtering Firewalls. Filtering firewalls are the most basic. They filter information packets based on a destination or a source address. A filtering firewall could be used to only allow outbound traffic from a certain IP or range of IP addresses. It could only allow inbound traffic from trusted IP domains. While filtering firewalls were useful when remote access and router networks were first established, they have become a bottleneck in today's network. With Dynamic Host Configuration Protocol (DHCP), remote access via an ISP, and other dynamic IP assignments, it is cumbersome — if not impossible — to update configurations to take into account all these new source or destination addresses. Most new solutions combine packet filtering with more advanced rules to protect the network.

Application Firewalls. A more advanced filtering firewall is an application firewall. Application firewalls are designed to only allow certain application traffic to pass its boundaries. They can be configured to allow different access, depending on whether the information is inbound or outbound. For example, a company may allow its employees to send files using File Transfer Protocol (ftp), but not allow anyone to receive files by the same manner. Other common application filters include telnet access, e-mail, and HTTP traffic. Each application can have a separate set of access rules applied to it. Application firewalls can prevent an unauthorized user from gaining access via a custom protocol. Users are forced to work within the confines of an allowed application protocol. Application firewalls can include the source or destination address as part of the rules-based access. Exhibit 49.1 shows an example of an application firewall in which business partners can access the company's mainframe via TN3270, but internal users cannot access external mainframes. External users can also send and receive mail (SMTP) and do web transactions (HTTP), but they cannot telnet into the mainframe or do file transfers to or from it. Application firewalls typically do not provide authentication services.

Proxy Firewalls. Proxy firewalls begin to provide a level of authentication. Users are normally asked to provide an ID and password to access systems via the proxy server. The main function of a proxy server is to hide

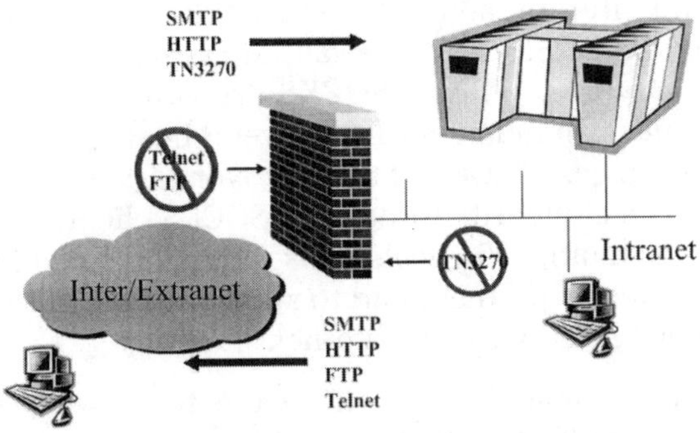

Exhibit 49.1 Example of an Application Firewall.

the IP addresses of internal users accessing outside systems and the IP addresses of internal systems. Most attempts at breaking into a system require the knowledge of its IP address. IP addresses are like telephone numbers: one needs to know how to contact the machine or person one wants to speak with.

Assume that a company allows its employees to send and receive files over the Internet using ftp via a proxy server. While each internal user has a unique IP address, all users collectively appear as a single IP address, that of the proxy server, to the outside world. Users requesting to do an ftp transfer are routed to the proxy server; the proxy server takes the user's requested transfer and replaces the source address with its own. The proxy server then sends the packets on to the destination. When the remote host sends the information back to the user, it uses the proxy server's IP address as the destination. The proxy server receives the packets and then forwards them on to the user's PC. Throughout the entire ftp transmission, the remote machine has never established a direct communications link with the internal user machine. The remote machine not only does not know the IP address of the actual user, but cannot "piggyback" an attack on the communication because the proxy server is only sending valid ftp packets back to the user.

Proxy firewalls support most common applications, including ftp, telnet, HTTP, SMTP, etc. Companies can require their users to "log on" to the proxy server, or preconfigure the connection in the browser. Both Netscape and Microsoft browsers support proxy configurations.

SOCKS Firewalls. Each successive generation of firewalls continues to build upon previous security systems. SOCKS version 5 is the latest generation of firewall technology. Not only does it support application, filtering, and proxy support, it also adds authentication services, encryption capabilities, and interoperability between different vendors' SOCKS firewalls based on RFC 1928. SOCKS firewalls are typically used to deploy Virtual Private Networks (VPNs). SOCKS firewalls can also be used to establish extranet communications between business partners. One of the biggest advantages of SOCKS firewalls over earlier implementations is that SOCKS v5 is a standard. Authorized SOCKS clients can dynamically negotiate through multiple firewalls. This reduces the complexity; prior to the standard, business partners had to statically configure their firewalls to match the configurations of each other.

Suppose there is a manufacturer that wants to allow its suppliers to access a mainframe application that tracks the purchase orders for the supplier. Instead of the supplier having to call the manufacturer every time to check on order status, the supplier accesses the application and looks up the information directly. In order to protect their networks, both companies have implemented a SOCKS firewall. Each company has its own access and control policies, and each firewall is configured differently. The supplier employees who need access to the manufacturing application are given SOCKS client software. This client software allows the supplier to negotiate an outbound connection through the supplier's firewall and then negotiate an inbound connection through the manufacturer's firewall. No configuration details are needed for either firewall. The SOCKS protocol handles the negotiation and authentication process transparently to the end user or the mainframe application.

While SOCKS firewalls are the latest generation of technology to protect the integrity of the network, they still have some drawbacks. SOCKS firewalls are not yet widely deployed. They make up about ten percent of the installed base. In order for the client negotiation to work, SOCKS client software must be installed on every desktop that needs access. This can cause a maintenance and distribution headache for network administrators because they now have an additional software component on every desktop that must be maintained. A number of application vendors have recognized this and are adding SOCKS client support directly into their applications, providing users with the benefits of a SOCKS firewall without requiring separate installation and maintenance of SOCKS client software.

Firewalls are the most common and one of the best ways to protect the integrity of one's network. In order to facilitate access, one needs to allow authorized users to access only those resources for which they have permission. This brings us to the topic of authentication and access controls.

Authentication/Access Controls

Access controls are used to authenticate users and allow them to access only certain resources. These resources could be an internal ftp server, a specific mainframe, or a specific mainframe application. Or, one might want to restrict access to standard working hours. Although one might allow business partners or customers access to certain applications or resources, one does not want them to have full reign over one's internal systems. Access controls involve two major components: authentication and resource permissions. Both of these areas were solved long ago in the IBM mainframe environment. Unfortunately, tools like RACF or ACF-2 cannot really be used to control and authorize users for network-based resources.

RACF is one of the most well-known systems for providing authentication services and access controls for mission-critical systems. When a user is authenticated by RACF, that user has a permission profile that is activated. This profile not only contains what areas or applications on the mainframe system(s) the user is allowed to access, but also what resources and portions of data records that the applications the user is running can access. While this system provides a high degree of control over the environment, a user must log on to the mainframe for it to be active. RACF cannot protect or control access to nonmainframe applications. And, because the mainframe applications are mission critical, one will want to make sure an unauthorized user does not even get to the mainframe logon screen. RACF, while very powerful, is still a user-ID-and-password-based system. Once a password is compromised, the security is gone. What one needs to implement is an RACF-style authentication system for one's network. The two functions that must be provided by the network are authentication services and access controls.

Authentication Services. Most early solutions solely relied on user-IDs and passwords. This is authentication in its simplest form. This is also the easiest type of system to break. Due to the large number of systems that a typical user needs to access via a password, most pick short, easy-to-remember passwords. For example, many choose a child's name, date of birth, or other common item. Trying to remember which password was for e-mail, which was for mainframe access, etc. was a chore the common user reluctantly was forced to deal with. Some companies are successful in forcing their users to change their passwords every 30 to 90 days, while others gave up that requirement due to user revolt.

There have been several improvements to simple ID-and-password-based systems. One of the popular methods is to use a hardware token or card to generate a new password every time. One of the popular implementations of this was SecureID. SecureID consists of an authentication server at the central office and a hardware card. This hardware card is synchronized with the server and generates a four-digit number every 30 to 60 seconds. This number is the user's password. A log-on dialog box is displayed to the user. The user enters the correct ID and the password displayed on the SecureID card. The SecureID server validates the user and current password, and then grants access to the network.

While this method eliminates the problem of easily guessed passwords, it does have some drawbacks. For starters, users have to rely on a hardware card. While the cards are not expensive, if a user loses the card, he loses network access. This does happen, especially with traveling users. Also, the default for most token-based systems is to generate a new password every 30 seconds. If the server takes longer than 30 seconds to authenticate the user, or the user attempts to log on after 29 seconds have passed, the user may have to attempt a connection several times before he is successful. While MIS departments may feel this is a small price to pay for security, tell that to the CEO or Vice President of Sales after having to make several attempts after a long day of meetings. One final drawback with this system is that the network-based application server must support the system. While most popular systems such as Windows NT and UNIX support SecureID, not all operating environments or web servers do.

The newest authentication systems involve a digital signature. A recognized certificate authority, such as Verisign, can generate this digital signature or the company can generate it. The most common type of digital signature is an X.509 certificate (cert) used for secure socket layer authentication. This certificate is an electronic stamp or file that is stored on the local PC. The most common use of a cert is for browser authentication. Both Microsoft Internet Explorer and Netscape Communicator support authentication to a web server using certs. When accessing a network via the web server, the server queries the browser for its cert. The server looks up the cert to see if the user is authorized for access. This mechanism solves the password issues by storing the digital signature. The only drawbacks are that the cert is tied to a physical machine and that not many systems other than web servers support digital certificates yet.

Companies and vendors are addressing this latter issue by utilizing enterprise directory services. The directory service that is quickly becoming the most popular is Lightweight Directory Access Protocol (LDAP), which falls into the category of access controls.

Access Controls. There are several network-based access control systems available today. These include RADIUS, TACACS+, and even some custom-developed systems. RADIUS servers have become very common for remote access authentication systems. They cannot only authenticate a user, but some control information can be configured. This includes the time of day or days of the week a user can dial in, how long the user can be connected, etc. RADIUS servers became very common in UNIX and remote access environments. A standard was developed so that systems can support RADIUS servers from multiple vendors. RADIUS was the first attempt at having a centralized authentication and access server. While RADIUS offers a number of good features for access controls, it is not flexible enough or widely supported enough to become a true enterprise authentication server. TACACS+ was Cisco's implementation of an authentication and access control server for its remote access solutions. TACACS+ offers the same level of services as RADIUS, but is specific to a Cisco environment.

The industry realized quite some time ago that an enterprise directory would be needed. The first attempt at an enterprise standard was the X.500 directory standard. This was a very complex standard that attempted to define all possible options. It was very complex, added a lot of overhead, and was very difficult to implement and administer. As a result of this complexity and the lack of real requirements at that time, most applications did not offer X.500 support. Those that did were unwieldy, slow, and not well received by the user community.

In addition to the complexity, X.500 directory services did not gain popularity because it was not until recently that the network has moved beyond its role as a communications infrastructure to become the focal point for access. Users now wish to be able to log on once and then have access to all the applications for which they are authorized. This desire, coupled with new developments in directory services, now make "single signon" a reality.

LDAP. LDAP defines an interface to locate network services. Authentication and access controls are part of network services. As mentioned earlier, LDAP is not the first network directory to be introduced. Along with X.500, Novell has developed Network Directory Services (NDS) and Microsoft has been promoting its own undelivered Active Directory. Both offerings from Novell and Microsoft are designed to work best in their specific network environments. To be fair to Novell, it has gone to great lengths to make NDS available for non-NetWare platforms. The proprietary nature of these directory services is one of the main reasons that they have not gained popularity. User organizations have decided to follow standards-based approaches whenever possible. This is especially true for web-based solutions. LDAP is a public standard. There are a variety of places that one can even get the source code to LDAP royalty free. This gives organizations the choice of purchasing a commercial LDAP server or maintaining their own source code.

Because LDAP defines an interface and not necessarily the underlying structure, organizations can define what information they want to store in their central repositories. User information can be stored in a central LDAP server. Applications can be "LDAP-aware." With LDAP, organizations can finally begin to regain the centralized control they used to have via mainframe systems like RACF. A user can log on to the network and gain authentication via an LDAP server. That logon process could also connect authorized users to their main applications by reading the user ID and passwords for those applications from the LDAP directory. One of the

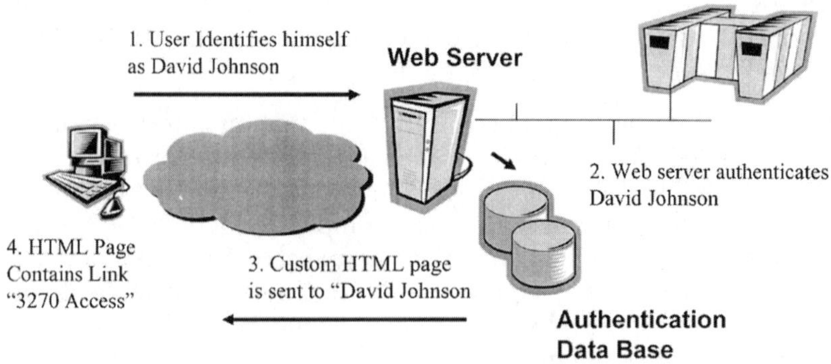

Exhibit 49.2 Example of a User.

biggest benefits to the user community is the ability to achieve single logon. From an MIS point of view, there is only one place to administer and maintain user information. Instead of going to the e-mail server, authentication server, file server, etc. to add, change, or delete a user, all this can be done from one central location.

User profiles can be stored in this central repository and utilized by LDAP-aware applications. In a mainframe access environment, information such as the SNA gateway address, destination mainframe, and logical unit name are used to provide connectivity. The user's mainframe password and ID can be sent to the mainframe automatically. Other user-configured items such as emulator keyboard and color map can also be stored and managed centrally. Exhibit 49.2 shows an example of a user David Johnson gaining authentication and access control using LDAP servers. In this case, a customized HTML page is built and sent based on David Johnson's unique set of parameters stored in the LDAP server.

LDAP not only provides a standards-based method for authentication and access control, it allows administrators the ability to recentralize user profiles and other information. Not all applications are LDAP-aware. Most successful network application vendors have realized the power of centralized management. More and more solutions that support LDAP are available. With the finalization of LDAP 3.0, most vendors will soon be offering LDAP support in their applications. IBM has even announced a version of LDAP 3.0 that will run on OS/390.

Application Integrity

The last major difference between enterprise mainframe systems and the Inter/intranet is the method of communication. 3270 applications were designed with a terminal and controller at one end of a leased line and the mainframe via a front-end processor at the other, using SNA as the communication protocol. SNA is a state or persistent type of connection. The communication link is established for the duration of the application session. This is analogous to a standard telephone conversation — one calls somebody and then has an established connection until hangup. Most 3270 applications are long-running conversations.

The HTTP protocol, utilized by the World Wide Web and corporate intranets, is a stateless protocol with the browser controlling the conversation. The browser asks the web server for some information and then disconnects. Using the phone call analogy, this would be similar to:

1. Dialing the number, asking how the person is doing, getting a response, and then hanging up
2. Dialing the number, asking if the person is available for dinner, getting a response, and hanging up, etc.

This is an awful lot of overhead for a conversation and, worse, the potential is there for someone else to easily dial up and break into the conversation.

Providing secure browser-based access to SNA applications must take this issue into account. It is possible, using technology such as Java or ActiveX, to recreate the SNA-style persistence and supply either a character or graphical interface to the existing SNA application. By using TCP socket technologies, applications developed in Java or ActiveX can create a state-oriented connection. Both Java and ActiveX allow a character-style interface or a full graphical interface to be used. Both solutions can also encrypt all the information that is flowing to and from the SNA Web server. This ability allows one to explore the possibility of the public Internet as an access method or a low-cost remote access network for employees, business partners, and even customers.

Data Protection. The ability to keep corporate information safe from prying eyes while traveling across a TCP/IP infrastructure is paramount in today's exposed environment. This is one area where web-based solutions have completely surpassed traditional emulation. TN3270(E) has exposed a potentially serious security hole in the enterprise. TN3270(E) sends all information in simple clear text. It would not take long for a user with a shareware network "sniffer" program to collect mainframe user IDs and passwords. Exhibit 49.3 shows a network trace done using a standard shareware program from the Internet. The first is a trace of a standard TN3270E connection to the mainframe, clearly showing the user ID as AU09615 and password as TSCHAULE. Exhibit 49.4 shows the same connection using a web-based solution that provides built-in encryption. The user ID and password are impossible to decipher.

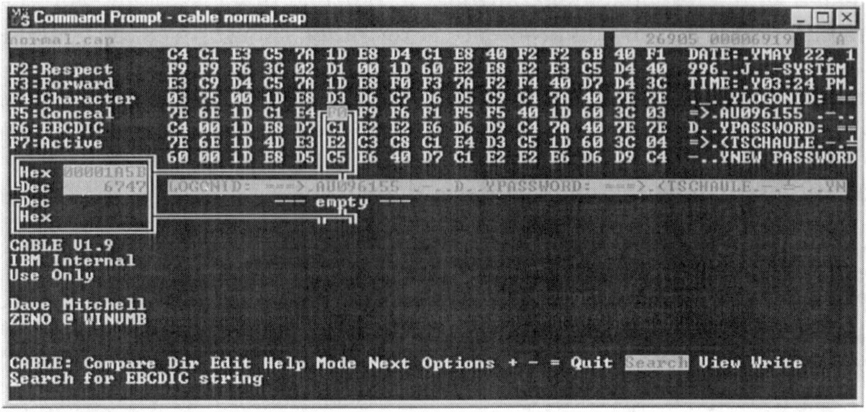

Exhibit 49.3 Network Trace Using a Standard Shareware Program from the Internet.

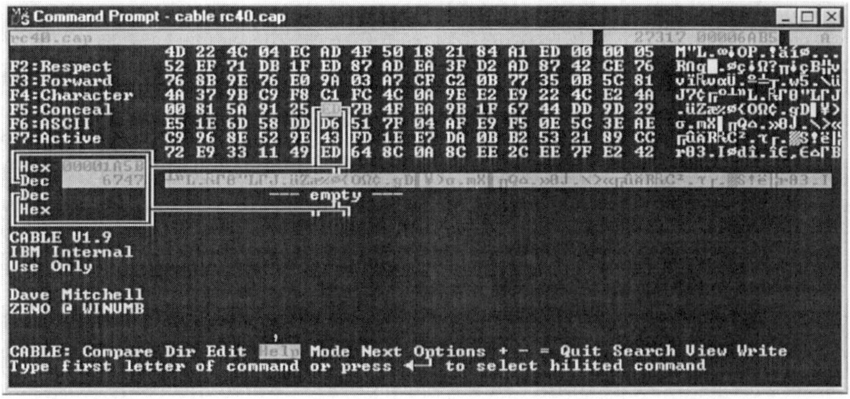

Exhibit 49.4 Connection Using a Web-Based Solution.

One used to frequently hear the phrase, "It's our internal network; security is not an issue." Most people do not realize that research shows 80 percent of all break-ins are by internal users. Who would have more to gain from access to your mainframe, the 13-year-old hacker looking for a cool game or the disgruntled employee who "sniffs" the accounting user IDs, passwords, and account numbers, looking for an unauthorized pay raise?

Exhibit 49.5 Access Control Checklist.

✔	Type of Security	Why a Requirement?
___	Firewalls in place providing at least filtering	Prevent unauthorized external access
___	User authentication and access controls in place, at a minimum user ID and password	Ensure that only authorized users gain access to the systems and network required to do their jobs
___	Plan in place to integrate network-based authentication/access control (e.g., LDAP) with legacy systems	Provide comprehensive and seamless single logon with reduced administration and high level of security
___	Web-to-host solutions ensure SNA-style persistent sessions	Prevent unauthorized break-in to the middle of an SNA session
___	All new mainframe access is encrypted	Prevent internal and external attacks on mainframe systems

CONCLUSION

This article has discussed several options to help secure one's network but still provide access to these mission-critical applications for authorized users such as mobile employees, business partners or suppliers, and even customers. Implementing security in areas such as network integrity, authentication/access control, application integrity, and data protection can protect one's network and systems, but still provide the flexibility needed for authorized access. When evaluating options for providing Internet access to mainframe systems, use Exhibit 49.5 as a set of minimum requirements.

With adequate protection in these areas, an organization can begin to utilize the Internet to be more competitive. Benefits can range from offering 7·24 customer service applications to realtime dissemination of information to business partners and suppliers. Financial benefits consist of both internal cost reductions and increased revenue, and customer satisfaction that comes with realtime information access. Secure, controlled access to "glass house" applications can bring enormous benefits to an organization. There are solutions and systems available today to make that secure access a reality.

Chapter 50 Applets and Network Security: A Management Overview

Al Berg

APPLETS, NETWORK-BASED PROGRAMS THAT RUN ON CLIENT SYSTEMS, ARE ONE OF THE NEWEST SECURITY CONCERNS OF NETWORK MANAGERS. This chapter describes how applets work, the threats they present, and what security precautions network managers can take to minimize the security exposures presented by applets.

INTRODUCTION

Applets are small programs that reside on a host computer and are downloaded to a client computer to be executed. This model makes it very easy to distribute and update software. Because the new version of an application only needs to be placed on the server, clients automatically receive and run the updated version the next time they access the application.

The use of applets is possible because of the increasing bandwidth available to Internet and intranet users. The time required to download the programs has been decreasing even as program complexity has been increasing. The development of cross-platform languages such as Sun Microsystems, Inc.'s Java, Microsoft Corp.'s ActiveX, and Netscape Communications Corp.'s JavaScript has made writing applets for many different computers simple — the same exact Java or JavaScript code can be run on a Windows-based PC, a Macintosh, or a UNIX-based system without any porting or recompiling of code. Microsoft is working to port ActiveX to UNIX and Macintosh platforms.

APPLETS AND THE WEB

The World Wide Web is the place that users are most likely to encounter applets today. Java (and to a lesser degree, JavaScript) have become webmasters' tools of choice to add interesting effects to their web sites or to deliver applications to end users. Most of the scrolling banners and other special effects found on today's web pages depend on applets to work. Some Web pages use applets for more substantial applications. For example, MapQuest(http://www.mapquest.com) uses Java and ActiveX to deliver an interactive street atlas of the entire U.S. *Wired* magazine offers a Java-based chat site that, when accessed over the Web, allows users to download an applet that lets them participate in real-time conferencing.

The Security Issue

Every silver lining has a cloud, and applets are no exception. Applets can present a real security hazard for users and network managers. When web pages use applets, the commands that tell the client's browser to download and execute the applets are embedded in the pages themselves. Users have no way of knowing whether or not the next page that they download will contain an applet, and most of the time they do not care. The Internet offers an almost limitless source of applets for users to run, however, no one knows who wrote them, whether they were written with malicious intent, or whether they contain bugs that might cause them to crash a user's computer.

Applets and computer viruses have a lot in common. Both applets and viruses are self-replicating code that execute on the user's computer without the user's consent. Some security experts have gone as far as to say that the corporate network manager should prohibit users from running applets at all. However, applets are becoming an increasingly common part of how users interact with the Internet and corporate intranets, so learning to live safely with applets is important for network managers.

What Are the Risks?

According to Princeton University's Safe Internet Programming (SIP) research team, there have been no publicly reported, confirmed cases of security breaches involving Java, though there have been some suspicious events that may have involved Java security problems. The lack of reported cases is no guarantee that there have not been breaches that either were not discovered or were not reported. But it does indicate that breaches are rare.

As web surfing increasingly becomes a way to spend money, and applets become the vehicle for shopping, attacks on applets will become more and more profitable, increasing the risk. Sun, Netscape, and Microsoft all designed their applet languages with security in mind.

JAVA: SECURE APPLETS

Java programs are developed in a language similar to C++ and stored as source code on a server. When a client, such as a web browser, requests a page that references a Java program, the source code is retrieved from the server and sent to the browser, where an integrated interpreter translates the source code statements into machine-independent bytecodes which are executed by a virtual machine implemented in software on the client. This virtual machine is designed to be incapable of operations that might be detrimental to security, thus providing a secure sandbox in which programs can execute without fear of crashing the client system. Java applets loaded over a network are not allowed to:

- Read from files on the client system
- Write to files on the client system
- Make any network connections, except to the server from which they were downloaded
- Start any client-based programs
- Define native method calls, which would allow an applet to directly access the underlying computer

Java was designed to make applets inherently secure. Following are some of the underlying language security features offered by Java:

- All of an applet's array references are checked to make sure that programs will not crash because of a reference to an element that does not exist.
- Complex and troublesome pointer variables (found in some vendors' products) that provide direct access to memory locations in the computer do not exist in Java, removing another cause of crashes and potentially malicious code.
- Variables can be declared as unchangeable at runtime to prevent important program parameters from being modified accidentally or intentionally.

Java: Holes and Bugs

Although Sun has made every effort to make the Java virtual machine unable to run code that will negatively impact the underlying computer, researchers have already found bugs and design flaws that could open the door to malicious applets.

The fact that Sun has licensed Java to various browser vendors adds another level of complexity to the security picture. Not only can security be compromised by a flaw in the Java specification, but the vendor's implementation of the specification may contain its own flaws and bugs.

Denial-of-Service Threats. Denial-of-service attacks involve causing the client's web browser to run with degraded performance or crash. Java does not protect the client system from these types of attacks, which can be accomplished simply by putting the client system into a loop to consume processor cycles, creating new process threads until system memory is consumed, or placing locks on critical processes needed by the browser.

Because denial-of-service attacks can be programmed to occur after a time delay, it may be difficult for a user to determine which page the offending applet was downloaded from. If an attacker is subtle and sends an applet that degrades system performance, the user may not know that their computer is under attack, leading to time-consuming and expensive troubleshooting of a nonexistent hardware or software problem.

Java applets are not supposed to be able to establish network connections to machines other than the server they were loaded from. However, there are applets that exploit bugs and design flaws that allow them to establish a back-door communications link to a third machine (other than the client or server). This link could be used to send information that may be of interest to a hacker. Because many ready-to-use Java applets are available for download from the Internet, it would be possible for an attacker to write a useful applet, upload it to a site where webmasters would download it, and then sit back and wait for information sent by the applet to reach their systems.

What Kind of Information Can the Applet Send Back? Due to another implementation problem found in August 1996 by the Safe Internet Programming research team at Princeton University, the possibilities are literally endless. A flaw found in Netscape Navigator 3.0 beta 5 and earlier versions, and Microsoft Internet Explorer 3.0 beta 2 and earlier versions, allows applets to gain full read and write access to the files on a web surfer's machine. This bug means that the attacker can get copies of any files on the machine or replace existing data or program files with hacked versions.

Giving Java applets the ability to connect to an arbitrary host on the network or Internet opens the door to another type of attack. A malicious applet, downloaded to and running on a client inside of a firewalled system, could establish a connection to another host behind the firewall and access files and programs. Because the attacking host is actually inside the secured system, the firewall will not know that the access is actually originating from outside the network.

Another bug found in August 1996 by the Princeton team affects only Microsoft Internet Explorer version 3.0 and allows applets (which are not supposed to be allowed to start processes on the client machine) to execute any DOS command on the client. This allows the applet to delete or change files or programs or insert new or hacked program code such as viruses or backdoors. Microsoft has issued a patch (available on its web site at http://www.microsoft.com/ie)to Internet Explorer that corrects the problem.

Princeton's SIP team also found a hole that would allow a malicious application to execute arbitrary strings of machine code, even though the Java virtual machine is only supposed to be able to execute the limited set of Java bytecodes. The problem was fixed in Netscape Navigator 3.0 beta 6 and Microsoft Internet Explorer 3.0 beta 2.

JAVASCRIPT: A DIFFERENT GRIND

Netscape's JavaScript scripting language may be named Java, but it is distinct from Sun's applet platform. JavaScript is Netscape Navigator's built-in scripting language that allows webmasters to do cross-platform development of applets that control browser events, objects such as tables and forms, and various activities that happen when users click on an object with their mouse.

Like Java, JavaScript runs applications in a virtual machine to prevent them from performing functions that would be detrimental to the operation of the client workstations. Also like Java, there are several flaws in the implementation of the security features of JavaScript. Some of the flaws found in JavaScript include the ability for malicious applets to:

- Obtain users' E-mail addresses from their browser configuration
- Track the pages that a user visits and mail the results back to the script author
- Access the client's file system, reading and writing files

A list of JavaScript bugs and fixes can be found on John LoVerso's web page at the Open Software Foundation (http://www.osf.org/[sim]loverso/javascript/).

ACTIVEX: MICROSOFT'S VISION FOR DISTRIBUTED COMPONENT COMPUTING

Microsoft's entry in the applet development tool wars, ActiveX, is very different from Java and presents its own set of security challenges. ActiveX is made up of server and client components, including:

- Controls, which are applets that can be embedded in web pages and executed at the client. Controls can be written in a number of languages, including Visual Basic and Visual C++
- Documents that provide access to non-HTML content, such as word processing documents or spreadsheets, from a web browser
- The Java virtual machine, which allows standard Java applets to run at the client

- Scripting, which allows the web developer to control the integration of controls and Java applets on a web page
- The server framework, which provides a number of server-side functions such as database access and data security

Java applets running in an ActiveX environment (e.g., Microsoft's Internet Explorer web browser) use the same security features and have the same security issues associated with JavaScript. Microsoft offers a Java development environment(i.e., Visual J++) as well as other sandbox languages (i.e., VBScript, based on Visual Basic and JScript, Microsoft's implementation of Netscape's JavaScript) for the development of applications that are limited as to the functions they can perform.

When developers take advantage of ActiveX's ability to integrate programs written in Visual Basic or C++, the virtual machine model of Java no longer applies. In these cases, compiled binaries are transferred from the server to the web client for execution. These compiled binaries have full access to the underlying computing platform, so there is no reason that the application could not read and write files on the client system, send information from the client to the server (or another machine), or perform a destructive act such as erasing a disk or leaving a virus behind.

Using Authenticode for Accountability

Microsoft's approach to security for non-Java ActiveX applications is based on the concept of accountability — knowing with certainty the identity of the person or company that wrote a piece of software, and that the software was not tampered with by a third party. Microsoft sees the issues related to downloading applets from the Web as similar to those involved in purchasing software; users need to know where the software is coming from and that it is intact. Accountability also means that writers of malicious code could be tracked down and would have to face consequences for their actions.

The mechanism that Microsoft offers to implement this accountability is called Authenticode. Authenticode uses a digital signature attached to each piece of software downloaded from the Internet. The signature is a cryptographic code attached by the software developer to an applet. Developers must enter a private key (known only to them) to sign their application, assuring their identity. The signature also includes an encrypted checksum of the application itself, which allows the client to determine if the applet has changed since the developer released it.

ActiveX: The Downside

This approach provides developers and users with access to feature-rich applications, but at a price. If an application destroys information on a user's computer, accountability will not help recover their data or repair damage done to their business. Once the culprit has been found, bringing them to justice may be difficult because new computer crimes are developing faster than methods for prosecuting them.

Microsoft acknowledges that Authenticode does not guarantee that end users will never download malicious code to their PCs, and that it is a first step in the protection of information assets.

Further information on ActiveX can be found on Microsoft's web site (http://www.microsoft.com/activex)and at the ActiveX Web site run by CNet Technology Corp. (http://www.activex.com).

AN OUNCE OF PREVENTION

So far, this chapter has discussed problems posed by applets. Following are some steps that can be taken to lessen the exposure faced by users.

Make Sure the Basics Are Covered

Users need to back up their data and programs consistently, and sensitive data should be stored on secure machines. The surest way to avoid applet security problems is to disable support for applet execution at the browser. If the code cannot execute, it cannot do damage.

Of course, the main downside of this approach is that the users will lose the benefits of being able to run applets. Because the ability to run applets is part of the client browser, turning off applets is usually accomplished at the desktop and a knowledgeable user could simply turn applet support back on. Firewall vendors are starting to provide support for filtering out applets, completely or selectively, before they enter the local network.

Users Should Run the Latest Available Versions of Their Web Browsers

Each new version corrects not only functional and feature issues, but security flaws. If an organization is planning to use applets on its web pages, it is preferable to either write them internally or obtain them from trusted sources. If applets will be downloaded from unknown sources, a technical person with a good understanding of the applet language should review the code to be sure that it does only what it claims to.

Mark LaDue, a researcher at Georgia Tech, has a web page (available at http://www.math.gatech.edu/[sim]mladue/HostileApplets.html) containing a number of hostile applets available for download and testing. Seeing some real applications may help users recognize new problem applets that may be encountered.

CONCLUSION

IS personnel should monitor the Princeton University Safe Internet Programming group's home page (located at http://www.cs.princeton.edu/sip)

for the latest information on security flaws and fixes (under News). It is also a good idea to keep an eye on browser vendors' home pages for news of new versions.

Applets offer users and network managers a whole new paradigm for delivering applications to the desktop. Although, like any new technology, applets present a new set of challenges and concerns, their benefits can be enjoyed while their risks can be managed.

Chapter 51
Intranet Security

John R. Vacca

ENTERPRISE COMPUTER NETWORKS ARE HARDLY NEW. Nevertheless, when setting up linked enterprisewide computer systems, enterprises are increasingly taking their cues from the Internet's multimedia wing — the World Wide Web. Servers connected to PCs across vast distances, as shown in Exhibit 51.1, have migrated into what are now known as intranets.

Intranets are private networks that mirror the structure of the Web. Using a standard Internet language called TCP/IP (Transmission Control Protocol/Internet Protocol), this emerging intranet technology provides an alternative method for disparate computer systems (such as Macintosh, Windows, or UNIX workstations) that are within enterprises to communicate with one another. This technology has resulted in improved enterprise communications. Employees using PCs equipped with web browsers can now access databases housed on servers anywhere in the enterprise network. This same technology also enables users to collaborate on documents and other projects. While e-mail is available in other ways, one primary use for intranet technology is e-mail. E-mail greatly speeds up the distribution of information to employees, whether from one person to another or from headquarters to an enterprise's thousands of employees. Generally, these intranet systems are also connected to the Internet (the Net); thus, enterprise users can also have access to the vast amount of resources it offers — for enterprise purposes only, of course.

Unfortunately, the world and the people who inhabit it are not perfect. Nor are all those people honest and incorruptible. For that reason, intranets cannot be left open to just anyone who wants access to them. In order to let authorized users in and to keep outsiders out, enterprises secure their private data on the intranet by means of firewalls.

Enterprise interest with regard to intranets is increasing substantially for several reasons:

- First and foremost, intranets enable enterprises to receive a single, consistent flow of information, which is an enterprise lifeline that provides data on everything from departmental budgets to 401(k) plans. In other words, with an intranet, one makes all the data accessible to

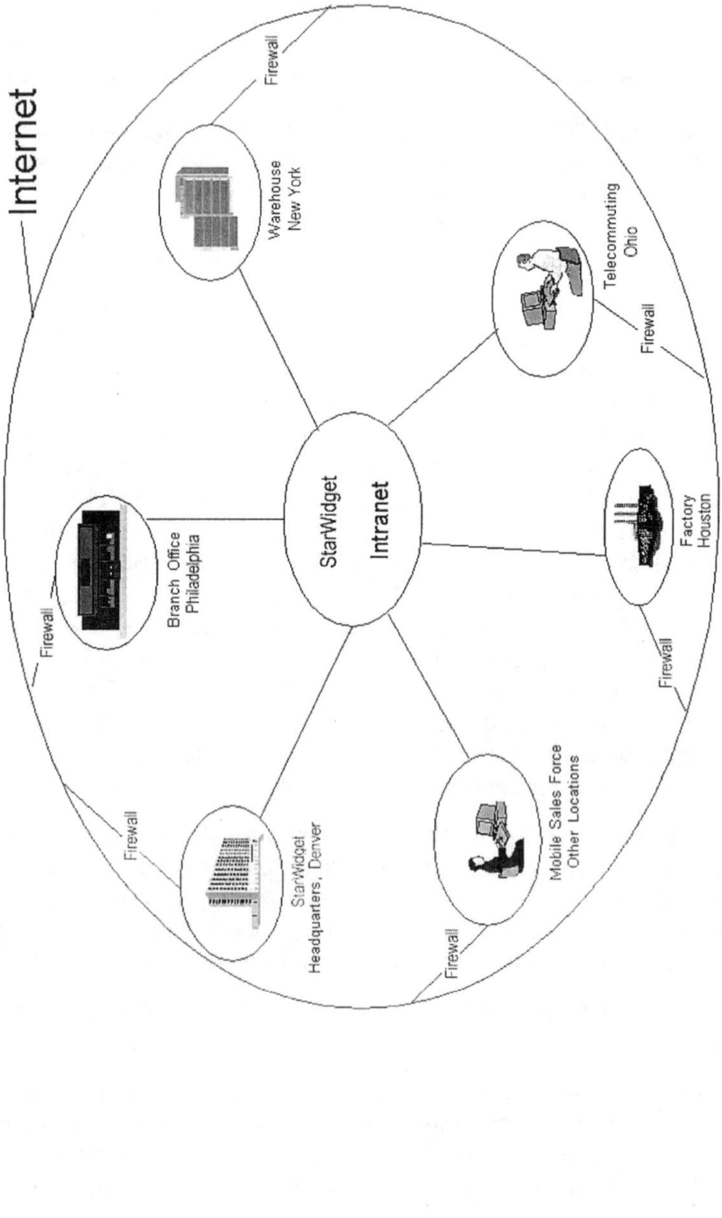

Exhibit 51.1 An Intranet.

a web browser and — boom! — it is there for users to use. However, a major concern that must be dealt with is security.

- The Web's intuitive environment provides tremendous ease of use: users just point and click on highlighted words and pictures to navigate and access information.
- Systemwide deployment can take place quickly. Most enterprises already have intranet building blocks in place, with various types of computers and servers hooked up to local area networks (LANs) and wide area networks (WANs).

The benefits of intranet technology can deliver greatly enhanced communications to an enterprise, as well as improving the enterprise environment as a whole. However, intranets bring with them an entirely new arena of security issues. Dealing with these issues effectively is imperative — the alternative could be devastating to the enterprise structure.

With this in mind, intranet security can be defined as protection against unauthorized physical or electronic intrusions, manipulations, or attacks, preserving end-to-end integrity of the intranet and transmitted information. It requires defenses against several types of disruptions, including electronic attack, physical attack, human error, and natural disaster.

As government, the private sector, and the general public become increasingly dependent on information services to support the nation's economic well-being and lifestyle, the reliability and security of intranets that provide and deliver information services will become increasingly critical. These intranets must have provisions to protect against catastrophic failure or loss of service from a range of threats or conditions. There also must be provisions to monitor the health of the intranets and rapidly restore services when failures do occur. This article provides an assessment of the capabilities of an enterprise's resources to implement an effective intranet security strategy. The rationale for the assessments provided in the article will examine the threats faced by intranets which comprise most enterprises, evaluate their vulnerability to these threats, and identify the shortfalls in the countermeasures currently employed — as well as learn what enterprises can do to avoid them in the future.

The rest of this article will identify some possible security breaches of a technical nature, and then take a look at some methods that can be used to interrupt or detect these breaches when or if they occur. Take a look at an aerial view of this "security battlefield" so that IT management and computer system and intranet security administrators understand the many dangers that lie ahead.

SECURITY TRENDS AND ISSUES

In recent years, technological advancements in today's world have come fast and furious. It seems that developments have gained momentum in the 1990s. Cases in point include computing, telecommunications, applications, and specific functions. This has resulted in expanded tools and services that are now available for information technology (IT) systems. A huge number of users can now use computers for countless tasks because of the proliferation of computers in the world. Also, hardware and software enterprises have developed computing devices and application programs suited for particular users in an effort to meet demands. Thus, with the growing number of products for personal computing, specialized IT systems, and other computing facilities for distributed processing and application tasks, this trend has become most evident.

Due to the nature of distributed processing, more dependence on the use of telecommunication lines, communication services, and computer intranets is required. In other words, the use of communication facilities contributes to the integration of newly developed computer devices and user-specific applications. It thus makes possible the efficient distribution of computing resources and functions.

Also, an increasing dependence on the capabilities of computer systems by enterprises has resulted in their storing and processing higher proportions of confidential, financial, and proprietary data. This information is available by remote access through private or public communication lines or organized in intranets. Advances in computer technology have resulted in extremely complex IT environments. These are environments where many users access available systems from remote locations — and where many different applications, jobs, programs, and procedures are executed at the same time.

So, particularly applicable to enterprise information technology are the complex structuring and use of computer systems. In other words, users cannot get detailed insight into their internal operational rules because many software products are presented as black boxes. Thus, operating any system (other than a PC) involves many components with internal interactions, control data, and thread of control, which are difficult to follow, analyze, or modify.

Look at it this way: if everything is working well, there is no problem. Nevertheless, if something does go wrong, regular users are often unable to efficiently modify their systems. The combination of "bad guys" who want access to a system, the potential for computer failure (whether by accident or by natural disaster), and the sensitive nature of the information computers process leads to one conclusion: computer system administrators must protect their charges against the possibility of undesirable events.

This means that, to be on the safe side, system administrators should use intranet security and disaster recovery features and products.

Also, with the distribution of resources and use of communication links, an increase in security threats and vulnerabilities is inevitable. This results in the grouping of link distances and types of computing structures into categories, which can then be easily implemented into enterprise intranets.

For example, the intranets may be standalone or part of a LAN. Because the enterprise totally owns the system and can inspect it (depending on the distance between each communicating device), the communications links used within a building are the easiest to control. Thus, the most common transmission media in these systems are twisted pair wire, coaxial cable, and fiberoptic lines. Nevertheless, the intrabuilding link security problem increases when an enterprise leases lines or uses voice dialup circuits. The security problems, therefore, become similar to those of LANs in these situations because the organization has relinquished control of its computing facilities. One is basically looking at a terminal network here.

On the other hand, the LAN communications can be user-owned cables (including fiberoptic), microwave circuits, leased lines, and voice dialup. Thus, the possibility for communications interference or line taps is greater than in intranets.

One can go up another notch or two on the security problems ladder with nationwide intranets. These intranets use private or public types of communication lines and are very large; data is transported totally by vendor media if public communication lines are used. This may be advantageous if packet-switching intranets route various pieces of information via different routes. However, OSI structures are generally considered the most vulnerable computing systems with this method.

Now, if one really wants to fall off the security problems ladder, try worldwide intranets. These intranets have all the security vulnerabilities of nationwide intranets, but do come with additional problems: they are subject to limitations on the flow of data across borders. Also, because of various national limitations, data encryption is sometimes neither allowed nor feasible.

Therefore, the first step in establishing a strategy for intranet security (regardless of the size of the intranet) is for IT management to set a security policy. The basis of all other criteria is formed by this security policy. Keep in mind, however, that the security policy's adequacy may not be completely known when it is first defined. In other words, an ongoing part of computer security is proper evaluation and testing with respect to the overall system operation, current computer and communications technologies, and the enterprise's evolving intranet growth.

In order to conform to predefined rules, regulations, and user requirements, current security trends require the overall operation of each complex system. This property of each complex computing system is what is known as *dependability.* Dependability is a key issue today. It involves the total system. It also includes privacy of information, protection of data and programs, integrity of system resources, fault avoidance/tolerance, special security protocols and applications, etc.

This part of the article deals with one specific but broad area of dependability: regular operation of computer systems or intranets. Basically, this operation includes only two main aspects: integrity of computer data, and programs and confidentiality. Therefore, in accordance with user specifications, this article's main topic is safeguarding and protecting system resources from possible threats. The article also does not detail any other dependability properties of information use that involves compliance with ethical and legal principles of privacy protection because they are beyond the scope of this article.

Thus, intranet security is mainly concerned with various operational problems that are the consequences of potential threats to the operations of the system. These problems include:

- threats due to accidental activities and operations that are destructive to the regular status of the system — and, that may also be treated as efforts or mistakes
- threats due to intentional illegal activities and operations, which may be carried out in order to modify, destroy, or disclose computer resources, or simply to create problems in regular systems operations

Here, *integrity measures* or *integrity constraints* are procedures and methods for the elimination of accidental efforts and mistakes appearing during regular system operations. On the other hand, for intentional destructive activities, protective methods are usually called computer protection. Thus, the logical structures of each security system comprise protection and integrity of system resources. A distinction is made in this section (although the two are not always treated separately) when considering particular security mechanisms and procedures, and when discussing integrating individual security mechanisms into a complete intranet security system.

Also, within the enterprise intranet, various strategies protect computer resources. Most of these strategies are designed and implemented as so-called "security services." Each strategy is concerned with specific aspects of computer operations. Thus, the first security service is user identification and authentication, which verifies the user's identity.

Confidentiality of users and data is among other important security services. This confidentiality is achieved primarily by the application of cryptography and access control. It guarantees that all system resources are used only by authorized persons in authorized ways. In other words, cryptography converts data into text that is unintelligible (cipher text) until it is decoded by the computer that receives it. Thus, the purpose of cryptography is to prevent unauthorized people from gaining access to the information.

On the other hand, access control is concerned with access to the physical system or intranet, system software, applications, and data. One aspect of this service may be information flow control. This is the attempt to control the flow of information between two (or more) users within the single computer system or within the communication network. In addition, access control also encompasses inference control, where information generation may be the result of combining statistical data with individual data.

DESIGNING SECURITY MECHANISMS

Design, implementation, and use of security functions in a computer system can be globally structured into four groups of activities: technical, physical, administrative controls, and legal and social issues. The nucleus of design and application of security mechanisms (for users and resources of computer systems and intranets) are methodology and practical implementation activities. These mechanisms take the form of software or hardware modules to protect computer resources. This aspect of intranet security, usually called the technical aspect, is this article's primary focus. There will also be some consideration of administrative controls and the legal and social issues that have an impact on intranet security.

Contrary to what many people think, design and implementation of tightly controlled computer systems or intranets is not enough. Layers of control external to the system design must supplement the technical aspect.

The next level of technical measures is physical security. This refers to procedures aimed at the physical protection of computer installations such as locks, guards, alarms, fire precautions, protection of stored data files, etc. Of course, these measures cannot be applied to communication facilities (satellites, public switched network (PSN), etc.) that extend beyond the control of the organization. Physical security for these communications facilities depends on owner organization of the software and hardware.

Administrative controls that ensure correct management and use of the system or intranet are the next layer of intranet security. These administrative controls include programmers, systems engineers, and IT managers. They also extend to user departments, auditors, and management.

Last, but very important, are legal and social regulations. Each person or enterprise responsible for security must follow some well-defined procedures in order to design and implement an efficient intranet security system. First comes the recognition and identification of all threats to the global computing information infrastructure or to the computer installation, often called vulnerability analysis.

The determination and evaluation of which potential threats may actually occur, and what the consequences might be, is the next step. This process is called risk analysis or risk assessment. It determines the requirements and scope of intranet security services. Each security service can be implemented by different mechanisms in different ways. In other words, in order to develop the most efficient, effective security system for a particular installation, security personnel must rate and verify all available security mechanisms.

Finally, users must always consider the needs and steps for the disaster recovery process. Of course, one assumes that no intranet security system is perfect — both for the security system itself, and for computer system and intranet resources.

Thus, in order to restructure or redefine the implemented security system in case of its malfunction or compromise, this disaster recovery process may be necessary. It is especially necessary in order to recover system resources whose security or integrity have been violated. In sum, assessment of intranet security consists of the following mutually related steps:

1. identifying potential threats
2. analyzing vulnerability
3. assessing risks
4. rating/evaluation of security mechanisms
5. verifying security mechanisms
6. identifying procedures for recovery

Therefore, in order to reach a reliable security system for their situations, all users, intranet security administrators, or IT management bodies must go through such a sequence of decisions. The set of decisions resulting from these steps constitutes the security policy. In other words, it actually represents the plan of action and rules for applying security methods. Essentially, during the general operation of a given system, a security policy states what is and is not permitted in the field of security.

In essence, security policies (together with applicable legislation and regulations) determine a number of things:

- to what extent data communications will be used by the enterprise
- the type of data that will be transmitted
- acceptable behavior for system users
- responsibility for intranet security
- security-related budgetary constraints

All these factors are elements of the intranet operation or total system.

This article section addresses the analysis and determination of security policies, and also contains the survey of security services and mechanisms that represent basic components of each secure intranet. The final decision within each security policy is which security services and mechanisms will be used, and how they will be implemented, used, controlled, and managed.

Analyzing Vulnerability

Normally, if all threats and risks have been eliminated, a computer system is considered secure. Threats and risks are activities or situations that may be dangerous or undesirable for regular operation of a system or intranet. Therefore, one cannot determine the seriousness of a potential danger out of context. A problem that is serious for one installation may not be serious for another.

The risk and vulnerability assessment procedure defines the vulnerabilities, assesses potential consequences if those risks are exploited, identifies threats to a computer or intranet, and identifies and evaluates the already-implemented security methods. Both the communications intranet and the computers (hosts) must be included in this analysis.

Threats and risks are dependent on operations, users, and resources of computer systems or intranets. Each operation represents a source of potential vulnerabilities and problems. The seriousness of each potential problem depends on the users and resources involved. It also depends on the sensitivity of their content and status. The following sequence of conclusions may be established:

1. You must first establish a list of expected threats to define a secure computer system.
2. The set of operations available in the system is related to the list of threats.
3. Each threat and its corresponding operation must be related to a particular resource, its importance, and its content.

In sum, the more potential threats and problems, the larger the set of operations. This also provides the basis for a more efficient and better-integrated security system. To establish a large list of operations, the author assumes that the computer system is a general system attached to an intranet, with all kinds of devices, data, programs, users, and operations. Such an environment contains numerous chances for errors, illegal activities, system malfunctions, and other vulnerabilities, as shown in Exhibit 51.2. This list of threats can be extended, restructured, or modified. The main point stressed here is that an efficient integrated security system must cope with all these threats.

Exhibit 51.2 System Vulnerabilities.

Intranet Environment Vulnerabilities

Users
- illegal identification
- illegal authentication
- invalid authorization
- subtle modification of software

Remote Locations
- attachment of signal recorders
- illegally planted bugs
- radiation

Communication Subsystem
- radiation
- telephone taps
- crosstalk.
- improper connections
- cross-coupling
- insertion of illegal messages
- deletion of regular messages
- unauthorized access to messages
- association monitoring

Personnel (Operators, System Programmers, Maintenance)
- replacement of authorized software by unauthorized software
- revealing protective measures
- disabling hardware-protection devices
- use of standalone programs
- disabling software protection features
- private access to the system

Processor
- failure of protection circuits
- contribution to software failures
- failure of software protection features
- invalid access control
- invalid user identification
- invalid access control

Files
- theft of data
- illegal copying
- unauthorized access
- destruction or deletion

Furthermore, all problems or sources of threats can be placed into four main groups. The problems in each group are mutually related. In other words, their sources and operations are similar. Thus, problems in one group may also relate to problems in other groups:

1. user interactions
2. data communications
3. database systems
4. data processing

All further discussions, problems, protection methods, and evaluation and verification procedures in this article are organized according to these four aspects.

Performing Risk Analysis

Security personnel must perform risk analysis and assessment before any intranet security system is developed or implemented at any installation. Risk analysis has two main components:

- analyzing threats to regular system operations
- identifying undesirable events that can result from those threats

Risk analysis also identifies system security weaknesses. These weaknesses may be the reasons for some threats. Thus, risk assessment must be performed after risks have been analyzed. This means that undesirable events are rated and ranked in some kind of priority order according to their severity and possible consequences. Therefore, the likelihood of each event is then related to its acceptability. Security officials can then develop, modify, or implement a security measures program. This is done by comparing the operational cost of each method to the losses it is designed to prevent.

Rating Protection Methods and Security Mechanisms

The next step is rating various protection methods, and includes intranet security mechanisms or packages with respect to their scope and efficiency for eliminating anticipated problems. A user must determine relevant properties of each security mechanism; establish desired protection levels; determine the required or expected efficiency of each security mechanism; and, finally, evaluate all available security methods, modules, products, and packages.

The definition of ideal security requirements for the list of expected threats is the first step in the rating procedure. After that, by determining the existence of particular methods in that system, security personnel must relate each specific security mechanism or complete security system to the ideal concept. In other words, compared to the ideal one, this gives the user a relative quantification of the actual security system.

The implementation of the security system is the second step. Implementation can take place by developing and writing the system in-house, by adopting and customizing, or by modifying an existing product.

RATING, REQUIREMENTS, AND VERIFICATION PROCEDURES

Rating procedures vary from empirical and descriptive to completely analytical and exacting. The following list summarizes the steps to determine quantitative characteristics of several security mechanisms and to gage their effectiveness in given situations:

1. Establish the set of requirements (specifications) of the ideal security system.
2. Assign a constant value, based on importance, to each security service and mechanism of the ideal system.
3. Evaluate the presence of each security mechanism in some security system and determine its scope and efficiency.
4. Determine a quantitative parameter (by some formula) that will specify the characteristics (rating) of the security system.
5. Compare various security systems and determine their effectiveness for the specific situation.

Requirements for evaluation of security systems vary depending on each user, computer installation, available operations, risks, and threats, but those elements may include:

- All data, programs, users, and hardware devices (system resources) must be protected by default.
- Access to the system must be based on temporary user IDs and passwords for user identification and authentication.
- There must be a possibility for an audit trail with tracing and logging features.
- Security measures should have a minimal effect on system performance.
- All security mechanisms must be transparent to their users to as large an extent as possible.
- The security system must provide simple backup, restore, and recovery procedures.

In addition, the following questions might be asked:

- How well are the interfaces to other systems defined?
- Is data in transit protected?
- What protection is provided for programming subsystems?
- Is there default protection on production programs?

- Is program pathing available?
- What are the database protection features?
- What is the lowest level for database protection?
- Are subjects structured according to their capabilities and objects classified according to their sensitivity level?
- What are the cost factors and expenses?
- Is the command language simple and easy to use?

Reporting and Auditing Security-Related Events

The key to any effective security system are the functions of auditing and reporting intranet security-related events. Qualified personnel must thoroughly audit the security of the entire IT system, including the intranet, in order to achieve and maintain integrity and reliability.

The audit can be performed by the enterprise's auditors (if they are qualified), the data security director, or a consultant. Nevertheless, the performance of the audit is only the first part of the process. Notifying management (and sometimes law enforcement or regulatory officials) of the findings in a comprehensive audit report is the second part.

When a security violation results in broken government regulations or laws, reporting is also necessary. Enterprise unwillingness to report and prosecute computer abusers costs stockholders and customers money. It also does little to encourage lawmakers to introduce remedial legislation. Prosecution, where warranted, serves to deter those considering computer abuse. It can also provide compensation to victims of abuse. Legislators are also provided with an impetus by prosecutions to develop appropriate laws to define and deal with the abuse of computer technology.

Disaster Recovery

Reestablishing the correct status and functioning of system resources is what is known as the process of recovery. Because no intranet security system is perfect and problems usually arise, the recovery process is sometimes necessary. Therefore, the reasons for the violation of regular procedures and for illegal manipulation with system resources fall into three main groups:

- illegal use or modification of security services and mechanisms
- illegal access to the system by somehow avoiding the rules of security services
- external problems (fire, earthquake, sabotage, etc.)

In each of the preceding cases, the recovery procedure to use is specific. First of all, recovery must be primarily concerned with implemented security mechanisms, especially if someone illegally uses or modifies the security system. In order to avoid repeating the same or similar problems, they must be analyzed, redesigned, and tested again.

Second, recovery must be primarily concerned with programs and data (system resources), especially if someone has entered the system by avoiding security mechanisms. This is also necessary in the first case. However, accessing the system through the security barriers, even illegally, gives the intruder more limited scope of activities and available resources than in this second instance.

Finally, by using so-called backup installation (backup disks and the like), problems external to the system can be solved quickly. Therefore, the recovery process may be concerned with the global security system itself; with computer programs; with data and other resources; and with the computer system as a whole.

A SURVEY OF INTRANET SECURITY SERVICES

Basic tools that various security mechanisms in computing environments can implement are also known as security services. In principle, security services are the functions of a security system. Security mechanisms, on the other hand, are particular implementations (procedures, algorithms, hardware devices, etc.) of those services.

This section contains a broad list of various security services for general computing structures (intranets and open systems) which are available today as international standards, commercial products, theoretical solutions, or just initial proposals. The purpose of listing those services is twofold: first, to define security methods appearing today, and second, to define a concept of a secure system.

Many approaches, suggestions, and proposals are relevant for listing and structuring security services. The structure is based on some international standards and original research results. References with individual security services are related to materials where additional information about a particular service can be found.

Security Services for Users

This group of security services is applicable to any type of user in a computer system or an intranet. Therefore, an entity whose use of system resources must be controlled is known as a user. A user can also be a person using the system, a program, a procedure, a hardware device, or a process. With that concept in mind, the International Standards Organization's (ISO) *Information Processing Systems Manual* gives all such users a common name: entities.

Unique User Identification. Here, a process is in place for verifying information that uniquely identifies a user in the system. This user identification may be a number, a character string, or some other type of information.

User Authenticity. User authentication is usually activated after user identification. This is the process of establishing that a would-be user is in fact the entity he or she claims to be.

Access Control. The security service that controls which users may access which system resources, in what way, and under what circumstances is known as access control. This access control appears in all four groups of security services because it has such broad interpretation and applicability. Users and processes are treated as subjects of this service (active components). On the other hand, communication lines and database segments are considered objects (passive components).

Peer-to-Peer Authentication. Here, the user's activities are usually executed as interactions after user identification and authentication, either with some local system or with another user in the intranet. Thus, most user operations are executed as a sequence of commands issued by or sent to the user. Security purposes require authentication of the communication channel. This is called peer-to-peer authentication.

Nonrepudiation. In the stage of exchanging messages (after an association has been established), each user needs a nonrepudiation security service. This consists of the following functions:

1. the proof to the receiver of the origin of data (sender's identity and authenticity)
2. the proof to the receiver of the original content of data
3. the proof to the sender of delivery of data
4. the proof to the sender of the original content of the delivered data (receipt)

The first function alone is called a digital signature, and the second function is called a digital seal. The third and fourth functions together are called certified mail. In other words, the third function is the digital signature of the confirmation reply message. Also, the proof to the sender that the original content of the delivered message (the fourth function) contains some form of digital seal of the received message is what confirms the original content.

Secure Electronic Mail. An application in which authenticated messages are sent to a receiver is what is known as secure electronic mail. The sender, therefore, needs a receipt for the content and delivery — even while the receiver may not be logged into the system at the time of message delivery. This means that the only security service feasible is the local user authentication — not peer-to-peer authentication. It also means that verification of the digital seal and digital signature on the delivered mail cannot be performed upon delivery. Therefore, it must be postponed until the receiver logs into the system.

Simultaneous Nonrepudiation Contract Signing. A contract signing between two mutually suspicious users is known as simultaneous nonrepudiation. This is the case when neither user trusts the other. Therefore, no private information should be exchanged without reciprocity.

Disclosure Protocol. The disclosure protocol is needed when one user is willing to disclose some information to another user. It also means that the recipient confirms that he or she will not divulge that information. Again, this is a parallel protocol between two mutually suspicious users. The previous security service is the parallel exchange of digital seals and digital signatures. However, this service exchanges a secret message for confirmation of its content and delivery.

Security Services for Communications

The most general concept of data transfer between any two systems is known as data communications. Passive and active attacks are two types of threats to regular operations that exist in such a situation.

In a passive attack (without interfering with the content of individual messages or sequences of packets), the illegal user observes or records messages passing on a communication channel. In active attacks, on the other hand, illegal users modify, delete, delay, reorder, or duplicate legal messages or insert false messages. The security services discussed next can be used to eliminate such threats and various threats of attacks.

Data Confidentiality. To prevent unauthorized individuals, entities, or processes from gaining access to information in messages, one may want to use the data confidentiality security service. This service is usually implemented via cryptography. It can also be used for database or process protection.

Data Integrity. The security service that is concerned with data that has not been altered or destroyed accidentally or in an unauthorized manner is known as data integrity. This service also provides continuous availability of data in high-quality form. That means that the content and structure remain unaltered.

Access Control. Access control with regard to security services for communications is concerned with preventing unauthorized use of communication resources and illegal access to remote resources in host systems via communication lines. This service is also applicable to control or database security.

Preventing Analysis of Message Patterns. Intruders can examine the internal structure of transmitted messages if data is encrypted for content protection. Original data being transmitted can be discovered by the structure of the message in this situation. By preventing message pattern analysis, this problem is eliminated. This service can also be used by security personnel to protect records in databases.

Information that might be illegally derived from the observation of traffic flow is protected by traffic flow confidentiality. The intruder may see various message control fields even if transmitted data is encrypted, and thus be able to obtain information on identities and locations of communicating entities. The intruder can also get information about the nature of the data by examining the lengths and frequencies of transmitted messages. Such passive attacks are usually referred to as traffic analysis or violation of transmission security.

Putting a Stop to Message Stream Modification. Message stream integrity is concerned with the quality of transmission. It is not concerned with individual messages, but with the complete sequence of message packets. Message stream modification, on the other hand, includes attacks on the authenticity, integrity, and ordering of individual messages transferred on the association. Authenticity here means that the source of each message can be reliably determined. Thus, integrity means that the message content has not been modified during transmission; and ordering means that the sequence of received messages is identical to the sequence of messages delivered into the communication system.

Continuous Communication. Disruption of continuous communication among users may occur accidentally or by illegal actions. Denial of message service (one category of active attacks in a communication system) includes all attacks in which the intruder either discards some or all messages. Here, the intruder also passes messages on to an association or delays messages in one or both directions.

Authenticity of Associations. Here, authenticity is concerned with the identity of participating users, communication channels, and all messages being transferred with so-called spurious associations. In other words, spurious association is the situation in which the intruder can establish an illegal association by one of the following means:

- evading, somehow, the initial user identification and authentication procedures
- taking advantage of the possible malfunction of the security system
- using a previous legitimate association initiation sequence to gain illegal access by playing back the recording
- using a trial-and-error method to find a false identity that can be used to establish an association

Security Services for Databases

Database security services are concerned with data stored in data banks on external storage devices. Protection is provided by them during storage, retrieval, organization, deletion, and other such operations. Some other aspects of database protection, related to data protection during program execution, are considered later in the article in the section "Security Services for Process Control."

Data Confidentiality. Data confidentiality here is the same as for data in transfer. It prevents unauthorized inspection of data contents if data is in transit or in computer databases. The service is concerned only with external retrieval and data storage.

Data Integrity. Again, this type of data integrity is also the same as for data in transfer. It is concerned with the correct content of data and their continuous availability in databases. Often included as part of the security system is the protection of data content (in the sense of correctness) and continuous availability in high-quality form.

Access Control. This type of access control, as with data in transfer, prevents unauthorized access to segments of databases. Here, access control can be considered a static security service; that is, the protection states of the system are changed externally by execution of certain control commands or in some similar way. During regular system operations, this method usually does not reflect (and therefore does not depend on) the changes. Furthermore, access control is equally concerned with processes in the system, as well as with data (as system resources). There are, however, some modifications of this security service when compared with the protection of data in transit. This is especially true when considering dynamic security schemes. These extensions consist of the following four security services:

- flow control
- context protection
- prevention of inference
- prevention of information generation

Flow control. The flow control security service (also known as consistency control) regulates the dissemination of values among the data sets accessible to a user by copying either complete files or certain records. All such transfers must be in accordance with user- or system-imposed security specifications, rules, and constraints. If this service is successfully implemented in an intranet, it means that the data is always logically consistent in the databases. Thus, sensitive data values are disseminated only under special authorization (called value-dependent restrictions).

Context protection. The context protection security service is based on dynamic database protection schemes. Protection with this service is based on an individual database object in a particular instant of time. It is also based on the overall security environment and previous operations executed with the same history-dependent restrictions.

Prevention of inference. Here, the deduction of confidential individual values from data stored in databases is blocked by the prevention of inference security service. The service accomplishes this task by posing specially designed sequences of queries and correlating the responses.

Prevention of information generation. If an individual does not have direct access to classified information in the database, the prevention of information generation security service can prevent an entity from being able to generate data above his or her security level. This is the result of certain logical relations and dependencies between data values.

Security Services for Process Control

Process control security services are concerned with the protection of processes in the system (as system resources) during their execution. They are also concerned with the protection of all system resources against invalid or unauthorized access by executing processes. Authentication and process identification are equivalent to the same services mentioned earlier for users.

Data Integrity. Here, the data integrity security service provides integrity of data in main memory during program efforts, wrong updating, or against inconsistencies due to incorrect program scheduling. Regardless of malfunctions or misuse of an individual process, this service keeps data continuously available to all processes and always in correct form.

Access Control. The most commonly used service for process control and security is access control. Thus, various activities with system resources can be seen as accesses to resources during the processing of information. In other words, access control must control access to resources and their use. It has basically two aspects:

- protection of intranet resources against access by illegal processes and unauthorized requests of legal processes
- integrity of resources against invalid process scheduling, wrong operations, or incorrect process synchronization

Protection of Software. The protection of software security service guards software resources (programs, routines, macros, control modules, etc.) from unauthorized inspection, copying, or usage. This service may also include special requirements for secure distribution of software, such as controlled licensing schemes and copyrights.

Verification of Processes. The establishment of integrity and protection against invalid processes can also be accomplished through a confidence in quality of operations (verification and certification of programs and processes before their execution). Also, the correctness of programs and processes must be tested, verified, and guaranteed. In this way, various problems inherent to the internal program structure, such as viruses, can be eliminated.

CONCLUSION

Despite all the excitement about privacy, virus infections, and the legalities of just being online, intranets are not a communications panacea. As these intranets expand, they could become big and anarchic. Yet, in the short run, intranets may be more enticing as the promise of cheaper and higher-speed Internet connections grows. Regardless of how information travels, enterprises will still seek new ways of communicating. For now, the hot model is the intranet, and it must be secure!

Chapter 52
A Look at Java Security

Ben Rothke

WHY SHOULD JAVA SECURITY CONCERN YOU? Many push-based applications are being ported to Java. In addition, Java is one of the cornerstones of active content, and an understanding of Java security basics is necessary for understanding the implications of push security issues.

A lot of people ask: "Why do I need Java security? I thought it was safe." Java as a language is basically safe and is built on top of a robust security architecture. But security breaches related to bugs in the browser, poorly written Java code, malicious Java programs, poorly written CGI scripts and Javascript code, and more often occur. Moreover, placing the enforcement of a security policy in the browser, and thus in the hands of end users, opens up many opportunities for security measures to be defeated. In addition, many push vendors are relatively new startups that do not always understand mission-critical software and security needs. Such circumstances only exacerbate the security predicament.

While some people might opine that Java is too insecure to be used in production environments and that it should be completely avoided, doing so creates the situation where a tremendous computing opportunity is lost. While the company that decides to bypass Java alleviates itself of Java security worries, that means that they also relinquish the myriad benefits that Java affords. In addition, a significant amount of cutting-edge Internet-based activities such as E-commerce, online trading and banking, and more are all written in Java. Also, many firewall and router vendors are writing their management front-end application in Java. By cutting off from Java, an organization may likely cut itself off from the next generation of computing technology.

The end result is that push-based programs are powerful and flexible web tools, and where the Web is directed. But these programs, by their nature, are inherently buggy and untrustworthy. Now take a look at the Java security model.

A QUICK INTRODUCTION TO THE JAVA PROGRAMMING LANGUAGE

The essence of Java is to be a portable and robust programming language for development of write-once programs. Java was created to alleviate the quandary of writing the same applications for numerous platforms that many large organizations faced in developing applications for large heterogeneous networks. To achieve this, the Java compiler generates class files, which have an architecturally neutral, binary intermediate format. Within the class file are Java bytecodes, which are implementations for each of the class's methods, written in the instruction set of a virtual machine. The class file format has no dependencies on byte-ordering, pointer size, or the underlying operating system, which thereby allows it to be platform independent. The bytecodes are run via the runtime system, which is an emulator for the virtual machine's instruction set. It is these same bytecodes that enable Java to be run on any platform. Finally, two significant advantages that increase Java's security is that it is a well-defined and openly specified language.

While many systems subscribe to the *security through obscurity* model, Java achieves a significant level of security through being published. Anyone can download the complete set of Java source code to examine for themselves. In addition, numerous technical security groups and universities have done their own audits of Java security.

The second area where Java security is increased is through its architectural definitions. Java requires that all primitive types in the language are guaranteed to be a specific size and that all operations defined must be performed in a specified order. This ensures that two correct Java compilers will never give different results for execution of a program, as opposed to other programming languages in which the sizes of the primitive types are machine- and compiler-dependent, and the order of execution is undefined except in a few specific cases.

OVERVIEW OF THE JAVA SECURITY MODEL

The Java applet[1] security model introduced with the 1.0 release of Java SDK considers any Java code running in a browser from a remote source to be untrusted. The model anticipates many potential attacks, such as producing Java code with a malicious compiler (one that ignores any protection boundaries), tampering with the code in transit, etc. The goal of the Java security model is to run an applet under a set of constraints (typically referred to as a sandbox) that ensures the following:

- No information on the user's machine, whether on a hard disk or stored in a network service, is accessible to the applet.
- The applet can only communicate with machines that are considered as trusted as itself. Typically, this is implemented by only allowing the applet to connect back to its source.

- The applet cannot permanently affect the system in any way, such as writing any information to the user's machine or erasing any information.

From a technical perspective, this sandbox is implemented by a layer of modules that operate at different levels.

Language Layer

The language layer operates at the lowest layer of the Java language model and has certain features that facilitate the implementation of the security model at the higher levels.

Memory Protection. Java code cannot write beyond array boundaries or otherwise corrupt memory.

Access Protection. Unlike C++, Java enforces language-level access controls such as private classes or methods.

Bytecode Verifier. When a Java applet is compiled, it is compiled all the way down to the platform-independent Java bytecode where the code is verified before it is allowed to run. The function of bytecode verification is to ensure that the applet operates according to the rules set down by Java and ensures that untrusted code is snared before it is able to be executed.

While the language restrictions are implemented by any legal Java compiler, there is still the possibility that a malicious entity would craft its own bytecode or use a compromised compiler. To deal with this possibility, Sun Microsystems architected the Java interpreter to run any applet bytecode against a verifier program that scans the bytecode for illegal sequences. Some of the checks performed by the verifier are done statically before the applet is started. However, because the applet can dynamically load more code as it is running, the verifier also implements some checks at runtime.

The bytecode verifier is the mechanism that ensures Java class files conform to the rules of the Java application. While not all files are subject to bytecode verification, those that are have their memory boundaries enforced by the bytecode verifier.

Security Manager. The function of Java security manager is to restrict the ways in which an applet uses the available interfaces and the bulk of Java's security resources are implemented via the security manager.

At the highest level, the security manager implements an additional set of checks. The security manager is the primary interface between the core Java API and the operating system and has the responsibility for allowing or denying access to the system resources it controls.

This security manager can be customized or subclassed, which allows it to refine or change the default security policy. Changing the security manager at runtime is disallowed because an applet could possibly discover a way to install its own bogus security manager. All of the Java class libraries that deal with the filesystem or the network call the security manager to ensure that accesses are controlled.

From a technology perspective, the security manager is a single interface module that performs the runtime checks on potentially dangerous methods that an applet could attempt to execute.

Security Package. The security package is the mechanism that allows for the authentication of signed Java classes. Those are the classes that are specified in the *java.security* package.

Signed applets were introduced in version 1.1 of the Java SDK and specifically are collections of class files and their supporting files that are signed with a digital signature.

The way in which a signed applet operates is that a software developer obtains a certificate from a certificate authority (CA) and uses that certificate to sign their applications. When an end user browses a web page that the developer has signed, the browser informs the end user who signed the applet, and allows the user to determine if he wants to run that applet.

Key Database. The key database works with the security manager to manage the keys used by the security manager to control access via digital signatures.

The Java Standard Applet Security Policy

The exact set of policies that are enforced by Java in a specific environment can be modified by creating a custom version of the security manager class. However, there is a standard policy that has been defined by Sun and is implemented by all web browsers that implement Java applets. The standard policy basically states.[2]

- An applet can only connect back to its source. This means, for example, that if the applet source is outside a company firewall, the applet is only allowed to talk to a machine that is also outside the firewall.
- An applet cannot query system properties because these properties could hold important information that could be used to compromise the system or invade the user's privacy.
- An applet cannot load native libraries because native code cannot be restricted by the Java security model.
- An applet cannot add classes to system packages because it might violate some access-control restrictions.

- An applet cannot listen on socket connections. What this means is that an applet can connect to a network service (on its source machine), but it cannot accept connections from other machines.
- An applet cannot start another program on the client workstation. This way, an applet cannot then spawn some other program or rogue process on the workstation. From a programming perspective, an applet is not allowed to manipulate threads outside its own thread group.
- An applet cannot read or write to any files on the user's machine.
- An applet can only add threads to its own thread group.

Java Language Security

This is not the place to detail the security features of the Java programming language, but a few of its most significant security-based features include the following.

Lack of Pointer Arithmetic. Java security is extended through lack of pointer arithmetic since Java programs do not use explicit pointers. Pointers are simply memory locations in applications. Consequently, no one can program (either maliciously or accidentally) a forged pointer to memory. The mishandling of pointers is probably one of the largest sources of bugs in most programming languages. To get around the lack of pointers, all references to methods and instance variables in the Java class file are via symbolic names.

Garbage Collection. Java garbage collection is the process in which Java deallocates memory that it no longer needs. Most languages such as C and C++ simply allocate and deallocate memory on the fly. The use of garbage collection requires Java to keep track of its memory usage and to ensure that all objects are properly referenced. When objects in memory are no longer needed, the memory they use is automatically freed by the garbage collector so that it can be used for other applets. The Java garbage collection engine is a multithreaded application that runs in the background and complements the lack of memory pointers in that they prevent problems associated with bad pointers.

Compiler Checks. The Java compiler checks that all programming calls are legitimate.

E-Commerce and Java

Sun Microsystems has entered the E-commerce arena in a big way and envisions having Java at the forefront. To assist in that attempt, Sun has created a Java E-commerce architecture to promote that.

Components of the architecture are the Java Wallet, Commerce Client, Commerce API, and Commerce JavaBeans.

The Java Wallet is a family of products written in Java that enable secure electronic commerce operations. The Java Wallet combines the Java Commerce Client, Commerce JavaBeans components, the gateway security model, and the Java Commerce Messages to create a single platform for E-commerce. It should be noted that the components can be used independently of one another. The Java wallet is written in Java; thus, it can run in any Java-capable browser.

THREATS

McGraw & Felten describe in *Java Security: Hostile Applets, Holes & Antidotes,* four classes of threats that Java is susceptible to:

- System modification: This is the most severe class of threats where an applet can significantly damage the system on which it runs. While this threat is the most severe, the defenses Java has to defend against it are extremely strong.
- Invasion of privacy: This is the type of attack where private information about host, file, or user is disclosed. Java defends against this type of attack rather well because it monitors file access, and applets can only write back to the channel in which they were originally opened.
- Denial of service: Denial of service attacks are written to deny users legitimate access to system resources. Denial of service attacks take many forms, but are primarily applications or malicious applets that take more processes or memory allocation area than they should use, such as filling up a file system or allocating all of a system's memory. Denial of service attacks are the most commonly encountered Java security concern and, unfortunately, Java has a weak defense against it.
- Antagonism: An antagonistic threat is one in which the applet simply annoys the user, such as playing an unwanted sound file or displaying an undesired image. Many antagonistic attacks are simply programming errors. Most denial of service attacks can be classified as antagonistic threats, but the ones defined here are less annoying than their denial of service counterpart. Like their counterpart, Java has a weak defense against them.

USING JAVA SECURELY

By following some generic guidelines, and then customizing those guidelines for an environment's unique needs, Java can be safely used in most environments. Java security, like most computer security, is built on a lot of common sense. A few of the major issues are:

- Make sure that your browser is up to date: Many Java vulnerabilities have originated in browser design flaws. Staying with a relatively new release of a browser ensures that discovered security flaws have hopefully been ameliorated.

- Stay on top of security alerts: Keep track of advisories from CERT (www.cert.org), CIAC (www.ciac.llnl.gov), and the appropriate browser vendor.
- Think before you visit a web site: If visiting www.whitehouse.gov, chances of downloading a hostile Java applet are much less than if visiting www.hackers.subterfuge.org. The bottom line, use your head when surfing the Web.
- Know your risks: Every company must access its risks before it can really understand how to deal with the security risks involved with Java. If the risk of Java is too great (i.e., nuclear control centers), do not use Java; if the risks are more minimal (i.e., home), one can pretty much use Java with ease.

THIRD-PARTY SOFTWARE PROTECTION

There are numerous third-party software tools available to further secure Java and add protection against the potential security threats that Java can produce. Such products are a necessity for running push and active content applications.

- Finjan — SurfinGate & SurfinGate (www.finjan.com)
- eSafe Technologies — eSafe Protect (www.esafe.com)
- Digitivity — Cage (www.digitivity.com)
- Security7 — SafeGate (www.security7.com)

CONCLUSIONS ABOUT JAVA SECURITY

Java has an impressive security architecture and foundation, but one cannot rely on the sandbox model exclusively. Combined with poorly written PERL and CGI scripts, browser vulnerabilities, operating system holes, web server holes, and more, there are plenty of potential openings in which a malicious or poorly written application could wreak havoc.

But by knowing what one's risks are, combined with an understanding of what the Java vulnerabilities are, in association with active content protection, Java security is not an oxymoron.

Notes

1. An applet is defined as a Java program that is run from inside a web browser. The html page loaded into the web browser contains an <applet> tag, which tells the browser where to find the Java .class files. For example, the URL http://cnn.com/TECH/computing/JavaNews.html starts a Java applet in the browser windows since the source code contains the entry <applet code=Ticker.class>.

2. This article cannot list all of the details of the standard policy. For a thorough listing, view the Java SDK documentation set.

References

Frequently Asked Questions–Java Security, http://java.sun.com/sfaq/index.html.

Under Lock and Key: Java Security for the Networked Enterprise, http://java.sun.com/features/1998/01/security.html.

The Java Commerce FAQ, http://java.sun.com/products/commerce/faq.html.

The Gateway Security Model in the Java Commerce Client, http://java.sun.com/products/commerce/docs/whitepapers/security/gateway.pdf.

Low Level Security in Java by Frank Yellin, http://www.javasoft.com/sfaq/verifier.html

Chapter 53

Creating Firewall Policy Following a Functional Model

John P. Ceraolo, CISSP

THE SECURITY POLICY OF ANY BUSINESS IS THE FOUNDATION FOR A SECURITY PROGRAM. While this is not news to the security practitioner, the concept of creating a policy specifically for the firewall may be. The standard security policy covers the "don'ts" of information resource usage, generally prefaced with a bold statement of how the information *is* the company and we all need to protect it. Within this policy usually lies the "acceptable usage" directives of the company-supplied Internet connection, with the details of what are the business needs initially funding it. After the security group has laid down what a user can and cannot do via the Internet, an equal policy needs to exist to regulate the most basic Internet security component, the firewall.

In most corporate environments, it is difficult enough to get a regular corporate security policy approved. Acceptance by management and employees is usually prefaced with resistance. Final policy approval can take months and even years. So why add another policy in an already difficult process? The answer lies in remembering why companies have the firewall in the first place. Many media outlets offer frightening accounts of Internet-based break-ins or mayhem. Maybe some of those stories have been in your bag of tricks when getting the firewall acquisition approved. Now with the firewall in place, management believes that, based on what has been told to them and everything they have seen in the trades, they are safe. Because it takes a lot more than that, one must remember that the firewall can create its own share of problems for the business. Problems such as line performance issues, a single-point of failure, a perception of the big brother concept and, more importantly for those in security, another area of responsibility. Therefore, the firewall policy is born with the core focus of satisfying the business needs while still protecting the corporate assets.

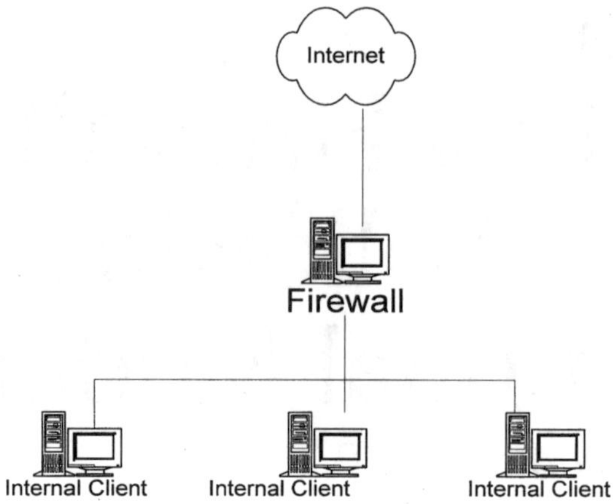

Exhibit 53.1 Dual-Homed Bastion Host.

The policy is used to document the inward and outward access of an Internet connection to and from your organization.

The standard configuration of any firewall architecture places the firewall between two different areas: the unprotected network, and the protected network. The firewall sits on the fence of the perimeter, acting as the gatekeeper and choke point for all traffic to and from the protected network. The dual-homed bastion host scenario is a common configuration (see Exhibit 53.1).

As the prevalence of outside services has grown, so has the DMZ concept. This is an architecture whereby, either through a separate NIC card (see Exhibit 53.2) or using two firewalls, a sacrificial network is established. Not that the security practitioner feels that any device should be left up for grabs, but the servers in a DMZ network are intended for inward access from users on the unprotected network. Because of their isolation into a distinct network, a compromise from unsavory users on the unprotected side would yield a more minimal risk to the business. In the scenario of Exhibit 53.2, it is understood that DMZ servers provide the information to the outside while being supplied the data from the inside. This redundancy also protects against the more common occurrence of denial of service. Service can be denied during an outside attack, as well as something as simple as a hard drive failure. To have the original data source online and protected within the internal corporate network certainly buys time and integrity during such an event.

Financial constraints may not allow for repetitious servers in a DMZ configuration. Firewall protection coupled with expanding to a *Client-Side*

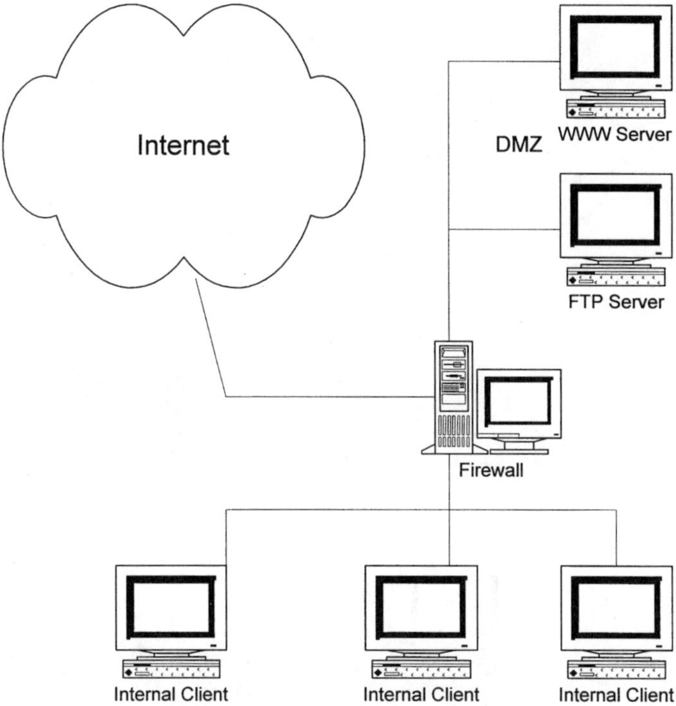

Exhibit 53.2 DMZ Firewall Set-Up.

Network configuration is an inexpensive way to provide secure access (see Exhibit 53.3). An additional network is created by placing a separate NIC card into the internal server. A separate NIC card of the firewall addresses only this client network. Along with application proxy access and server security, a legitimate user will only gain access to the required data source, and internal users should not be allowed outside access to the Internet via these components.

These are only a few types of security architectures that can work for a company firewall strategy. The basic premise remains clear: protection of the inside corporate network from the evils of the unknown that may lurk on the Internet. This is the rule understood that most likely got the financial and management backing for a firewall. What also needs to extend to this rule is the threat that employees are to the other companies on the Internet and, unfortunately, to themselves. Therefore, the firewall policy must begin with the concept that it should be split into two separate and distinct phases: how the *inward network traffic* and the *outward network traffic* are regulated. Before this is examined in more detail, there are some basic tenets to follow in the policy creation.

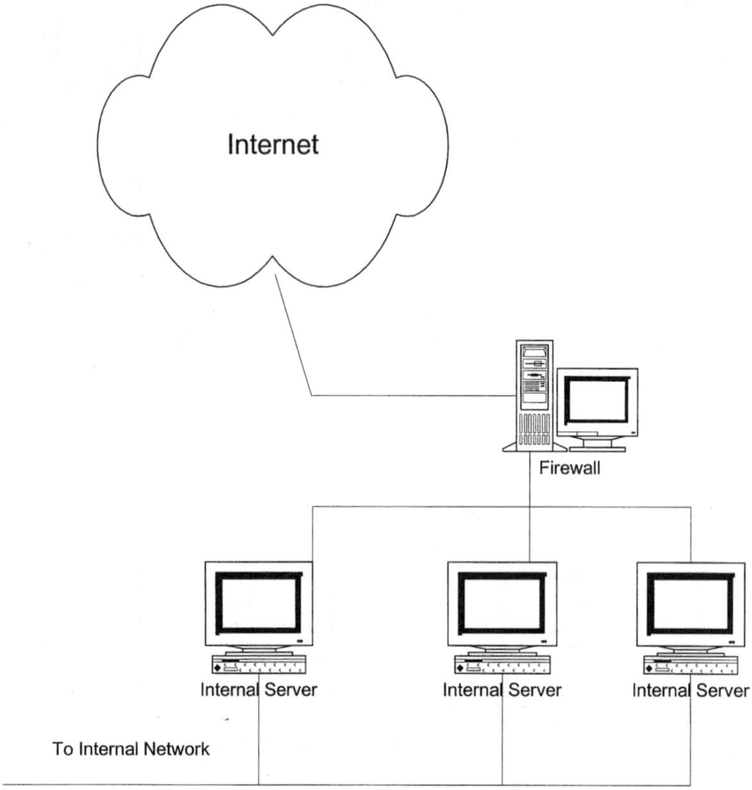

Exhibit 53.3 Client Side Network Configuration.

FIREWALL POLICY FOUNDATION

The firewall becomes a player in the security architecture the moment it is provided electrical power. This is the reason it is important that the very basis of the firewall policy should be that *no services except those that are specifically authorized are allowed into the protected network or out to the unprotected network*. Everything relating to firewall access should build from this foundation upward. This is the concept that all services are strictly prohibited unless an exception is created. All exceptions should be based on a business need as requested by business units or management. This ensures that the flow of information in either direction is from directives of the business and not from the firewall administrator. While the business leaders are forging these decisions, the first role of the firewall administrator is to analyze the associated risks of the request.

For example, a request to allow all employees HTTP access to the Internet would be very common. The business argument may be that all users have the right to use the Web as a resource in their daily functions. The risks of such a request start with the idea that every browser-enabled workstation will have the ability to access the Internet. When the original proposal was created for Internet access, did the networking group have these kinds of numbers in mind? A specification may have been made that a T-1 line would be capable of handling all the Internet-related business to a DMZ-based Web or ftp site only. Additionally, all browser-enabled workstations could access sites deemed inappropriate in the "acceptable usage" section of the corporate policy. This could create a sexually hostile environment if an employee decides to view an adult-oriented site in plain view of others, opening the organization up to sexual harassment issues. From these risks, a control is recommended such as a separate server that monitors and blocks content. Additional time will be needed by the security department to audit these logs daily. The control would further recommend that evidence of violation of "acceptable usage" is presented to the human resources department and appropriate action taken with the offending employee. Combining all of these factors presents both the business leaders and the firewall administrator with this equation:

> (The likelihood for the risk of inappropriate/excessive usage for Web access) multiplied by the (cost of upgrading the Internet connection speed, purchasing a content filtering solution, and code of conduct inquiries involving HR and security) divided by the business need.

It is often likely that the business need as a denominator will far outweigh the costs of such a request, and the first exception to "no services are allowed" is born. Keeping the business unit leaders focused on these issues will smooth the process that creates the policy bullet points. Now the policy creation effort needs to view the access in its two separate domains, inward and outward traffic.

INWARD NETWORK TRAFFIC

The services allowed into the protected network by the firewall should be the minimal, and afford the maximum protection. Since TCP/IP is the protocol of the Internet, one is dealing with the vulnerabilities through the specific IP ports of the firewall (maintain your own awareness of vulnerabilities by using the resources provided by CERT, CIAC, and COAST, to name a few). A risk analysis is important for every request to open a hole in the firewall. Requests for services such as Telnet and ftp should be weighed not only on their merits of the unit making the request, but also the risk they may pose for the entire internal network. This is especially important in an age where costs have driven servers to do more than just one business task. All traffic coming in from the unprotected network should be viewed as suspect. While it can be a daunting task, the firewall logs are one's best ally in getting a feel for who is coming into the network. In an environment (such as in Exhibit 53.2) where a separate NIC controls the DMZ network, ftp and HTTP traffic could be a normal business day. Attempts on the ftp ports (21,22) and HTTP (80) should raise concerns when dealing with the log entries for the internal network. In an application proxy environment, one may have the freedom to choose a nonstandard port for external Internet clients coming into the protected network. This can be used to one's advantage against port scanning attacks. For example, if one creates a database application and houses it internally on a server, a nonstandard port may reduce the risk of a standard attack. For example, a client starts an application front-end developed by your business. With the firewall as the gateway to your internal network, they request the firewall's outside interface address and Port 5988. In the application proxy environment, one can plug Port 5988 to an internal server and the standard port number normally used for this type of application. This will allow monitoring of the logs for access on nonstandard Port 5988, and attempts at the standard port (which may signify an attack). If an attacker spends the time to probe all of your ports and gets some kind of response from 5988, they now must figure out what it is used for. In advance of this access, it becomes imperative that the database server handles the authentication of the user making the access, and that these logs are also reviewed in a diligent manner by the DB administrator. This can become a bargaining point during the risk analysis phase. If the business need presses for opening this hole in the firewall, then security measures on the targeted internal server must be increased. This can range anywhere from longer password lengths to smart card authentication. A good audit of the server's security controls is recommended prior to opening access. From this example, two policy bullet points can be created:

- Any access from the Internet into our internal network must use nonstandard IP ports whenever possible.
- Any server receiving access requests from the Internet via the firewall must have a full security review prior to setting up the access. Both the systems and DB administrators must perform regular log audits of this access.

Another issue that will impact firewall policy with regard to Internet users driving inward traffic may still come from employees. Employees with their own Internet accounts may decide it is time to test the merits of the security program and run SATAN against your firewall one evening. In a worse scenario, they may try something less organized and possibly devastating, such as trying to telnet into your DNS port (53). If this policy point does not already exist in corporate policy, the time may be right to add it here:

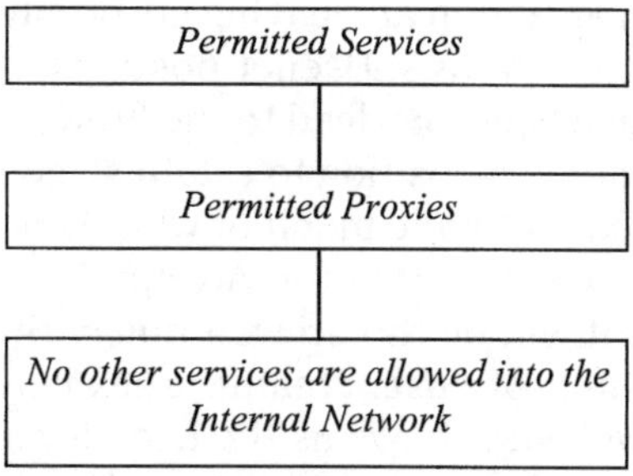

Exhibit 53.4 Decision Tree for Inward Access.

- Testing of security measures installed for the corporation must be authorized by the security department.

Other uses of the Internet can have an impact on the overall performance of both the ISP connection and the firewall. Carefully review the need to have channel technology constantly running across the firewall. Is it necessary in a parts plant to have stock updates going to every employee? What about continual software pushes or newsletter subscriptions? Make sure that in log reviews these uses are sorted out and weighed against their impact on performance. This may create a need for a further dialogue with business unit leaders about acceptable usage.

INWARD NETWORK POLICY MODEL

When reviewing a request for inward access from the Internet, it should fall into a category of either an allowed service or, in the case of an application proxy gateway, an allowable proxy. All other nonrequired services should remain blocked until an analyzed business need grants justification (see Exhibit 53.4).

In the firewall policy, document what has been allowed, the requested business group, IP port, and date of access. For example:

> Lotus Notes Access has been granted via a plug proxy for Marketing, requested by John Doe on 7/2/98. The port 1352 is being used.

OUTWARD NETWORK TRAFFIC

The services allowed outside to the Internet can be more flexible because strictly browsing outwardly has less of an impact on your internal business. But there are inherent dangers with granting access to the Internet that can be strictly controlled from the firewall. These access controls should not only serve to protect what is brought back to the screen or disk of a user, but also what a user can do to other systems. Users accessing the

Internet from your organization are leaving corporate footprints wherever they go. Whether on the routers, Usenet postings, or in probe attacks of other systems, the trace back may lead to the firewall IP address. This puts the security practitioner in a position to establish controls on how the outside world is to be treated. While much of what is needed for external access policies should be covered in the Acceptable Usage portion of any good security policy, it will be the firewall policy that is needed to enact them. The earlier example of a user creating a sexually hostile environment by viewing an adult web site shows us the need for a control such as content filtering, or outright blocking from sites. This leads to the following policy bullet point:

- Access to nonbusiness-related sites can be blocked and will be audited by the security staff.

As in any policy statement, wording is important. To say simply "adult-oriented" would leave room for argument, as would the word "inappropriate." To blanket the statement with "All access to nonbusiness …" would be a difficult point to enforce. Does the company come to a halt if someone reads the news from a cable news web site? Probably not. But the stakes are higher if the site offends, or in some states violates the law. When using a content-filter server, or even if blocking access on the firewall, a stern warning message screen upon an access violation is useful in putting a stop to the behavior:

> WARNING — You have entered a site that is nonbusiness-related, and the action has been logged. Move to a business-related site immediately.

It is important as well to follow up this warning with action so that it is taken seriously. Finding the sites to block can take a considerable amount of time and is sometimes post facto; check with your firewall vendor on the most comprehensive solution.

Electronic mail will also have an impact on outbound Internet access. A comprehensive e-mail policy, or subsection within the corporate policy, should cover most issues. From a performance side, consider the limitations needed for e-mail attachments. This applies to both mail received and sent. Some firewalls will work on a "store-and-forward" method that will use valuable disk space on the firewall for unreachable mail servers. Large attachments could cause problems. Verify what the limits are on your firewall and establish this as a policy point:

> Mail either sent or received can have attachments no larger than five megabytes. Files of a larger size must use alternate means of delivery.

Follow this with a disclaimer statement (that may or may not already exist in the e-mail policy) regarding mail or Usenet postings:

> All mail messages and postings must have the following disclaimer — "These opinions are my own and not that of my employer."

General services should be listed as a policy bullet point, and who they apply to:

> HTTP and ftp access will be allowed all employees for business purposes only.

If one has become granular in his or her access to particular networks, document this point as well:

> All users in the 192.20.10.x through 192.20.22.x subnets only are allowed Internet access.

As in the policy statements relating to internal access, it is imperative to cover the issue of "testing" the security of outside servers on the Internet. An ambitious employee who would like to see how secure a governmental server is, or test an extranet partner's security, would cause problems for any business:

> Testing of security measures installed for any other network devices is prohibited.

Remember that, in scenarios involving a DMZ or client side-style of network (Exhibits 53.2 and 53.3), one will need to make accommodations for providing data to these systems, as well as management functions. For example, if one is going to be supplying pages to a web server, then the ftp service will need to be used to provide current data. The firewall will need to accommodate this access, and the server will need to be configured to accept ftp from the firewall. When it involves SNMP monitoring devices, place one within the DMZ or on the client side network to avoid SNMP traffic over the firewall. A network monitoring station that is isolated within the DMZ or client side network will provide alerts for these specific devices and will not open another hole in the firewall.

Finally, document in the firewall policy what has been allowed out, the requested business group, IP port, and date of access. For example:

> HTTP access has been granted for the Marketing subnet, per a request by John Doe on 7/3/98.

OUTWARD NETWORK POLICY MODEL

A request for outward access to the Internet should be carefully reviewed as to its impact on the performance of the firewall and network, legal implications, and overall business need (see Exhibit 53.5). Follow the model in Exhibit 53.5 when defining outward access policies.

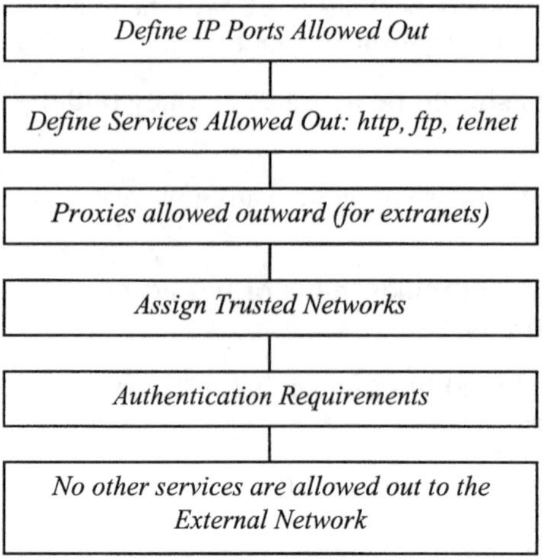

Exhibit 53.5 Decision Tree for Outward Access.

OTHER ISSUES WITH OUTWARD NETWORK ACCESS

Some businesses running IP networks never figured on an Internet presence and quite possibly could be using internal addresses that are registered to someone else. To attempt to use an already registered address will cause problems in routing and other network issues. Before ramping up to the Internet, ensure the use of your registered InterNIC address. Use either a legal, nonbroadcast IP address or something from the nonroutable set of IP addresses (172, 192, 10). In addition, this can be confusing when reviewing firewall logs. If using another company's registered address, your firewall will attempt a DNS resolution (more overhead) and create larger, less-reliable logs. Also remember that an eye towards legal issues is helpful in creating the policies. Beyond what has already been mentioned about hacking and sexually hostile environments, there are also issues with online wagering and child pornography that have extensive liability implications. Involve your organization's legal department in establishing and enforcing the policies.

GENERAL INFORMATION SECTION

Include a section in the firewall policy relating to general information. This can be a subset that is just supplied to technical staff, but should tie in with this policy. It should include all information about the firewall hardware itself. Include the model, type, CPU, serial numbers, power requirements, and required backup media. Be certain that, at the very least, the main configuration files are backed up regularly. Compile a list of all warranty information, technical, and emergency contacts. If using token cards, be certain that all users who have these cards are documented and required to return them upon termination or resignation. It helps to know the battery type needed for the cards as well, in case of failure.

Outline the network requirements — is 10 or 100 base-T running? Are the NIC cards set to a maximum MTU (maximum transfer unit, the size limit for a protocol or network) speed? Be certain that the firewall is included in any recovery plans for the data center in the event of a disaster. Finally, minimize the number of individuals who have accounts on the firewall. These should be individuals in a trusted class who have direct responsibility for the device.

CONCLUSIONS

It is important to continually monitor for unusual access, as well as the expected access (such as seen in proxies). Also, performing a penetration test from the outside of the firewall is a good way to determine the effectiveness of the structure and the technical setup. Observe how this affects the log files so the patterns that indicate trouble can be recognized. When using plug proxies that are not transparent, provide a legal banner regarding authorized access (like the one that should be on the login screen of all servers). Avoid revealing too much about the organization in any banners or connections made via the firewall. Customers should already know where they are going.

When initiating a firewall policy, view the process in two phases — access out and access in. Remember that any policy is a living and growing document that will often evolve. It needs to remain flexible in its approach because it is business-driven. Security is always the main focus for this device.

When it comes to the policy, require a well-planned collaborative process among security, systems, business unit leaders, and network personnel. Document by date all changes made to the rules files of the firewall, and provide an explanation for any of the changes in writing. Finally, always focus the request for access via the firewall from a business-need perspective. Business should be the only issue strong enough to drive a hole into any firewall.

Chapter 54
Assessing and Eliminating Virus Threats in Distributed Networks

Frank Horwitz

COMPUTER VIRUSES COST USERS BILLIONS OF DOLLARS PER YEAR IN LOST DATA, LOST PRODUCTIVITY, AND CLEANUP COSTS. This chapter examines the problem by defining viruses, why they matter, how they infect systems, and how to discover whether they have infected a system. Standard approaches to fighting viruses are explained, weaknesses in some commonly used virus-fighting techniques are illustrated, and the ideal virus defense system is discussed.

PROBLEMS ADDRESSED

Technically speaking, a computer virus is similar to a biological virus: it wants to reproduce itself. A virus does not necessarily inflict any damage. Industry experts define viruses differently. A virus can be described as a piece of code that attaches itself to a file, critical disk sector, or memory location for the purpose of replicating. Another definition describes a virus as a program designed to replicate and spread, generally with the victim being oblivious to its existence. A more complete definition says that a virus is a program that replicates itself, attaches itself to other programs, and performs unsolicited, if not malicious, actions. By any definition, reproduction is the common theme.

Unless it is deflected or killed, a virus usually spells difficulty and expense for network administrators whose task is to eliminate them. This chapter provides practical information for preventing, discovering, and eliminating viruses.

How Pervasive Is the Viral Threat?

One of the most damaging effects a virus can have on a corporate LAN or WAN is a drain on system resources. A very destructive virus, such as the Byway Virus, can reproduce rapidly enough to fill a multigigabyte hard drive overnight, and can cause an entire system to crash. Others, such as Junkie Virus, fill memory and cause system response times to slow drastically. In either situation, the best case is a loss of productivity; the worst case is the systemwide loss of data.

Loss of data is another specter of the viral threat. One of the most common DOS viruses, Jerusalem, is designed to erase any program executed using the DOS execute program call. All of the programs users try to run suddenly cease to exist. A variant of this virus (known as the 1704-Format), when activated, attempts to reformat part of the hard drive. Another common virus, Disk Killer, attempts to scramble all data on an infected disk or diskette. These and many other viruses can cost days of cleanup and restoration in a well-maintained network, or wipe out months of productivity in a poorly backed-up network.

Another problem created by viruses is the cost of cleaning them off infected systems. A survey of corporations with more than 1,000 PCs reported that the average cost of cleanup can be as high as $254,000, a figure that includes only the direct labor expense for system recovery and data backup. The indirect expense of lost productivity is much higher. One estimate states that viruses cost American businesses $2.7 billion in 1994. In addition, the average recovery time required to clean up an organization having more than 25 PCs is four days. Even worse, 25 percent of those experiencing a virus attack suffered a reinfection by the same virus within 30 days.

Even one virus incident can potentially cost a company millions of dollars. Although budgets often place computer security low on the priority list, the cost of prevention seems almost negligible when compared to the potential loss of time and money.

The odds of being infected with a virus are getting worse every day. Consider this progression of averages:

- In 1986, one new virus came into existence every one and a half months (there were eight known viruses; four of them existed only in computer laboratories).
- In 1989, one new virus came into existence every week.
- In 1990, one new virus came into existence every two days.
- In 1991, six new viruses came into existence every day.
- In 1994, approximately 7,000 known viruses existed.
- In 1995, approximately 15,000 known viruses existed.

Currently, the number of viruses doubles every eight to eight and a half months. Hackers and virus authors are working cooperatively. Electronic bulletin boards allow them to share not only new viruses, but virus-creating engines. A would-be virus author can learn from books, virus kits, the Internet, and even CD-ROMs. However, antivirus companies, in order to maintain profitability, work alone, unwilling to share source code. The result is that there are 1,200 known virus authors but only 200 virus researchers. At that 6:1 ratio, the virus authors are getting more done than the researchers. Only 38 percent of corporate users consistently apply workstation antivirus products. As a result, more than 40 percent of all networks have viruses.

How Viruses Infect Systems

Usually, a virus enters a system through an intrusion point such as floppy drives on user workstations. On a network, intrusion points include E-mail, modem pools, and gateways to other networks. Approximately 87 percent of viruses enter systems from floppies, and 43 percent of those are brought from home by unsuspecting users. Once on a system, a virus usually either attaches itself to an executable file so that whenever that file is executed the virus is too, or the virus infects the boot sector of the PC so that from there it can travel to other floppies or logical disks.

Major Types of Viruses

The following sections discuss the most prevalent types of viruses, including file, boot sector, multipartite, file overwrite, stealth, polymorphic, and macro-based.

File. File viruses usually attach themselves to an executable file, such as .EXE and .COM on DOS machines. The virus can insert its code into the host program's code so that when the program executes, the virus executes first. Most of the thousands of viruses known to exist are file viruses. Windows 3.1 barely runs in the presence of a file virus. If a file virus is resident in the memory of a DOS system (which is exactly where file viruses like to reside), in many cases Windows cannot even start. This generally causes the user running Windows to eliminate the virus, perhaps unwittingly, as they attempt to fix their system. A growing trend toward Windows 95 and 32-bit operating systems may signal a resurgence of file viruses.

Boot. Boot sector viruses cause the vast majority of actual attack incidents. Each of the top 12 viruses reported last year were boot sector viruses. Whenever a computer is booted up, it looks for instructions about how to operate and what to do. It finds those instructions in the boot sector of a hard drive or floppy disk. Boot viruses insert themselves into boot sectors so that the virus executes first and gains control of the system, even before the operating system is loaded.

Boot viruses are especially dangerous because they can spread from anything that has a boot sector. Any floppy disk — even an allegedly blank one — can spread boot viruses. If a boot virus on a floppy disk is inserted into a computer, the virus goes into RAM and infects every disk that computer accesses until the computer is rebooted, which wipes the boot virus from memory.

Multipartite. Multipartite viruses combine characteristics of file and boot viruses. Multipartite viruses can spread as easily as a file virus, yet still insert an infection into a boot sector, making them very difficult to eradicate.

File Overwriters. File overwriters are file viruses that link themselves to an executable program but keep the program intact. Executing the program also executes the virus, which attempts to add itself to as many files as possible. File overwriters often have no purpose other than to replicate, but even then they take up space and slow performance. They may damage or destroy files inadvertently.

Stealth. Stealth viruses are engineered to elude detection by traditional antivirus checkers. The virus may target and eliminate the detection function of a commercial antivirus product. Stealth viruses reside in memory, intercepting the system's MS-DOS calls in order to make infected files appear uninfected. The stealth virus can then infect every floppy diskette and logical drive the system accesses. Some antivirus scanners help propagate stealth viruses because they open and close files to scan them, giving the virus additional chances to spread.

Polymorphic. Polymorphic viruses include a mutation engine that makes the virus change minor parts of its code each time the virus is executed. Different encryption algorithms are nested within a polymorphic virus to help it hide from scanners. A decryption routine included in the virus allows it to return to a normal state when it executes. The stable bytes (the decryption algorithm) become shorter with repeated executions of the virus. This defeats first-generation virus scanners, which operate by checking code for any matches with virus code.

Virus authors can access polymorphic engines, which can take a non-polymorphic virus as input and output the virus with polymorphic qualities. The availability of such engines has made the authoring of polymorphic viruses a simple, straightforward task. As a result, the number of polymorphics has doubled about every eight months. Today, more than 200 polymorphic viruses produced by these engines exist, and another 50 polymorphic viruses are known to exist that do not use the engines. The latest generation, the superfast polymorphic infector, can lay waste to every executable in every directory on a PC's hard disk without requiring that .COM and .EXE files launch first. Running a directory listing is enough to trigger the virus.

Macro-Based. Macro-based viruses are the newest innovation. A macro virus is unusual because it can infect documents instead of programs. It is the first virus that can cross platforms, infecting both PCs and Macintoshes. The one known form of the virus, written in Word Basic and referred to by Microsoft as the Prank Virus, infects only Microsoft Word 6.0 files. The virus is not destructive; it simply adds nonsense Word macros to documents that end with .DOC or .DOT. Although Prank is not really destructive, its implications for the future are disturbing because it has introduced an entirely new method for viruses to spread.

Common Spread Scenarios

Viruses spread through organizations several ways, including through the use of shared machines, shared diskettes, popular programs, and LAN servers.

Shared Machines. Viruses spread throughout an organization most commonly through shared machines. A computer used by many different people can serve as a center of infection. If a user runs an infected program on the machine, the infection has probably spread to programs on the machine's hard disk. If other users bring their own diskettes to run on the machine, the diskettes and any programs on them are likely to become infected. The diskette will probably carry the infection to other machines.

Shared Diskettes. Many diskettes, such as diagnostic diskettes, product demos, or company manuals, are routinely carried from machine to machine. If such a diskette becomes infected, the infection can quickly spread to many machines.

Popular Programs. Popular games, demos, or animations often cause the user who obtains a copy to want to pass it on to other people. If one of these programs becomes infected, the infection can spread quickly to many machines.

LAN Servers. If a program on a LAN server used by many workstations becomes infected, a large percentage of the LAN workstations can become infected very quickly (sometimes within an hour or two). One common mistake is to have the LAN logon program in a place where anyone on the LAN can write to it. This setup means that if any workstation on the LAN becomes infected, the logon program quickly becomes infected, and then every workstation that logs on to the LAN immediately becomes infected.

HOW TO DISCOVER A VIRUS

Viruses can continue replicating until they are detected. The most well-crafted viruses show no symptoms to reveal their presence. However, many viruses are flawed and betray their presence with some of these indications:

- Changes in the length of programs.
- Changes in the file date or time stamp.
- Longer program load times.
- Slower system operation.
- Reduced memory or disk space.
- Bad sectors on a floppy diskette.
- Unusual error messages.
- Unusual screen activity.
- Failed program execution.
- Failed system bootups when booting or accidentally booting from the A: drive.
- Unexpected writes to a drive.

Instead of waiting for a sign, network managers should use the appropriate tools to seek out viruses before they get far enough to compound problems. The ideal is to repel them before they infect the system.

STANDARD APPROACHES TO FIGHTING VIRUSES

There are several ways to combat viruses. Computer viruses have become increasingly cunning in their programming and ability to avoid detection or eradication. However, virus-fighting tools have also grown through several generations to meet the challenge. Some of the various approaches are described in the following sections.

Signature-Based Scanners

Traditionally, virus scanners look for known virus code and when they find a match, they alert the user. The leading scanners are signature-based. Signatures are strands of code unique to a single virus, analogous to DNA strands in a biological virus. Virus researchers and antivirus product developers catalog known viruses and their signatures. Scanners use these catalogs to search for viruses on a user's system. The best scanners have an exhaustive inventory of all viruses known to exist and examine all possible locations for infection, including boot sectors, system memory, and files.

Multilevel Generic Detection

Generic detectors are used to eliminate unknown viruses. This method performs integrity checking using checksums.

A checksum is created when an algorithm reads a file's bytes sequentially, creating a unique numeric code based on the file itself. Generic antivirus detectors then compare checksums recorded when the system was in a known clean state with checksums recalculated subsequently. If a virus has attached itself to a file, the bytes will add up differently and the new checksum will no longer match the old (i.e., clean) checksum.

Using this method, it is not necessary to know anything about a virus; instead, the system focuses on what the clean file should look like. The Secret Service uses the same method when teaching agents how to spot counterfeit currency. New agents receive extremely detailed training on what a real dollar should look like rather than on what various counterfeits look like.

The other techniques used in generic detection enable antivirus programs to distinguish between normal, legitimate writes to a file in contrast to viral additions. Expert systems test a system's software by examining code flows, calls, executions, and other functions to spot viral activity. Sophisticated versions of this approach not only spot viruses, but clean them automatically.

TSR Monitoring

Terminate-and-stay-resident (TSR) programs stay in memory but operate in the background while other programs run. Because most viruses are essentially TSR, it makes sense to combat them with a TSR. Antivirus TSR programs can provide real-time monitoring of disks and files, expert systems analysis of virus-like behavior and code, and may even detect stealth and polymorphic activity. Rather than only working when invoked, TSRs stay on in automatic mode whenever the workstation is in use. Instead of looking for code that matches memorized patterns, as scanners do, antivirus TSRs attempt to catch viruses "in the act." On a network, antivirus TSRs can download from a server to each client as it logs on so that users do not need to remember to activate antivirus tools.

Behavior Blocking

This is the only defense that can prevent viral infection, rather than merely detecting viruses after they have infected. Behavior blocking performs on-the-fly code analysis, monitoring the sequence of code behavior until it can distinguish whether the code is safe or harmful. Harmful code is not permitted to execute. Instead, the behavior blocker notifies the user. Behavior blocking programs use some or all of the following techniques.

File Attribute Monitors. A virus cannot infect (i.e., write to) an executable that is marked read only. Many viruses work around this by first modifying the file's attributes so that the file is now a read-write file. Behavior blockers can intercept code that attempts to change or delete the attributes of files.

Intercept Reboot. Some behavior blocking intercepts Ctrol+Alt+Del warm reboots and checks any inserted floppy for viruses before allowing the computer to warm-boot off that floppy. If the floppy has a virus, the behavior blocker warns the user that the floppy is infected. This technique can halt boot viruses.

Smart Blocking. This term refers to very sophisticated behavior blockers that are able to distinguish complex virus behaviors from the complex behaviors of a user running complex software. Smart behavior blockers can analyze detailed sequences of behavior, using statistical analysis to determine the probability that a particular sequence is a virus.

Rescue Disks. Rescue disks are used to salvage data once a virus has infected a PC. It is important that each PC have its own rescue disk. During the installation, an operator must be present to put in the diskette — there is no automatic installation. Users must keep track of their rescue disks. If the disk is lost, there is no way to rescue the PC from the virus infection.

Physical Access to PCs. One simple but important technique for defeating viruses is to control who is able to use the computers. Despite the rise of the Internet, most viruses still enter machines through floppy disks. Although the majority of infections come through the hands of unwitting employees, a percentage of attacks emanate from hostile intent. Therefore, some viral attacks can be deflected simply by deterring unauthorized personnel from using machines. Besides taking measures such as securing physical access to computer rooms, a manager can also use security products that render physical and logical drives invisible to certain users or user groups on a network. Thus, fewer personnel have the opportunity to hack those drives.

Drawbacks of Signature Scanning

Despite the existence of sophisticated antivirus tools, many organizations rely almost entirely on signature scanning to detect viruses. In light of the virus boom, signature scanning alone is a mediocre defense, at best. Some of the drawbacks of this commonly used approach are described in the following sections.

Passivity. The most profound flaw in relying on signature scanners is that they are reactive, or passive. The goal of scanning is to detect a virus that has already infected a file or a boot sector. The ideal method is to prevent viruses from infecting the system at all, not merely to be informed of the problem after the fact.

Incomplete Checking. A polymorphic virus, which produces varied but fully operational copies of itself, can deceive signature scanners by altering or encrypting its signature. Signature scanners have attempted to address this by including several signatures for a given virus, one for each possible encryption method or iteration of the signature. As polymorphic viruses become increasingly sophisticated, the brute force method of including more signatures in the scanner will not be able to keep up with all the possible variants of all the polymorphic viruses. Many polymorphs already evade detection by interspersing noise instructions or by interchanging mutually independent instructions within the code to continually modify the signature. A simple signature-based scanner cannot reliably identify this type of code.

Failure to Scan for Newer Viruses. Scan strings can only be extracted and cataloged if the antivirus vendor has a sample of the virus. In the recent past, it took the most common viruses six months to three years to become prevalent, giving vendors enough time to send out regular updates of known viruses and head them off. The exponential growth in viruses has increased the likelihood of a new virus reaching the LAN or PC before the update from the antivirus company does. Besides creating a chance of missing an unknown virus, signature-based scanners require constant updating. If the signature scanner is not centrally administrated, it slows productivity and drains resources because of the management tasks needed to install each successive enterprisewide update.

Insufficient Scanning Frequency. In theory, a virus infecting a system at 8:59 a.m. could be caught one minute later if the network is routinely scanned at 9:00 a.m. However, the opposite scenario is just as likely. A network may be scanned at 9:00 am and become infected at 9:05 am. If the virus is a fast infector such as Dark Avenger or Frodo, once it is in memory it can infect not only executed programs, but even those that are merely opened. Such a virus has almost 24 hours of free time to wreak havoc in the network. Even worse, because many signature scanners open files in order to scan them, the very act of using the scanner can allow the virus to infect all programs at once.

Slow Scanning. Any scanner takes a finite amount of time to scan a machine for viruses — perhaps five minutes or more. If the 70 million U.S. employees who use PCs spend five minutes a day scanning, and earn $15 an hour, the annual cost of scanning(260 days a year) is more than $22 billion. The costs of scanning exceed the purchase price of antivirus software after just a few weeks of scanning. More sophisticated tools can cut this time drastically by scanning checksums instead of the entire contents of every file. The more viruses a scanner must search for, the more places within a file it must search, and the more files it must search across, the slower the search must be. Because strings must be stored in memory, and memory is limited, there will soon be two-pass products that load one set of strings, scan, then load a second set and scan. Although computers are faster now, hard drives are also getting larger.

Dependence on User Compliance. Traditional scanners do not work unless employees remember to use them. Some users are inclined to value their own productivity and convenience more than their employer's security concerns, and thus are not motivated to consistently scan. Even diligent users tend to get lax if scanning every day for a month produces no alarms.

RECOMMENDED COURSE OF ACTION

As long as there are hackers inventing new forms of maliciousness, no antivirus vendor can guarantee that their products will completely eliminate viruses. However, there are advanced products that come very close to providing the ideal defense. Knowledgeable implementation of advanced protection strategies and products can prove an effective deterrent to viruses in the short and long term.

Strategies for Virus Prevention

The first priority for an antivirus strategy is that any defenses put in place must be used. Many approaches emphasize end-user convenience to the point of rendering defenses useless. A company can, however, set up antivirus software on its LAN servers so that each time a user logs in, the program checks for its own presence on the user's workstation. If such antivirus software is not present on the workstation, the program loads itself onto the PC and scans the PC's hard drive before allowing the user to continue. If the program finds an earlier version of itself, or a modified version of itself on the workstation, it loads the newer, clean version onto that workstation and scans. The entire process happens rapidly enough not to harm user productivity. Many users do not even notice it happening.

This approach is far preferable to that of programs that depend on users remembering to scan periodically. Such programs leave holes in a system's defenses every time even one user forgets to scan. Users are often tempted to skip scanning, especially if the scanning process is slow. This adds an even more haphazard quality to network defense. Antivirus software should offer an unobtrusive way of forcing users to keep their machines clean.

Repelling Viruses Proactively. An antivirus strategy should be proactive. It should detect and repel viruses before they infect anything on the system. A signature scanner working as the sole defense of a network can do nothing more than occasionally report bad news. The ideal system must be able to stop boot viruses before they infect and must be able to remove all viruses without necessarily knowing the virus. Proactive antivirus software provides signature scanning as well as multilevel generic detection, a TSR approach, and behavior blocking to remove viruses that are known and unknown.

Comprehensive Security. Some antivirus software scans only for the 200 most common viruses, which account for the majority of infections. Protecting a system from these common viruses may offer sufficient protection because the likelihood of infection by another virus is quite slim. However, the ideal system is not one that usually works, or hardly ever misses a virus, but one that seals off every conceivable intrusion point.

In addition, viruses tend to spread in a regional fashion, turning up much more frequently in one particular country or geographical area than other areas. If a virus common in a particular region is one that the software perceives as uncommon, the scanner could miss the virus. This is especially threatening in companies that have international offices. Effective antivirus software uses a combination of traditional and proprietary heuristic techniques to ferret out even the trickiest viruses, Trojan Horses, and logic bombs. Scanning alone is not sufficient. The most effective antivirus system should use the latest generation of defenses in concert.

Automatic Logging. Antivirus systems should document any security events that occur so that managers can stay informed about threats to their defense system. Documentation should include logins, logoffs, program execution, and a separate log of failed login attempts. Effective antivirus software should also require password entry upon any bootup and prevent access to hard disks any other way. After a period of inactivity at the keyboard, a time-out feature should inhibit input from the keyboard and mouse. Documentation and automatic logging requirements help management restrict physical access to workstations, which is vital to maintaining a protected environment.

Chapter 55
Mobile User Security

Ralph R. Stahl, Jr.

INFORMATION TECHNOLOGY PROFESSIONALS TODAY ARE UNSURE OF THEMSELVES IN A STRANGE, NEW ENVIRONMENT. However, end users are telling security practitioners that they can no longer perform optimally and beat the competition if their access to information and processing power is restricted by the mainframes in the corporate data center.

The days when security practitioners arrived at the office to find a stack of computer printouts on their desks are gone. Paper has been replaced by computers on the desktop in the environmentally correct and secure office. In addition, users today expect to be able to connect their notebooks by modem from any location to the server at headquarters. Tomorrow, mobile users may expect connectivity for their notebooks in the air as they fly, and on the ground as they drive to their next destination. The business traveler may anticipate that all applications will behave in exactly the same manner on the road as in the office.

Large companies like AT&T are encouraging employees to telecommute for three major reasons:

- State and local governments are requiring companies to take action to reduce air pollution and traffic congestion.
- Office sharing allows companies to reduce their real estate expenses.
- Telecommuting benefits employees by allowing them more flexibility in managing their professional and personal lives.

LINK Resources Corp., a New York City consulting firm, reports that Americans bought for home use a record 5.85 million microcomputers last year. One out of three American households already has a microcomputer. BIS Strategic Decisions estimates that 45 million workers in the United States are considered part of the mobile work force. Other surveys estimate that, in addition to the time spent in the office, the average white-collar worker spends six hours a week working at home.

Against this backdrop, the challenge for the application developer is to develop systems that may be used in any environment. The information architecture for the enterprise must also accommodate many methods of

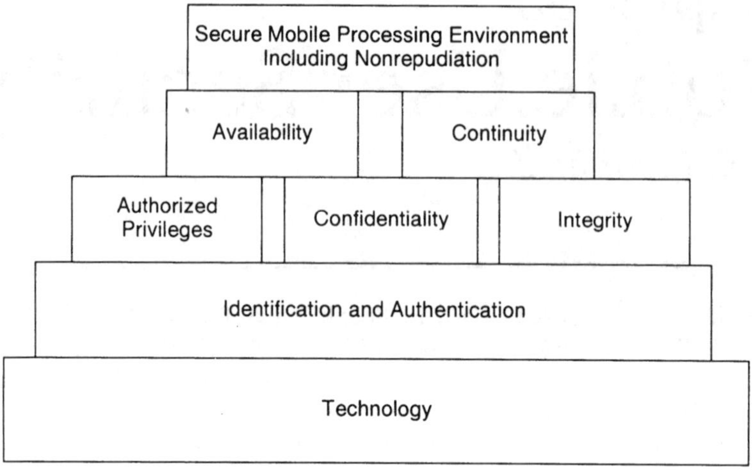

Exhibit 55.1 Model of Security Services.

remote connectivity (i.e., dialup, Integrated Services Digital Network [ISDN], Cellular Digital Packet Data [CDPDI, Internet, wireless, video, and image transmission) in addition to the traditional local area network and wide area network connectivity.

This chapter is divided into four major sections: availability and continuity; integrity; confidentiality; and new technology considerations, which briefly reviews the security implications for some of the emerging technologies. The architectural model of the security services in Exhibit 55.1 provides a high-level view of the interdependence of identification and authentication, authorized privileges, availability, continuity, integrity, and confidentiality in providing a trustworthy environment that supports nonrepudiation and mobile power user security.

AVAILABILITY

In this chapter, availability is defined as the assurance that an authorized user's access to an organization's resources will not be improperly impaired. Achieving such assurance involves properly categorizing information privilege keys and ensuring that the mobile user's authorized privileges are properly associated with these privilege keys. Availability also involves physical considerations (e.g., theft prevention, device identification, mobile uninterrupted power supply), notebook connectivity (e.g., a power source, telephone communications tools), and miscellaneous toolkit necessities.

Scheduling Considerations

Information availability is an operations scheduling issue, although some organizations believe that all availability needs are covered by their business resumption practices. The security practitioners must be aware of the need to maintain operational schedules. If the backup and batch processing is scheduled to end at a precise time so that the online or remote transaction processing may start, then the credibility of the central staff to meet their commitments to the field are tested every day. Although capacity planning is not a security issue, the complete information protection plan will make sure that the topic is adequately addressed by the appropriate operational staff members.

Physical Considerations

Concerns associated with the desktop microcomputers in the corporate office also apply to notebooks for the mobile user. However, with respect to mobile computing, security practitioners may need to be more creative to achieve the desired results.

Theft-Prevention Devices. Such theft-prevention devices as cabling and bolting plates can be used to minimize the potential of notebook theft by opportunity. The cables are designed so that they may be looped through an opening in a stationary object to tie the laptop down while the user is traveling. Resistance to these devices exists because many users feel that having these devices gives the impression of not trusting coworkers or business associates. However, security administrators who use theft-prevention devices in their companies indicate that they have experienced a significant decrease in loss. Although the products are effective, corporate procedures with strong enforcement practices are usually required before these products are put into use.

Device Identification. Device identification is critical to the ability to identify a misplaced or stolen notebook. In addition to traditional identification methods (e.g., serial number registers, tags, labels, and engraving), microcomputers can be marked by using invisible ink to record the company's name and the notebook's serial number on the inside of the lid just under the monitor display area. The invisibly inked number must match the serial number recorded in the corporation's asset inventory register. This practice can also be used to resolve disputes associated with ownership of the microcomputer.

Mobile Uninterrupted Power Supply. Mobile uninterrupted power supply implies that each mobile user should have a portable surge protector with sufficient electrical outlets for each device that is connected to the microcomputer or notebook. Electricity follows the path of least resistance, and it will reach the microcomputer through any device cable if the power source for the device is not protected. Surge protector plugs are available at most hardware and electronics retailers. It is also recommended that the user carry a fully charged spare battery pack for the notebook. Usually the battery can be purchased from the dealer that sells the notebook.

Notebook Connectivity

To ensure the notebook's power source, the electrical wall connection should be used so that the notebook's battery can be conserved or recharged. For proper grounding, the electrical code requires that all computer male plugs be three-pronged. In some facilities, the female electrical wall receptacles may accept only two-pronged male plugs. In such cases, the problem can be averted by carrying a female/male converter plug that converts three-pronged plugs to two-pronged plugs and has a grounding wire that may be attached to the wall receptacle's holding screw. This type of converter is available in hardware stores. If the user travels internationally, the toolkit must also include an international voltage adapter that eliminates the need to carry different converter plugs.

Telephone Communications Tools

Offices and hotels are updating their telephone PBXs to digital service, but modems and the majority of the PBXs are still analog systems. To ensure connectivity in either environment, a converter should be purchased that can covert the phone line to analog at the modem connection. The complete converter kit should include alligator clips for phones that do not have RJ-11 jacks. The phone line converter requires an AC adapter as the power source; the full-functioning kit will have the capability to use a 9-volt battery when an electrical wall outlet is not available.

Data transmission may be interrupted if a phone system with call-waiting capability is used. The feature can be suspended during the data transmission call by adding *70 (occasionally #70) at the beginning of the dial string. As a general rule, this is probably not required. However, if a transmission session is interrupted for an unknown reason, call waiting may be responsible.

A miscellaneous toolkit should include the following items:

- A small screwdriver with both a flat and a Phillips head.
- An extra-long telephone cord with male RJ- 11 connectors on both ends.
- A connector with two female RJ- 11 receptacles.

CONTINUITY

Continuity is defined in this section as the processes of preventing, mitigating, and recovering from service disruption. The terms business resumption planning, disaster recovery planning, and contingency planning may also be used in this context; they concentrate on the recovery aspects of continuity that ensure availability of the computing platform and information when needed.

Recovery diskettes will reduce the user's lost time when access to information and remote computing resources is lost due to a major breakdown of the notebook. Recovery diskettes should be customized to the exact configuration of the notebook and should contain the following files:

- DOS system files (COMMAND.COM and the two hidden files), which make the recovery diskettes bootable.
- PARTNSAV.FIL, which contains the hard drive partition table, boot sector, and CMOS information.
- CONFIG.SYS, which contains the appropriate values for files, buffers, stacks, and hard disk drivers.
- REBUILDOOM, which restores the CMOS.
- SYSCOM, which enables the transfer of the operating system to another disk.
- Copies of any special drivers that are necessary to meet the standard operational configuration of the organization's protocol stack.
- Communication programs that support emergency downloading of files and programs.

Creating Mobile Backup

The fastest and easiest backup medium is a cassette tape; however, this represents an additional cost and bulky devices to transport. Most (if not all) mobile users want to travel as lightly as possible. The diskette then becomes the most acceptable medium, but care must be taken to minimize the number of diskettes needed.

The proper organization of the notebook's hard drive will minimize the time and number of diskettes required to create a backup. A directory called *data* can be created with all of the subdirectories necessary to easily organize, track, and access data unique to the user. Application software, databases, or operating software should not be included in the data directory. These files should be stored in other appropriately established directories. When necessary, these software and data base files may be downloaded from the office server.

The backup diskettes should be kept in a different place than the notebook. If the notebook is lost, the backup diskettes will not be lost.

Loss of Computing Resources

Loss of computing resources as a result of the loss or theft of a notebook or its mechanical failure is the most difficult problem to deal with. The user is away from the office yet may need to repair or replace the notebook immediately, and canceling the next few days' appointments is not an acceptable option.

The recovery process requires planning and discipline on the part of the notebook's user. The data files on the notebook should be backed up regularly. If any ingredient of the business resumption plan is missing or incomplete, lengthy delays are unavoidable.

Mechanical Failure. When the problem is a mechanical failure, the existence and awareness (by the user) of a national maintenance agreement with a rapid response clause can ensure fast repair. The remote user should be able to easily obtain the location of the nearest repair location. This can be handled by the organization's help desk service, which should provide 24-hour accessibility. After the notebook is repaired, the user must determine if any data or application programs were lost or damaged. Lost or damaged software may be replaced by using the emergency recovery diskettes to download the needed data from the office server. Although this may be a long transmission session, it is preferable to getting on an airplane and flying back to the office. After this is accomplished, the notebook owner can restore individual files from the backup diskettes. This process may consume a full day and may require the active participation of the remote user, but after it is done, the user's machine has been restored with the least amount of lost time.

Lost or Stolen Notebook. When the notebook is lost or stolen, the plan must provide rapid delivery of a new notebook to the user in the field. Spare notebooks with the standard operating software, application software, master files, and current databases should be available at the data center. After a call is received to ship a backup machine, the only step necessary is to find the quickest method. Airlines and bus depots should be called to determine whether they provide shipment that is faster than that provided by the standard 24- hour service providers. After the shipment arrives, the next step is for the user to restore personal data from the separately stored backup diskettes.

If backup procedures for mobile users are to be effective, they should be tested and adjusted frequently. The recovery process may never be needed, and these procedures may be regarded as taking a lot of time. However, without such procedures, it may take the better part of a week just to obtain a replacement microcomputer. After that, it must be determined what applications and databases must be loaded. Finally, if backup procedures are not enforced, how will the user's personal information be restored to a new notebook?

INTEGRITY

In this section, integrity is defined as the process of ensuring that the intended meaning of information is maintained. Information integrity is provided by allowing only authorized persons and processes to perform only those tasks that they are authorized to perform. Everything else is prohibited.

Software Considerations

Virus Protection. Remote users are no more or no less susceptible to viruses than their office-based counterparts. Therefore, they are likely to experience a virus infection of their computer within the next few years. Proper procedures can prevent the virus from attaching itself to any of the hard drive's files. A corporate contract should be purchased for one of the leading virus detection and eradication software products. The cost can be surprisingly reasonable, in some cases less than $15 per user per year, although some very good products may be even cheaper. The secret of success is to have the detection software active in memory at all times as a terminate-and-stay-resident (TSR) program. Some vendors can relocate the TSR so that base memory is not used at all, while other TSR programs may take as little as 5K bytes of base memory. With the detection software active in memory, the computer's user may remain passive. Many organizations' detection programs have limited success because they require the user to execute a scan program to check diskettes or hard drives after disconnecting from a bulletin board. During a normal day's workload, users may often overlook this program. If TSR software is used, however, the microcomputer should lock up when it detects a virus, and it should not allow the user to proceed. The eradication or cleaning program should be run immediately to remove the virus. The remote user should be provided with a system-bootable diskette that contains a virus detection and eradication program. If additional directions are required to remove the virus, the mobile user should contact the organization's help desk for assistance.

Notebook Configuration Integrity. Although the name personal computer may have been appropriate at one time, today it is a misnomer because the microcomputer has become such an integral part of information processing for the business community. By personalizing the notebook to the configuration of their choice, users may cause incompatibility with their organizations' requirements.

As part of the processing infrastructure, it is important that the information technologist control the protocol layers to ensure proper connectivity, memory management, and execution of the company's processes. This is not to say that mobile users are not allowed to install some software of their choice. An example of coexistence on the microcomputer would be to control the autoexec.bat by not allowing anyone except the systems administrator to modify the file. However, the last line of the autoexec.bat is a call to an autoexec.bat that gives the user the ability to add activity to the booting process. Today, a number of client-based products exist that are designed to establish administrative control over the configuration and accessibility of directories and files on the hard drive of notebooks. These products support multiple-user confidentiality and several levels of administrators.

The content of certain directories on the hard drive should also be under the control of the systems administrator. Again, policies and guidelines should allow a section of the hard drive to be used at the discretion of the mobile user. Procedures and guidelines should clearly state the areas that may not be altered, and the latitude the mobile user has to customize the microcomputer.

These same robust access control products provide complete control over the DOS computing environment. Access to printers, ports, disk drives, modems, files, and directories may be constructed so that several users of the same microcomputer may not have the same authorized use of these facilities. In addition, passive DES (or proprietary) encryption and decryption of files or directories may be established. The major vendors provide the ability to configure directories on the hard drive to be encrypted or decrypted on the fly. When a file is written to the designated directory, it is automatically encrypted. Conversely, when a file is read into memory by the authorized owner of the directory, it is automatically decrypted. This prevents unauthorized access to information even though several users may share the same microcomputer.

Nonrepudiation. In this section, nonrepudiation is defined as the process of ensuring that a user — either the originator or the recipient — can be identified as having engaged in a particular transaction. This facility may also be used during the normal course of business activity to identify the originator of information. Nonrepudiation involves the following procedures:

- Electronic identification of the sender is accomplished by digital signaturing. Depending on the nature of an organization's business, a choice of standards can be followed.
- Message hashing ensures that the content of the message is not altered. This is accomplished by cycle redundancy checks which sum up a total value of the message's bits and stores that hash with the digital signature in an encrypted envelope for the message. Two major hashing algorithm standards exist: MD5, which is supported by RSA and the Internet (ANSI X930 part 2 and RFC 1321), and secure hash standard, which is supported by NIST (FIPS 180 and 180-1).

- A copy of the message (the hash) with the digital signature of the originator is sent to the message archive. Each message must have an established retention date that will vary according to the message content. The records retention policy should serve as the guide for establishing the date. The message should be automatically removed from the archive database when this date is reached.
- Proof that the message was delivered requires an electronic acknowledgment containing the date stamp of the activity to be sent to the archive and matched to the message.
- Proof that the message was opened requires an electronic acknowledgment containing the date stamp of the activity to be sent to the archive and matched to the message.
- When the importance of a transaction dictates that nonrepudiation is required, a utility should monitor the message activity to ensure that the message was received and opened. Electronic message status should be returned to the sender for appropriate follow-up as required (e.g., to determine why the message was not received or opened).
- The trustworthy information processing infrastructure must provide assurances that the message and audit details cannot be altered.

Many business uses exist for nonrepudiation in the mobile world (e.g., purchase orders, expense statements, strategic management directives, conflict of interest forms, and other important documents) that can provide technology an opportunity to reduce today's manual administrative efforts. However, nonrepudiation does not address confidentiality of the message; this is accomplished through encryption.

Remote Access Authentication. After the decision is made to allow modem, Internet, CDPD, or ISDN access to the infrastructure, the risk of unauthorized access to the infrastructure increases. Call forwarding and other advances in technology have eliminated the security effectiveness of dial-back modems. The Gartner Group considers dial-back modems to have limited effectiveness.

The most effective way known to authenticate a remote user is through two-phase authentication: with something that the remote user knows and something that the remote user possesses. Another means of authentication is through the use of biometrics, which includes voiceprint, fingerprint, or retinal scan. Currently, the cost and technical problems associated with remote biometrics scans render them impractical for common use.

When selecting one of the myriad products that support two- phase authentication, the following items should be added to the functional requirements list:

- It must provide the capability for centralized administration of access system controls (e.g., personal identification numbers, passwords, alarms, use analysis), while the actual authentication platforms may be decentrally deployed.
- It must function independently from the network infrastructure. The product should be independent of modem type, BAUD rate, or any other characteristics related to transmission of data. When changes are made to the network infrastructure, the product should not require modification.
- It must function independently from the hardware infrastructure. The product must function with all hardware platforms and operating systems. When changes are made to the hardware infrastructure, the product should not require modification.
- The product must function independently from all application software. Changes to the application should not dictate changes to the product. A one-time modification to the application software may be required to request the user's identification to the product.
- It must provide a random-number challenge (algorithm) to the product making the call that is in the possession of the caller. The challenge response (one-time password) must be unique for each authentication session. This ensures that the caller has the product in his or her possession each time a call is placed.
- The product must allow encrypted data to be processed. It is not a requirement for this product to perform the encryption.
- The product must accommodate caller mobility. The caller may need to call different processors or locations that are not part of the infrastructure. In addition, the caller may want to place calls from different devices (e.g., a different microcomputer in a different location); therefore, the authentication process must be capable of being relocated.
- The product must provide magnetic and printed reports of audit trail activity. The following data should be included in the audit log: date and time for all access attempts, the line on which the call entered, entry time, disconnect time, reason for disconnect, caller associated with the call, and system violations or other unusual occurrences.
- If the product in the caller's possession fails, a backup capability should exist that will grant the requester access to the infrastructure. The backup process must be available at all times. The most economical way would be to place a call to the network help desk; the help desk will then grant one-time access on verbal authentication of the requester. This requires a process that will mitigate social engineering.
- The process device must provide controlled one-time access for some individuals (e.g., vendors or customers) that is granted by a remote authority. An example of this feature would be a one-time password generator that would relay the challenge and response over the

Exhibit 55.2 Safe Practices for Mobile Users.

1. Be aware of the surroundings.
2. Make portable devices inconspicuous.
3. Shred confidential documents before discarding them.
4. Lock portable devices out of sight when leaving them unattended.
5. Hide physical security tokens; do not carry them in the same case with the notebook.
6. Establish a regular schedule for performing appropriated backup practices.
7. Select nontrivial passwords.
8. Change passwords frequently.
9. Follow the common-sense rule to question whether you are in an appropriate place and time to be working with the company's information assets.
10. Most important of all: treat sensitive company information as if it were the combination to your personal safe.

---

phone. The central unit issues a random challenge number, the handheld password generator calculates a response through an algorithm using the personal identification number of the requestor, and this unique response is compared to the central unit's response for that user. If the two responses agree, then access is granted.

- It must support the establishment of alternate dial authentication hot sites when the primary site goes off line for any reason.

CONFIDENTIALITY

Confidentiality in this section refers to the facilities by which information is protected against unauthorized reading. To facilitate establishing adequate levels of protection for information, data trustees (owners) must provide a classification to all information. This classification is based on the level of damage to the enterprise that may result from allowing individuals to gain access to information that they do not need.

Mobile Employee Information Security Recommendations

Ensuring compliance with appropriate security procedures and practices depends not on the security tools that are provided, but on the effectiveness of the security awareness program. Awareness contributes to the success of the effective information protection program. Exhibit 55.2 provides a baseline of awareness requirements for the mobile user. Posters in the organization's facilities are very effective and are an important part of the overall awareness program. However, use of E-mail and articles in the organization's internal communication media are more effective than posters with mobile users.

Software Considerations

Version Management. If the application software or database on the notebook is not the current version, the mobile user may create and transmit incorrect data into the organization's record of reference databases. If this happens, the integrity of the databases may be damaged and inaccurate information may be given to a business partner.

Each time the mobile user connects into the organization's network servers, a process should be performed to ensure that all of the software (both application and operating system) on the mobile client is current. The same is true for all database subsets that are mirrored on the notebook. The organization's change control process should notify the synchronization process when production environment changes are made, and should have the updates available when a remote connection is made. If the notebook does not have all the current software and database data, the infrastructure servers must not accept information uploads until the software and database are synchronized. The mobile user must recreate the information before attempting to upload using the current versions of the processing environment.

When the mobile user connects into the infrastructure and determines that a download of updates is required, the user should have an option to delay the download. The user is not allowed to upload information to the server, but queries may be made. This is important if the user is with a customer and wants to obtain status information; if a potentially lengthy download takes place, the user would waste the customer's time.

Encryption and Decryption. To date, encryption is the most effective security measure to ensure information confidentiality. One type of technology uses a two-part key in which the private key is kept by the owner and the public key is published. The recipient's public key is used to encrypt the data, which can only be decrypted by the recipient's private key. To reduce the computational overhead, encryption is often used to create a digital envelope that holds a DES encryption (symmetric) key and DES-encrypted data. Message nonrepudiation uses document hashing and digital signature as a means of verifying the message sender. This is accomplished by encrypting a message with the sender's private key and letting others decrypt the message with the sender's public key.

The major security concern is maintaining integrity and confidentiality of the keys. Each organization must devise a process to distribute the public keys to everyone who is involved in the encrypted messaging process, including customers and other business partners. The recommendation is to establish a comprehensive public- key database on a central server that may be accessed by everyone (this means that it is located outside the security firewall) and to have each mobile user keep a subset of public keys on his or her notebook for major business partners.

Another concern is protecting corporate equity. Consideration must be given to the necessity for the corporation to decrypt messages when the owner of the private key is not available. One method may be to include the symmetric DES key (discussed in the previous encryption section) in an extractable format in the message archiving facility (discussed in the non-repudiation segment). A tightly controlled process to extract the DES key would allow the message to be decrypted without compromising the private key of the originator. Right-to-privacy concerns are outweighed by corporate equity considerations, because company resources were used to create the messages.

NEW TECHNOLOGY CONSIDERATIONS

It is important to have an appreciation for new connectivity technologies so that users may determine their potential threats and vulnerabilities. By looking at what is coming, users should be able to develop the mitigating security measures before deploying the technology. Many security concerns exist, but very few proven answers are associated with emerging technologies. However, the technologist (and to some degree the mobile user) often wants to implement the technology quickly, before the technology itself has reached commercial strength.

PC Card. The unified standard that combines Personal Computer Memory Card International Association (PCMCIA) standard and the Japan Electronic Industry Development Association (JEIDA) standard is called PC Card. The credit-card-sized devices take the form of memory cards, modems, and disk drives that can be plugged into slots in computers. The card's security measures should be the same as those for the hard drive. Experience indicates that many users do not take the card out of the drive when it is not in use; therefore, passive encryption is recommended. For this reason, cards may be ineffective if used as a removable security lock. The card is effectively used in several applications, most notably as a removable modem.

Smart Cards. Although they are the size of an ordinary credit card, smart cards use an embedded processor that gives both the system designer and the system user a powerful authentication tool. Smart cards are a subset of the rapidly growing integrated circuit card industry.

Two types of smart cards exist: contact cards and contactless cards. The contact-type interface uses an eight-position contact located at one corner of the card. A contact card reader also uses a matching set of contact points to transfer information between the card and the reader. The contactless card does not come in direct contact with the card reader, but uses an inductive power coil and transmit and receive capacitor plates to transfer information to the contactless reader. AT&T's contactless card product is essentially an 8-bit computer with a proprietary operating system and either 3 Kbytes or 8 Kbytes of user-accessible, nonvolatile memory inside the smart card.

Many applications take advantage of the strong authentication capabilities of smart cards. The most common application is electronic money. Another example of smart card flexibility is in a building security system protected by card access that otherwise requires a large network of cables connecting the door reader, door controllers, and host computer. By converting systems to a smart card system and using the onboard database and encryption capabilities, the miles of network cabling and host computer may be eliminated. However, the need for a special card reader coupled with the mobile user's desire to travel light all but eliminates the smart card as a practical security mechanism in today's mobile arena.

Cellular Digital Packet Data. Cellular digital packet data (CDPD) technology is rapidly building its infrastructure right alongside the traditional analog cellular infrastructure. In fact, any of the existing analog base stations are being used. Similar to most public communications efforts, a coalition of common carriers has cooperated to ensure interoperability. CDPD was developed by a group of major cellular communications companies. This prestigious group makes it clear that by leveraging existing technologies and infrastructures, they will have a nationwide network available within a very short time.

As the caller moves from one analog cell to the next, the ability to transmit data digitally means faster transmission speeds and a solution to the current problem of lost or repeated transmission. The user will be able to send and receive data while in a moving car.

Technologists believe that because existing common protocols are used, today's applications can use CDPD without modification. The one drawback that may impede CDPD deployment is the development of effective modems that support CDPD.

Three types of CDPD services are provided. CDPD Network allows subscribers to transfer data through their applications as an extension of their internal network. In CDPD Networked Applications Services, the cellular provider provides specific application services like E-mail, directory services, and virtual terminal services to subscribers. CDPD Network Support Services provides network management, use accounting, and network security. The security practitioner should not assume that cellular carrier's security interpretation or objectives are the same as those of the organization. The carrier's direction is intended to protect its investment in the

CDPD Network and principally to ensure that only authorized paying subscribers use the network. A secondary concern is providing data privacy. It is not practical for the carrier to comply with each subscriber's security policies.

The user should already be aware of the security risk associated with cellular transmissions. Cellular (analog or digital) transmission is a miniature radio station broadcasting to everyone who has the receiving equipment. Digital transmission will be able to scramble the transmission by channel hopping, which makes interception more difficult but not impossible for the motivated eavesdropper. Therefore, the best solution to maintain data integrity and confidentiality is through digital signaturing and encryption. Because of the administrative overhead associated with key management of symmetric keys (the same passphrase is used to encrypt and decrypt the message), public and private key encryption is recommended.

Wireless Communication. Although cellular communication is a desirable tool for mobile users, wireless communication may be a valuable capability for those roving from place to place within the confines of their own building. Wireless networking employs a number of different methods. One of the most popular of these methods is spread-spectrum radio. Wireless adapters connect into a computer either internally or through the parallel port, and they then communicate to a base station or what could be called a wireless hub. The communication receiving area of the base station is usually several hundred feet, and the microcomputer and adapter may be placed anywhere within that radius. Other methods use a line-of-sight transmission technology.

Because the transmission takes place within the confines of the company's buildings, the security requirements may be fewer. However, depending on the organization's type of business and the need for confidentiality, consideration might be required to neutralize the motivated eavesdroppers who may be within the company or stationed outside the facility.

Other technologies use radio frequencies to forward messages to a central station, which in turn (when requested) sends the message to the recipient. This method is known as store and forward, and it is not normally used for interactive messaging. The best answer to ensure confidentiality in all wireless applications is encryption.

The World Wide Web. The World Wide Web (known as the Web or WWW) provides an infrastructure for accessing information. The Web provides a simple means of attaining almost any type of information that is available on a web server that is attached to the Internet. All information on the Web is stored in pages, using a standardized hypertext language. Many companies use the Web to provide timely information to their customers.

Exhibit 55.3 Breakdown of Mobile Work Force.

26%	Sales
23%	Field Services
18%	Administrative
15%	Field Engineering
10%	Senior Management
8%	Other

1997 Projections by The Yankee Group

They typically provide information about products, upgrades and patches, and feedback areas. The Web is considered a companion to E-mail because it provides information by using an interface. However, the Web is designed to accommodate limited two-way communication.

Security considerations must be the same as with all other Internet connections. The web server should be placed outside of the firewall. If confidential information is placed on the Web, then encryption should be used. In general, the web architecture does not provide for the integrity and confidentiality of information. Access to the infrastructure should not be allowed through the web server.

SUMMARY

Security procedures, guidelines, and practices accepted by end users must enable them to do their jobs. If end users interpret security to be a roadblock, they will often find ways to circumvent security requirements. To ensure that this does not happen, the security practitioner should spend time learning the problems and security concerns of users. The practitioner should consider scheduling one day per month to stay at home and telecommute in addition to dialing in while on business trips. This practice enhances understanding of the remote access conditions, and assists the security practitioner in developing more effective security practices. Exhibit 55.3, published by the Yankee Group, provides a breakdown of the professions that make up the mobile community.

Information today is stored not only in the data center but also on desktops, notebooks, and home computers; it is stored wherever mobile users have taken the data. By understanding the implications of this fact of business life, practices can be better established to secure assets while supporting employees' requirements to perform optimally and competitively through having access to the most current information and computing power.

Chapter 56
An Introduction to IPSec

Bill Stackpole

THE IP SECURITY PROTOCOL WORKING GROUP (IPSEC) WAS FORMED BY THE INTERNET ENGINEERING TASK FORCE (IETF) IN 1992 to develop a standardized method for implementing privacy and authentication services on IP Version 4 and the emerging Version 6 protocols. There were several specific goals. First, for the architecture to be widely adopted it would have to be flexible. It must be able to accommodate changes in cryptographic technology as well as the international restrictions on cryptographic use. Second, the architecture must support all the client IP protocols (i.e., transmission control protocol or TCP, user datagram protocol or UDP) in standard or cast (i.e., multicast) modes. Third, it must be able to secure communications between two hosts or multiple hosts, two subnets or multiple subnets, or a combination of hosts and subnets. Finally, there had to be a method for automatically distributing the cryptographic keys. This article will cover the key features of the IPSEC security architecture, its major components, and the minimum mandatory requirements for compliance.

FEATURES

The goals of IPSEC were transformed into the following key architectural features.

Separate Privacy and Authentication Functions with Transform Independence

IPSEC privacy and authentication services are independent of each other, which simplifies their implementation and reduces their performance impact upon the host system. It also gives end users the ability to select the appropriate level of security for their transaction. The security functions are independent of their cryptographic transforms. This allows new encryption technologies to be incorporated into IPSEC without changing the base architecture, and avoids conflicts with location-specific use and exportation restrictions. It also makes it possible for end users to implement transforms that best meet their specific security requirements. Users can select authentication services using hashed cryptography, which have low implementation costs, minimal performance impacts, and few international use restrictions. These implementations can be widely distributed, and they provide a substantial improvement in security for most of today's Internet transactions. Users can also select privacy functions based on private-key cryptography. These are more difficult to implement, have higher performance impacts, and are often subject to international use restrictions, so although they provide a much higher level of security, their distribution and use is often limited. Alternatively, users can combine these functions to provide the highest possible level of security.

Network Layer (IP) Implementation with Unidirectional Setup

Introducing security functionality at the network layer means all the client IP protocols can operate in a secure manner without individual customization. Routing protocols like exterior gateway protocol (EGP) and border gateway protocol (BGP), as well as connection and connectionless transport protocols like TCP and UDP, can be secured. Applications using these client protocols require no modifications to take advantage of IPSEC security services. The addition of IPSEC services makes it possible to secure applications with inherent security vulnerabilities (e.g., cleartext password) with a single system modification. And this modification will secure any such application regardless of the IP services or transports it utilizes.

This capability even extends to streaming services using multicast and unicast packets where the destination address is indeterminate. IPSEC makes this possible by using a unidirectional initialization scheme to set up secure connections. The sending station passes a setup index to the receiving station, and the receiving station uses this index to reference the table of security parameters governing the connection. The receiving station does not need to interact with the sending station to establish a secure unidirectional connection. For bidirectional connections, the process is reversed. The receiving station becomes the sender, passing its setup index back to the originator. Sending and receiving stations can be either hosts or security gateways.

Host and Gateway Topologies

IPSEC supports two basic connection topologies, host-to-host and gateway-to-gateway. In the host (sometimes called end-to-end) topology, the sending and receiving systems are two or more hosts that establish secure connections to transmit data among themselves. In the gateway (also called subnet-to-subnet) topology, the sending and receiving systems are security gateways that establish connections to external (untrusted) systems on behalf of trusted hosts connected to their own internal (trusted) subnetwork(s). A trusted subnetwork is defined as a communications channel (e.g., Ethernet) containing one or more hosts that trust each other not to be engaged in passive or active attacks. A gateway-to-gateway connection is often referred to as a tunnel or a virtual private network (VPN). A third scenario, host-to-gateway, is also possible. In this instance the security gateway is used to establish connection between external hosts and trusted hosts on an internal subnet(s). This scenario is particularly useful for traveling workers or telecommuters who require access to applications and data on internal systems via untrusted networks like the Internet.

Key Management

The ability to manage and distribute encryption keys effectively is crucial to the success of any cryptographic system. The IP security architecture includes an application-layer key management scheme that supports public and private key-based systems and manual or automated key distribution. It also supports the distribution of other principal session parameters. Standardizing these functions makes it possible to use and manage IPSEC security functions across multiple security domains and vendor platforms.

Two other key features of the IPSEC security architecture are support for systems with multilevel security (MLS), and the use of IANA (Internet Assigned Numbers Authority) assigned numbers for all standard IPSEC-type codes.

IMPLEMENTATION AND STRUCTURES

The IPSEC security architecture is centered around two IP header constructs, the Authentication Header (AH) and the Encapsulation Security Payload (ESP) header. To understand fully how these mechanisms function, it is first necessary to look at the concept of security associations. To achieve algorithm independence, a flexible method for specifying session parameters had to be established. Security Associations (SA) became that method.

Security Associations

An SA is a table or database record consisting of a set of security parameters that govern security operations on one or more network connections. Security associations are part of the unidirectional initialization scheme mentioned above. The SA tables are established on the receiving host and referenced by the sending host using an index parameter known as the Security Parameters Index (SPI). The most common entries in an SA are

- The type and operating mode of the transform, for example, DES in block chaining mode. This is a required parameter. Remember IPSEC was designed to be transform-independent, so this information must be synchronized between the end points if any meaningful exchange of data is going to take place.

- The key or keys used by the transform algorithm. For obvious reasons this is also a mandatory parameter. The source of the keys can vary. They can be entered manually when the SAS is defined on the host or gateway. They can be supplied via a key distribution system or — in the case of asymmetric encryption — the public key is sent across the wire during the connection setup.
- The encryption algorithm's synchronization or initialization vector. Some encryption algorithms, in particular those that use chaining, may need to supply the receiving system with an initial block of data to synchronize the cryptographic sequence. Usually, the first block of encrypted data serves this purpose, but this parameter allows for other implementations. This parameter is required for all ESP implementations but may be designated as "absent" if synchronization is not required.
- The life span of the transform key(s). The parameter can be an expression of duration or a specific time when a key change is to occur. There is no predefined lifespan for cryptographic keys. The frequency with which keys are changed is entirely at the discretion of the security implementers at the end points. Therefore, this parameter is only recommended, not required.
- The lifespan of the security association. There is no predefined lifespan for an SA. The length of time an SA remains in effect is at the discretion of the end point implementers. Therefore, this parameter is also recommended, but not required.
- Source address of the SA. An SA is normally established in one direction only. A communications session between two end points will usually involve two SAs. When more than one sending host is using this SA, the parameter may be set to a wildcard value. Usually, this address is the same as the source address in the IP header; therefore, this parameter is recommended, but not required.
- The sensitivity level of the protected data. This parameter is required for hosts implementing multilevel security and recommended for all other systems. The parameter provides a method of attaching security labels (e.g., Secret, Confidential, Unclassified) to ensure proper routing and handling by the end points.

SAs are normally set up in one direction only. Before a secure transmission can be established, the SAs must be created on the sending and receiving hosts. These SAs can be configured manually or automatically via a key management protocol. When a datagram destined for a (secure) receiving host is ready to be sent, the sending system looks up the appropriate security association and passes the resulting index value to the receiving host.

The receiving host uses the SPI and the destination address to look up the corresponding SA on its system. In the case of multilevel security, the security label also becomes part of the SA selection process. The receiving system then uses those SA parameters to process all subsequent packets from the sending host. To establish a fully authenticated communications session, the sending and receiving hosts would reverse roles and establish a second SA in the reverse direction.

One advantage to this unidirectional SA selection scheme is support for broadcast types of traffic. SAs can still be established even in this receive-only scenario by having the receiving host select the SPI. Unicast packets can be assigned a single SPI value, and multicast packets can be assigned on SPI for each multicast group. However, the use of IPSEC for broadcast traffic does have some serious limitations. The key management and distribution is difficult, and the value of cryptography is diminished because the source of the packet cannot be positively established.

Security Parameters Index

The SPI is a 32-bit pseudorandom number used to identify an SA uniquely. The source of SPIs can vary. They can be entered manually when the SA is defined on the host or gateway, or they can be supplied via an SA distribution system. Obviously, for the security function to work properly, the SPIs must be synchronized between the end points. SPI values 1 and 255 have been reserved by the IANA for use with openly specified (i.e., standard) implementations. SPIs require minimal management, but some precautions should be observed to ensure that previously assigned SPIs are not reused too quickly after their associated SA has been deleted. An SPI value of zero (0) specifies that no SA exists for this transaction. On host-to-host connections, the SPI is used by the receiving host to look up the security association. On a gateway-to-gateway, unicast, or multicast transaction, the receiving system combines the SPI with the destination address (and in an MLS system with the security label) to determine the appropriate SA. Now we will look at how IPSEC authentication and privacy functions utilize SAs and SPIs.

Authentication Function

IPSEC authentication uses a cryptographic hashing function to provide strong integrity and authentication for IP datagrams. The default algorithm is keyed Message Digest Version 5 (MD5), which does not provide nonrepudiation. Nonrepudiation can be provided by using a cryptographic algorithm that supports it (e.g., RSA). The IPSEC authentication function does not provide confidentiality or traffic analysis protection.

The function is computed over the entire datagram using the algorithm and key(s) specified in the SA. The calculation takes place prior to fragmen-

IPv4 Header	AH Header	Upper Protocol (e.g. TCP, UDP)

Exhibit 56.1 IPv4 Placement Example.

Next Header	Length	RESERVED
Security Parameter Index		
Authentication Data (variable number of 32-bit words)		
1 2 3 4 5 6 7 8 1 2 3 4 5 6 7 8	1 2 3 4 5 6 7 8	1 2 3 4 5 6 7 8

Exhibit 56.2 IP Authentication Header Structure.

tation, and fields that change during transit (e.g., ttl or hop count) are excluded. The resulting authentication data is placed into the AH along with the SPI assigned to that SA. Placing the authentication data in its own payload structure (the AH) rather than appending it to the original datagram means the user datagram maintains its original format and can be read and processed by systems not participating in the authentication. Obviously there is no confidentiality, but there is also no need to change the Internet infrastructure to support the IPSEC authentication function. Systems not participating in the authentication can still process the datagrams normally.

The AH is inserted into the datagram immediately following the IP header (IPv4) or the hop-by-hop header (IPv6) and prior to the ESP header when used with the confidentiality function, as seen in Exhibit 56.1. The header type is IANA-assigned Number 51 and is identified in the next header or the protocol field of the preceding header structure. There are five parameter fields in an authentication header, four of which are currently in use. These are shown in Exhibit 56.2.

- *The next header field* — To identify the IP protocol (IANA-assigned number) used in the next header structure
- *The payload length* — The number of 32-bit words contained in the authentication data field
- *The reserved field* — Intended for future expansion, this field is currently set to zero (0)
- *The SPI field* — The value that uniquely identifies the SA used for this datagram
- *The authentication data field* — The data output by the cryptographic transform padded to the next 32-bit boundary.

IP Version 4 systems claiming AH compliance must implement the IP AH with at least the MD5 algorithm using a 128-bit key. Implementation of AH is mandatory for all IP Version 6 hosts, and must also implement the MD5 algorithm with a 128-bit key. All AH implementations have an option to support other additional authentication algorithms (e.g., SHA-1). In fact, well-known weaknesses in the current MD5 hash functions (see Hans Dobbertin, *Cryptanalysis of MD5 Compress*) will undoubtedly lead to its replacement in the next version of the AH specification. The likely replacement is HMAC-MD5. HMAC is an enhanced method for calculating hashed message authentication codes that greatly increases the cryptographic strength of the underlying algorithm. Because HMAC is an enhancement rather than a replacement, it can be added easily to existing AH implementations with little impact upon the performance of the original algorithm. Systems using MLS are required to implement AH on packets containing sensitivity labels to ensure the end-to-end integrity of those labels.

The calculation of hashed authentication data by systems using the AH does increase processing costs and communications latency; however, this impact is considerably less than that of a secret key cryptographic system. The AH function has a low implementation cost and is easily exportable because it is based on hashing algorithms. Nevertheless, it would still represent a significant increase in security for most of the current Internet traffic.

Confidentiality Function

IPSEC confidentiality uses keyed cryptography to provide strong integrity and confidentiality for IP datagrams. The default algorithms use the cipher block chaining mode of the U.S. Data Encryption Standard (DES CBC), which does not provide authentication or nonrepudiation. It is possible to provide authentication by using a cryptographic transform that supports it. However, it is recommended that implementation requiring authentication or nonrepudiation use the IP AH for that purpose. The IPSEC confidentiality function does not provide protection from traffic analysis attacks.

There are two modes of operation — tunnel and transport. In tunnel mode the entire contents of the original IP datagram are encapsulated into the ESP using the algorithm and key(s) specified in the SA. The resulting encrypted ESP along with the SPI assigned to this SA become the payload portion of a second datagram with a cleartext IP header. This cleartext header is usually a duplicate of the original header for host-to-host transfers; however, in implementations involving security gateways, the cleartext header usually addresses the gateway whereas the addressing point of the encrypted header is to the end-point host on an interior subnet. In transport mode, only the transport layer (i.e., TCP, UDP) portion of the frame is encapsulated into the ESP so the cleartext portions of the IP header retain their original values. Although the term *transport mode* seems to imply a use limited to TCP and UDP protocols, this is a misnomer. Transport-mode ESP supports all IP client protocols. Processing for both modes takes place prior to fragmentation on output, and after reassembly on input.

IPv4 Header	AH Header (optional)	Encapsulated Security Payload

Exhibit 56.3 IPv4 Placement Example.

Security Parameter Index		
Initialization Vector Data (variable number of 32-bit words)		
Payload Data (variable length)		
...Padding Data	Pad Length	Payload type
1 2 3 4 5 6 7 8 1 2 3 4 5 6 7 8	1 2 3 4 5 6 7 8	1 2 3 4 5 6 7 8

Exhibit 56.4 IP ESP Header Structure.

The ESP header can be inserted anywhere in the datagram after the IP header and before the transport layer protocol. It must appear after the AH header when used with the authentication function (see Exhibit 56.3).

The header type is IANA-assigned number 50 and is identified in the next header or the protocol field of the preceding header structure. The ESP header contains three fields (Exhibit 56.4):

- *The SPI field* — The unique identifier for the SA used to process this datagram. This is the only mandatory ESP field.
- *The opaque transform data field* — Additional parameters required to support the cryptographic transform used by this SA (e.g., an initialization vector). The data contained in this field is transform specific and therefore varies in length. The only IPSEC requirement is that the field be padded so it ends on a 32-bit boundary.
- *The encrypted data field* — The data output by the cryptographic transform.

IPv4 or IPv6 systems claiming ESP compliance must implement the ESP supporting the use of the DES CBC transform. All ESP implementations have an option to support other encryption algorithms. For instance, if no valid SA exists for an arriving datagram (e.g., the receiver has no key), the receiver must discard the encrypted ESP and record the failure in a system or audit log. The recommended values to be logged are the SPI value, date/time, the sending and destination addresses, and the flow ID. The log entry may include other implementation-specific data. It is recommended that the receiving system not send immediate notification of failures to the send system because of the strong potential for easy-to-exploit denial of service attacks.

The calculation of the encrypted data by systems using the ESP does increase processing costs and communications latency. The overall impact depends upon the cryptographic algorithm and the implementation. Secret key algorithms require much less processing time than public key algorithms, and hardware-based implementations tend to be even faster with very little system impact.

The ESP function is more difficult to implement and subject to some international export and use restrictions, but its flexible structure, VPN capabilities, and strong confidentiality are ideal for businesses requiring secure communications across untrusted networks.

Key Management

Key management functions include the generation, authentication, and distribution of the cryptographic keys required to establish secure communications. The functions are closely tied to the cryptographic algorithms they are supporting but, in general, *generation* is the function that creates the keys and manages their lifespan and disposition, *authentication* is the process used to validate the hosts or gateways requesting keys services, and *distribution* is the process that transfers the keys to the requesting systems in a secure manner.

There are two common approaches to IP keying — host oriented and user oriented. Host-oriented keys have all users sharing the same key when transferring data between end points (i.e., hosts and gateways). User-oriented keying establishes a separate key for each user session that is transferring data between end points. The keys are not shared between users or applications. Users have different keys for Telnet and FTP sessions. MLS systems require user-oriented keying to maintain confidentiality between the different sensitivity levels, but it is not uncommon on non-MLS systems to have users, groups, or processes that do not trust each other. Therefore, the IETF Security Working Group strongly recommends the use of user-oriented keying for all IPSEC key management implementations.

Thus far, only traditional cryptographic key management has been mentioned. However, traditional key management functions are not capable of supporting a full IPSEC implementation. The IPSEC transform independence requires that all the elements of the security association, not just the cryptographic keys, be distributed to the participating end points. Without all the SA parameters, the end points would be unable to determine how the cryptographic key is applied. This requirement led to the development of the Internet security association and key management protocol (ISAKMP). ISAKMP supports the standard key management functions and incorporates mechanisms to negotiate, establish, modify, and delete SAs and their attributes. For the remainder of this section, the terms "SA management" will be used to refer to the management of the entire SA structure (including cryptographic keys) and "key management" to refer to just the cryptographic key parameters of an SA. It is important to note that key management can take place separately from SA management. For example, host-oriented keying would use SA management to establish both the session parameters and the cryptographic keys; whereas user-orientated keying would use the SA management function to establish the initial session parameters, and the key management function to supply the individual-use session keys.

The simplest form of SA or key management is manual management: the system security administrator manually entering the SA parameters and encryption keys for their system and the system(s) it communicates with. All IPv4 and IPv6 implementations of IPSEC are required to support the manual configuration of SAs and keys. Manual configuration works well in small, static environments but is extremely difficult to scale to larger environments, especially those involving multiple administrative domains. In these environments the SA and key management functions must be automated and centralized to be effective. This is the functionality ISAKMP is designed to provide.

Internet Security Association and Key Management Protocol

The ISAKMP protocol provides a standard, flexible, and scalable methodology for distributing SAs and cryptographic keys. The protocol defines the procedures for authenticating a communicating peer, creating and managing SAs, techniques for generating and managing keys and security associations, and ways to mitigate threats like replay and denial of service attacks. ISAKMP was designed to support IPSEC AH and ESP services, but it goes far beyond that. ISAKMP has the capability of supporting security services at the transport and applications layers for a variety of security mechanisms. This is possible because ISAKMP separates the SA management function from the key exchange mechanism. ISAKMP has key exchange protocol independence. It provides a common framework for negotiating, exchanging, modifying, and deleting SAs between dissimilar systems. Centralizing the management of the SAs with ISAKMP reduces much of the duplicated functionality within each security protocol and significantly reduces the connection setup time because ISAKMP can negotiate an entire set of services at once.

A detailed discussion of ISAKMP is beyond the scope of this article, so only the operations and functional requirements of an SA and key management system will be covered. An SA and key management system is a service application that mediates between systems establishing secure connections. It does not actively participate in the transfer of data between these systems. It only assists in the establishment of a secure connection by generating, authenticating, and distributing the required SAs and cryptographic keys.

Two parameters must be agreed upon for the system to work properly. First, a trust relationship must be established between the end point systems and the SA manager. The SA manager can be a third-party system — similar to a Kerboros key distribution center (KDC) — or integrated into the end point IPSEC implementation. Each approach requires a manually configured SA for each manager and the end points it communicates with. The advantage is these few manual SAs can be used to establish a multitude of secure connections. Most vendors have chosen to integrate ISAKMP into the end point systems and use a third-party (e.g., certificate authority or CA) system to validate the initial trust relationship. The second requirement is for the end points to have a trusted third party in common. In other words, both end points must have an SA management system or CA they both trust.

The operation is pretty straightforward. Here, systems with integrated SAs will be used for this scenario. System A wishes to establish a secure communications session with System B, and no valid SA currently exists between them. System A contacts the SA management function on System B. The process then reverses itself (remember, SAs are only established in one direction) as System B establishes a secure return path to System A. ISAKMP does have the capability of negotiating bidirectional SAs in a single transaction, so a separate return path negotiation is usually not required.

The ISAKMP protocol has four major functional components. They are

- Authentication of communications peers
- Cryptographic key establishment and management
- SA creation and management
- Threat mitigation

Authenticating the entity at the other end of the communication is the first step in establishing a secure communications session. Without authentication, it is impossible to trust the identification of an entity, and without a valid ID, access control is meaningless. What value is there to secure communication with an unauthorized system?

ISAKMP mandates the use of public key digital signatures (e.g., DSS, RSA) to establish strong authentication for all ISAKMP exchanges. The standard does not specify a particular algorithm. Public key cryptography is a very effective, flexible, and scalable way to distribute shared secrets and session keys. However, to be completely effective there must be a means of binding public keys to a specific entity. In larger implementations, this function is provided by a trusted third party (TTP) like a CA. Smaller implementations may choose to use manually configured keys. ISAKMP does not define the protocols used for communication with TTPs.

Key establishment encompasses the generation of the random keys and the transportation of those keys to the participating entities. In an RSA public key system, key transport is accomplished by encrypting the session key with the recipient's public key. The encrypted session key is then sent to the recipient system, which decrypts it with its private key. In a Diffie–Hellman system, the recipient's public key would be combined with the sender's private key information to generate a shared secret key. This key can be used as the session key, or for the transport of a second randomly generated session key. Under ISAKMP these key exchanges must take place using strong authentication. ISAKMP does not specify a particular key exchange protocol, but it appears that Oakley will become the standard.

SA creation and management is spread across two phases of connection negotiation. The first phase establishes an SA between the two end point SA managers. The second phase establishes the SAs for the security protocols selected for that session. Phase one constitutes the trust between the managers and end points; the second phase constitutes the trust between the two end points themselves. Once phase two has been completed, the SA manager has no further involvement in the connection.

The ISAKMP protocol integrates mechanisms to counteract threats like denial of service, hijacking, and man-in-the-middle attacks. The manager service sends an anticlogging token (cookie) to the requesting system prior to performing any CPU-intensive operation. If the manager does not receive a reply to this cookie, it assumes the request is invalid and drops it. Although this certainly is not comprehensive anticlogging protection, it is quite effective against most common flooding attacks. The anticlogging mechanism is also useful for detecting redirection attacks. Since multiple cookies are sent during each session setup, any attempt to redirect the data stream to a different end point will be detected.

The ISAKMP protocol links the authentication process and the SA/key exchange process into a single data stream. This makes attacks that rely on the interception or modification of the data stream (e.g., hijacking, man-in-the-middle) completely ineffective. Any interruption or modification of the data stream will be detected by the manager and further processing halted. ISAKMP also employs a built-in state machine to detect data deletions, thus ensuring that SAs based on partial exchanges will not be established. As a final antithreat, ISAKMP specifies logging and notification requirements for all abnormal operation, and limits the use of on-the-wire error notification.

SUMMARY

As a standard, IPSEC is quickly becoming the preferred method for secure communications on TCP/IP networks. Designed to support multiple encryption and authentication schemes and multivendor interoperability, IPSEC can be adapted to fit the security requirements of large and small organizations alike. Industries that rely on extranet technologies to communicate with their business partners will benefit from the IPSEC flexible encryption and authentication schemes; large businesses will benefit from the IPSEC scalability and centralized management; and every company can benefit from the IPSEC VPN capabilities to support mobile workers, telecommuters, or branch offices accessing company resources via the Internet.

The IPSEC architecture was designed with the future in mind, and is garnering the support it deserves from the security and computer communities. Recent endorsements by major manufacturing associations like the Automotive Industry Action Group, product commitments from major vendors like Cisco Systems, as well as the establishment of a compliance certification program through the International Computer Security Association are clear signs that IPSEC is well on its way to becoming the industry standard for business-to-business communications in the 21st century.

Section IX

Network Management

Chapter 57
Network Baselining as a Planning Tool

Gilbert Held

BASELINING PROVIDES A MECHANISM FOR DETERMINING THE LEVEL OF UTILIZATION OF A NETWORK, including its computational and transmission facilities. As such, it plays a central role in a network manager's capacity planning effort because the baseline shows whether or not there is sufficient capacity available, as well as providing a foundation for future network measurements that can be compared to the baseline to indicate the direction of network utilization. Thus, the network baselining effort represents the first major step in the capacity planning effort.

In addition, baselining enables network managers and administrators to identify and respond to network capacity requirements before they become an issue, in effect providing a mechanism to head off network-related problems.

BASELINING TOOLS AND TECHNIQUES

There are a variety of network baseline tools and techniques that can be used to facilitate an organization's capacity planning effort. The actual techniques employed are commonly based on the type of tool used. This article focuses on a number of commercially available network baselining tools, and discusses appropriate techniques concerning their use.

SimpleView

SimpleView is an easy to use and relatively inexpensive Simple Network Management Protocol (SNMP) management platform from Triticom, Inc. of Eden Prairie, MN. Through the use of SimpleView, users can retrieve statistical information maintained by Remote Monitoring (RMON) network probes. SimpleView supports a Management Information Base (MIB) walk capability, shown in the MIB Walk window, that lets a user click on an MIB group to select the group starting point, or double-click on the group to explode its elements, enabling a specific element from the group to be selected for retrieval.

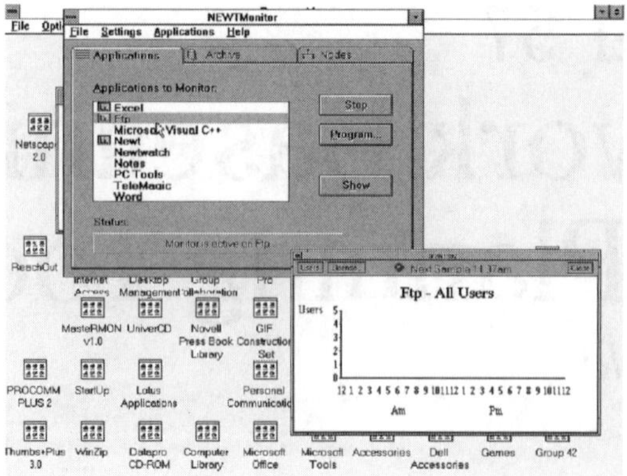

Exhibit 57.1 The NetManage NEWMonitor program provides the capability to examine the activity on a network based upon certain types of predefined applications.

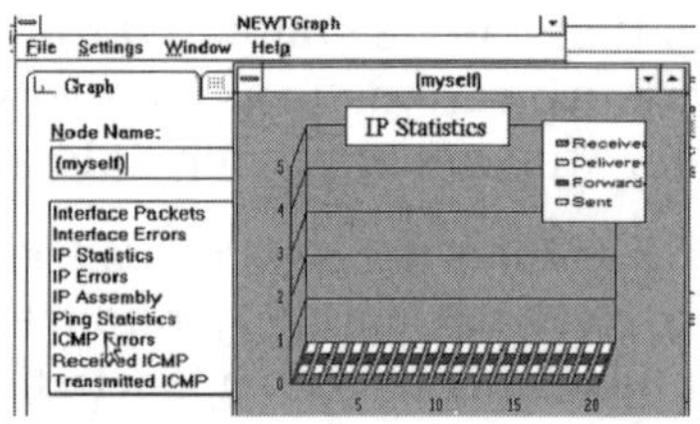

Exhibit 57.2 Through the use of the NetManage NEWTGraph program, graphs of different types of TCP/IP statistical information can be displayed.

NEWT

NetManage of Cupertino, CA, well known for its Chameleon suite of Internet applications, also markets a program called NEWT that can be used to monitor the use of desktop applications as well as to provide statistics on network activity associated with individual users. Exhibit 57.1 illustrates the use of NEWTMonitor on the author's computer to monitor the number of simultaneous FTP sessions occurring over a period of time. Doing so can be extremely important, especially when used in conjunction with normal RMON traffic statistics that do not look beyond the data link layer. NEWT-Monitor enables the use of specific types of TCP/IP applications. In comparison, if the network probes and network management system support RMONv2, or can be upgraded to this new version of RMON, it can be used to obtain a distribution of traffic through the application layer.

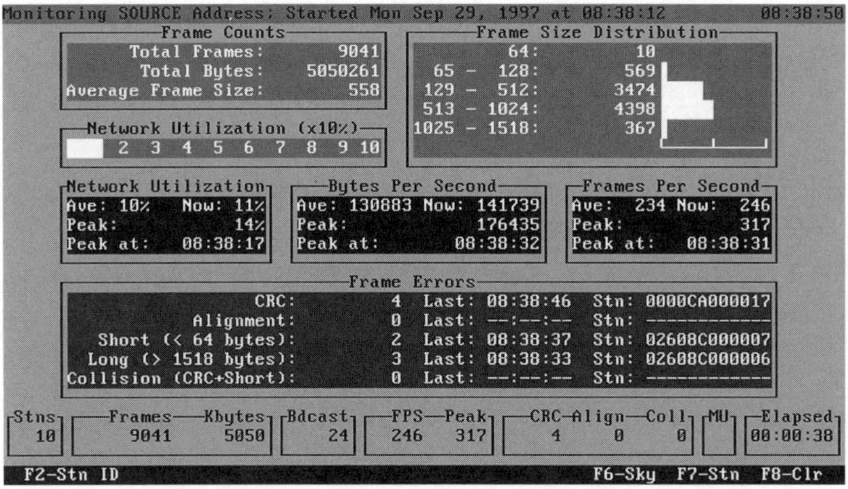

Exhibit 57.3 The Triticom EtherVision statistics summary display can be used to obtain information about network utilization and frame distribution.

Exhibit 57.2 illustrates the use of NEWTGraph to display different TCP/IP statistics by node. In the example shown in Exhibit 57.2, the author displayed Interface Errors for his node.

EtherVision

When checking the activity associated with an individual network, users can choose from a variety of network monitoring programs. One such program is EtherVision, also from Triticom, Inc. of Eden Prairie, MN.

Exhibit 57.3 illustrates the statistics summary display based on the monitoring of frames using their source address for constructing a statistical baseline. EtherVision supports monitoring by either Source or Destination address, enabling users to build two baselines. In examining Exhibit 57.3, note that the statistics summary presented indicates the frame count over the monitored period of time, current network utilization in the form of a horizontal bar graph, and a summary of "average," "now" or current, and "peak" utilization displayed as a percentage, as well as the time peak utilization occurred. The latter can be extremely handy as it allows a user to run the program on a workstation connected to an Ethernet LAN and return at the end of the day to determine the peak percentage of network use as well as when the peak occurred.

Although not shown in Exhibit 57.3, an EtherVision user can also set the program to generate a report that will log each period of activity over a

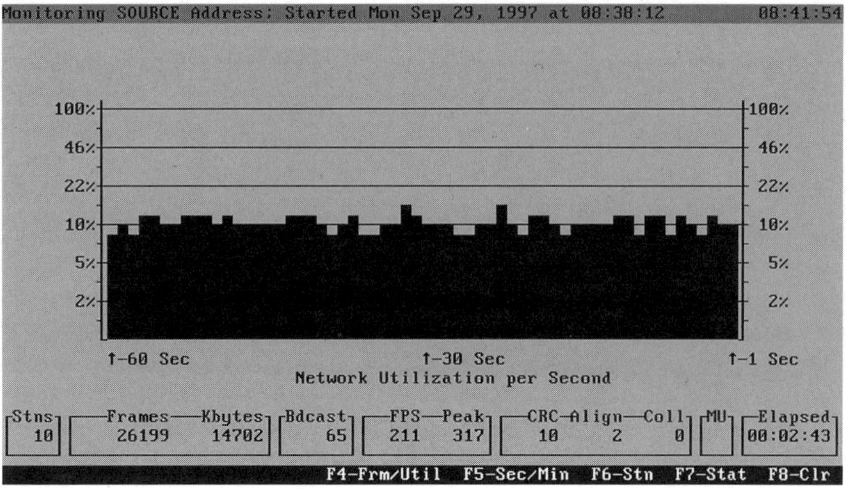

Exhibit 57.4 EtherVision supports the display of network utilization over a period of time, which facilitates observing the changing state of this important baseline metric.

certain percentage of network activity. Then, using the logged report, a network manager or LAN administrator can easily determine the distribution of network utilization throughout the monitoring period.

In the upper right corner of Exhibit 57.3, note that EtherVision maintains a distribution of frames transmitted on the network based on their size or length, falling into five predefined intervals. By examining the distribution of frames based on their length, users can determine the general type of traffic flowing on a network. This is possible because interactive query-response applications are generally transported in relatively short frames. In comparison, file transfers, such as web-browser pages containing one or more images, commonly fill frames to their full length. In examining the distribution of frame sizes shown in Exhibit 57.3, note that there are relatively few full-sized Ethernet frames in comparison to the total number of frames encountered during the period of monitoring. This indicates a low level of file transfer and web browser activity occurring on the monitored network.

Although EtherVision provides numeric information concerning network utilization, many users prefer to work with charts that note trends at a glance. To accommodate such users, EtherVision includes a number of built-in displays such as the one shown in Exhibit 57.4, which plots network utilization over a period of time. By examining a visual display, users can immediately note any potential capacity-related problems. In the example shown in Exhibit 57.4, the maximum level of network utilization is

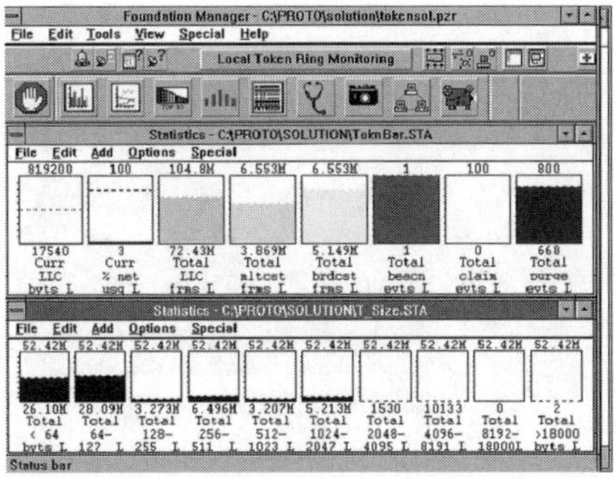

Exhibit 57.5 Using Network General's Foundation Manager to monitor the distribution of frames by length and type on a Token Ring network.

---

slightly above 46 percent. However, based on the monitored period, network traffic rose from 22 to 46 percent numerous times during the monitoring period. Since an Ethernet LAN gets congested at utilization levels above 50 percent due to its CSMA/CD access protocol, and the effect of the delay associated with the use of a random exponential backoff algorithm after a collision occurs, Exhibit 57.4 indicates a baseline of network utilization that justifies careful attention and a scheduled remonitoring effort to ensure traffic on the network does not turn into a bottleneck.

Foundation Manager

Foundation Manager, a product of Network General Corporation, is a sophisticated SNMP Network Management System (NMS) platform that operates on Intel-based computers using different versions of Microsoft's Windows operating system.

Foundation Manager was recently upgraded to support the emerging RMONv2 standard. When used to gather statistics from an RMONv2-compatible probe, it can provide a summary of statistics through the application layer, allowing it to replace the use of multiple products to obtain equivalent information.

Exhibit 57.5 illustrates the use of Foundation Manager to monitor a local Token Ring network. In the example, two buttons under the Local Token Ring Monitoring bar were pressed to initiate two displays of information from the Token Ring Statistics Group that an RMON probe on the local network accumulates. The first button clicked on is the bar chart icon to the right of the icon with the upraised hand in the form of a stop sign. Clicking on the bar chart icon results in the display of the top row of eight bar charts that indicate the total number of different types of frames and level of network utilization.

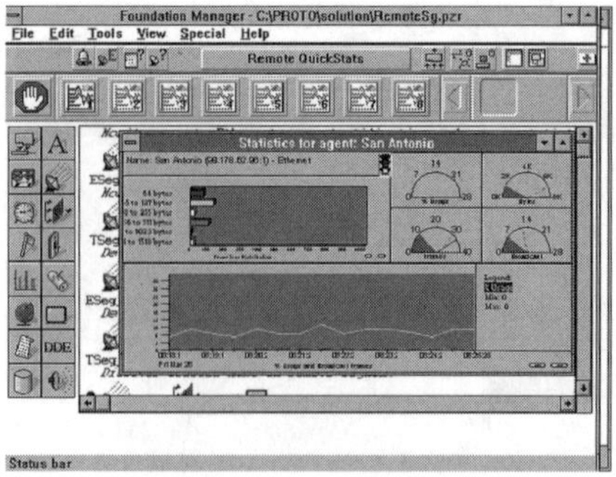

Exhibit 57.6 The Foundation Manager QuickStats display provides users with the ability to visually note important network baseline parameters both in real-time and over a period of time.

For example, the second bar chart located on the left side of the top display indicates that network utilization is at 3 percent on a 100 percent basis. Other bar charts on the top row indicate the current number of logical link control (LLC) bytes and frames, multicast frames, broadcast frames, beaconing frames, purge events, and claim events. The second row of bar charts resulted from clicking on the third icon to the right of the raised hand icon. This sequence of ten bar charts indicates the distribution of Token Ring frames in a manner similar to the method that EtherVision used to summarize Ethernet frame sizes. Foundation Manager follows the RMON standard and provides a more detailed breakdown of the distribution of Token Ring frames by their length.

Similarly, when using Foundation Manager to monitor Ethernet networks, the program retrieves RMON probe-kept frame distribution information that is more detailed than that kept by EtherVision. However, it is important to note that the retail price of EtherVision is under $500 and it can operate by itself. In comparison, the retail price of Foundation Manager is approximately $5000 and a single probe can cost approximately $1000, requiring an investment of an additional $5500 to obtain an enhanced level of frame size distributions as well as some additional features.

Two of the more interesting features of Foundation Manager are its QuickStats and discovery, and baselining capabilities. Exhibit 57.6 illustrates the use of the Foundation Manager Quick Stats feature to display a quick set of statistics for a remotely monitored network. In the example, statistics for an RMON probe connected to a network located in San Diego are displayed.

Foundation Manager is capable of displaying up to eight Quick Stats graphical reports at one time, with each report generated by clicking on an

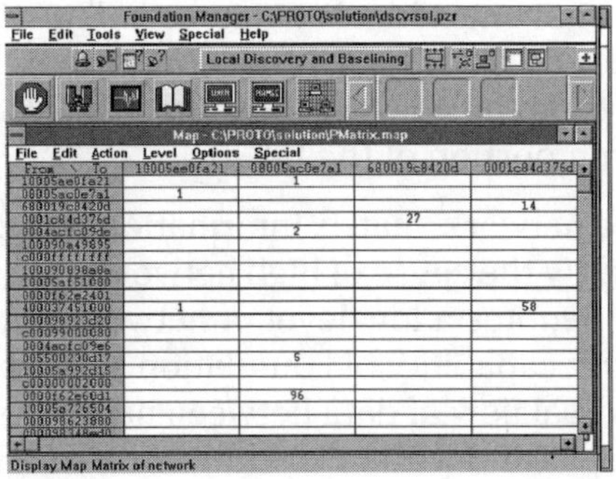

Exhibit 57.7 The Local Discovery and Baselining capability of Foundation Manager enables the flow of data between stations to be identified.

---

appropriate icon to the right of the icon with the raised hand in the form of a stop sign.

Each Quick Stats display presents summary information about a monitored network in a similar manner. In examining the statistics display for the network located in San Diego, the upper left display presents a distribution of frame length for the monitored LAN as a horizontal bar chart. The upper right portion of the display contains four gauges that provide a real-time view of network utilization, bytes transmitted, broadcast traffic, and frame rate. The lower half of the display shows a real-time plot over a period of predefined length for any two of the gauge values. Thus, the use of the Foundation Monitor Quick Stats display provides users with the ability to visually note important network baseline parameters both in real time and over a period of time.

A second interesting feature built into Foundation Manager is its discovery and baselining capability. This capability is available for both local and remotely located networks being monitored, and provides the ability to gather pattern flow information that can be extremely valuable when attempting to determine if the cause of a high level of network utilization results from the activity of one or a few stations on the network.

Exhibit 57.7 illustrates the Foundation Manager local discovery and baselining display as a matrix map of network activity. The first portion of the title of the display, "Discovery," results from the fact that the probe examines each frame flowing on the monitored network and discovers its source and destination by examining the source and destination addresses contained in the frame. The second portion of the title of the display, "Baselining," results from the fact that Foundation Manager extracts information from a matrix table maintained by the probe that denotes the number of frames transmitted from one address to another. Thus, in examining Exhibit 57.7, such numerics as 1, 2, 27, 14, 58, 5, and 96 represent the number of frames transmitted from the address located in the row in the table to the address in the column portion of the table.

When baselining a network, matrix information should be considered as a mechanism to identify the cause of high network utilization. If Quick Stats or a similar display denotes a low level of network utilization, there is no need to use the matrix capability of Foundation Manager or a similar product to identify the actual flow of data between network stations. This is because even if the user can locate a station using too much bandwidth, a modification of the operation of the station will, at best, have a negligible effect upon improving network performance if the network already has a low level of utilization.

CONCLUSION

Baselining is an important process that enables the communications manager and LAN administrator to quantify the status and activities of a network. In doing so, it provides a base of information that enables network trends to be identified, which can allow changes to be made to a network infrastructure prior to network capacity becoming an issue. Thus, network managers and administrators should consider using the tools and techniques described in this article as a mechanism to baseline their network infrastructures.

Chapter 58
Working with TCP/IP Utilities

Gilbert Held

THERE ARE FOUR KEY TCP/IP UTILITY APPLICATION PROGRAMS ONE CAN CONSIDER USING TO OBTAIN INFORMATION ABOUT NETWORKING DEVICES, NETWORK USAGE, AND THE GENERAL STATE OF A NETWORK. The utilities include PING, TRACEROUTE, FINGER, and NSLOOKUP. Of the four utility application programs, this article primarily focuses on the first three. The reason for this focus is that NSLOOKUP operates on the DNS database and the operation of the utility program requires a considerable amount of knowledge concerning the different types of records in the DNS database.

It is important to note that the ability to use one of the four previously mentioned utility programs depends on the manner by which a TCP/IP protocol stack developer includes or excludes certain utility programs. It is common that not all of the four programs are supported by a specific operating system. For example, while the programs PING and TRACEROUTE, implemented under the name TRACERT, are supported by Windows 95 and Windows 98, FINGER and NSLOOKUP are not. Although many UNIX operating systems support all four utility programs, certain flavors of that operating system lack support for one or more of the programs. Similarly, it is important to note that although one or more of the TCP/IP utility programs covered in this article may be supported by several protocol stacks built into different operating systems, the manner in which they operate and the parameters they support can differ between protocol stacks. In fact, as noted later in this article, a popular implementation of the PING utility program includes an option that can be employed — either intentionally or nonintentionally — with the resulting effect equivalent to an unsophisticated denial of service attack. Now that there is an appreciation for the fact that the four key TCP/IP utility programs covered in this article may not be supported by all operating systems, and if supported can be implemented with different options, attention can be turned to each utility program. In doing so, this article examines the basic use of each program and the basic parameters supported under one or more operating systems. When applicable, potential security hazards that can result from the improper use of the program, and the manner by which network managers and LAN administrators can protect their networks from those hazards, will be examined.

PING

PING is a TCP/IP utility application program that is included in every protocol stack with which this author has worked. Those protocol stacks include NetManage and FTP Software TCP/IP protocol stacks developed for use with Windows 3.1, several versions of UNIX, Windows 95, Windows 98, Windows NT, and the beta version of Windows 2000.

The name PING can be traced to two separate historical developments. First, according to some, the similarity in operation between PING and shipboard sonar determined the name. PING operates by generating an Internet Control Message Protocol (ICMP) echo message which results in the destination address issuing an echo response message. Shipboard sonar is similar in that the sonar unit generates a sonar signal and uses the ping response from the target to obtain a bearing and distance to the target. The distance is obtained by timing the delay between transmitting the signal and receiving its pinged reflection. That time represents the round-trip delay. Thus, dividing the time by two represents the time to the target. Because the velocity of propagation of a sonar wave is known, the time to the target, multiplied by the velocity of propagation, results in the distance to the target. Thus, according to legend, the echo response generated by an echo request was viewed similar to sonar pinging, resulting in the term PING being used for the TCP/IP application.

A second commonly told tale concerning the name of the PING application program reports that the name resulted as an acronym for Packet InterNetwork Groper. In this tale, the echo request message generated by the client is used to elicit, or grope, a response from the destination. Because the application was designed to cross network boundaries, it supported internetworking. Thus, the name PING was used as an acronym for Packet InterNetwork Groper.

Regardless of the actual manner by which the name PING evolved, this application utility program typically represents the first tool commonly used to test and troubleshoot network devices. For example, one can use PING to determine if a workstation, server, or another TCP/IP network device recently connected to a network is correctly cabled to the network and the protocol stack is operational and correctly configured. That is, after a device is connected to a network and its TCP/IP protocol stack is configured, one can test the cabling to the network and the operational state of the protocol stack by PINGing the device from another station on the network or located on a different network. If the PING elicits a response, the response informs one that the device is correctly cabled to the network and that the TCP/IP protocol stack is operational. Based on the fact that PING will also return the round-trip delay between source and destination, this application can also provide indirect information about network activity. That is, if round-trip delay time is relatively long in comparison to previously obtained round-trip delay times, the cause would normally be attributable to network congestion.

If one receives a timeout in response to issuing a PING that denotes the absence of an echo response message after a predefined period of time, one should not automatically assume the destination is not available or improperly cabled. If the destination is on a different network, it is possible that one or more intermediary devices, such as a router or gateway, is not operational and the PING cannot reach its destination. As noted later in this article, if one obtains a timeout response to the use of PING when attempting to access an address on a different network, one should turn to the use of the TRACEROUTE utility program prior to concluding that the destination is not operational.

Command Format

Although most modern operating systems are based on the use of a graphic user interface (GUI), most implementations of the PING utility program require the use of a command line interface. One common command line interface is:

```
PING [-q] [-v] [-c count] [-s size] {Host_namelIP_Address}
```

Where the command line parameters are as follows:

- q implements the quiet mode, resulting only in the display of summary lines at startup and completion
- v implements verbose mode, resulting in the listing of ICMP packets received in addition to each echo response message
- c specifies the number of echo requests (count) sent prior to concluding the test
- s specifies the number of data bytes (size) transmitted with each echo request message

Host_name or IP_Address Identifies the Target to be PINGed

To illustrate an example of the use of the PING utility, assume that a computer with the IP address 198.78.46.8 has just been installed and the host name is gil.feds.gov. To test the cabling of the computer to the network and the operation of the protocol stack on the computer, one can go to another station on the network and enter the following command:

```
PING-c 5 gil.feds.gov
```

This command line would result in the transmission of five echo request messages to the host address gil.feds.gov. Assume that the response was five PING timeout messages, each indicating the lack of a response from the destination host. Although one might assume that there is a problem with the destination, it is important to note that routing is based on the use of IP addresses. Because TCP/IP-compliant devices respond to IP addresses, it is quite possible that the host is up and operational but, for some reason, an appropriate entry was not made into the local DNS server to enable the host name to be resolved into an appropriate IP address. Recognizing this potential problem, one would then retry PING using the IP address of the destination as follows:

```
PING–c 5 198.78.46.8
```

The response to the PING would appear similar to the following:

```
72 bytes from 198.78.46.8 time = 12.2 ms
72 bytes from 198.78.46.8 time = 12.2 ms
72 bytes from 198.78.46.8 time = 12.2 ms
72 bytes from 198.78.46.8 time = 12.2 ms
72 bytes from 198.78.46.8 time = 12.2 ms
```

If one does not specify the number of data bytes to be used in each echo request, a default size of 64 is used. Because the ICMP header adds 8 bytes to the data size, the result in the count of echo response packet size being displayed is 72 bytes. Although the round-trip time for each echo request to echo response message is shown as being the same at 12.2 ms, it is important to note that this is not always true. When one uses a host address instead of an IP address, it is important to note that most times the issuance of multiple echo requests will result in the first round-trip delay having a longer time value than succeeding times. The reason is that the first PING is delayed by the address resolution process. During the address resolution process, the hostname is converted into an IP address to enable the echo request to be routed toward its destination. If one is using PING to obtain a precise round-trip delay time — such as to determine if a network structure can support voice over IP — it is important to recognize that the first round-trip delay time may significantly exceed succeeding delay times. Thus, it is a good and recommended procedure to issue at least two separate PING commands, permitting the first one to obtain the host IP address via the Address Resolution Protocol (ARP) and place that entry in the ARP table maintained by the PINGing device. Then, the second PING command to the same destination will reflect round-trip delay times that do not include the address resolution process.

In concluding this examination of the use of the PING application utility program, one can turn attention to the format of the program supported by Microsoft Corporation in Windows 95 and Windows 98. Exhibit 58.1

Exhibit 58.1 Microsoft Windows 95/98 PING Utility Program Command Line Format Microsoft® Windows C:\WINDOWS>ping 95. Copyright Microsoft Corp. 1981–1996.

```
Usage:    ping [-t] [-a] [-n count] [-l size] [-f] [-i TTL] [-v TOS]
          [-r count] [-s count] [[-j host-list] I [-k host-list]]
          [-w timeout] destination-list

Options:  -t             ping the specifed host until interrupted
          -a             resolve addresses to hostnames
          -n count       number of echo requests to send
          -l size        send buffer size
          -f             set Don't Fragment flag in packet
          -i TTL         Time To Live
          -v TOS         Type Of Service
          -r count       record route for count hops
          -s count       timestamp for count hops
          -j host-list   loose source route along host-list
          -k host-list   strict source route along host-list
          -w timeout     timeout in milliseconds to wait for each reply
```

illustrates the command format for PING displayed in response to entering the command without any command line parameters.

In examining the command line entries supported by Microsoft's implementation of PING, note the -t option. The use of that option results in a continuous PINGing of the destination until the CTRL-BREAK key combination is pressed to interrupt the program.

Continuous PINGing requires a continuous response from the destination. Because the destination must suspend what it is doing to respond to the PING, its ability to perform its intended function, such as a web server responding to HTTP queries, is degraded. In effect, continuous PINGing is often considered an unsophisticated hacker denial of service attack.

Another area of concern with respect to PING is that it can be used to discover network devices. That is, a sophisticated programmer can use PING within a shell program in an attempt to learn the addresses of network devices and then target those devices. To protect against the adverse effects of PING, some organizations use either a firewall or program a router's access list to prevent PINGs from reaching their network. If a router's access list is programmed, the denial of PINGs is usually expressed on a universal basis. Because a firewall has a more sophisticated filtering capability, many network managers and LAN administrators will configure the firewall to permit up to five PINGs per source IP address within a predefined period of time. This action permits a reasonable level of PINGing for a legitimate purpose to occur from users outside the organization's network while precluding denial of service attacks based on the use of PINGs.

An alternate option some network managers use is to program a firewall to allow employees behind the firewall to issue outbound PINGs but to bar inbound PINGs. The problem with this solution is the fact that, if every organization implemented it, nobody would be able to legitimately use PING. Having an appreciation for the operation and utilization of PING, one can now focus on the utility program that should be considered as a follow-on to its use — the TRACEROUTE program.

TRACEROUTE

The TRACEROUTE command invokes a program by that name which provides information about the route packets take from source to destination. If one PINGs a device on a different network, the lack of a response via a timeout does not necessarily mean that the destination device is not operational. Instead, it is possible that one or more intermediate routers or gateways that provide a connection between networks are not operational. Thus, TRACEROUTE can be used to examine the route from source to destination to determine if a path to the destination is available. If a path is available but the destination cannot be reached, this would then indicate that there is a problem with the device previously PINGed.

Operation

TRACEROUTE operates by transmitting a series of User Datagram Protocol (UDP) datagrams to an invalid port address at the destination device. First, the program transmits three datagrams in sequence, with each datagram having its time-to-live (TTL) field value set to unity. With this setting, each datagram will timeout as soon as it is processed by the first router in the path between the source and destination networks.

After the first router decrements the TTL field value by one and notes its value is zero, it generates an ICMP time exceeded message (TEM) response that indicates the datagram expired and sends it to the great bit bucket in the sky. The TEM response enables the device issuing the TRACEROUTE to compute the round-trip delay to the router for each of the three datagrams in the sequence. Next, the issuing device sets the TTL field value to 2 and transmits another sequence of three UDP datagrams. This sequence of three datagrams flows through the first router, which decrements their TTL field values to 1. However, the next router decrements their TTL field values to zero and returns TEM responses to each datagram, enabling the round-trip delay to the second router in the path to be determined.

This process continues until the datagrams either reach their destination or encounter a broken path. If the datagrams reach their destination, the use of an invalid port number in the UDP header results in the destination device generating an ICMP destination unreachable message in response to each of the three datagrams. This message tells the TRACEROUTE program that the destination was reached and it should terminate its operation. In the event an open path is encountered due to a router or gateway failure, TRACEROUTE terminates its operation when a timer expires. By noting that the last router that responded was not connected to the destination network, one can observe that a wide area network connection or router failure is the reason behind the inability to access a destination device.

Command Format

A commonly used TRACEROUTE command line format is shown below.

```
TRACEROUTE [-m ttl] [-q packets] {IP_Address| Host_name}
```

Where:

- -m represents the maximum allowable time-to-live (ttl) value, which is the number of hops allowed prior to the program terminating. The typical default ttl value used by most implementations of TRACEROUTE is 30.
- -q represents the number of UDP packets transmitted with each ttl setting. The common default value is 3.

Exhibit 58.2 illustrates the use of TRACERT from a Windows 95 client to determine the path to a device whose IP address is 205.131.176.15. It should be noted that the author used a commercial Internet service provider's network to access the Internet. From the access point, a total of 13 hops were required to reach the destination. In examining the entries in Exhibit 58.2, one will note that routers are typically configured to return a description in addition to the TEM message, and that description typically provides an indication of the location and organization operating the router. For example, hop 8 represents a router operated by BBN Planet in Vienna, VA. Because BBN Planet was acquired over a year ago by GTE, it is also apparent that the description of the router was not updated.

Although the Microsoft implementation of TRACEROUTE, which is known as TRACERT, uses a default of three datagrams to obtain three round-trip delay times per hop, it does not average those times. Other implementations of TRACEROUTE provide a statistical summary of round-trip response time. Similarly, other implementations may have one or more optional parameters that allow for control of timeout delays. However, because all versions of TRACEROUTE work in a similar manner, it is important to note that regardless of implementation, its use provides the ability to determine if a path is available to the destination.

FINGER

The FINGER utility program provides the ability to determine who is logged onto another computer and information about one or more users. The most common format of FINGER is as follows:

Exhibit 58.2 Using Microsoft's Windows 95 Version of TRACERT C:\WINDOWS>tracert 205.131.176.15.

Tracing route to www5.opm.gov [205.131.176.15] over a maximum of 30 hops:

1	446 ms	410 ms	542 ms hil-c45-068-eth00.as.wcom.net [209.154.36.97]
2	411 ms	407 ms	407 ms hil-ppp1-fas1-1-0.wan.wcom.net [209.154.36.65]
3	405 ms	400 ms	403 ms hil-core1-fas4-0-0.wan.wcom.net [205.156.214.145]
4	436 ms	421 ms	444 ms hyt-core1-atm1-0-3.wan.wcom.net [205.156.223.134]
5	466 ms	451 ms	414 ms hyt-peer1-fdd4-0.wan.wcom.net [205.156.223.68]
6	481 ms	440 ms	462 ms hyt-mae-east-pos3-0-0.wan.wcom.net [205.156.223.98]
7	453 ms	445 ms	445 ms maeeast2.bbnplanet.net [198.32.186.2]
8	453 ms	431 ms	538 ms p2-2.vienna1-nbr2.bbnplanet.net [4.0.1.93]
9	456 ms	455 ms	559 ms p1-0.vienna1-nbr3.bbnplanet.net [4.0.5.46]
10	480 ms	473 ms	460 ms p1-2.atlanta1-nbr1.bbnplanet.net [4.0.5.142]
11	442 ms	465 ms	480 ms p2-0-0.atlanta1-cr3.bbnplanet.net [4.0.5.113]
12	449 ms	459 ms	500 ms s0-1-3.usopm.bbnplanet.net [4.0.156.110]
13	473 ms	573 ms	610 ms www5.opm.gov [205.131.176.15]

Trace complete

FINGER [username]@[host_name | IP_address]

If one enters a username, the remote system will, if configured to respond to FINGER, return the username, the user's real name, his process identifier, application being accessed, and terminal port number. If one enters the command without username, some destinations will provide the previously mentioned information for all users on the system.

In an era of privacy issues, many colleges and universities where FINGER was previously enabled have turned it off. In addition, its use enables a hacker to attempt to gain a person's password by noting persons on a system and calling them. Pretending to be the administrator, a common ploy is to tell them that their use of a specific application noted from FINGER caused a lockup and they need the user's password to unlock a process. More often than not, a hacker was able to use FINGER as a starting point for gaining illegal access to different computers. Although Windows NT supports FINGER, neither Windows 95 nor Windows 98 support it.

NSLOOKUP

The last TCP/IP utility program discussed in this article is NSLOOKUP. Here, NS stands for the term "name server" and NSLOOKUP provides the ability to look up information from an organization's DNS server.

NSLOOKUP operates in two modes: interactive and noninteractive. If the program is entered without parameters, the user is placed in its interactive mode of operation. One would then enter the IP address of a DNS server, which would be used for querying DNS records concerning a target destination. Through the use of a range of DNS record types, one can obtain information about individual stations on a network, an organization's mail exchangers, and the set-up of a distant DNS. While the actual query types one is able to use can vary considerably between different implementations of the NSLOOKUP program, a few points concerning its use can illustrate how it can indicate the reason for certain problems. Assume that the use of a host name always results in the inability to reach a network device, yet one can ping it using its IP address. Through the use of NSLOOKUP, one can determine if an appropriate A record is in the DNS server. If not, this action informs one that a call to the administrator of the DNS is in order.

SUMMARY

TCP/IP utility programs can provide a significant insight to potential and actual network problems. By understanding how the programs operate and using them at appropriate times, one can isolate network problems and initiate action to correct noted problems. Although the use of TCP/IP utility programs will not allow one to note software problems that only a protocol analyzer can denote, their use represents the first line of investigation readers should consider when operations do not appear to be correct.

Chapter 59
Using RMON to Manage Remote Networks

Gilbert Held

BY STANDARDIZING THE MANAGEMENT INFORMATION BASE (MIB) FOR ETHERNET AND TOKEN RING LOCAL AREA NETWORKS (LANS), the management platform obtained from one vendor can be used with Remote Network Monitoring (RMON) probes obtained from other vendors. Thus, remote monitoring provides interoperability between network management products that monitor activity on distant networks, which may be extremely valuable in alleviating network problems.

This chapter reviews the Simple Network Management Protocol (SNMP) on which RMON is based, and describes the use of Ethernet and Token Ring RMON probes. Screens captured from a personal computer executing a management program illustrate how RMON provides information concerning activity on a remote network.

SIMPLE NETWORK MANAGEMENT PROTOCOL

The Simple Network Management Protocol was originally developed as a mechanism for managing Ethernet gateways in the Internet. The SNMP architecture is based on three components: an agent, a manager, and an MIB.

The agent is a software program or read-only memory (ROM) firmware code that operates in a managed network device, such as a computer, bridge, or router. The manager represents another software program, however, this program operates on a computer that becomes the network management station. The manager interoperates with agents through the use of SNMP commands. The third component of SNMP is the MIB. The MIB denotes a database of standardized elements that is queried by SNMP agents and provided to the manager.

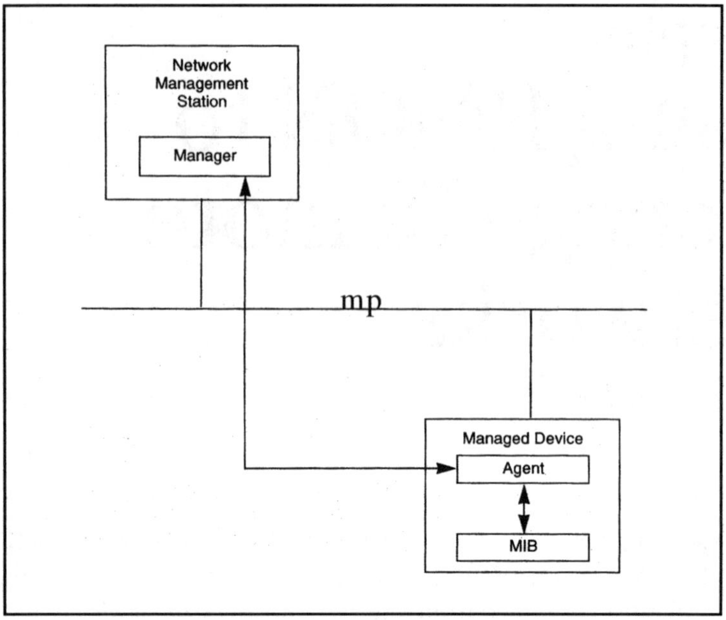

Exhibit 59.1 Basic Components of SNMP.

Exhibit 59.1 illustrates the relationship of the three basic components of SNMP on a bus-based Ethernet LAN. The manager operating on a network management station queries an agent operating on a managed device. The agent, in turn, depending on the manager's command, can query its database, returning requested information to the manager.

SNMP Commands

SNMP specifies five commands that are referred to as verbs and protocol data units: GetRequest, GetNextRequest, SetRequest, GetResponse, and Trap.

The GetRequest command is issued by the manager to retrieve a single value from the agent. By comparison, the GetNextRequest is used by the manager to examine an agent's MIB, one element at a time. The agent replies to both the GetRequest and GetNextRequest commands through the use of the GetResponse command.

The SetRequest command is used to modify or reset one or more configuration parameters of a managed device. Because of the lack of security in the first version of Simple Network Management Protocol, many vendors did not include recognition of the SetRequest command in their agents.

Under SNMP Version 2, which included security to prevent unauthorized managers from using the SetRequest command, that command is now commonly supported by agents.

The Trap command represents an unsolicited command. A Trap command is transmitted from an agent to a manager to alert the manager to a predefined condition, such as a circuit failure.

RMON Probes

As previously mentioned, SNMP was originally developed as a mechanism for managing TCP/IP and Ethernet gateways. A logical extension of SNMP was to permit the monitoring of remote networks. This was accomplished by the development of RMON probes based on request for comment (RFC) 1271 and 1513 for Ethernet and Token Ring networks, respectively. The term "probe" represents a host computer operating RMON software, or a hardware device that is operating ROM code that supports RMON.

As an extension to SNMP, an SNMP manager can be used to interact with RMON probes. In fact, even the RMON MIB can be considered an extension of the originally defined Ethernet and Token Ring LAN MIB.

ETHERNET AND TOKEN RING RMON MIBS

The Ethernet RMON MIB consists of managed objects arranged in nine groups, each representing a basic unit of conformance. If an RMON device implements a group, it must implement all objects in the group. The following sections briefly describe each group; the statistics and alarm groups are discussed at greater length later in the chapter.

Statistics. This group contains statistics measured by the remote probe for each monitored interface on the network management tool.

History. This group contains periodic samples that are stored for later retrieval.

Alarm. This group periodically retrieves statistical samples from variables in the probe and compares them to predefined thresholds, generating an event if the threshold is exceeded.

Host. This group contains statistics associated with each host discovered on the network.

HostTopN. This group is used to prepare reports for a list ordered by the top N statistics for a host.

Matrix. This group stores statistics about conversations between two network addresses.

Filter. This group enables packets to be matched based on a predefined filter equation. Matched packets may be used to generate events, or they can be captured.

Packet Capture. This group enables packets to be captured based on predefined filters.

Event. This group controls the generation and notification of events.

The Statistics and Alarm Groups

For many network management applications, the statistics and alarm groups are especially important. The statistics group contains statistics measured by the RMON probe in real time for each monitored interface on the probe. Thus, viewing the statistics group provides the network manager or administrator with information concerning the utilization level of a monitored network, the type of activity being performed, and similar information that may indicate the cause of a network problem.

Through the use of the alarm group, statistical samples from variables in the RMON probe are compared to previously configured thresholds. Thus, users obtain a proactive network management capability through the use of the alarm group because they can configure thresholds below the level where serious degradation occurs. This results in the generation of alarms that can then be used as signals to carefully monitor network activity, such as information generated by the statistics group. Many times the network manager will focus primarily on the statistics and alarm groups, using the information obtained from the other groups only periodically.

Statistics Group Counters

Because of the importance of the statistics group, the counters for Ethernet and Token Ring for that group are listed. Although Ethernet has one statistics group, Token Ring has two such groups.

The first Token Ring statistics group consists of a series of counters that track Media Access Control packets. The second Token Ring statistics group consists of a series of counters that track all frames that flow on the network with the exception of MAC frames, which are tracked by the general statistics group. Because this statistics group tracks all non-MAC frames, it is referred to as a "promiscuous statistics group." Under RFC 1271, the implementation of the promiscuous statistics group is optional.

Exhibit 59.2 lists the Ethernet RMON statistics group counters, Token Ring MAC, and Token Ring promiscuous statistics group counters. For each type of network the RMON statistics group tracks error conditions applicable to each type of LAN. In addition, the Ethernet statistics group and the Token Ring promiscuous statistics group track the distribution of frame lengths.

Exhibit 59.2 RMON Statistics Groups for Ethernet and Token Ring.

Ethernet RMON Statistics Group	Token Ring MAC Statistics Group	Token Ring Promiscuous Statistics Group
Dropped events	Dropped events	Dropped events
Octets	MAC Packet Octets	Data Octets
Packets	MAC Packets	Data Packets
Broadcast Packets	Ring Purge Events	Data Broadcast Packets
Multicast Packets	Ring Purge Packets	Data Multicast Packets
CRC Alignment Errors	Beacon Events	Data Packets 18 to 63 Octets
Undersize Packets	Beacon Time	Data Packets 64 to 127 Octets
Oversize Packets	Beacon Packets	Data Packets 128 to 255 Octets
Fragments	Claim Token Events	Data Packets 256 to 511 Octets
Jabbers	Claim Token Packets	Data Packets 512 to 1023 Octets
Collisions	NAUN Changes	Data Packets 1024 to 2047 Octets
Packets<=64 Octets	Line Errors	Data Packets 2048 to 4095 Octets
Packets 65 to 127 Octets	Internal Errors	Data Packets 4096 to 8191 Octets
Packets 128 to 255 Octets	Burst Errors	Data Packets 8192 to 18000 Octets
Packets 256 to 511 Octets	AC Errors	Data Packets>18000 Octets
Packets 512 to 1023 Octets	Abort Errors	
Packets 1024 to 1513 Octets	Lost Frame Errors	
Congestion Errors		
Frame Copied Errors		
Token Errors		
Soft Error Report Frames		
Ring Poll Events		

FOUNDATION MANAGER: AN EXAMPLE OF REMOTE NETWORK MANAGEMENT

In an SNMP environment, the Foundation Manager program (developed by ProTools, Inc., a subsidiary of Network General Corp.) is a network management platform with a built-in SNMP manager. The network management platform interacts with the SNMP manager to retrieve data from remote probes and then uses its built-in display capability to present the retrieved information in a format that facilitates the understanding and comparison of multiple RMON agent variables.

Foundation Manager can monitor both Ethernet and Token Ring networks, either singularly for a local network or simultaneously for multiple networks. Although Network General markets a probe for use with its Foundation Manager, any vendor's RMON probe can be used as long as it is compliant with RFC 1271 or 1513.

Remote Ethernet Monitoring

Under the main window menu is the program's ribbon bar, which contains a series of icons in two sections separated by a pulldown menu in the center. The icons in the ribbon bar provide quick access to various program options, such as starting or stopping monitoring by an agent, viewing outstanding alarms, and configuring SNMP communications settings. Under the ribbon bar is the solution bar, represented by a series of larger icons. Those icons represent functions associated with the pulldown menu labeled "Remote Ethernet Monitoring," which is displayed in the ribbon bar.

Foundation Manager permits the monitoring of one local network to which the computer running the software is directly connected, as well as the monitoring of multiple remote networks. In addition, this network management platform can perform functions beyond those standardized by RMON, such as packet decoding. Through the selection of a pulldown ribbon bar entry, users can initiate the basic operation of the management platform.

The Icon Panel

Users can construct paths of predefined functions to analyze a network by moving icons into the main portion of the window that is the program's path area. For example, the icon at the top left column in the icon panel is the "Acquire" icon. By moving that icon into the path area, users can acquire frames from the network. Then, by clicking on a predefined icon in the path area, users can display a specific RMON group of information. As an alternative, users can directly select a solution bar icon to rapidly display different types of network statistics.

Exhibit 59.3 illustrates the resulting display obtained from selecting the second and third icons from the left in the program's solution bar. The second icon, which appears as a bar chart labeled, "Net," produces the display of network activity that is shown in the top portion of the program's main window area. The labels of each bar correspond to Ethernet RMON statistics group counters. The display method of those counters is not standardized, so management platform developers can select any method they desire.

Displaying Network Information in a Line Graph. If selected, the third icon in the solution bar displays Ethernet use and broadcast-packets-per-second in a line graph format. The result of the selection of this line graph icon in the solution bar is displayed in the lower half of the Foundation

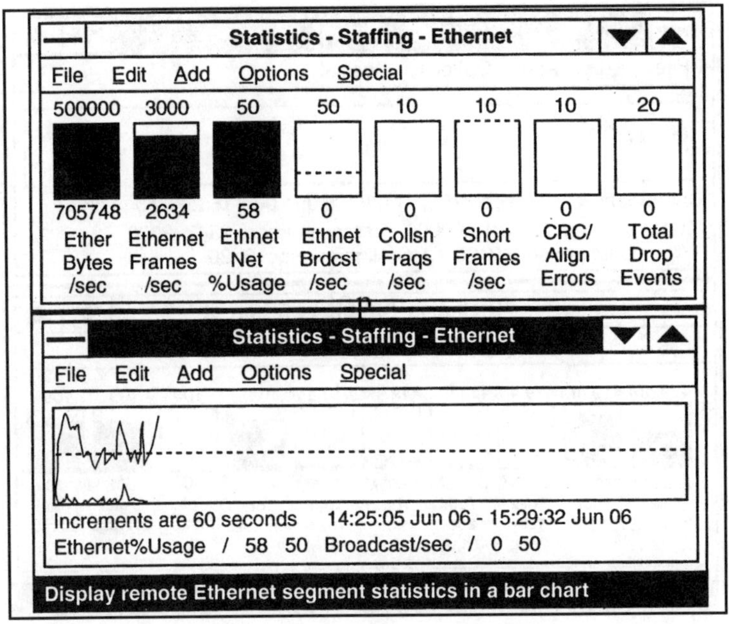

Exhibit 59.3 Network Statistics Displayed in Bar Chart and a Line Graph of Network Utilization and Broadcast Packets per Second.

Manager's main window area. Although the percentage of network use is not included in the Ethernet statistics group, it can be easily calculated because the maximum number of frames or packets that can flow on an Ethernet LAN is known, and by counting packets actually transmitted, the management platform can compute the utilization of the network. Similarly, although there is no broadcast-per-second counter in the Ethernet statistics group, there is a broadcast counter. By dividing the value of that counter by one-second time intervals, a management platform can track and display broadcast packets per second.

Displaying Network Information in a Bar Graph. Through the selection of the fourth and fifth icons in the solution bar, users can display another line graph of information, and a bar graph showing the distribution of packets by size as well as packet errors. Exhibit 59.4 shows the line chart display of four variables on a per-second basis (shown in the top portion of the main window area) and the display of packet-size distribution in a bar chart format in the lower portion of the main window area. The line graph provides a visual display of bytes, fragments, frames, and cyclical redundancy checking (CRC) and alignment errors on a per-second basis, as well as a digital display of a running total. Thus, users can view this display for an

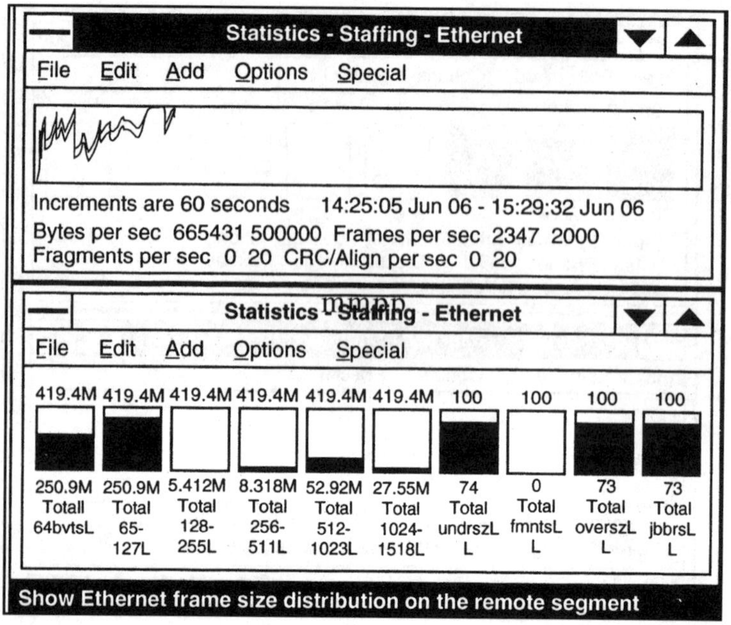

Exhibit 59.4 Viewing Four Variables in a Line Graph and Packet-Length Distribution in a Bar Graph.

indication of the transmission activity on the remote network as well as the state of two error conditions—fragments and CRC/alignment errors.

In the lower portion of Exhibit 59.4, the bar chart displays the distribution of packets by their size as well as four types of packet-size or packet-length error conditions. By examining the distribution of packets by length, users can determine, without packet decoding, the general type of activity occurring on the monitored network. Relatively short packets are used to transport interactive query traffic, and relatively long packets are used for file transfer operations.

Remote Token Ring Monitoring

The use of Foundation Manager to monitor a remote Token Ring network produces a display similar to that obtained from the program with an Ethernet RMON probe. The key difference between monitoring an Ethernet and a Token Ring network involves the display of statistical group counters that differ between RMON probes.

Selecting icons labeled "Net" and "Size" from the Foundation Manager screen and the alarm icon from the program's solution displays three windows in the program's main window area. The upper left window shows network usage information in a bar chart format. This window provides an overview of network activity on the monitored Token Ring network including network use and various data rates, and a soft error rate.

The window labeled “Alarm” displays current alarm activity as well as the thresholds currently set to generate alarms. As indicated in the window, a ring usage exceeding 50 percent in 30 seconds or more than ten ring purges in 30 seconds are currently defined alarm generation criteria. The RMON alarm group contains other alarm criteria in addition to the two just mentioned. Through Foundation Manager or another management platform, users can set the thresholds for various alarms. Then, the occurrence of a threshold can be used to generate a visual and audio alert as well as log the alarm to a logfile.

Selection of the bar chart icon labeled “Size” in the solution bar results in the display of a bar chart. Similar to the Ethernet bar chart, it denotes the distribution of packets by length on the monitored remote Token Ring networks, providing users with an indirect mechanism to note the type of LAN activity on the network.

CONCLUSION

The exhibits in this chapter showing the display of counters maintained by Ethernet and Token Ring RMON probes indicate how RMON can provide the network manager and administrator with valuable information. Through the use of a network management platform and one or more remote probes, users can monitor the vital signs of remote networks. In addition, by setting appropriate alarm threshold values, users can be alerted to deteriorating conditions. This allows time to reconfigure a network or perform another type of network modification to alleviate problems before the end user notices poor network performance. Because of this, network administrators should consider RMON a proactive network management tool that can be extremely valuable for isolating, reducing, and eliminating remote network problems.

Chapter 60
Proactive Performance Management

Tim Clark

ONCE VIEWED AS INTERNAL UTILITIES FOR INTERORGANIZATION COMMUNICATION, THE NETWORKS OF TODAY ARE BEING BUILT TO DELIVER ESSENTIAL BUSINESS PROCESSES TO PARTNERS AND TO CUSTOMERS. It is critical for organizations to be able to proactively manage the available bandwidth and systems for effective overall performance management.

Network management is a term that encompasses many complex, interrelated disciplines. Many organizations define the elements of network management as FCAPS — fault management, configuration management, accounting/billing, performance management, and security management — and try to address each element. However, the primary emphasis in many organizations is on fault determination and event isolation and correlation. These are critical elements, obviously, but the changing nature and complexity of the network are elevating the importance of performance management.

Enterprise networks are increasingly becoming a critical delivery mechanism for the products and services of a company. Corporate intranets are being accessed by business partners and even the public to perform basic business transactions. The purchase of books, the tracking of packages, and home banking are just a few examples of goods and services being delivered over the Internet.

With this trend, the enterprise network is both more critical to the day-to-day business operations of an entity and under more stress. The bandwidth demands are skyrocketing as the usage of the new web-based applications grows. The potential for bottlenecks grows. To meet these increasing demands, LAN and WAN switches are being deployed. Service providers are augmenting the private network. The performance of this new network is much more complex to manage than its relatively static and stable, in-house, router-based predecessor.

THE CURRENT STATE OF PERFORMANCE MANAGEMENT

Organizations today often have a variety of different tools to help manage the performance of the network. Those that have mainframes deployed are running a version of TME 10 NetView. SNMP (Simple Network Management Protocol) management applications by HP, IBM, Sun, Cabletron, and others are being widely deployed. Applications from router and switch vendors help to manage those elements and devices. A wide range of specific point tools provide information limited to a particular facet of performance management.

These tools are utilized to conduct many different types of network performance analysis:

- monitor real-time network performance
- network performance threshold alarms
- historical network performance reports
- network performance service level agreements (SLAs)
- LAN capacity planning
- WAN capacity planning
- server capacity planning
- network modeling
- baselining
- benchmarking

The analysis is performed using performance reports from a very wide range of network components:

- routers
- serial links
- frame relay CIR
- Ethernet segments
- Token Ring segments
- hubs
- LAN switches
- WAN switches
- servers
- workstations
- applications

Despite the plethora of tools and the variety of tasks that are undertaken in the typical enterprise network, comprehensive and proactive performance management still eludes many organizations. The barriers are numerous. In many cases, there is insufficient staff available that have the experience and training necessary to configure and administer the system,

Exhibit 60.1 Performance Management Checklist.

1.	Do you know the exact state of the network across the enterprise at any given time?	Yes___ No___
2.	Do you know the demand on network resources?	Yes___ No___
3.	Do you know how and where things have changed on the network?	Yes___ No___
4.	Do you know where problems are emerging — or are about to emerge?	Yes___ No___
5.	Do you know who is using the network?	Yes___ No___
6.	Do you have instant accessibility of real-time performance data?	Yes___ No___
7.	Do you have real-time trending statistics from across the enterprise?	Yes___ No___
8.	Do you have reliable access to critical information for making business decisions?	Yes___ No___
9.	Are you able to identify over/underutilized resources?	Yes___ No___
10.	Can you spot trouble before it starts?	Yes___ No___
11.	Are you able to plan your network's growth with accurate useable information?	Yes___ No___
12.	Can you see, analyze, and baseline all ports on all routers, hubs, and switches?	Yes___ No___
13.	Can you provide management with the network reports it wants?	Yes___ No___

write applications to enhance the system, and maintain the system. As a result, organizations often resort to reacting to network performance problems rather than proactively managing the network's performance.

Exhibit 60.1 should be used to evaluate the current state of performance management within an organization.

PROACTIVE PERFORMANCE MANAGEMENT

The goal of proactive performance management is to ensure that SLAs are being met and will continue to be met. To accomplish this goal, it is necessary to identify current and potential bottlenecks, inefficient or poorly performing components, and potential failures. Proactive performance management is the gathering of statistical data for monitoring and correcting the operation and measuring the overall effectiveness of the network, as well as aiding in the network planning and analysis process. Performance management is not performed in a vacuum, however. It must be integrated with other network management functions, like fault management and event correlation, to be maximally effective. Performance and fault management should be tightly integrated to ensure NMS and operations staff are aware of potential errors or trends that could impact client services.

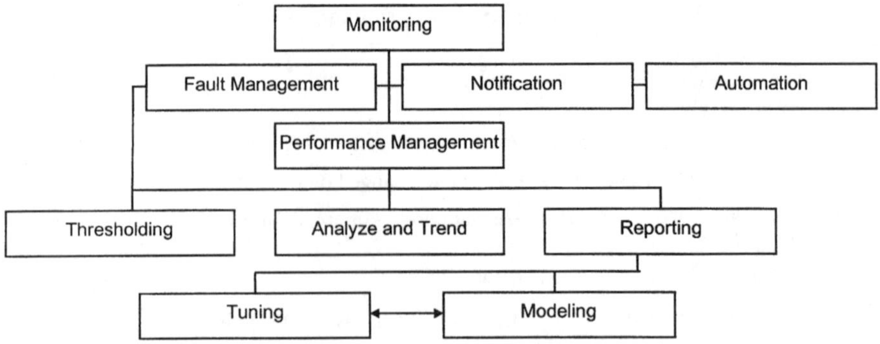

Exhibit 60.2 Proactive Performance Management Activities.

Performance management involves establishing a set of performance objectives for resources and addressing the following steps:

1. *Monitoring:* tracks resource and communication activities in order to gather the appropriate data for determining performance.
2. *Thresholding:* sets and modifies the operational standards (e.g., threshold values) by which the system performance is monitored and judged.
3. *Reporting:* involves the presentation of gathered data.
4. *Analyzing and trending:* consists of trending and statistical analysis, which *was* performed by network operators and planners to assess the results of performance measures. New software systems now do this for you.
5. *Modeling:* imports performance data to predict where additional resources will be required to meet SLAs, and to understand what impact a new application will have on current resources.
6. *Tuning:* focuses on the adjustment or reconfiguration of resources and communications activities to improve performance.

Exhibit 60.2 illustrates the relationships between these various activities, and the key interaction between performance management steps and fault management.

Exhibit 60.3 lists the required elements of a proactive performance management system or set of systems.

DEVELOPING A BASELINE

In order to begin a process of enhancing the performance management capabilities of an organization, one must begin with a baseline. With today's technology, few problems are caused by equipment failure. Most are caused by the increasing demands being placed on large and complex networks. To

Exhibit 60.3 Required Elements of a Proactive Performance Management System.

Required Elements	Benefits
Monitor hardware, software, remote links, and systems	Enables shift from reactive to *proactive* network performance management
27 · 7 collection of SNMP and RMON statistics	Enables staff to quantify network performance; improves understanding of network health and performance
Automatic notification of performance alerts	Increases end-user satisfaction and productivity; avoids additional headcount and capital expenditures
Predictive trend analysis	Predicts when next trouble spots will occur and identifies cause; solves problems before they impact network service levels
Daily reports and reviews of network performance trends and capacity trends	Enables true proactive management; provides necessary factual data and frees up IT personnel to focus on more strategic objectives

get a jump on these problems, one needs a comprehensive picture of the network that can be used for the analysis and modeling efforts that are part of any ongoing network design task.

One must have a baseline of past activity by which to judge present and future performance. With a baseline established, network hot spots can be acted on quickly or can be predicted and prevented altogether. Developing a baseline requires the focus on the first three activities in performance management: monitoring, thresholding, and reporting.

Most organizations have implemented products and systems that are already gathering information about network performance. SNMP MIB performance metrics represent specific performance characteristics of network elements. SNMP MIB attributes are periodically polled and trend data is collected over a period of time. Some metrics are requested on a demand-only basis, while others are collected only when an alarm condition exists on the network. Networks contain a wealth of information in their SNMP MIBs and Remote Monitoring (RMON) data. However, collecting and analyzing the data can be an arduous, resource-intensive process. The development of a network baseline should pull together the relevant SNMP and RMON data that are indicators of network performance. Exhibit 60.4 illustrates these indicators.

Resource availability statistics, derived from the fault management system, are the actual measure of the accessibility of a network resource when compared to the time the resource should be available. It must be noted

Exhibit 60.4 Indicators of Network Performance.

Network	Resource Availability	Quality of Service	Capacity	Utilization
Physical	Link status	Disrupted seconds %	Bandwidth	Bandwidth usage %
Routing	Device/ interface status	Packets dropped %	Packets per second	Packet per second %
LAN (Ethernet, Token Ring)	Status	Error packets	Bandwidth	Bandwidth usage %

that there is a clear distinction between reliability and availability. Reliability is the measure of the network's ability to function without failure. Availability is the measure of the network's ability to serve the purpose for which it was installed (from the user perspective). Availability depends on the reliability of the network, as well as the efficiency of the network control staff to restore services after a failure. As noted previously, clients will primarily be concerned with service availability, whereas the network control center will also be concerned with actual component reliability.

- *Quality of service* is the measure of service delivery in terms of number of errors, packets or circuits dropped, etc.
- *Capacity* or *throughput* is the measure of the network size and can be measured in terms of bandwidth, number of WAN links in the physical layer, or packet processing speed in the routing layer.
- *Utilization* is the measure of the actual use of a resource when compared to its theoretical throughput.

Once the appropriate monitoring is set in place, thresholding allows network technicians to monitor, identify, analyze, and react to changes in system and network performance. When conditions occur that cause monitored parameters to fall outside of predetermined thresholds, the management system can be set to trigger alarms or monitor and report changes in system status. For each indicator of interest, thresholds must be carefully selected. Thresholds that are too rigid will result in frequent alarms, while thresholds that are too relaxed can result in performance degradation that may be overlooked by support personnel. In the development of a baseline, thresholds can identify current hot spots and potential problems.

The final step in developing a network baseline is the development of pertinent reports for communication to various constituents. This then becomes an ongoing process, particularly with new tools that allow the easy publishing of the information via corporate intranets. Performance and capacity reports can be generated using a variety of reporting tools. These tools provide an automated process to easily summarize SNMP MIB element data that has been collected from network objects. The reporting features of these tools allow for report generation as well as immediate online viewing of the data.

Performance reports are useful to clients, the network operations staff, and the network systems planning group. These reports will provide the vehicle to assist network planners in preventing future outages and drops below the minimum service level agreements (SLAs), as well as allowing sufficient time for the network system planning to provide necessary network resources for the future. An excellent medium for providing personnel with reports is via HTML web pages over the corporate intranet. Performance problems may be due to inadequate resources (e.g., bandwidth), inefficient processes (e.g., unnecessary encapsulation of protocols), or an inefficiently used resource (e.g., excessive noise on a circuit requiring retransmission). Some sample performance and capacity reports are described below.

- *Resource availability report:* The minimum/average/maximum availability of the network resource reported for a day/week/month; 99.91 percent availability over a period of time
- *Resource maintainability report:* the mean time to repair (MTTR) for the network resource
- *Resource failure report:* the number of resources reported to be unavailable for more than "n" time

Online Utilization Reports

The following reports can be provided online to the user community:

- LAN utilization
- interface utilization statistics
- environmental reporting

Automated Trend Analysis

A statistical analysis is needed for graphs of collected data. The analysis could contain a description of the data's behavior, including any trends. It may contain an opinion with a recommendation to change. It can provide the statistics derived from the data for those who like such numbers. The enclosed statistics should include data set minimum, maximum, average, standard deviation, and linear regression parameters. For example, the bandwidth utilization graph with an analysis tells one if the usage trend is increasing to the point of degraded performance. If so, it will predict a theoretical date that performance will hit 100 percent and it will advise one to upgrade the bandwidth speed. Separate advice should be provided to accommodate serial links, Ethernet, and Token Ring. However, if bandwidth utilization is normal, it should tell one that. All this information allows one to predict when upgrades will be required, if needed, rather than passively wait for some device to fail due to saturation.

Trouble Ticket Reports

Clients need to have confidence that their trouble ticket report is seriously considered as a source of identifying and attacking the problems that they are encountering. They also need to know who is handling their problem, that the problem is understood from their perspective, some projected time frame when can it be resolved, and when the problem is considered resolved by the management system. Some trouble ticket reports will be generated by conditions within the network system itself and routed to operations prior to client reports. In many cases, problems will be resolved automatically as they occur or prior to client reporting.

Reports issued for user feedback should include the following:

- number of complaints by region, system, and user group
- number of trouble tickets generated by region, system, and user group
- number of trouble tickets resolved daily, weekly, and monthly
- average time to resolve trouble ticket per category
- ratio of trouble tickets per category
- ratio of resolution per category
- types and numbers of recurring trouble tickets by region, system, and user group

A PROACTIVE PLAN

Network systems planning is the process of using historical operating data, as well as estimated future resource requirements, to determine the optimal design of the network. The ultimate goal of network systems planning is to make optimal use of network resources to ensure network services are provided at a level necessary to meet client demands and SLAs while maintaining cost-effectiveness.

An extremely important aspect of Network System Planning is network optimization. Optimization involves balancing the efficient use of network resources, client requirements, and costs. Effective network optimization requires a detailed understanding of the following areas:

- client business requirements
- client priorities (i.e., response time vs. availability)
- disaster recovery requirements
- network protocols (WIN NT, TCP/IP, AppleTalk)
- transmission methods and media (analog, digital, fiber, copper, microwave)

- comparison of vendor-provided services vs. in-house supported services
- application design and implementation
- component capacity and throughput (client systems as well as switches)
- component compatibility
- industry trends
- available technology and future directions
- standards efforts
- network metrics/indicators and their interpretation
- network modeling
- costs and return on investment (ROI)

The importance of having a planning group that understands all of the above items and their interrelationships cannot be overstated. A network design omitting even one of these items could result in undesirable consequences.

Input to the planning process can come from any area of network management. Specifically, these can include, but are not limited to:

- performance reports
- trouble ticket reports
- component configuration
- client resource usage estimates
- security requirements
- costs

Component configuration refers to understanding the limitations of each component in the network with regard to capacity as well as physical limitations (number of available slots, power requirements, environmental limitations, etc.). In addition, planners must be aware of future changes to vendor-supplied components such as incompatible hardware upgrades or software that will no longer be supported.

The most difficult piece of planning data to gather is an accurate list of future client resource requirements. In many cases, end users may not have the technical skills or tools to estimate network resource requirements.

This process of optimizing the network design as performed in the network systems planning function is a key to turning the performance management function from being reactive to being proactive. This is the critical function that will allow organizations to prevent problems from occurring. By utilizing the reports generated from the monitoring and thresholding activities, potential bottlenecks can be identified and the resolution of those bottlenecks can be set in place before they hamper network performance.

SUMMARY

Today's networks are becoming very critical and very visible elements in the day-to-day operation of a business. The complexity of these networks will continue to grow as the demands on them increase over time. Network performance management is an essential component of an overall strategy for these new networks. If you are a bank and your customers become frustrated with your home banking service because of network bottlenecks and long response times, they will find somewhere else to bank. It is that easy.

Today, many organizations are forced into reacting to performance issues rather than anticipating them. Often, they simply throw bandwidth at the problem although a bottleneck is caused by overutilization of a device, system, or component. Putting in place a proactive performance management system that assists in identifying current and future trouble spots will be key to the organization's ability to respond to new demands on the network.

Chapter 61
Using Intelligent Agents to Manage Multivendor Environments

Nathan J. Muller

THE CHALLENGE OF MANAGING MULTIVENDOR SYSTEMS IS INCREASINGLY BEING MET THROUGH THE USE OF INTELLIGENT AGENTS, special software installed on managed stations on the local or wide area network that collects performance information in a standard format, implements preemptive actions based on predefined policies, and automates routine, repetitive tasks. These capabilities help minimize traffic on the network, increase systems availability and reliability, and reduce the overall cost of systems management.

INTRODUCTION

Today's distributed computing environment usually includes a mix of UNIX, MVS, VMS, MS-DOS, NetWare, Windows, and other operating systems from multiple vendors. Managing a multivendor environment is difficult, especially because standards for systems management and underlying technologies are lacking. The OSF's DME effort has failed largely because it attempted to address everything from applications to infrastructure all at once instead of setting priorities and proceeding incrementally.

Open systems management is compounded by the fact that every vendor defines the term differently and therefore takes a different approach to systems management. Some vendors offer integrated suites of applications; others merely provide platforms or frameworks and rely on third parties for specific applications that plug into them. Because vendors offer overlapping but not identical sets of products, users find product comparisons and selection difficult. In many cases, this difficulty has delayed full-scale implementation of systems management, especially for client/server environments.

The most promising approach to the challenge of managing multivendor environments is embodied in the concept of intelligent agents. Intelligent agents are special software installed on managed stations on the LAN/WAN that collects performance information in a standard format and implements preemptive actions based on predefined policies. Because filters and thresholds are used to send the management console specific items of information only, the software provides a more efficient process than continuous polling of every station to notify the central management station of all events. Using intelligent agents thus minimizes traffic on the network which, in turn, consumes less bandwidth and contains costs. It also has the following business benefits:

- By detecting exceptional events and triggering automated actions to correct problems before they affect systems services, intelligent agents increase systems availability and reliability.
- By automating routine, repetitive tasks, intelligent agents relieve the systems administrator of a heavy work burden and reduce the overall cost of systems management. They speed up problem resolution while permitting businesses to get the most out of their technology investments.

Organizations that use intelligent agents to continuously monitor systems and applications for problems, and carry out corrective actions on an automated basis, thus achieve a competitive edge over companies that react to problems with manual processes.

KEY CONCEPTS IN SYSTEMS ADMINISTRATION

Manager and Agent

The key concepts in systems (and network) administration are the manager and agents. The manager is the workstation that is set up to view information collected by the agents. The agents are special programs that are designed to retrieve specific information from systems on the network. An application agent, for example, works on each workstation to log application usage. Workstation users are not aware of the agent and it has no effect on the performance of the workstation or the applications running on it. The collected information is retrieved from one or more databases for viewing on the systems administrator's console.

The use of agents is not new, of course. The manager-agent relationship is intrinsic to most standard network management protocols, including SNMP. SNMP agents are now used routinely in all kinds of internetworking devices including bridges, routers, hubs, multiplexers, and switches. Exhibit 61.1 depicts the interactions between an SNMP manager and agent.

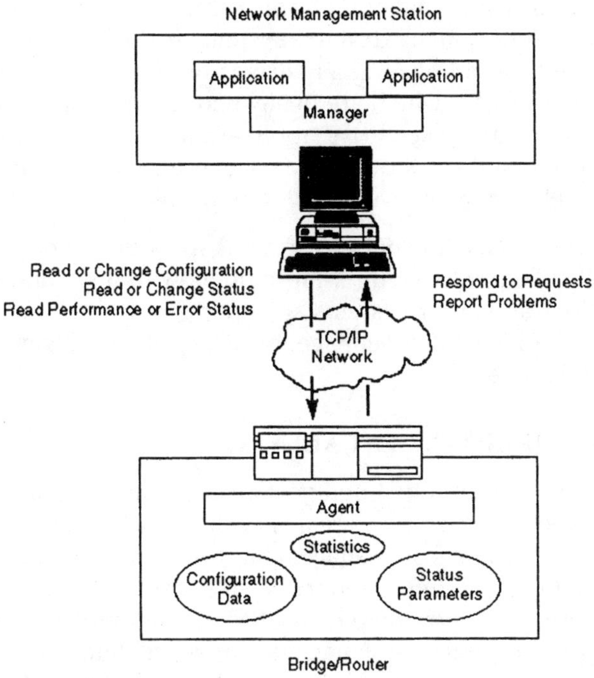

Exhibit 61.1 Interactions Between an SNMP Manager and Agent.

What is relatively new is the use of intelligent agents to manage systems-level activities. The difference between network-level (i.e., SNMP) agents and systems-level agents lies in their degree of intelligence.

Agent IQ

In the SNMP world, agents respond to polls from an SNMP management station that requests information on the operational status of network devices. Based on that information, agents are then directed by the management station to get more data, set management variables, or generate traps when significant events occur. By necessity, SNMP agents are dumb; that is, they act merely as passive data collectors that provide information only when they receive a request from the central management station. That such agents cannot perform filtering or processing on their own is acceptable, though, because bridges, routers, and other internetworking devices have limited memory and processing resources.

The idea of using SNMP to manage UNIX and other systems started to catch on a few years ago, but it quickly became apparent that dumb agents were not very efficient for this purpose. Dumb agents must be polled continuously by central management software, a process that increases network traffic. Today, many desktop systems reside at remote sites, and polling over wide-area links quickly drives up operating costs. Even on higher-bandwidth LANs, which are being increasingly burdened with multimedia and other delay-sensitive applications, traffic from continuous polling and the resultant data transfers can easily affect performance thresholds. Intelligent agents address these problems.

What makes intelligent agents so smart is the addition of programming code that tells the agent exactly what to do, how to do it, and when to do it. Because UNIX and other systems are equipped to run complex computations anyway, adding intelligent agent software does not significantly tax a machine's resources.

CAPABILITIES OF INTELLIGENT AGENTS

The information collected by intelligent agents from multiple sets of data can be displayed in several ways (e.g., in cells, charts, and text), analyzed for such purposes as license management or inventory management, and printed as a detail or summary report. Agents are becoming an increasingly popular method of carrying out repetitive, programmable tasks. In essence, they play the dual role of manager and agent. Intelligent agents that reside on UNIX systems are capable of tracking hundreds of parameters. They are used mostly for scheduling, controlling, and monitoring the execution of processes across the network, and ensuring their successful completion.

A comprehensive agent-based management application allows the systems administrator to perform the following main functions:

- View and manipulate network data
- Automate file distribution
- Maintain hardware inventory
- Manage installed software, including application usage
- Receive notification of network events
- Establish and manage network printer support
- Automate network processes, such as backup and virus detection
- Monitor disk and file usage
- Create task lists
- Manage storage

All of the agents that collect information in support of these functions are configured at the management console using commands selected from the menu bar. Once configured, each type of agent can be assigned an icon that launches its associated viewer for displaying collected information. The many types of agents available include those for monitoring disk, application, and file usage; distributing files; tracking network activity; and process scheduling. The following sections explain how intelligent agents work in the distributed environment and aid in multivendor systems management.

DISK MONITORING AGENTS

To remain productive, network users must have an adequate amount of disk space for storing programs and data. A disk monitoring agent that reports volume capacity helps the systems administrator manage use of disk space and meet user demand. Current information on disk use helps the systems administrator make decisions about reallocating resources or purchasing additional drives or servers.

The disk monitoring agent scans the server volumes to collect disk use information. The collected information includes a list of volumes that were scanned, total volume capacity, and space used. In addition, the time and date disk data was collected is recorded for comparing disk use over a period of time. Some vendors' disk monitoring tools also record disk-usage data based on directory and file owner.

Depending on the product selected, one or more of the following categories of information can be collected by a disk monitoring agent:

- Volumes. Includes the date and time data was collected, server name, volumes scanned, capacity, space used, and space available.
- Directories. Includes the date and time data was collected, server volume and directory names, creation date and time, file count, directory size (in bytes), owner name, and groups to which the owner is a member.
- Directory and file owners. Includes the date and time data was collected, server and volume names, owner name, groups to which the owner is a member, total number of files, and total space used (in bytes).

A disk monitoring agent provides automatic notification if disk usage reaches a specified level. In accordance with organizational needs, notification can be sent to the individual responsible for responding to the situation through a console, network, pager, or E-mail message. The console provides view formats and types that allow the systems administrator to analyze current disk use as well as disk use patterns over a period of time.

Usually, the disk monitoring agent can process data in the background while other programs are running. It also can be made to run interactively, with progress displayed in a status window of the console as volumes are being scanned. A scheduling agent can be tasked to run the disk monitor automatically in batch mode to collect data at night. This feature is particularly helpful because collecting data from some volumes takes a significant amount of time. Using the scheduling agent, the systems administrator can regularly collect data each night, once a week, or once a month, depending on organizational requirements. This process ensures that information is collected consistently.

APPLICATION MONITORING AGENTS

The ability to track software usage helps ensure that the organization complies with vendor licensing agreements and that network users have access to needed applications. Tracking software usage also helps reduce software acquisition costs, because accurate usage information helps managers determine which applications are run most before deciding on upgrades and how many copies to buy.

An intelligent agent is used to meter software usage. Metering allows the systems administrator to control the number of concurrent users of single applications as well as those in a software suite. Metering of a software suite tracks each application in the suite individually to automatically allow the correct number of concurrent users. In addition, the application agent collects information on which applications have been used and identifies the users who have accessed them.

The systems administrator can also choose to be notified of the times when users are denied access to applications because all available copies are in use. If work must be tracked by project, the intelligent agent can also be used to automatically apply application usage to a particular project code. The systems administrator views and analyzes the collected metering data at the console.

For each application, the intelligent agent meters use of the main executable file as well as all associated executable files. For example, metering use of Micrografx, Inc.'s Designer can also track use of associated executable files, including Slideshow and Batch Print, because these applications are a part of the Designer product.

Examples of the types of files that are automatically recognized by the intelligent agent for metering are:

- .com (command)
- .exe (program)
- .bat (batch)
- .pif (starts Disk Operating System applications in the Windows environment)
- .bin (program)
- .scr (screen savers)

Because some executable files use file extensions other than those in the list, the intelligent agent provides the option of allowing the systems administrator to identify additional files and extensions for metering software usage.

Each package in the software inventory has a unique name (i.e., tag) that is used for metering. When a user launches a metered application, the tag name is checked to determine if there are available copies (i.e., licenses), and a record is generated. The intelligent agent checks software licenses in and out — when users start an application, one less copy is available to run; likewise, when the application is closed, the copy becomes available again. Before users are granted access to metered software, the software inventory is checked to determine whether copies are available. An error message indicating that there are no available copies is displayed to a third user if two users are running an application that has only two licenses.

FILE MONITORING AGENTS

A file monitoring agent is a tool that is used to collect data about file access, such as which files are being accessed and by whom. Tracking file access is important for:

- Recording file access to a sensitive file or group of files
- Diagnosing file access problems
- Reviewing access times for application bottlenecks

Monitoring file access data also helps the systems administrator maintain network security, and it provides useful information for tracking the source of security breaches.

The file agent monitors access to specified files and provides the following information:

- Who accessed the files
- The complete path for the file
- When the files were accessed and for how long
- What operations were performed

The systems administrator sets up the file monitoring agent to automatically collect this kind of information, after which it can be viewed and analyzed at the console.

FILE DISTRIBUTION AGENTS

A file distribution agent is a tool that is used to automate the process of distributing files to particular groups or workstations. A file distribution job can be defined as software installations and upgrades, startup file updates, or file deletions. Using a file distribution agent allows these types of changes to be automatically applied to each workstation, or to a specified group of workstations.

The agent can be set up to collect information on file distribution status, which the systems administrator can view at the console to determine if files were distributed successfully. The console provides view formats and types that allow the administrator to review status data, such as which workstations are set up for file distributions, the stations to which files have been distributed, and the number of stations waiting for distributions.

Because users can be authorized to log in at one or more workstations, the file distribution agent determines where to distribute files based on the primary user (i.e., owner) of the workstation. The owner is established the first time an inventory is taken of the workstation. Before automated file distributions are run, the hardware inventory agent is usually run and distributions made only when the owner is actually working at the station.

Using scripts, the systems administrator defines distribution criteria, including the group or station to receive files and the day or days on which files are to be distributed. Scripts usually identify the files for distribution and the hardware requirements needed to run the file distribution job successfully. Many vendors provide templates to ease script creation that are displayed as a preset list of common file tasks. Using the templates, the systems administrator outlines a file distribution script and then uses the outline to actually generate the script.

EVENT MANAGEMENT AGENTS

The ability to monitor network activity allows the systems administrator to ensure the effectiveness and efficiency of network services. An event manager agent is used to track network activity (i.e., events), log network activity, and automatically alert the individual responsible for responding to certain network occurrences. Any network occurrence can be considered an event, from running the virus scan program, to logging in to the network, to performing a network backup.

The systems administrator specifies the network activity to be tracked, such as the times when users log in and out of the network or when programs are run. Network activities that may require immediate attention are also specified. A notification feature can be set to alert the systems administrator, or the person responsible, of the times when these events occur. Different notification methods can be set as appropriate for each network event.

Event notifications can be processed based on priority level. For example, if three network events occur simultaneously, notification of the event with highest priority is sent first. Priority levels range from a low of one to a high of nine. When the systems administrator specifies the network activity to monitor, a priority level for each event is assigned based on how critical the activity is and whether someone has to be notified when the event occurs.

The following methods of event notification are typically available:

- Console messages. Text messages display the name of the event and color-coded views indicate the priority level of the event.
- E-mail messages. The event level and name are sent in an electronic mail message.
- Pager messages. A phone number, the event name, or both, are sent to a pager.

The systems administrator chooses one or more contacts to receive notification of each event level. For example, a technician can be specified to receive a pager message when high-priority events occur, and a help desk operator can be specified to receive an E-mail message when routine application-related events occur. An acknowledgment of receipt for event data can be sent to the console to help ensure a proper response to events.

PROCESS SCHEDULING AGENTS

A scheduling agent is an application used to organize and run jobs. A typical job can be running a command for network backup, or sending a broadcast message to users before the backup begins. Instead of having to keep a written log of jobs and remembering to run them, the systems administrator uses a scheduling agent to specify that the jobs run automatically at a certain time. For example, the systems administrator could schedule the system backup process for 2 A.M. — a time when the system is not likely to be heavily used. At the specified time, the scheduling agent automatically starts jobs unattended. The job specifications can be saved and updated whenever necessary, and jobs can be changed or deleted at any time.

The scheduling agent can also provide job completion information, such as whether the job ran successfully, when it ran, and the next time it is scheduled to run. Following are examples of jobs a scheduling agent streamlines:

- Performing network backup
- Running computer virus scan
- Processing overnight mailings
- Performing file copying and distribution
- Printing and processing reports
- Sending network messages

A scheduling agent can implement jobs that run once or at these regular intervals: hourly, daily, weekdays, weekly, monthly. Instead of being run individually, jobs can be grouped to accomplish multiple tasks. An entry can be posted in the event log after jobs are completed that tells the systems administrator when the job started and ended. For example, with regard to scheduling the network virus scan process, the scheduling agent can automatically add a record to the log of network events provided by the event agent.

SYSTEMS MANAGEMENT PRODUCTS

Some systems management products that make extensive use of intelligent agents include Computer Associates' CA-Unicenter, Hewlett-Packard Co.'s OperationsCenter, and Saber Software Corp.'s LAN Management System (a suite of systems management products). The agent software is written for each vendor's system platform. The extent to which agent software works with other hardware platforms differs according to vendor. Hewlett-Packard's agent software, for example, works on the following UNIX platforms:

- AT&T Systems 3000
- Sun Solaris
- IBM RISC/6000
- Sun Microsystems SPARCstation
- BULL HN Information Systems DPX/20
- HP 9000 and 3000 Series systems

Other companies that have awakened to the management possibilities offered by intelligent agents include Apple Computer, BMC Software, Candle Corp., Ki Networks, and Sun Microsystems. IBM Corp. has announced that it will revamp its entire family of management products with a multivendor intelligent agent architecture. The company's Common Agent Architecture, being implemented in phases, uses the DMTF DMI technology which lets systems administrators gather a variety of desktop system data such as disk usage and PC configuration, and monitor the performance and status of individual applications. APIs let management consoles, such as HPOV and IBM's NetView for AIE platforms, share event and topology information.

INTEROPERABILITY AND OTHER ISSUES

The term "agent" implies interoperability with other vendors' management software, at least on a basic level. This interoperability usually takes the form of SNMP support. The expectation is that if agents are sitting out there on various stations, multiple managers from different vendors will be able communicate with them. But in the distributed environment, the applications are not designed to interoperate with other vendors' applications. So what appears as agent technology might really be just a piece of a distributed application.

However, some agents are interoperable. IBM's Systems Monitor/6000, for example, is a true intelligent agent. It provides an MIB interface to UNIX information by exposing UNIX kernel information in SNMP format. Still, products that offer such interoperability are few and far between.

Agent Management

The issue of agent management also needs to be addressed. A large network can have thousands of agents that need to be monitored to find out what agents are actually in the network as well as what they are doing and what has changed. One tool that does monitor agent activity is Legent Corp.'s AgentWorks, which is available for several UNIX platforms, as well as for Windows NT and OS/2.

AgentWorks uses icons to represent nodes on the network and allows an administrator to view the overall system, an individual site, a particular network segment, or a particular node. If the systems administrator is looking at an overall view and an error has occurred in a database process, the node where the error occurred appears in red. As the user refines the view to just the node, the database manager agent uses the color red to further pinpoint the problem. If an individual forgets to restart a process, such as an interrupted backup, AgentWorks automatically displays the error.

In addition to allowing systems administrators to create their own intelligent agents, AgentWorks can run as a standalone network management system or interoperate with other management platforms such as SunConnect's SNM, HPOV, and IBM's NetView for AIE. AgentWorks is also one of the few products that integrates current SNMP platforms and intelligent agents in a distributed client/server architecture. AgentWorks even allows systems administrators to choose among agent modules for systems, network, application, and database management. Within the basic modules, systems administrators can gradually add management functions as the need arises.

The concept of modular agents is not without its share of potential drawbacks. Because it is possible, for example, to have multiple agents in the same system, more than one agent might respond to a command from the management console. In addition, agents may contend for the same output port when attempting to respond.

This issue is addressed by designating one agent the master and the others as subordinates. The master agent reconciles requests and responses and multiplexes data from subordinates over the output port. However, this scheme introduces a new level of complexity into multivendor environments. Because the masters and subordinates of different vendors are not compatible, they will not work together. Although many vendors have been coalescing around a few approaches for defining master-subordinate agents, such as SNMP Research Inc.'s Emanate extensible agent architecture, many users are still forced to buy agents from the same vendor from which they bought the management system.

DMI was designed to promote multivendor interoperability by addressing standardization of the agent interface. To date, however, one of the biggest problems with DMI has been the lack of guidelines on how to retrieve information remotely. The Desktop Management Task Force recently completed its remote specification, which should make it easier to add management capabilities to computers, software, and peripheral devices.

CONCLUSION

Intelligent agents have a myriad of other uses. BeyondMail from Banyan Systems, Inc. is an example of agent-based messaging. BeyondMail's intelligent agents filter, sort, and prioritize mail by sender or topic on behalf of users. A number of sophisticated messaging products have become endowed with agents to help users manage and access data. For example, intelligent agents are used on AT&T's wireless Personalink network to perform a variety of tasks for mobile users, such as sorting and directing E-mail according to priorities. Among other things, important messages can be sent to the user's PDA, and routine messages can be output to a fax machine back at the office.

Intelligent agents offer several compelling advantages for systems management. They detect exceptional events and trigger automated actions to correct problems before they affect systems services, thereby increasing systems availability and reliability. An intelligent agent architecture obviates the need for expensive network polling, because only critical alerts are sent to the management console.

Finally, intelligent agents can be used to automate routine, repetitive tasks, relieving the systems administrator of a heavy work burden. This reduces the overall cost of systems management by speeding problem resolution and permitting businesses to get the most out of their technology investments. This ability to carry out corrective actions on an automated basis, and to keep business systems and applications up and running, can give a company a competitive edge in an economy where information availability and the speed of delivery are critical.

Index

A

A/D, *see* Analog-to-digital converters

AAL, *see* ATM Adaptation Layer

AAL1, 175, *see also* Asynchronous Transfer Mode

AAL3/4, 175–176, *see also* Asynchronous Transfer Mode

AAL5, 176–177, 180, *see also* Asynchronous Transfer Mode

Abstractions, high-level, 533

Accept message, *see* Messages

Acceptable usage, 623, 627, 630

Access

FINGER, 695

  FTS2000 Offnet Calling, 292–293

  protection and Java security model, 617

  restricted of legacy systems, 422

  router, 226–228

Access control

  Intranet security, 600–601, 608, 609, 610, 612, 613

  mission-critical applications, 574, 578–579, 580–582

  mobile user security, 656, 662

  Virtual Private Networks, 266–267

  virus fighting, 642

Access charges, 320

Access codes, 288–289

Access control lists (ACLs), 225–226, 547

Access logic, 474

Accountability concept, 592

ACD, *see* Automatic call distribution

ACE SecureID, 97

ACLs, *see* Access control lists

Acoustic feedback, 84

Active Server Page (ASP), 113

ActiveX, 24, 477, 591–593

Address resolution protocol (ARP), 690

Addresses

  destination

    Cellular Digital Packet Data, 308

    frame relay, 340, 353

    IPv6 protocol, 38, 41, 43, 44, 48

  extranet security and application layer gateway firewall, 102

  fixed, 575

  IP

    identification and PING, 689–690

    Java programming language, 529

    outward network traffic and firewall creation, 632

  proxy firewalls and mission-critical applications, 577

  IPv6 protocol, 41–44

  media access control, *see* Media access control

  mapping and ATM network operation, 180, 181

  OSPF versus EIGRP protocol, 57

  source, 666

  unicast address, *see* Unicast address

  Virtual Private Networks, 261

Addressing

  ATM network operation, 177–178

  building the IP foundation for one network, 53–56

  frame relay, 342–343

IPv4 versus IPv6 protocol, 38
Telnet access to router, 227
Administrative controls, 601, *see also* Internet
ADSL, *see* Asymmetrical digital subscriber lines
ADSU, *see* ATM Data Service Unit
Advanced Peer to Peer Networking (APPN), 30–33, 34, 35, 441
Agent IQ, 719–720, *see also* Intelligent agents
Agent management, 726–727
Agents, systems administration, 718–719
AgentWorks, 727
AIW, *see* APPN Implementers Workshop
Alarm, 705, 712
Alarm group, 699, 700
ALF lightweight session, 198–199
All nodes' address, *see* Addresses, destination
All routers' address, *see* Addresses, destination
AlphaNet, 81
Analog-to-digital (A/D) converters, 236
Animation, 108, 111
Antagonistic attacks, 620
Anticlogging mechanism, 674
Antivirus
scanners, 638
software, 653
strategies, 644–645
tools, 640–644
ANX, *see* Automotive Network Exchange
Anycast address, *see* Addresses, destination
AnyNet, 67
API, *see* Application Programming Interface
Applets
advantages of IP-based protocols, 21–22
data center integration with intranets, 67, 69, 73–76
integrated SNA applications with Web, 472–473, 474
Java security model, 616–617, 618–619, *see also* Java, security
network security
ActiveX, 591–593
Java: secure applets, 589–591
JavaScript, 591
lessening user exposure, 593
World Wide Web, 588
remote method invocation, 535
TCP/IP–SNA transitions, 24
untrusted and JavaBeans, 542
Application firewalls, 576, *see also* Firewalls
Application layer, 530, *see also* Java
Application monitoring, 722–723, *see also* Intelligent agents
Application Programming Interface (API)
JavaBeans, 540, 541
Internet telephony, 81
handheld network devices, 515
Java enterprise server platform, 543, 547
Java Naming and Directory Interface, 536
Applications
Internet telephony, 78–82
mission-critical and data center integration with intranets, 65
transaction-based and Cellular Digital Packet Data, 300
APPN, *see* Advanced Peer to Peer Networking
APPN Implementers Workshop (AIW), 30–31
Arbitrated loop topology, 467
Architecture
Cellular Digital Packet Data, 301–305

changes and enterprise network, 7–11
high-speed and cable modems, 253–254
Intranet, 596
Java security, 616
Parallel Sysplex, 410, 412
router-based extranet, 95–96
xswatch, 400–405
Archive management, 398
Archive transparent restore, 398
Archiving, 398, 458
Area 0, 57
ARP, *see* Address resolution protocol
ARPANET, 37
AS/400 Networks
practical applications of frame relay, 357–358, 359
tn3270/tn5250 Internal standards, 482–483, 489
ASCII–EBCDIC conversion, 432
ASL-Connect, 517–518
ASM, *see* Automated storage management
ASP, *see* Active Server Page
Asymmetrical digital subscriber lines (ADSL)
additional digital subscriber lines (DSL), 245–246
carrierless amplitude/phase (CAP) modulation, 244–245
developments and field tests, 245
discrete multitone (DMT) operation, 242–244
Asynchronous technologies, 112, *see also* Individual entries
Asynchronous Transfer Mode (ATM)
access
key technologies, 278–281
network core, 271–274
should you start now, 281–282
Wide Area Network, 274–278
backbone and Ethernet/Token Ring networks, 136
implementation
concepts, 171–180
different types of switches, 169–171
integrated solutions, 167–169
migration: integration with current technology, 180–183
role in internetworks, 163–167
multiservice networks, 9–10
synchronous optical network, 375, 381
Token Ring traffic switching, 119, 120, 121, 124–125, 127–128
Voice over Internet Protocol, 218
voice/video on Local Area Networks, 208–209
Web-enabling networking trends, 472
ATM Adaptation layer (AAL), 175
ATM Data Service Unit (ADSU), 179
ATM layer, 174–175, *see also* Asynchronous Transfer Mode (ATM)
ATM, *see* Asynchronous Transfer Mode
Attachment unit interface (AUI), 142–143
Attachments, e-mail, 630
Audio streaming technology, 111
Audioconferencing, 85
Audits, 607, 628
AUI, see Attachment unit interface
Authenticating Web server, 99, 100
Authentication
authorization systems in enterprise network, 11
Cellular Digital Packet Data, 307, 308
Intranet security, 600–601, 609
inward network traffic and firewall creation, 628
IP Security Protocol Working Group, 663–664, 667–669

IPv4 versus IPv6 protocol, 39
Java enterprise server platform, 547
RADIUS-based and Virtual Private Networks, 268
remote access and mobile user security, 655–657
mission-critical applications, 575, 578–582
network computers, 504
peer-to-peer, 609
smart cards and mobile user security, 660
3270-to-HTML conversion, 473
tunnel, 264
two-phase, 655
user and extranet security, 97, 103
Virtual Private Networks, 262, 265
Authenticity of association, 611
Authenticode, 592
Autoexec.bat file, 654
Auto-GUI, 72, see also Graphical user interface
Automated storage management (ASM), 397–398
Automatic call distribution (ACD), 219
Automatic logging, 645
Automatic transparent restore (IXR), 398
Automotive Network Exchange (ANX), 101
Auto-negotiation, 148–149
Availability
mobile user security, 648–650
proactive performance management, 712
VM/ESA operating system, 422
Available copies, monitoring, 722–723
AvantGo Desktop, 516
Awareness, 657, see also Security, mobile user

B

Back end tools, 516–520
Backbone
Advanced Peer to Peer Networking, 33
building and corporate readiness for one network, 52–53
Cellular Digital Packet Data, 306–307
multicast networks, 198–199
Back-door communications, see Communications, back-door
Backup
data center security, 397–398
Ethernet versus Token Ring, 132
installations and Intranet security, 608
mobile user security, 651, 652
storage area networks, 453
Systems Network Architecture and frame relay, 356
two-phase authentication, 656
Backward explicit congestion notification (BECN), 343, 356
Bandwidth
Advanced Peer to Peer Networking, 33
architectural changes and enterprise network, 10
Asynchronous Transfer Mode, 167, 275
cable modems, 252
Data Link Switching, 30
frame relay, 352–353, 355
Internet
near zero cost, 476
problems, 105
reduction and site development, 113–114
telephony, 86, 87
IPv4 versus IPv6 protocol, 39
100BASE-T, 141
utilization

frame on gigabit Ethernet, 159
maximizing mainframe-based WAN resources, 361, 362, 367–368
proactive performance management, 713–714
voice/video traffic and high-speed Token Rings, 126, 127
Bar charts, 705
Bar code scanning, 514
Bar graphs, 703–704
Base Sysplex, 410
Baseline
development and proactive performance management, 710–714
simple and maximizing mainframe-based WAN resources, 369–370
tools and techniques, 679
Baud signaling rate, 251
BECN, see Backward explicit congestion notification
Behavior blocking, 641
Behavioral subsets, 464
Bell System divestiture, 317
Bell System standards, modems, 283
BER test, 345
Best-of-breed purchasing decisions, 365
Best-of-breed network services, 266
BGP, see Border gateway protocol
Bidirectional amplifiers, 251–253
Billing, Virtual Private Networks, 267–268
BIOS, 493
Biphase shift key (BPSK), 252
Bipolar semiconductor technology, 407–408
Bipolar violations (BPV), 345, 346
Bit error rate testing, 345
Boot sector viruses, 637–638
Boot service, 503
Border gateway protocol (BGP), 664
BPSK, see Biphase shift key
BPV, see Bipolar violations
British Telecom (BT) TYMNET, 288
Broadband integration, 164
Broadband PCS, 521
Broadcast address, see Addresses, destination
Broadcast domain, see Virtual Local Area Network
Broadcast service, 196–202
Broadcast storms, 193
Browser
–based access, 20–22, 24
current and Java security model, 620
data center integration with intranets, 68
incompatibilities and Internet site development, 114–115
BT, see British Telecom
Bugs, 589–591
Burst rate, 352
Business drivers, 4–7
Bytecode, 430, 502, 539, 616, 617, see also Java

C

C code, 430
C plus language, 502
CA, see Certificate authority
Cable modems, preparation
course of action, 255
fundamentals, 250–251
high-speed architecture, 253–254
on the web, 254–255
problems addresses, 249–250
quadrature amplitude modulation technique, 251
unidirectional to bidirectional system, 251–253

Cable television, 317

Cableport cable modem system, 255, see also Cable modems

Cabling, migration, 135

CAC, see Call Admission Control

Cached applets, 74–75, see also Applets

Caching, 29

Call Admission Control (CAC), 182

Call center, 221–222

Call control, 212, 218

Call waiting, 320, 650

Calling circles, 78

Calling line identifier, 208

Campus-class switches, 170

CAP, see Carrierless amplitude/phase modulation

Capacity, 712, see also Proactive performance management

Capacity planning, 649

Capital expenditures, 381, 382, 384, see also Synchronous optical network

Carrier class networks, 271

Carrier extension technology, 155, 156, 159, 160

Carrier sense multiple access with collision detection (CSMA/CD) protocol

- 100BASE-T systems, 141, 142, 149, 151
- gigabit Ethernet, 155

Carriers, negotiating price, 322–323

Carrierless amplitude/phase (CAP) modulation, 244–245, 246

Carterphone decision, 316

Cascading style sheets (CSS), 106

CASL, see Compact Application Solution Language

Catalyst 5000/5500, 170

CATV, see Cable television

CDPD, see Cellular Digital Packet Data

Cell-in-Frame, 210, 211

Cell relay technology, 169, 170

Cell switching (CS), 166, 169

Cellular Digital Packet Data (CDPD), mobile networks

- applications, 298–299
- architecture and protocol, 301–305
- backbone, 306–307
- benefits to users, 300–301
- emerging class of remote uses, 299
- equipment requirements, 301
- fundamentals, 298
- is there a mass market, 310–312
- security, 660–661
- sending data, 308–310
- service pricing, 299–300
- transparent operation, 307–308

Cellular telephony, 77–78, 521, 522

Central Management Administrator Kit (CMAK) wizard, 266

Central processing unit (CPU), 131, 134, 448

Central site error, see Errors, central site

Central site routers, see Routers

Certificate, 618

Certificate authority (CA), 673

Certification program, 507

Certified mail, 609

Certs, see X.509 certificate

CF, see Coupling Facility

CF Control Code (CFCC), 417

CFCC, see CF Control Code

CGI, see Common Gateway Interface

Channel-attached controllers, see Controllers, channel-attached

Channel-attached gateways, see Gateways, channel-attached

Channel-attached router, see Routers, channel-attached

Channel-attached switches, see Switches

Channel hopping, 298

Channel interface processor (CIP), 363

Channel networking, 463

Channel-to-channel connections, 409–410

Chat rooms, 111, 112

Checksum, 592, 640–641

Chip coding, 463–464

CIP, see Channel interface processor

CIR, see Committed information rate

Circuit size, 375, 380

CIXC, see Competitive interchange carrier

Class D addressing, multicast networking, 196–197, 202

Class of Service (CoS), 30, 31, 68, 164

Clean up program, 636, see also Viruses

Cleartext header, 669

Cleartext password, 664

CLECs, see Competitive local exchange carriers

Click-to-call, 221–222

Client resource requirements, 715

Client socket, 531

Client/server technology

- changing environment of mainframes, 437–438
- communications and Java enterprise server platform, 545
- enterprise network, 4
- source routing and Ethernet versus Token Ring, 133
- synchronous optical network, 377–378
- tn3270/tn5250 Internet standards, 483, 485–487

Client-to-server virtual private network, 100

Client-side network, 624–625, 626

Clients, single, 266

Cluster controller, 438, 439–440

CMAK, see Central Management Administrator Kit wizard

CMOS technology, 407–408

Coaxial cable, 251, 252

Codewarrior, 514–515

Coding system, 143, 146–147

Collision domain, 151, 155, 159

Command formats, 689, 693, 698–699

Command/response field, 343

Command set recognition, 287

Commissioners, negotiating price, 322

Committed information rate (CIR), 340–341, 352

Common Gateway Interface (CGI), 534

Communications

- back-door, 590
- client/server and Java enterprise server platform, 545
- Internet, 599, 610–611
- layered, 528

Community technologies, 111–113

Compact Application Solution Language (CASL), 516

Compatible addresses, see Addresses, destination

Compatibility issues, 238

Competition, 238, 316–318

Competitive interchange carrier (CIXC), 272–273, 277

Competitive local exchange carriers (CLECs) , 272–273, 277, 317

Compiler checks, 619

Compliance plan, Y2K, see Y2K, desktop

Compliance, user, 644

Component broker, 434–435

Component configuration, 715

Compressed address form, 41

Compressed video, 110–111

Compressed voice, 217

Compression rate, 251

CompuServe, 289–290

Computing resources, loss, 651–652, see also Security, mobile user

Concurrent Garbage Collection, 434

Conduit Development Kit (Java edition), 517

Confidentiality

- Intranet security, 600–601
- IP Security Protocol Working Group, 669–671
- mobile user security, 657–659

Configuration

- change and router security in organizations, 225, 226, 228–229
- Integrity and mobile user security, 653–654

Conflict, Internet telephony, 83–84

Congestion, 33, 343, 344, 688–689

Connectionless networking, 532

Connection-oriented networking, 531

Connectivity

- data center security, 393, 394
- mobile user security, 653–654
- synchronous optical network, 380

Consistency control, see Flow control

Console management, automated, 396

Console messages, see Messages

Console security, see Security

Contact cards, 659–660

Contactless cards, 659–660

Context protection, 613

Continuity, 650–652, see also Security, mobile user

Continuous PINGing, 691

Controllers, channel-attached, 439–440

Convergence sublayer, 175, 176, 177

Conversation on Internet, 78, 79, see also Internet telephony

Conversions, digital/analog signals, 236, 238

Cookie, see Anticlogging mechanism

CORBA, 544, 545

Cordless telecommunications, 522, 523

Corporate intranet, 5–6, 29

Corporate management, 364

CoS, see Class of Service

Cost

- Asynchronous Transfer Mode switching, 165, 169, 273
- Cellular Digital Packet Data, 299
- Fibre channel technology, 467
- frame relay, 340, 353
- Linux operating system, 555
- maintenance and xswatch, 405
- TCP/IP–SNA transitions, 16, 17
- Virtual Private Networks, 267–268
- voice and data network integration, 219
- voice/video on Local Area Networks, 207
- VM/ESA operating system, 422
- wireless communications, 327
- Y2K compliance on desktop, 496

Cost of ownership, 509–510, see also Network computers; Total cost of ownership

Cost plus formula, 319

Counter, value of NT-based systems, 566, 568

Coupling Facility (CF), 417

CPE, see Customer premise equipment

CPU, see Central processing unit

Crashing, 589

CRC, see Cycle redundancy check

Critical paths, 56

Cross-platform viruses, 639

Cross-System Coupling Facility (XCF), 410
Cryptography, 600–601, 610, see also Encryption
CS, see Cell switching
CSMA/CD, see Carrier sense multiple access with collision detection protocol
CSS, see Cascading style sheets
CTS/RTS, 287
CU-seem, 85
Custom tags, 109
Custom tariff, see Tariffs
Customer premise equipment (CPE)
Asynchronous Transfer Mode, 170, 278–279
wireless communications, 328
voice and data network integration, 220
Voice over Internet Protocol, 218
Customization, JavaBeans, 541
Cycle redundancy check (CRC) errors, 345, see also Errors

D

Data
- caching, 417
- compression, 286–287
- confidentiality, 610, 612
- corruption, 453
- field, 156, 160
- files, 494, 495
- integrity, 610, 612, 613
- link layer, 39, 528
- locking, 417
- loss, 636
- migration, 454, 456
- path, 308, 309
- protection, 583–584
- rates, 178, 179
- sharing, 456–458
- storage, 598, see also Storage
- stream flow, 489, 490, 674
- traffic, 120, see also Traffic
- transfer, 237, 561, 660
- transmission, 339, 374
  - Cellular Digital Packet Data, 308–310
- warehousing, 456

Data center
- Intranet integration
  - technologies, 66–68
  - too much choice, 68–76
- security
  - automated production control solutions, 395–397
  - automated storage management software, 397–398
  - connectivity factor, 393
  - intranet security and systems management workstation, 391–393
  - intranet security software and systems management, 390–391
  - multimedia storage management, 398–399
  - powerful solutions, 393–395
  - xswatch, 399–405
- Y2K problems, 469

Data communications equipment (DCE), 180
Data Exchange Interface (DXI), 178–180
Data link connection identifier (DLCI), 342, 352, 353
Data Link Switching (DLSw)
- Advanced Peer to Peer Networking comparison, 34, 35
- concept, strengths, weaknesses, 27–30
- data center integration with intranets, 67, 68
- TCP/IP–SNA transitions, 17, 24

Data networks, see also Voice/data networks

group-oriented and multicast networking, 193
information communication, 216–217
past comparison with now, 3–4
Data terminal equipment (DTE), 180
Data/voice
integration, see Voice/data integration
consolidation
building the IP foundation, 53–56
making the IP infrastructure corporate ready, 52–53
routing protocols, 56–64
Databases, 108, 203, 495, 561, 618
Dataconferencing, 202
Datagram, 532
Datagram Socket, 532
Date stamp, 655
DCE, see Data communications equipment
DCL, see Digital control language
DCRT, see Digital cellular radio technology
DE, see Discard eligibility
Dead time, 157, 160
Debit cards, 299
Decision making, Internet, 602
Decision trees, 629, 631, 632
Decryption, 658–659
DECT, see Digital Enhanced Cordless Telecommunications
Dedicated lines, 259–160
Deencapsulation, 310
Degradation, 32–33
Delays
Internet telephony, 83, 87, 90
variable and disadvantages of frame relay, 341
voice/video on Local Area Networks, 209–210
Denial-of-service, 589–590, 620, see also Java, security model
Dense mode topology, 201
Department of Transportation (DoT), 203
Depreciation, 319
Design
issues for wireless communications, 335–336
mechanisms for Intranet security, 601–606
Design time, 542–543
Deskpro 4000N, 508
Desktop Management Interface (DMI), 505, 727
Destination address, see Addresses, destination
Destination unreachable, see Addresses, destination
Deutsche Telekom AG, 88, 89
Development cycle, 403–404, see also xswatch
Device identification, see Identification
Devise sharing, 450–453
DHCP, see Dynamic Host Configuration Protocol
DHCP address, see Addresses, destination
Diagnostics, 375
Dial backup circuits, 356
Dial-up line, 260
Dial-up network technology (DUN), 266
Dial-up user, 259
Diffie–Hellman system, 674
Digital cellular radio technology (DCRT), 298
Digital Cellular Telephone standard, 522
Digital certificates, 103
Digital control language (DCL), 392
Digital Enhanced Cordless Telecommunications (DECT), 522, 523

Digital seal, 609

Digital signaling, 238

Digital signature

- ActiveX and network security, 592
- Intranet security, 609, 673
- mobile user security, 655
- network computers, 504
- secure access to mission-critical applications, 580, 661

Digital subscriber lines (DSL), 241

Digital transmission technology, 298

Directories, 31, 212, 222, 401–402,

Disaster recovery

- automated and data center security, 395–396
- Intranet security, 602, 607–608
- Parallel Sysplex, 407, 415–416

Discard eligibility (DE), 343

Disclosure protocol, 610

Discrete multitone (DMT), 242–244, 246

Discrete signaling, 143–144

Discrete wavelet multitone coding, 246

Disk mirroring, 454, 455

Disk monitoring, 721

Distributed processing, 598

DLCI, see Data link connection identifier

DLSw, see Data Link Switching

DMI, see Desktop Management Interface

DMT, see Discrete multitone

DMZ network, 624–625, 631

DNS, see Domain Name Service

Domain Name Service (DNS)

- building the IP foundation for one network, 53, 54, 55
- Java programming language, 536
- NSLOOKUP, 695

Domain Naming Structure, 102

DoT, see Department of Transportation

Dotted decimal notation, 41, 43, 529

Dow Jones News/Retrieval, 290–292

Downloading, faster, 113–114

Downstream transmission, 242, 244–245 255

Downtime, 167

Drag and drop, 107

Dribbling errors, see Errors, dribbling

Drivers, Linux, 554

Drop/insert capability, 375

DS3 circuits, 280–281

DSL, see Digital subscriber lines

DSLw technology, 353–354

DTE, see Data terminal equipment

Dual-homed bastion host, 624

Dumb agents, 719, see also Intelligent agents

DUN, see Dial-up network technology

Duplex operation, 142

DVMRP, 199–200

DXI, see Date Exchange Interface

Dynamic fonts, 106

Dynamic groups, 202

Dynamic Host Configuration Protocol (DHCP), 53, 54, 55, 576

Dynamic HTML, 106–107, 109

Dynamic positioning, 106

E

Ease-of-use, 335

Eavesdropping, 307

Echo response message, 688, 689, 690

Echo test, 346–347

E-commerce

- advantages of IP-based protocols, 20
- competition, 419
- initiatives and TCP/IP, 16
- Internet telephony, 79–80

Java security model, 619–620
tn3270/tn5250 Internet standards, 483, 490
web-based, 6–7
Y2K problems affecting mainframe shops
integrating SNA applications with the Web, 472–474
overview, 469–472
programmatic access, 474–476
what to expect, 476–477
Economies of scale, 382
e-Control, 18
Edge switches, 210, 211
EDI, see Electronic Data Exchange
Efficiency
Cellular Digital Packet Data, 300
frame relay, 353, 355–356
TCP/IP–SNA transitions, 18
EGP, see Exterior gateway protocol
802.1p, 209–210, 211
802.1Q Standard, 124
8B6T coding, 146–147
8-bit/10-bit methodology, 463–464
EIGRP protocol, 57, 61–62
EJBs, see Enterprise JavaBeans
Electromagnetic spectrum, 521, see also Personal Communications Systems
Electronic Data Exchange (EDI), 3
Electronic software, see Software
E-mail
event management monitoring, 724
Cellular Digital Packet Data, 299, 300, 305
Intranet security, 609–610
intranet technology, 595
outward network traffic and firewall creation, 630–631
Employee information, 657
EN, see End nodes
Encapsulation Security Payload (ESP) header, 47–48, 669–670
Encryption
Cellular Digital Packet Data, 307
end-to-end and 3270-to-HTML conversion, 473
extranet security and application layer gateway firewall, 101
IP Security Protocol Working Group, 666
IPv6 protocol, 47
Java enterprise server platform, 547
mobile user security, 654, 658–659, 661
secure access to mission-critical applications, 583, 584
security and architectural change, 11
Virtual Private Networks, 264–265
Encryption key management, 261
End nodes (EN), 31
End users, 5, 380
End-point devices, 171, 172
End-to-end analysis, 347–348
End-to-end topology, see Host-to-host topology
End-user systems, 39
Engine, xswatch, 400, 402–403
Enhancements, tn3270/tn5250, 487, 488
Enterprise-class switches, 170–171
Enterprise Extender, 471
Enterprise JavaBeans (EJBs), 432, 433, 435
executive, 544
Enterprise networks
architectural changes, 7–11
Asynchronous Transfer Mode, 163, 164, 271, 274, 275, 277
bandwidth, 106
business drivers, 4–7

Intranet architecture, 596, 597
Enterprise server upgrade, 124
Entertainment companies, 317
Entropy, 404
Eradication/cleaning program, 653, see also Viruses
Errors
Advanced Peer to Peer Networking, 31. 32
Cellular Digital Packet Data, 298
central site, 32
dribbling, 346
data center security, 397, 399
frame relay, 341, 344, 353
FCS, 346
messages, 48, 379
modems, 284
pilot, 4
tracking using Remote Monitoring network probes, 700
ESCON channels
high-speed and Sysplex introduction, 410
Local Area Networks-to-mainframe connectivity, 440
TCP/IP–SNA transitions, 18, 19
Web-enabling networking trends, 471
ESF, see Extended super frame
ESP, see Encapsulation Security Payload header
eSuite Workplace, 508, 511
Ethernet, see also Fast Ethernet; Gigabit Ethernet
Foundation Manager, 701–704
networks
support of all applications, 134–138
Token Ring comparison, 130–134
Remote Monitoring network probes, 699–700, 701
switch
bandwidth in enterprise network, 10
establishment for virtual Local Area Network, 185, 186, 187
voice/video on Local Area Networks, 209–211
Ethertype field, 190–191
EtherVision, 681–681
Event handling, 541, see also JavaBeans
Event management, 724–725
Explicit tagging, see Virtual Local Area Network
Extended data storage management, 399
Extended Recovery Facility (XRF), 416
Extended super frame (ESF), 374
eXtensible markup language (XML), 108–109
Extension headers/options, 44–46
Exterior gateway protocol (EGP), 664
Extranets
connections using Virtual Private Networks, 257
enterprise-to-enterprise transactions, 15
supporting and implementing
application gateway firewalls, 97–103
router-based architectures, 95–96
EZTrak, 515–516

F

Fabrics, Fibre channel technology, 466–467
Failure
link, 30, 31
loosely coupled systems, 410
mobile user security and mechanical, 652
OSPF versus EIGRP protocol, 58, 61
Parallel Sysplex, 414, 415–416
S/390 systems, 408

Fast Ethernet, see also Ethernet; Gigabit Ethernet
  adapter card and Cableport cable model system, 255
  bandwidth and architectural changes in enterprise network, 10
  delays in voice/video on Local Area Networks, 209
  high-speed Token Ring, 121
  100BASE-T
    100BASE-T2, 147–152
    100BASE-T4, 145–147
    100BASE-TX, 145
    how does it work: ISO view, 142–145
    impacts on network design, 152–153
  Web-enabling networking trend, 471, 476
Fast link pulses, 149
Fast Packet technology, 170
Fault masking, 413
Fault tolerance, 356, 375, 380
FCC, see Federal Communications Commission
FCS errors, see Errors
FCS field, 344
FCS, see Frame check sequence
FDDI, see Fiber Distributed Data Interface
FDM, see Frequency division multiplexing
FECN, see Forward explicit congestion notification
Federal Communications Commission (FCC), 334, 521
Feedback, user, 714
FEP, see Front-end processor
Fiber Distributed Data Interface (FDDI), 120, 121, 124, 143
Fiberoptic cables, 241
Fibre channel layers, 461–464
Fibre channel technology
  layers, 461–464
  message structure definition, 464–465
  scalability and guidelines for use, 465–467
Field tests, 245
File
  access and network security, 590
  attribute monitors, 641
  automated and data center security, 397
  format and Java programming language, 616
  intelligent agents, 723–724
  naming conventions and Y2K compliance on desktop, 496
  transfer, 229–230, 456
  Uniform Resource Locator-based programming, 533
viruses, 637
File overwriter viruses, 638
File Transfer Protocol (FTP)
  advantages of IP-based protocols, 20
  building the IP foundation for one network, 56
  Linux operating system, 554
  network computers, 501
  outward network traffic and firewall creation, 631
  secure access to mission-critical applications, 576, 577
Filter, Remote Monitoring network probes, 700
Filtering firewalls, 576, see also Firewalls
Filter-matching rules, 265
FINGER, 693–695
Firewall
  application layer gateway and extranet security, 97–103
  configuration and PING, 691–692
  creation

general information, 632–633

inward network policy model, 629

inward network traffic, 627–629

other issues with outward network access, 632

outward network policy model, 631

outward network traffic, 629–631

overview, 623–625

policy foundation, 626–627

modification and extranets, 93

Proxy Server and Virtual Private Network security, 267

Fixed address, see Addresses, fixed

Flags, 344, 353

Flash plug in, 107–108

Flaws

Java applets, 590, 591

JavaScript, 591

xswatch, 404–405

Floating Point Arithmetic, 432

Flooding, 199–200

Floppy disks, 637

Flow control

access control and Intranet security, 612–613

modems, 286–287

100BASE-T system, 149–150

Flow label, 39, 40, 46

Follow-me/find-me, 222

Foreign carriers, 318

Format prefix, 42

Forward explicit congestion notification (FECN), 343

Foundation Manager

network baselining, 683–686

Remote Monitoring network probes, 701–705

4B5B coding, 145

FRAD, see Frame relay access device

Fragment header, 45

Frame check sequence (FCS), 353

Frame relay

architectural changes and enterprise network, 10

multicast networking, 194

practical applications, 356–360

SNA transport strategies, 352

testing and training

advantages of packet-switching networks, 340–341

anatomy of frame, 342–344

disadvantages of, 341

importance of testing, 344–347

provider's perspective, 348

subscriber's perspective, 348–349

unique issues, 348

transporting Systems Network Architecture legacy traffic, 353–355

Frame relay access device (FRAD), 352, 353

Frames

flow

Ethernet versus Fast Ethernet networks, 156–158

gigabit Ethernet, 159–161

format, 121, 155

header, 342

message structure and Fibre channel technology, 464–465

size, 130–131, 136–137

support, 130

two-level model of virtual Local Area Network, 192

Framing type, 178, 179

FreeTel Communications, 88

Frequency allocation, wireless, 326, 334–335

Frequency division multiplexing (FDM), 242

Frequency hopping spread spectrum, 331, 336

Front-end processor (FEP), 28–29, 355–356, 438, 440–441

FTP, see File Transfer Protocol

FTS2000 Offnet calling, 292–293

Full duplex operation, 149, 151, 152

Full-duplex sound cards, 84

Functional layers, 173, 174

Functional support, 356

Funding, firewalls, 623

Future-proof technology, 182–183

G

G.723.1 voice code standard, 89–90

Garbage collection, 433–434, 619, see also Java

Gatekeepers, 212, 220

Gateways, 212, 219, 220, 442

Gateway-to-gateway topology, 664–665, 667, see also Virtual private network

General information, 632–633, see also Firewalls

Generation 5 notation, 414

Generic Resources capability, 414, 415

Generic Routing Encapsulation (GRE), 262

Geographically Dispersed Parallel Sysplex (GEOPLEX), 416, see also Parallel Sysplex

GEOPLEX, see Geographically Dispersed Parallel Sysplex

getContent, 534

GetNextRequest, 698

GetRequest, 698

GetResponse, 698

getTLL, 533

Gigabit Ethernet, see also Ethernet; Fast Ethernet

- availability and speed, 130
- backbone and Ethernet/Token Ring networks, 136
- frame calculation, 159–161
- frame flow calculation, 156–158
- 100BASE-T system, 151–152
- recommended course of action, 161

Gigabit high-speed Token Ring, 123

Gigacells, 384

Global uniformity, 326

GNU license, 549

Goals, proactive performance management, 709

Grammars, xswatch, 403

Graphical user interface (GUI)

- –based applications and enterprise network, 5
- intranet security and systems management workstation, 391–392
- JavaBeans, 540, 541
- PING, 689
- server connections and tn3270/tn5250 Internet standards, 487, 489

Graphics, 107–108, 113

GRE, see Generic Routing Encapsulation

Green-on-black emulation, 72, 73

Grounding, notebooks, 650

Growth, virus infections, 643

GSM compression, 86, 87

GUI, see Graphical user interface

GXC, 81

H

H.323 protocol, 205, 205, 211, 218

Hackers, 590, 636

Half-duplex sound cards, 84

Handheld network devices

- applications development, 514

back end tools that access and port data to/from, 516–520
development tools, 514–516
future, 523
implementation in personal communications systems, 520–523
types, 513–514
Hard drive, 503, 561, 654
Hardware
general information on firewall, 632
Internet telephony, 82–85
Private Branch Exchanges, 215
Hardware token, 579
Hashing, 654, 667, 669
Header
compression, Virtual Private Networks, 263
frame relay, 353
IP Security Protocol Working Group, 665, 668
IPv4 versus IPv6 protocol, 39–41
Heartbeat tracking scheme, 415
Hewlett-Packard, 508–509
High data-rate digital subscriber line (HSDL), 246
High performance routing (HPR), 31, 32, 67–68
High-Speed Serial Interface (HSSI), 179
High-Speed Token Ring (HSTR) , see also Token Ring
availability and speed, 130
defining requirements, 121–122
deployment, 123–124
emerging need, 120
meeting needs of voice/video in Local Area Networks, 127–128
options, 120–121
packet priority, 126–127
priority access for voice and video traffic, 125–126
quality-of-service, 124–125
technology and timescales, 122–123
virtual Local Area Networks, 124
Web-enabling networking trends, 471
High-Speed Token Ring alliance, 121–122
High-Speed Token Ring line cards, 124
HiPerLinks, 418
HMAC, 669
Holes, 589–591, 628
Home backbone router, 308, 310
Hop limit, 40–41
Hop-by-hop header, 44, 668
Host
authentication, 47, see also Authentication
IP Security Protocol Working Group, 671
IPv4 versus IPv6 protocol, 39
publishing, 71, 72–73
Remote Monitoring network probes, 699
Host-to-gateway topology, 665
Host-to-host topology, 664–665, 667
HostTopN, 699
Hot Spots, prediction, 711, 712
Hotels, access, 288–289
Hotlinks, 254
Hot-Plug-In mode, 413
HotSync technology, 517
HPR, see High performance routing
HSDL, see High data-rate digital subscriber line
HSSI, see High-Speed Serial Interface
HSTR, see High-speed Token Ring
HTML, see HyperText Markup Language
HTTP, see HyperText Transport protocol
HTTP/HTTP-S access, 99

Hush-a-phones, 316

Hybrid addresses, see Addresses, destination

Hybrid Asynchronous Transfer Mode Networks, 210–211, see also Asynchronous Transfer Mode

Hybrid network, 358, 359

HyperText Markup Language (HTML), 88

HyperText Transport Protocol (HTTP), 254, 530, 583

I

I/O, see Input/output connectivity

IAD, see Integrated Access Device

I-am-alive-and-well message, see Messages

IAP, see Internet access provider

IBM, see also Parallel Sysplex

- communications controllers and advantages of IP-based protocols, 22
- Java support, 429
- networking and Y2K problems, 470
- SNA revival, 19
- traditional devises and Local Area Networks-to-mainframe connectivity, 438–442

ICA, see Independent Computing Architecture

ICF, see Internal Coupling Facility

ICMP, see Internal Control Message Protocol

ICMPv6, 48

Icon panel, Foundation Manager, 702, 704–705

IDC, see International Data Corporation

Identification, see also Individual entries

- access to mission-critical applications, 576–577, 579
- mobile user security, 649, 654
- user and Intranet security, 608

IDL, see Interface Definition Language

Idle transmission patterns, 144–145

IDT, 80, 81

IEEE 802.11 standard, 331

IENF, see Integration Event Notification Facility

IETF, see Internet Engineering Task Force

IGMP dialog, 196–197

IIOP, see Internet InterORB Protocol

IMA, see Inverse multiplexing over ATM

Implicit tagging, see Virtual Local Area Network

IMTC, see International Multimedia Teleconferencing Consortium

Inbuilding wireless, 329–334

Inclusion, 276

Incomplete checking, 642–643

Independent Computing Architecture (ICA), 509

Independent functioning, 656

Independent software vendors (ISV), 488

Industry standards, 506–507

Information

- Cellular Digital Packet Data, 306
- disk monitoring by intelligent agents, 721
- Ethernet versus Fast Ethernet networks, 157–158
- Fibre channel technology, 464–465
- flow management and storage area networks, 447
- frame relay, 343
- Gigabit Ethernet, 160
- Intranet security, 613
- IPv6 protocol, 48
- multicast network applications, 202–203
- xswatch, 404

Information Technology Group (ITG), 265, 266, 267, 268, 269

Infrared technology, 330, 334

Infrastructure
- one network and corporate readiness, 52–53
- storage area networks, 452
- Voice over Internet Protocol, 218
- voice/video on Local Area Networks, 207–208

Input caching, 416
Input/output (I/O) connectivity, 448
Integrated Access Device (IAD), 279
Integrated services, 66–68, 167–169, 276
Integrated switching technology, 169–170
Integration Event Notification Facility (IENF), 394
Integrity
- mobile user security, 653–657
- network communications and Java enterprise server platform, 547
- secure access to mission-critical applications, 582–585

Intelligent agents
- application monitoring, 722–723
- capabilities, 720–721
- disk monitoring, 721
- event management, 724–725
- file distribution, 723–724
- file monitoring, 723
- interoperability and other issues, 726–727
- key concepts in systems administration, 718–720
- process scheduling, 725
- systems management products, 725–726

Intellisync, 518
Interactive contents, 107
Interactive operations, 251
Intercept robot, 641
Interchange carrier (IXC), 272–273
Interconnect Controller, 438, 441–442
Interconnection
- Ethernet and Token Ring networks, 136
- limitations of 56K-bps modems, 237–238
- telephone and telephone providers, 315

Interface, 5, 286, 379, 422
Interface Definition Language (IDL), 546
Interference, 335, 613
Interference control, 601
Intermediate session routing (ISR), 31
Internal consulting, 384
Internal Coupling Facility (ICF), 417
Internal security, see Security
Internal usage chargeback, 267–268
International Data Corporation (IDC), 5, 6–7
International Multimedia Teleconferencing Consortium (IMTC), 89–90
International Standards Organization (ISO), 188
International Telecommunications Union (ITU), 239
Internet access provider (IAP), 272
Internet, see also Web; World Wide Web
- banking, 20, 21
- connections and cable modems, 249–250
- e-commerce, 6, 15, 79–80
- enterprise network, 3–4
- firewall creation, 625, 627
- growth and background, 37–38
- layered communications, 528
- mission-critical applications and secure access, 573
- network computers, 502
- site development
  - low-bandwidth, content-rich, 106–108
  - XML, 108–115

telecommunications policy, 317
Virtual Private Networks, 259–260, 266–267
VM/ESA operating system, 421–422
Internet Control Message Protocol (ICMP), 688
Internet Engineering Task Force (IETF), 90, 485, 487
Internet Explorer, 115
Internet InterORB Protocol (IIOP), 545
Internet security association and key management protocol (ISAKMP), 671–674
Internet service providers (ISPs)
architectural changes and enterprise network, 10
56K-bps modem, 236, 239
Linux operating system, 549, 551
Internet Standards Track, 38
Internet telephony
applications, 78–82
future directions, 90–91
hardware requirements, 82–85
how it works, 78
software requirements, 85–88
standards, 89–90
synchronous technology and Internet site development, 111, 112
telco participation, 88–89
Internetwork
Asynchronous Transfer Mode role, 163–169
service testing, 346–347
Interoperability
Ethernet and Token Ring networks,135–137
frame relay efficiencies for Systems Network Architecture, 355–356
intelligent agents, 726–727
Internet telephony, 89
IPv4 and IPv6 protocol, 48–50
OSPF versus EIGRP protocol, 57
tn3270/tn5250 Internet standards, 485, 487–488, 489
virtual Local Area Network, 191–192
Interpretative mode, 430
Interswitch links, 124
Intersymbol interference, 250, 252, see also Interference
Interworking unit (IWU), 166
Intranet
computer interconnections and Virtual Private Networks, 260–261
corporate
business drivers of enterprise network, 5–6
network computers, 502
security
designing mechanisms for, 601–606
ratings, requirements, and verification procedures, 606–608
software and systems management, 390–391
survey of services, 608–614
trends and issues, 598–601
Intrasubscriber addresses, see Addresses, destination
Introspection, 542
Intrusion point, 637, 645
Inverse multiplexing over ATM (IMA), 280–281
Invisible ink, 649
Inward network traffic, see Traffic
Inward proxy mechanism, 98, 99
IP address, see Addresses, IP
IP Authentication header, 47
IP-based protocols
advantages of, 19–22

proved technology for SNA transition facilitation, 22–24
IP encapsulating security payload, see ESP
IP multicast, 133–134
IP Security Protocol (IPSec)
features, 663–665
implementation and structures, 665–674
Virtual Private Networks, 262, 264–265
IP Suite, 519
ip3270/ip5250, 66, 68, see also tn3270/tn5250 Protocol
IPng Working Group, 38
IPORTs, 266–267
IP-PBX, 221
IPSec standard, 101, 103
IPSec, see IP Security Protocol
IPv4, 37, 38–39, 668, 670
IPv6
addresses, 41–44
background and features, 38–39
extension headers and options, 44–46
header format, 39–41
IP Security Protocol Working Group, 668, 670
migration to, 48–50
quality-of-service parameters, 46–47
security, 47–48
IPX/IP protocol, 190
Iridium, 326–327
ISAKMP, see Internet security association and key management protocol
Islands of Computing, 3
ISO, see International Standards Organization
ISO seven-layer diagram, 142, 143
ISPs, see Internet Service Providers
ISR, see Intermediate session routing
ISV, see Independent software vendors
ITG, see Information Technology Group
ITU, see International Telecommunications Union
IWU, see Interworking unit
IXC, see Interchange carrier
IXR, see Automatic transparent restore

J

Java
applications access on network computers, 504, 511
data stream flow, 489, 490
enterprise connectors, 434–435
enterprise server platform, 543–548
handheld network devices, 519, 520
logistics, 435–436
network computers, 501, 502, 508
networking features
basics, 528–530
primitives, 530–537
security
applets, 24, 589–591
breaches, 589
introduction to programming language, 616
model overview, 616–620
third-party protection, 621
threats, 620
using, 620–621
support on OS/390, 429–431
threads and garbage collection, 432–434
Unix connection, 431–432
Java Database Connectivity (JDBC), 546–547
Java Development Kit (JDK), 429–431
Java Management API (JMAPI), 548
Java Message Service (JMS), 546
Java Naming and Directory Interface (JNDI), 536, 546

Java Server API, 548
Java Transaction Services (JTS), 548
Java virtual machine (JVM), 431, 432
JavaBeans
  characterization, 540–543
  component architectures, 540
  programmatic access from Web-to-data center, 475
  technology review, 539–540
JavaScript, 591
JavaTel, 81
JavaWallet, 620
JCL, see Job Control Language
JDBC, see Java Database Connectivity
JDK, see Java Development Kit
JetSend, 519–520
Jini, 519
JIT, see Just-In-Time Compiler
JMAPI, see Java Management API
JMS, see Java Message Service
JNDI, see Java Naming and Directory Interface
Job completion, 725
Job Control Language (JCL), 392, 397
JoinGroup, 533
JTS, see Java Transaction Services
Just-In-Time compiler (JIT), 429, 430
JVM, see Java virtual machine

K

K56 flex technology, 238
KDC, see Kerboros key distribution center
Kerboros key distribution center (KDC), 673
Key database, see Database
Key management, 665, 671–674
Keyed message Digest 5, 47
Kiosks, information delivery, 203

L

L2F, see Layer 2 Forwarding
L2TP, see Layer 2 Tunneling Protocol
Lafayette Life Insurance, 72, 73, 424–426
LAN emulation (LANE), 171, 181, 182, 210
LANDesk Client Manager, 505
LANE, see LAN emulation
Language layer, 617–618
Language security, 619
Languages, cross-platform, 587
LANs, see Local Area Networks
Laptops, 495
Last cable mile, 241
Layer 2 forwarding (L2F), 262
Layer 2 Tunneling Protocol (L2TP), 262–264
Layer 3 protocol, 136
Layer 7 protocol, 530
Layered communications, see Communications, layered
Layers, dynamic HTML, 107
LDAP, Lightweight Directory Access Protocol
LeaveGroup, 533
LEC, see Local exchange carrier
Legacy systems
  Asynchronous Transfer Mode, 272–273, 279–280
  Internet, 419–420
  Java enterprise server platform, 543–544, 546
Legal regulations, 601, 607, see also Intranet
Lempel–Ziv, 286
Light technologies, 335
Lights-out devices, 280
Lightweight Directory Access Protocol (LDAP), 580, 581–582
Lincoln National Reinsurance, 22, 23
Line graphs, 702–703

Line taps, 574, 599

Line-of-sight transmission, 661

Link failures, see Failure, link

Linux

- customers, 551
- drawbacks, 554
- features and interest in, 549–550, 552
- future outlook, 555–556
- guidance for usage, 555
- implications, 555
- operating requirements, 551–552
- related Web and file transfer protocol sites, 554
- resources, 556
- sources, 552–553
- vendor support, 550, 551

LLC, see Logical Link Control

LLC2 processing, 29

LMDS, see Local Multipoint Distribution Service

LMI, see Local management interface

Load balancing, 54, 55, 58

Local access phone numbers, 266

Local Area Networks (LANs)

- enterprise network, 4
- frame relay, 339, 347
- mainframe connectivity
  - change mainframe environment, 437–438
  - channel-attached gateways, 442–443
  - channel-attached routers and switches, 443–445
  - integrated mainframe adapters, 445
  - making the selection, 445, 446
  - traditional IBM and compatible choices, 438–442
- Virtual Private Networks interconnections, 261
- virus infections, 6379
- voice/video, 127–128, 209–210, see also Voice/video, Local Area Networks
- wired versus inbuilding wireless, 329–330

Local exchange carrier (LEC), 272–273

Local loop, 327–328

Local management interface (LMI), 344, 346

Local Multipoint Distribution Service (LMDS), 328, 329

Lock down, 498

Log

- file, 399
- firewall creation, 628, 629, 630
- watching program, 399–405

Logged report, 681–682

Logical link control (LLC), 126, 684

Logical partitioning (LPAR), 413–414

Logical subsets, 464

Logical Unit (LU) nailing, 574

Log-in procedures

- British Telecom TYMNET, 288
- CompuServe, 289–290
- Dow Jones News/Retrieval, 290, 292
- MCI Mail, 293–295
- U.S. Sprint Sprint/Net, 295–296

Logistics, Java, 435–436

Logon journaling, 415, 416

Logon process, 581–582

Logon test, 56

Long-distance competition, 316–317

Long-run incremental cost (LRIC)

- negotiating and synchronous optical network, 375–376, 382–383, 384
- ratemaking in telecommunications, 321

Loosely coupled configurations, 409–413

Lost frames analysis, 346
LPAR, see Logical partitioning
LRIC, see Long-run incremental cost
LU, Logical Unit nailing

M

MAC, see Media access control
Machine name, 533
Macro-based viruses, 639
MAE, see Multi-Access Enclosure
Mainframe
access and Y2K problems, 470
advantages of IP, 19–22
changing environment, 437–438
computing and Parallel Sysplex, see Parallel Sysplex
Internet use, 421
VM/ESA operating system, 423–424
Maintenance, 110, 499, 652
Malicious code, see Viruses
Managed network services, 10
Management Information Base (MIB), 679
Manager, systems administration, 718
MapQuest, 589
Market acceptance, 505–506, see also Network computers
Mass market, 310, 312
Massively Parallel Processing (MPP), 408
Master agents, 727
Matrix, 699
Maven software, 86
Mbone, 198–199, 200
MCA, see MicroChannel architecture
MCI Communications, 316–317
MCI Mail, 293–295
MD5, Message Digest Version 5
MDI, see Media-dependent interface
MD-IS, see Mobile data-intermediate systems
Mean time to repair (MTTR), 713
Media Access Control (MAC) address, see also Address
Asynchronous Transfer Mode networks, 178
Ethernet versus Token Ring, 132, 133, 135
multicast networking, 193–194, 196
100BASE-T systems, 142, 143
transporting SNA legacy traffic across frame relay, 354
virtual Local Area Network implicit tagging, 188–190
Media, Asynchronous Transfer Mode network operation, 178
Media-dependent interface (MDI), 143
Media-independent interface (MII), 143, 148
Memory
Internet telephony, 82–83
Java security model, 617, 619
performance of NT-based systems, 560
reclaiming unused and Java platform, 433–434
stealth viruses, 638
TFTP access to router, 227
Merced chip, 555–556
Messages, see also Individual entries
accept and Asynchronous Transfer Mode network operation, 173
console, 724
Fibre channel technology, 464–465
I-am-alive-and-well, 415
Intranet security, 610–611
pager and event management monitoring, 724
prune message and DVMRP multicast routing protocol, 199
rejection message, 173

time exceeded, 48, 692, 693
Message Digest Version 5 (MD5), 667
Metering software usage, 722–723
Metropolitan wireless communications, 327–329
MIB, see Management Information Base
MicroChannel architecture (MCA), 441, 442
Microsoft, 265–268
Microsoft Internet Explorer 3.0, 590, 591
Middleware servers, 67
Middleware solutions, 24
Migration issues, 135–137, see also Ethernet
MII, see Media-independent interface
Mission-critical applications, secure access
challenges, 574–584
why use Internet, 573
MMDS, see Multipoint Multichannel Distribution Service
MMX technology, 206
MNLP, see Mobile network location protocol
MNP protocol, 284, 285
MNPS, see Multinode persistent sessions
MNRP, see Mobile network registration protocol
Mobile database systems, 303
Mobile data-intermediate systems (MD-IS), 303
Mobile-end systems protocol, 303–305, 307
Mobile network location protocol (MNLP), 306
Mobile network registration protocol (MNRP), 306
Mobile user security, see Security, mobile user
Mobility management, 305
Modems
Cellular Digital Packet Data, 301
56K-bps
capabilities and operation, 235–237
limitations, 237–239
recommended course of action, 239–240
hardware for Internet telephony, 83
low-speed and Internet, 105
Monitoring
Cellular Digital Packet Data, 299, 300
proactive performance management, 712
real-time and virus detection, 641
simultaneous of LAN/WAN and frame relays, 347
wireless communications, 336
Mortgages, 206
MOSPF, see Multicast Open Shortest Path First
Motorola company, 317
MPOA, see Multi-Protocol Over ATM
MPP, see Massively Parallel Processing
MTTR, see Mean time to repair
Multi-Access Enclosure (MAE), 19
Multicast address, see Addresses, destination
Multicast connectionless networking, 532–533
Multicast file transfer, 202–203
Multicast networking
applications, 202–203
broadcast and multicast IP comparison, 196–202
Local Area Networks, 193–195
vendors, 203–204
Multicast Open Shortest Path First (MOSPF), 200
Multicast packets, 133–134
Multicast services, 466
Multicast Socket, 532–533
Multicell conflict resolution, 333

Multihomed Intranets, 102
Multilevel generic detection, 640–641
Multimedia information, see Information
Multimedia storage management, 398–399
Multinode persistent sessions (MNPS), 32, 414
Multipartite viruses, 638
Multiple processors, 32
Multiplexing connections, 467, 468
Multiplexing equipment, 374
Multipoint, multidrop network, 357–358
Multipoint Multichannel Distribution Service (MMDS), 328, 329
Multi-Protocol Over ATM (MPOA), 181, 182
Multiprotocol routers, see Routers, multiprotocol
Multiprotocol support, 261
Multiservice access concentrator, 279–280
Multiservice-class switches, 170
Multiservice networks, 9–10, 165–166
Multivendor environments, intelligent agents, see Intelligent agents
Mutation engine, 638

N

Narrowband PCS, 521
Natural disasters, 607
NCP, see Network Control Program
NDS, see Network Directory Services
Neighbors, 39
Net Vecta PC, 505, 508–509
Net2Phone, 80
NetBIOS, 136
NET-FONE, 84
netPC, 501–502, 508–509
NetREXX scripts, 435
Netscape Explorer 3.0, 590, 591
Netscape Navigator/Communicator, 115
NetShow, 111
Network
- adapter card, 156
- Asynchronous Transfer Mode role, 164–165, 169, 171, 172–173
- consolidation
  - Advanced Peer to Peer Networking, 30–33
  - Data Link Switching, 27–30
- EtherVision, 682–683
- integrity, 547, 575–578
- layer implementation, 664
- layer protocols
  - Java programming language, 528–529
  - Cellular Digital Packet Data, 303, 305
- optimization, 714–715
- PING, 688–689, 691
- planning, effectiveness, 714–715
- OSPF versus EIGRP protocol, 58
- synchronous optical network, 378
- Voice over Internet Protocol, 218

Network baselining
- EtherVision, 681–683
- Foundation Manager, 683–686
- NEWT, 680–681
- tools and techniques, 679

Network Computer Reference Profile, 506–507
Network computers
- advantages and challenges, 502–503
- cost of ownership, 509–510
- implementation, 510–511
- industry standards, 506–507
- management, 505
- market acceptance, 505–506
- operation, 503–504

security issues, 504
vendor approaches, 507–509
Network Control Program (NCP), 355, 363
Network Directory Services (NDS), 581
Network Interface Cards (NICs), 124, 130, 625, 633
Network Management System (NMS), 683
Network nodes (NN), 31, 32
Network Station Series, 507
Network-to-Network Interface (NNI), 171, 172
New Online Administrative System (NOLAS), 424–426
Newsgroups, 111, 112, 556
Newspapers, 203
NEWT, 680–686
Next header, 40
NICs, see Network Interface Cards
NMS, see Network Management System
NN, see Network nodes
NNI, see Network-to-Network Interface
No SNA-reengineering whatsoever, 68
Noise, 147
NOLAS, see New Online Administrative System
Nonbusiness sites, Internet, 630
Noncongestion-controlled traffic, see Traffic
Nondisruptive rerouting, 32
Nonrepudiation, 609, 654–655, 667
Nonroutable protocols, 136
NonY2K software, listing, 495, see also Y2K
Notebook connectivity, 650
Not-even-a-keystroke-dropped protection, 414
NSLOOKUP, 695
NT-based systems
changing the paging file, 561–563, 564
evaluating performance, 563, 565–569
performance effect, 560–561
virtual memory, 560
N-way multicast, 195
Nyquist theory, 250–251, 252

O

Object registry, 535–536
Off-board server, 65
OFX, see Open Financial Exchange
Ohio State University Program, 20
OHIO, see Open Host Interface Objects
OLAP, see Online Analytical Processing
10BASE-T systems, 141, 142, 255
100BASE-T system
attributes, 147
auto-negotiation, 148–149
flow control, 149–150
full duplex operation, 149
gigabit Ethernet, 151–152
media-independent interface, 148
repeaters and repeater connections, 150
topology rules, 151
100BASE-T4 system, 145–147
100BASE-TX system, 145, 146
One-way multicast, 194
Online Analytical Processing (OLAP), 474
Online catalogs, 113
Open database connectivity, 66
Open Financial Exchange (OFX), 109
Open Group survey, 503
Open Host Interface Objects (OHIO), 490
Open Shortest Path First (OSPF), 57, 59, 200
Open System Interconnection (OSI)
Asynchronous Transfer Mode network operation, 173, 174
Cellular Digital Packet Data, 303–305
enterprise network, 3

Fibre channel technology, 461
Intranet security, 599
Java programming language, 530
virtual Local Area Network implicit tagging, 188
Open Systems Adapter (OSA), 445
OpenCard framework, 504
OpenEdition, 432
openStream, 534
Operation
Linux operating system, 551–552
network computers, 503–504
Parallel Sysplex, 415–416
synchronous optical network, 378–382
threats and Intranet security, 600, 602, 603
TRACEROUTE, 692–693
Optical Carrier Level 3, 375, 384
Optimization protocols, 87
Oracle Consolidator/Oracle Lite, 518
OS/390, 429–431, see also Java
OSA, see Open Systems Adapter
OSI, see Open System Interconnection
OSPF, see Open Shortest Path First
Outbound proxy mechanism, 97–98
Out-of-the-box, 502
Outsourcing services, 276–278, 384
Outward network traffic, see Traffic, outward network

P

Packet
Cellular Digital Packet Data, 298
data link layer in Java programming language, 528
dropped and flow control in 100BASE-T systems, 149–150
Ethernet and voice/video, 210, 211
Fibre channel technology, 465
IPv4 versus IPv6 protocol, 39
multicast networking, 196
priority
end-to-end on Local Area Networks, 128
high-speed Token Rings, 127
Token Rings switching, 126–127
Remote Monitoring network probes, 700
transmission over data networks, 216–217
voice/data and Voice over Internet Protocol, 217
Packet InterNetwork Groper, 688
Packet too big error messages, 48
Packet-switched data networks (PSDN), 167
Packet-switching technology, 340–341
Packet-tagging scheme, 124
Pager messages, see Messages
Paging file, changing, 561–563
Palm Computing platform, 514
PalmIII, 514
PalmPilot, 514
Parallel sysplex
beyond loosely coupled, 409–413
coupling at its best, 417–418
divide and conquer, 413–414
nonstop working, 414–417
not your father's multiprocessors, 408–409
operation
Advanced Peer to Peer Networking, 32
TCP/IP–SNA transitions, 18
Parameter problem, 48
Partitioning, 448
Passivity, viral infections, 642
Password, see also Security

bootup and antivirus strategies, 645
router security, 227, 228–229, 230
secure access to mission-critical applications, 579
Patches, Y2K, 497, 499
Path changes, 131–132
Pattern/action pairs, 401
Pay phones, 288–289
Payload length, 40
Payload option, 44
Payload type (PT), 176, 177
PBX, see Private Branch Exchange
PC card, 659
PCM, see Pulse Code Modulation
PCS, see Personal Communications Systems
PDAs, see Personal digital assistants
PDU, see Protocol data unit
Peer-to-peer authentication, see Authentication
Peer-to-peer environment, 331–332, 333
Pen Dragon, 516
Penright!, 515
Performance
Advanced Peer to Peer Networking, 32
Asynchronous Transfer Mode, 167, 275
data center security, 399
Ethernet versus Token Ring, 130–134
extranets, 94–95
frame relay, 353
high-speed Token Ring, 120, 121
mainframe-based WAN resources, 365–369
NT-based systems, 563, 565–569
OSPF versus EIGRP protocols, 58, 60
virtual Local Area Network, 187
Performance management, proactive
current state, 708–709
developing a baseline, 710–714
goals, 709–710
plan, 714–715
PERL programming, 399–405
Permanent logical links (PLLs), 352
Permanent virtual circuit (PVC), 342, 345–346, 352, 355
Personal Communications Systems (PCS), 521–523
Personal computers, 207–208
Personal digital assistants (PDAs), 513
Personal Handyphone System (PHS), 523
Personal identification number (PIN), 547
Personalization technology, 112–113
PGML, see Precision Graphics Markup Language
Phone call, quality, 77
Phoneware, 77, 86–87
PHS, see Personal Handyphone System
Physical coding sublayer, 143
Physical design, Intranet security, 601
Physical issues, mobile security, 649–650
Physical layer, 173, 528
Physical medium sublayer, 173–174
Physical paths, 57
Physical subsets, 464
Physical units (PUs), 29
Pilot error, see Errors, pilot
PIM, see Protocol Independent Multicast
PIN, see Personal identification number
PING, 688–689
Platform portability, 430
Platform-lock issues, 543
PLLs, see Permanent logical links
PLS layer, 142
Plug compatible mainframes, 410
Plug-and-play operation, 168

PNNI, see Private Network-Network
Pointer arithmetic, 619
Points-of-Presence (POPs), 270
Point-to-Multipoint topology, 328, 331–332
Point-to-Point (PPP) frames, 262
Point-to-point topologies, 466
Point-to-Point Tunneling Protocol (PPTP), 262, 263–264
Policy foundation, firewalls, 626–627, 629, 631, 632
Polymorphic infectors, 638
Polymorphic viruses, 639, 643
POPs, see Point-of-Presence
Popular programs, virus spreading, 639
Port density, 150
Port number, 533
Port scanning attacks, 628
Portable network access
  British Telecom (BT) TYMNET, 288
  commercial dialing from hotels and pay phones, 288–289
  CompuServe, 289–290
  Dow Jones News/Retrieval, 290–292
  FTS2000 Offnet Calling, 292–293
  MCI Mail, 293–295
  modem terms, 283–287
  U.S. Sprint Sprint/Net. 295–296
PPP, see Point-to-Point frames
PPTP, see Point-to-Point Tunneling Protocol
Precision Graphics Markup Language (PGML), 108
Predictive analysis techniques, 87
Prevention, virus infections, 644
Primestar satellite, 317
Printing, 18
Priority issues
  access and high-speed Token Rings, 125–126
  event management monitoring, 724
  IPv6 protocol, 40, 46
  synchronous optical network, 380
  tags on Local Area Networks, 210
  Y2K compliance on desktop, 496
Privacy
  Cellular Digital Packet Data and mobile user security, 661
  FINGER, 693–695
  invasion of and Java security model, 620
  IP Security Protocol Working Group, 663–664
  IPv4 versus IPv6 protocol, 39
  Java enterprise server platform, 547
  networks and Internet telephony, 82
  synchronous technology and Internet site development, 112
  telecommunications, 316
  voice and data network integration, 220–221
Private Branch Exchange (PBX), 166, 215, 216, 219
Private Network-Network Interface (PNNI), 169, 171, 181–182
Privilege keys, 648
Privileged access, 230
PR/MS, see Processor Resource/Systems Manager
Process control, Intranet security, 613–614
Process scheduling, 725
Processor Resource/System Manager (PR/SM), 413
Processor Time Counter, 563, 565, 566
Production control documentation, 397
Productivity losses, 636
Programmatic server, 70
Programming, tn3270/tn5250 Internet standards, 488
Property control, 540–541, see also Java

Protection methods, rating, 605–606

Protocol data unit (PDU), 175, 176

Protocol identifier, 533

Protocol Independent Multicast, 201–202

Provided identifier, 42

Proxy firewalls, 576–577, see also Firewalls

Proxy mechanism, 97

Proxy Server, 267, see also Server

Prune message, see Messages, prune

PSDN, see Packet-switched data networks

PT, see Payload type

Public directories, 78, see also Directories

Public interactions, 112

Public networks, 163

Public service, voice/data network integration, 219

Public switched telephone network (PSTN), 236, 283, 284

Public voice services, 223

Pulse code modulation (PCM), 236, 239

Purchasing, decisions, 364–365

PUs, see Physical units

PVCs, see Private Virtual Connections

Q

QAM, see Quadrature amplitude modulation

QoS, see Quality-of-service

Quadrature amplitude modulation (QAM) technique, 251

Quality of Service (QoS)

- Asynchronous Transfer Mode, 164, 208–209, 275
- proactive performance management, 712
- high-speed Token Rings, 124–125
- Ipv6 protocol, 46–47
  - IPv4 comparison, 39

Queue scheduling, 210

Queuing mechanisms, 126–127, 417

Quick Stats, 684–685

R

RACF, Resource Allocation Control Facility

RAD, see Rapid Applications Development

Radio bands, 334

Radiofrequency (RF), 334

RADIUS services, 267, 580–581

Random-number challenge, 656

Range, expanded, 333

Rapid Applications Development (RAD), 515

RAS, see Remote Access Server

Rate base, 319

Rate design, 318

Ratemaking, 319–321

Rate-setting process, 318–319

Rating procedures, 606

RDBMS, see Relational database management systems

RDP, see Remote Desktop Protocol

Read-only memory (ROM) firmware, 697

RealNetworks, 111

RealPlayer, 110

Real-time packets, 46

Real-time protocol (RTP), 211–212

Realtime Transport Protocol (RPT), 90

Receive, connectionless networking, 532

Recentralization, 8–9

Reception, wireless communications, 333

Reconciliation Sublayer, 143

Recovery diskettes, 651, 652, see also Disaster recovery

Redirect procedure, 310, 311

Redundancy, 53, 624–625

Reed-Solomon error correction, 244, 245, see also Errors

Reflection process, 542
Registration procedure, 308, 310
Registry identifier, 42
Regulatory activism, 316–318
Regulatory body, influencing, 322–324
Rejection message, see Messages
Relational database management systems (RDBMS), 515
Relay-agent address, see Addresses, destination
Reliability
  Asynchronous Transfer Mode technology, 181–182
  Cellular Digital Packet Data, 301
  Fibre channel technology, 465
  proactive performance management, 712
  VM/ESA operating system, 422
  wireless communications, 335
Remote access
  mission-critical applications, 575, 576
  Virtual Private Networks, 258, 259, 268–270
Remote Access Server (RAS)
  56K-bps modem operation, 236
  Virtual Private Networks, 265, 266, 267–268
Remote Desktop Protocol (RDP), 509
Remote method invocation (RMI) , 534–535, 536
Remote Monitoring (RMON) network probes
  Ethernet and Token Ring Management Information Bases, 699–701
  EtherVision, 681
  Foundation Manager, 683, 684, 701–705
  network baselining, 679, 680
  Simple Network Management Protocol, 697–699
Remote object, 535
Remote office, 220
Remote sites, 416–417
Remote users, 299
Repeater connections, 150
Replacement, software, 497, see also Y2K
Report
  data center security, 396, 397
  multicast networking, 198
  proactive performance management, 712–714
  two-phase authentication and mobile user security, 656
Request for Comment (RFC), 699
Request for information (RFI), 322–323
Request for proposal (RFP), 322–323
Rerun, automated, 395–396
Rescue disks, 642, see also Disaster recovery; Viruses
Reserved multicast address, 44
Residual ratemaking, 320
Resolution, Y2K compliance, 495–498
Resource Allocation Control Facility (RACF), 574, 579
Resource Reservation Protocol (RSVP), 90, 211
Response time, 693
ResQ!Net, 474
Restart, automated, 395–396
Restore, 453
Restrictions, value-dependent, 612, see also Intranet security
Retail chains, wireless communications, 326
Retention date, 655
Retrieval, storage area networks, 458
Retrieval access codes, 291–292
Return on investment (ROI), 364
Revenue, rate setting, 318

Reverse path forwarding, 199

Reverse proxy mechanism, 97–98

Reverse-colocation, 376

REXX scripts, Java, 435

RF, see Radiofrequency

RFC, see Request for Comment

RFC 1112, multicast networks, 196

RFC1490, 354–355, 356, 357

RFI, see Request for information

RFP, see Request for proposal

RG-11 cable, 251, 255

RG-59 cable, 251

RIF, see Routing information field

Risk assessment

- extranet security, 94–95
- 56K-bps modems, 239–240
- firewall creation, 626–627. 628
- Intranet security, 602, 603, 605
- Java security model, 621

RMI, see Remote method invocation

rmic compiler, 535

RMON, see Remote Monitoring network probes

ROI, see Return on investment

ROM, see Read-only memory firmware

Router

- access, 226–228
- access lists, 231
- additional protective measures, 231
- central site, 29
- channel-attached, 32, 443–444, 472
- Data Link Switching communication, 28
- file transfer, 229–230
- internal security, 230
- IPv4 versus IPv6 protocol, 39
- multiprotocol routers, 30
- need for access security, 225–226
- securing console and virtual terminals, 228–229
- Systems Network Architecture
- transport strategies, 351–352
- –TCP/IP transitions, 18, 19, 24

Routing header, 44

Routing information field (RIF), 137

Routing-over-IP, 67

Routing protocols

- building one network, 56–64
- Data Link Switching, 29–30
- data networks, 216–217
- Ethernet and Token Ring networks,, 136
- multicast networking, 199–202
- network layer in Java programming language, 528

RP, 201

RPT, see Realtime Transport Protocol

RS/6000 Scalable Parallel systems, 408

RSVP, see Resource Reservation Protocol

RTP, see Real-time protocol

Rules

- –based access, 576
- –based policies, 393, 398–399, see also Multimedia storage management
- router-based extranet architecture, 95

Runtime, 542–543

S

S/390 Parallel Enterprise Server, 407

SA, see Security associations

Safe Internet Programming (SIP), 589, 590

Safety practices, 336, 657

Sales representatives, 79, 326

Sampling rate, 84

SAN, see Storage area networks

Sandbox, 616–617
SAP, see Service Advertising Protocol
SAR, see Segmentation and reassembly
Satellite arrays, 326
Satellite Forms, 515
Scalability
  Advanced Peer to Peer Networking, 32
  Asynchronous Transfer Mode, 167, 169, 182, 210, 272, 276
  channel-attached gateways, 443
  Data Link Switching, 29
  DVMRP multicast routing protocol, 200
  extranet security and application layer gateway firewall, 102
  Fiber Distributed Data Interface, 121
  Fibre channel technology, 465–
  high-speed Token Ring, 120
  OSPF versus EIGRP protocol, 58, 61, 63
  Parallel Sysplex, 413
  VM/ESA operating system, 422
  xswatch, 400, 402, 404
Scanning frequency, 643
Scheduling, 649, 721
Scout Server, see Server, scout
Scripts, 497, 724
SDLC devices, , 28, 29
SDSL, see Single-line digital subscriber line
SDUs, see Service data units
Searches, 108, 643
Secure machines, 593
Secure rooms, 574
Secure Sockets Layer (SSL), 473, 488
SecureID, 579–580
Security, see also Individual entries
  applets and World Wide Web, 589
  architectural changes and enterprise network, 11
  Asynchronous Transfer Mode technology, 182
  breaches and file monitoring, 723
  Cellular Digital Packet Data, 300
  console, 228
  communication level and mission-critical applications, 574
  enterprise network, 4
  extranets, 93, 94
  firewall creation, 631
  IPv6 protocol, 45, 47–48
  Java model, 617–618, 619, 621
    enterprise server platform, 547
    JavaBeans, 542
  mobile user
    availability, 648–650
    confidentiality, 657–659
    continuity, 650–652
    integrity, 653–657
    new technology, 659–662
  network computers, 504
  routers, 225–226, 230
  Secure Sockets Layer, 473
  software and data center, 390–391
  storage area networks, 453
  TCP/IP–SNA transitions, 17
  tn3270/tn5250 Internet standards, 488
  virtual Local Area Network, 188
  wireless communications, 333, 336
Security associations (SA), 665–667
Security parameters index, 667
Segmentation and reassembly (SAR), 176, 177, 180, 279
Self-monitoring analysis and reporting technology (SMART), 505
Self-Timed Interconnect bus (STI), 418
Send, connectionless, 532

Sequence number (SN) field, 175
Sequence number protection (SNP), 175
Server
- addressing and building one network, 54, 55–56
- data center integration with intranets, 68
- scout and handheld network devices, 517
- special application servers, 359–360
- tunnel server, 265
- Voice over Internet Protocol, 218–219
- Web, 22, 421–422

Server clustering technology, see Parallel Sysplex
Server Socket, 531
Service Advertising Protocol (SAP), 187
Service data units (SDUs), 175
Service-level agreements (SLAs), 275, 713
Service Location Protocol (SLP), 488
Service provider, 275, 277
Service Provider Interface (SPI), 536
Servlets, 548
SetRequest, 698, 699
setTLL, 533
Shannon's law235, 237
Shared diskettes, 639
Shared machines, 639
Shielded twisted pair (STP), 135, 178
Signal, fades, 298
Signaling protocol, 172–173
Signaling system
- 100BASE-T systems
  - impact on network design for 152–153
  - physical layer, 142, 143, 144
- 100BASE-T4 systems, 145–146

Signal-to-noise ratio, 235, 236, 237
Signature scanning, 640, 642–644
Signed classes, 618
Simple line coding, 246
Simple Network Management Protocol (SNMP), 548, 679, 697–699
Simultaneous nonrepudiation, 610, see also Nonrepudiation
Single signon, 581
Single System Image, 414
Single-line digital subscriber line (SDSL), 246
Single-pipe strategy, 279
Single-user interface, 99
SIP, see Safe Internet Programming
6bone trials, 50
SLAs, see Service-level agreements, 275
Slow scanning, 643
SLP, see Service Location Protocol
SMART, see Self-monitoring analysis and reporting technology
Smart bandwidth, 168–169, 182
Smart blocking, 642
Smart Cards, 103, 504, 547, 659–660
Smart networks, see Synchronous optical network
SmartTran, 518–519
SMDS, see Switched Multimegabit Data Service
SN, see Sequence number field
SNA, see Systems Network Architecture
SNA Network Interconnection (SNI), 22, 356
SNI, see SNA Network Interconnection
SNIA, see Storage Networking Industry Association
Snippets, 400
SNMP, see Simple Network Management Protocol
SNP, see Sequence number protection

SNPA, see Subnetwork point of attachment addresses
Social regulations, Intranet security, 601
Socket-and-port transmission, 529–530, 531
SOCKS firewalls, 577–578, see also Firewalls
Software
  antivirus, 653
  data center security and systems management, 390–391
  electronic, 203
  extranet security and application layer gateway firewall, 100
  installation on network computers, 504
  Internet telephony, 85–88
  Intranet security, 613–614
  mobile user security, 658–659
  Private Branch Exchanges, 215
  third-party and Java security model, 621
  Y2K compliance on desktop, 494
SONET, see Synchronous optical network
Sound cards, 83–84
Sound devices, 83–84
Source address, see Addresses
Source code, 590
Source–destination path, 692
Source Route Transparent (SRT) bridging, 132
Source route/translational (SR/TLB) bridge, 132, 136, 137
Source routing, 120, 124, 131–133
Source-route bridging (SRB), 354
Spanning tree protocol, 131–132, 136
Sparse mode topology, 201
Speak Freely, 86
Special application servers, see Servers
Speech, clipped, 77
Speed
  Ethernet versus Token Ring, 130
  Cellular Digital Packet Data, 300
  wireless communications, 331, 335–336
SPI, see Service Provider Interface
Spreading, viruses, 639
Spread-spectrum radio, 661
SR/TLB, see Source route/translational bridge
SRB, see Source-route bridging
SRT, see Source Route Transparent Bridging
SSL, see Secure Sockets Layer
SSL-enabled site, 99–100
Standards
  evolving and 56K-bps modems, 239
  inbuilding wireless communications, 330–331
  Internet telephony, 89–90
  IPSec and extranet security, 103
  network computers, 506–507
  Voice over Internet Protocol, 218
  voice/video on Local Area Networks, 211–212
Star-configured network, 357
Static, wireless communications, 335
Statistical multiplexing, 340
Statistics group, 699, 700
Status, message, 655, see also Messages
Stealth viruses, 638
STI, see Self-Timed Interconnect bus
Storage area networks (SAN)
  advantages, 450
  archive and retrieval, 458
  back/restore applications and increased complexity, 453
  data migration, 454, 456
  data sharing, 456–458
  disk mirroring, 454, 455
  getting started with device sharing, 450–453

new focus on storage, 447–448
what is it, 448–450
Storage Networking Industry Association (SNIA), 456, 458
Storage of state, 541
Store and forward, 333, 630, 661
STP, see Shielded twisted pair
Streaming media, 664, 110–111
Subfields, addresses, 41–42, 43
Subnet-to-subnet topology, see Gateway-to-gateway topology
Subnetwork point of attachment (SNPA) addresses, 178
Subscriber identifier, 43
Subscription-based delivery, 202–203
Support, 69, 554
Support libraries, xswatch, 402
Surge protector, 649–650
Surveys, 494
Swap file, see Virtual memory paging file
SWAT teams, 499
Switch configuration, 348
Switched cell network, 273
Switched Multimegabit Data Service (SMDS), 175, 177, 196
Switches, 169–171, 444–445
Switching, 205
Symmetric DES key, 658, 659
Symmetric Multiprocessor, 407, 408, 410, 411
Symptoms, 639–640
Synchronization process, 658
Synchronous optical network (SONET)
advanced computers to manage client/server computing, 377–378
how to begin, 376–377
new operating modes, 378–382
strategies, 382–385
technology, 375–376
Synchronous technologies, 111
Sysplex Timer, 410
System/390, 482–483, 489
System modification, 620, see also Java
System Properties dialog box, 561, 562
Systems administration, 718–720
Systems management products, 725–726
Systems Network Architecture (SNA)
address resolution and Ethernet versus Token Ring, 132
data center integration with intranets, 66, 67, 69
Data Link Switching, 27, 28, 30
enterprise network, 3
mission-critical applications and secure access, 574, 583
tn3270/tn5250 Internet standards, 483, 485, 487, 488
traditional devises and Local Area Networks-to-mainframe connectivity, 438
transition to IP
case against last-minute revival, 19
dispelling concerns, 17–18
primary advantages of moving to mainframe, 19–22
proven technology to facilitate, 22–24
TCP/IP, 8, 362–365
transporting legacy traffic
frame relay efficiencies, 355–356
how traffic is carried across frame relay, 353–355
practical applications of frame relay, 356–360
strategies, 351–352
why frame relay, 352–353

Web, 471, 472–474

Systems, how viruses infect, 637

T

T1, 374

T1/E1 speeds, 169

TACAS+, 580, 581

Tagging, 188–191, 209–210

Tape, obsolescence, 458

Tape library, 453

Tariffs, 321–324, 383

Task Control Block (TCB), 433

TCB, see Task Control Block

TCO, see Total cost of ownership

TCP/IP, see Transmission Control Protocol/Internet Protocol

TDM, see Time-division multiplexing network

Technology, decision making, 363

Telecommunications Act of 1996, 317

Telecommunications Industry Association (TIA), 239

Telecommunications policy

beginning of competition, 316–318

drafting a custom tariff or agreement with carrier, 321–322

influencing a regulatory body, 322–324

ratemaking, 319–321

rate-setting process, 318–319

who creates, 318

Teleconferencing, 196

Telephone communication tools, 650

Telephone companies, 88–89, 317

Telnet, 226–227

Telnet protocol, 501

Templates, 724

Terminal, virtual, 228

Terminate-and-stay resident (TSR) program, 641, 653

Testing, 344–347, 497, see also Individual entries

TFTP, see Trivial file transfer protocol

Theft, 652

Theft-prevention devices, 649

Theodore Vail concept, 315–316

Thin server, 502

Thin-client architecture

browser-based and data center integration with intranets, 73, 74

network computers, 501, 502, see also Network computers

software applications, 502–503

Web-enabling networking trends, 472, 474

Thrashing condition, 561

Threading scheme, 432–433

Threats

Identification and Intranet security, 600, 602, 603–604

Java security model, 620

pervasiveness of viral, 636–637

3270/5250 emulation, 73–76, see also tn3270/tn5270

3270/5250-to-HTML conversion, 67, 69, 70–73

3270-to-HTML conversion, 473, 476

Throughput, 18, 355

TIA, see Telecommunications Industry Association

TIC, see Token Ring interface coupler

TIME, see Time Interactive Multimedia Extensions

Time-division multiplexing (TDM) network, 166–167, 242, 245

Time exceeded error messages, see Messages, time exceeded

Time Interactive Multimedia Extensions (TIME), 111
Time synchronization, 529
Timeout response, 689, 690
Timer, report delay, 197–198
Timescales, 122–123
Time-to-Live (TTL) field, 692
Timing specifications, 142
TLS, see Transport Layer Security
tn3270/tn5250 protocol
    data center integration with intranets, 65, 67, 68
    gateways and IP-based protocols, 20, 22
    Internet standards
        enterprise data and logic, 481
        how does standardization help, 485
        integration, 489–491
        interoperability, 487–488
        newer platforms and devices, 482–484
        protocols, 485–487
        user productivity and confidence, 481
    Local Area Networks-to-mainframe connectivity, 443
    secure access to mission-critical applications, 583
    Systems Network Architecture clients and source routing, 132
    TCP/IP–SNA transitions, 23–24
    Token cards, 632–633
    Web-enabling networking trends, 472, 474
Token Ring, see also High-Speed Token Ring
    Data Link Switching, 28
    Foundation Manager
        monitoring, 683–684
        remote monitoring, 701, 704–705
    interface on front-end processor, 440
    voice/video on Local Area Networks, 209, 210
    Remote Monitoring network probes, 699–700
Token Ring interface coupler (TIC), 29
Token Ring LANs, 10
Token Ring statistics group, 700, 701
Topology rules, 151
Total cost of ownership (TCO), 367, 420, see also Cost of ownership
TP/PMD, see Transport protocol/physical medium dependent specification
TRACEROUTE, 692–693
TRACERT, 693, 694
Trade test, 56
Traffic
    data and need for high-speed Token Ring, 120
    firewall creation
    inward network, 627–629
    outward network, 629–631
    maximizing mainframe-based WAN resources, 367, 370, 371
    multihomed intranet and extranet security, 102
    noncongestion-controlled and quality-of-service parameters, 46
    TCP/IP–SNA transitions, 18
    type and Asynchronous Transfer Mode, 165
Traffic Class Expediting, 209
Training, 348–349, 392
Transform algorithm, 663–664, 666
Transmission Control Protocol/Internet Protocol (TCP/IP)
architectural changes and enterprise network, 8, 9–10
changing mainframe environment, 438
Data Link Switching, 29

Ethernet and Token Ring networks interconnection, 136, 137
installation on mainframes and SNA conversion, 16, 17
Internet telephony, 85–86
intranets, 595
Java programming language, 529
NEWT, 680
Local Area Networks, 190, 439
router-based extranet architecture, 95, 96
scaling, 29
tn3270/tn5250 Internet standards, 483, 485, 486, 487
utilities
PING, 688–692
TRACEROUTE, 692–693
FINGER, 693–695
NSLOOKUP, 695
Web-enabling networking trends, 470–471, 476
Transmission convergence sublayer, 174
Transparency, 307–308, 343
Transparent bridging, 131–132, 133
Transparent LAN services, 276
Transport layer, 529–530, see also Java
Transport Layer Security (TLS), 488
Transport mode, 669
Transport-mode ESP, 47
Transport protocol/physical medium dependent (TP/PMD) specification, 143, 144–145
Transport strategies, 351–352
Trap, 699
Trend analysis, automated, 713–714
Trivial file transfer protocol (TFTP), 227–228
Trouble ticket reports, 714
Troubleshooting, network devices, 688–689
TRU RADIUS, 270
TrueSpeech, 86
Trumpet software, 86
Trust relationship, 672
Trusted subnetwork, 665
Trusted third party (TTP), 673
TSR, see Terminate-and-stay resident programs
TSS modem standards, 283, 284
TTL, see Time-to-Live field
TTP, see Trusted third party
Tunnel authentication, see Authentication
Tunnel mode, 669
Tunnel-mode ESP, 47
Tunnel server, see Server, tunnel
TwinAxcess, 467
Twisted-pair wire, 242, 243
Two-part key system, 658–659
Two-phase authentication, see Authentication
Two-way multicast, 194, 195

U

U.S. Sprint Sprint/Net, 295–296
UDP, see User Datagram Protocol
UNI, see User-to-Network Interface
Unicast address, 41, 42–43, 194
Unicast protocol, 199
Unidirectional connection, 664, 667
Unification, 276
Unified messaging, 221
Uniform Resource Locator (URL), 254, 533–534, 536
Uninterrupted power supply, 649–650
Uniprocessor, 410, 411
Universal drivers, Linux operating system, 556
Unix, 431–432
Unlicensed band, 334–335

Unlicensed PCS, 521
Unshielded twisted pair (UTP), 135, 149, 178
Unsolicited calls, 78
UPC, see Usage Parameter Control
Upgrading, software, 497, 498
Upstream transmission, 242, 244–245, 255
URL, see Uniform Resource Locator
URLConnection, 534
Usage Parameter Control (UPC), 182
Usage Peak, 566, 567
User access, 230
User base, Linux operating system, 549, 551
User Datagram Protocol (UDP), 529, 692
User-oriented keying, 671–672
User-to-Network Interface (UNI), 171, 172
Utilization reports, 712, 713
UTP, see Unshielded twisted pair cabling
UUnet Technologies, 268

V

V.29 modulation, 284, 285
V.32 modulation, 284, 285, 286
V.34 modems, 235, 238
V.42 Recommendation, 284–285
V.42bis modems, 286–287
Value-added services, 219
Value-added networks (VANs), 3
Value of service, 320
VANs, see Value-added networks
Vat software, 86
VCC, see Virtual circuit connection
VCI, see Virtual channel identifier
VDOPhone, 85
Vector graphics, 107–108, 109
Vendors
    approaches to network computers, 507–509
    Linux operating system, 550
    multicast network applications, 203–204
    OSPF versus EIGRP protocol, 57
    purchasing decisions and SNA versus TCP/IP, 365
Venn diagram, 93, 94
Verification, 614
Verifier program, 617
Version management, 658
Vertical monopoly, 315–316
Very high data-rate digital subscriber line (VSDL), 246
Very large integration (VLSI) technology, 163
Very Small Aperture Terminal (VSAT), 326
Vestigial sideband, 253
Video, priority access, 125–126, 127
Video/voice, Local Area Networks, see Voice/video, Local Area Networks
Videoconferencing
    asymmetrical digital subscriber lines, 247
    audioconferencing combinations and Internet telephony, 85
    desktop, 206, 208, 222
    multicast network applications, 202, 203
Vienna Way, 81
Virtual channel identifier (VCI), 173
Virtual circuit connection (VCC), 176
Virtual Local Area Networks (VLANs)
    construction methods, 188
    definition, 185–186
    enhancing performance, 187
    explicit tagging, 190–191
    facilitating adds, moves, and changes, 187
    high-speed Token Ring, 124
    implicit tagging, 188–190

interoperability issues, 191–192
rationale, 186
security, 188
support and requirements, 186–187
Virtual memory, 560
paging file, 560–561, 563
Virtual networking, 164–165
Virtual path identifier (VPI), 173
Virtual Private Networks (VPNs)
architectural changes and enterprise network, 10
extranet security and application layer gateway firewall, 100, 101, 102–103
secure access over Internet
basic requirements, 261–265
common uses, 259–261
ease of management and use, 265–266
future planning, 268–269
Windows communication platform, 266–268
secure access to mission-critical applications, 578
Web-enabling networking trends, 472
Virtual second line, 222
Virtual storage, 418
Virtual telephone companies, 80–81
Viruses
applet similarity and network security, 589
mobile user security, 653
security on network computers, 504
threats
discovery, 639–640
problem addressed, 635–639
recommended course of action, 644–645
standard approach to fighting, 640–644
VLAN-ID field, 190–191
VLANs, see Virtual Local Area Networks
VLSI, see Very large integration technology
VM/ESA operating system
case study: Lafayette Life, 424–426
good news/bad news, 422–423
platform for the Internet, 421–422
unlocking power of mainframe applications, 423–424
VM:Webgateway, 72
VocalTec, 81, 86–87, 88, 89
Voice
consolidation, see Data/voice consolidation
high-speed Token Rings, 125–126, 127
Internet telephony, 80–81, 83, 86
traffic support and frame relay applications, 358–359
Voice/data networks
integration, 219–223, 276
private voice and data networks comparison, 215–216
Voice over Internet Protocol, 215, 216–219
applications, 220–223
Voice/video, Local Area Networks
technologies for integration, 208–212
value, 206–208
Voice over Internet Protocol (VoIP), see Voice, Internet telephony; Voice/data networks
Voice Telephony over ATM protocol (VTOA), 211
VoIP, see Voice over Internet Protocol
Voltage adapter, 650
VPI, see Virtual path identifier
VPNs, see Virtual Private Networks
VSAT, see Very Small Aperture Terminal

VSDL, see Very high data-rate digital subscriber line

VTOA, see Voice Telephone over ATN Protocol

Vulnerability, analysis, 602, 603–605

W

WANs, see Wide Area Networks

Warm boot, 641–642

WaveSync, 519

Weaknesses, assessment, 605

Web, see also Internet; World Wide Web

- application-specific solutions and TCP/IP–SNA transitions, 24
- browsers, 87–88, 593
- enabled legacy applications, 423
- gateway, 70
  - -to-data-center gateways, 67
- server, see Server, Web
- Linux operating system, 554
- surfing, 621

WebPhone, 86

WebSphere, 432, 476

WfM, see Wired for Management

Whiteboarding, 199, 206

Wide Area Networks (WANs)

- Asynchronous Transfer Mode, 169, 274–278
  - key technologies, 278–281
  - bandwidth cost reduction and use in time-division multiplexing network, 167
- frame relay, 339, 345–346, 347, see also Frame relay
- maximizing mainframe-based resources
  - SNA versus TCP/IP, 362–365
  - what does performance really mean, 365–369
  - building the models, 369–371
- multicast networking, 193
- outsourcing and architectural changes in enterprise network, 10

Wide area wireless communications, s325–327

Windows

- applications access on network computers, 504
- communication platform and Virtual Private Networks, 266–268
- NT, see NT-based systems
- running and viruses, 637
- software for Internet telephony, 86
- support of TCP/IP utilities, 687, 690–691, 693, 694
- terminals, network computers, 501, 509

Winterm 2000 series, 509

Wired for Management (WfM), 505

Wireless communications

- allocation of frequencies, 334–335
- design issues, 335–336
- inbuilding, 329–334
- metropolitan, 327–329
- mobile user security, 661
- outlook, 336–337
- wide area, 325–327

Wireless Local Loops (WLL), 520

Wiring, Private Branch Exchanges, 215

Wiring plan, 255

WLL, see Wireless Local Loops

Workgroup-class switches, 170

Workload

- balancing, 410, 413, 414
- management, 395
- traffic, 370

Workload Manager, 415, 432

Workstation

browser-enabled and firewall creation, 627
data center security, 391–393
virus spread, 639
World Wide Web, see also Internet; Web
addresses in IPv6 protocol, 43–44
applets and network security, 588
electronic commerce and enterprise network, 5
mobile user security, 661–662
Worldwide reach, 301
Write Once, Run Anywhere, 540, 543
Write-once programming, 616
Writes, legitimate, 641

X

X.400 Message system, 305
X.500 Directory services, 305, 581
X.509 certificate, 580
X2 technology, 238
XCF, see Cross-System Coupling Facility
XML, see eXtensible markup language
XON/XOFF, 287
XRF, see Extended Recovery Facility
Xswatch, 399–405
Xtails, 400

Y

Y2K, desktop
checklist, 500
control, 498–499
discovery, 494–495
maintenance, 499
reporting, 495
resolution, 495–498
mainframe retooling and readiness, 438

Z

Zenith cable modem, 252, 253, 255, see also Cable modems
Zero, addresses, 41